ANATOMY OF *Voice*

ANATOMY OF *Voice*

How to Enhance and Project Your Best Voice

Blandine Calais-Germain
and François Germain

Translated by Martine Curtis-Oakes

Healing Arts Press
Rochester, Vermont • Toronto, Canada

Healing Arts Press
One Park Street
Rochester, Vermont 05767
www.HealingArtsPress.com

Healing Arts Press is a division of Inner Traditions International

Originally published in French under the title *Anatomie Pour la Voix: Comprendre et améliorer la dynamique de l'appareil vocal* by Éditions Désiris
First U.S. edition published in 2016 by Healing Arts Press

Note to the reader: *This book is intended as an informational guide. The remedies, approaches, and techniques described herein are meant to supplement, and not to be a substitute for, professional medical care or treatment. They should not be used to treat a serious ailment without prior consultation with a qualified health care professional.*

Library of Congress Cataloging-in-Publication Data
Calais-Germain, Blandine, author.
[Anatomie pour la voix. English]
Anatomy of voice : how to enhance and project your best voice / Blandine Calais-Germain, François Germain ; translated by Martine Curtis-Oakes. — First U.S. edition
pages cm
Summary: "An illustrated guide to the dynamic physiological structures that create and individualize the voice"—Provided by publisher.
Includes bibliographical references and index.
ISBN 978-1-62055-419-7 (paperback) — ISBN 978-1-62055-420-3 (ebook)
1. Voice—Physiological aspects. 2. Human mechanics. I. Germain, François, co-author. II. Title.
RF511.S55C35 2016
612.7'8—dc23

2015010861

Printed and bound in India by Replika Press Pvt. Ltd.

10 9 8 7 6 5 4 3 2 1

Drawings by Blandine Calais-Germain

Text design and layout by Debbie Glogover
This book was typeset in Cronos with Helvetica and Garamond Premier Pro as display fonts

For further information about Blandine Calais-Germain's work go to **www.calais-germain.com**.

This book is dedicated to Laurent,
king of vocalizations, and to all of his
Down's syndrome brothers and sisters,
who bring nothing but sweetness
to this earth.

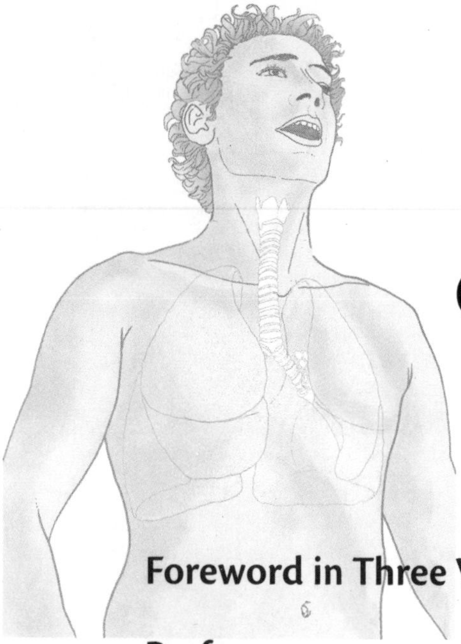

Contents

Foreword in Three Voices . . .

There is a time for the voice, and for some, there comes a time when it's necessary to understand the mechanisms behind the voice.

How, though, does one approach the Ariadne's thread of vocalizations while offering anatomical descriptions of the vocal apparatus? This is the challenge facing authors who need to make complex anatomical concepts accessible.

To accomplish this, Blandine Calais-Germain applies the effective approach she has used in her previous works, offering multiple drawings to show the anatomical structures from a variety of angles. To these she adds text that often acts like a virtual finger, tracing the structure and reinforcing the reader's understanding.

We find in this written work the same clarity of presentation that we experience in her workshops: all concepts are linked not only to the physiology of the vocal apparatus, but to movement as well. In many cases, François Germain elaborates on complex physical concepts while keeping the information accessible to a nonspecialized audience.

It's a pleasure to see how these two authors have presented the wonder of the voice and the anatomical structures that make it possible. This work gives us a glimpse of nature's capacity to produce a world of sounds.

Vicente Fuentes, Director; head of the Department of Voice and Language, Royal School for Dramatic Arts, Madrid; advisor to the National Theater of Spain

Dr. Guy Cornut, speech pathologist; former head of the Department of Phoniatrics, ENT Clinic, Faculty of Medicine, Lyon, France; choir director of the Vocal Ensemble of Lyon

Annie Trolliet-Comut, speech pathologist, Lyon; voice teacher specializing in vocal therapy for speech and song

Preface

Anatomy of Voice presents anatomical information and concepts as they relate to use of the voice. It is intended for those who regularly use their voices, sometimes in very extreme ways: singers, actors, lecturers, lawyers. It is also intended for anyone with an interest in the subject, for personal or professional reasons. Its purpose is to give readers an understanding of the anatomy of the vocal structures and the way these structures produce sound, so that they can then adapt and refine the way they use their own voice. This goal explains some of the technical choices and practices that are detailed here.

This book follows the structure of Blandine Calais-Germain's *Anatomy of Movement:* the text is often woven around the simple illustrations that it elucidates. This visual approach makes the subject easily accessible.

Also in the spirit of accessibility, in the anatomical descriptions the focus is on only those elements that pertain to the voice or that will be helpful in a vocal practice. Some structures, such as vessels and nerves, are not shown. Pathologies are not included either. Similarly, this book does not use the International Phonetic Alphabet when describing phonemes because this would have entailed a detailed description. Instead, it uses the common English alphabet.

The colors used in the illustrations were selected to facilitate easy reading. To this end, sometimes those colors aren't realistic, but they are consistent with the color code used throughout the book.

- As in many anatomy books, bones are presented as beige because it is easier to see against a white background than ivory, which would be more realistic.
- When bones and muscles are represented in a drawing together, bones are presented in gray.

- Cartilage is presented in light blue to represent the vitreous appearance of hyaline cartilage, although in reality much of the cartilage presented in the book is actually yellowish in color because of the fibers it contains.
- Muscles appear simply as solid or crosshatched red.

In order to keep the anatomical vocabulary accessible to a nonmedical audience, in many cases we have used the common nomenclature—such as *shoulder blades* for *scapula*. At other times we have used the scientific terminology, such as *apophysis* to describe certain bony prominences.

Some structures of the vocal tract have been described in other books by Blandine Calais-Germain: *Anatomy of Breathing, Anatomy of Movement, The Female Pelvis, No-Risk Abs.* In order to avoid repetition, we often refer the reader to these books for more detail.

There are many highlighted text blocks that present a detail regarding voice practice or the anatomy of the vocal apparatus. Among these, some explain how to palpate an area to find a specific anatomical structure. When you are asked to palpate an area, remember that this isn't a massage course—you need use only the lightest touch.

Acknowledgments

We wish to thank the people who assisted in the development of this book, those who posed for the drawings, and those whose debates, sometimes endless, helped bring it to fruition.

Pierre-Yves Binard
Bernard Coignard
Françoise Contreras
Anne Debreilly
Odile Dhénain
Benjamin Duluc
Gloria Gastaminza
Brigitte Hap
Francis Jeser
Allison Liddiart
Ibai Lopez
Jose Luis Marin Mateo
Etoile Mechali
Nicole Nussbaum
Julia Roux
Simone Ushirobira

Thanks as well to designers Marie-Luce Dehondt and Florence Penouty for their endless enthusiasm.

1 Introduction

Some Rules for Describing Movement

Whenever we talk about the voice, the subject of movement comes up. Naming and describing movement is a complex task because different movement techniques have different nomenclature systems, and there are always special cases. Movement in relation to the voice is no exception. Without going into too much detail, we've adopted some simple rules when talking about movement.

The Reference Position

In a standing position, the feet are parallel and the arms are at the sides. When we refer to a structure as vertical or horizontal, it is from this standing reference position.

The Planes of Movement*

Movements to the front and back are made in what we call the *sagittal* plane. In general, a movement to the front is *flexion,* and a movement to the back is *extension.*

Lateral movements are made in what we call the *frontal* plane. Lateral movements when referring to the spine are called *lateral inclination* or *lateral flexion.* When referring to the limbs or vocal cords, we speak of *abduction* (moving away from the center) or *adduction* (moving toward the center).

Movements around a vertical axis are said to be made in the *transverse* plane. Movements in this plane are typically called *rotations:* toward the centerline is *medial* or *internal rotation,* while away from the centerline is *lateral* or *external rotation.*

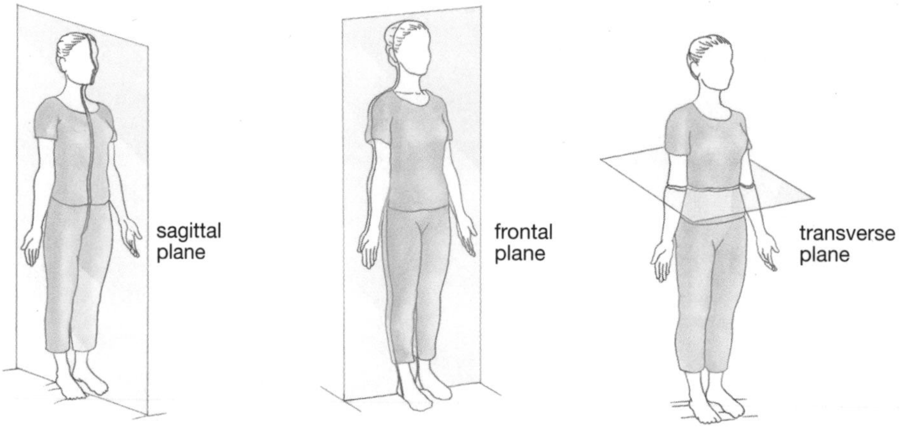

*In reality, most movements are made not exactly in the three orthogonal planes but in a combination of planes and directions. Just think of these planes as tools for description (a bit like a grid on a piece of paper), but know that movements are not limited to them.

The Vocal Apparatus

The voice is commonly known to be produced by an ensemble of regions in the body that we categorize under the term *vocal apparatus*. We distinguish three major functions and assign each to one of the three zones:

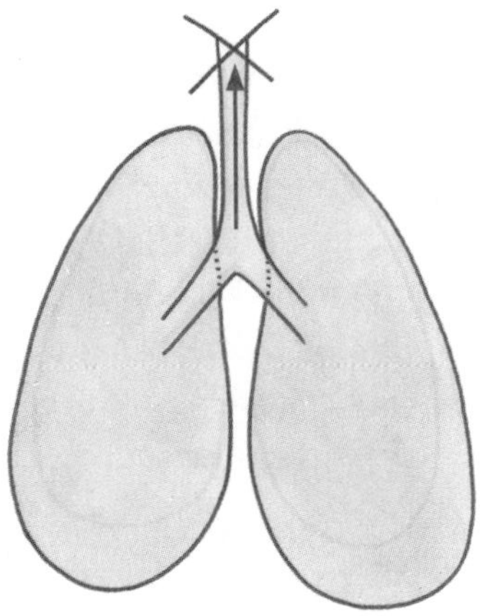

1. The first function involves sending air under pressure toward the vocal cords. This is done by what we call the **generator,** which corresponds to the respiratory part of the vocal apparatus.

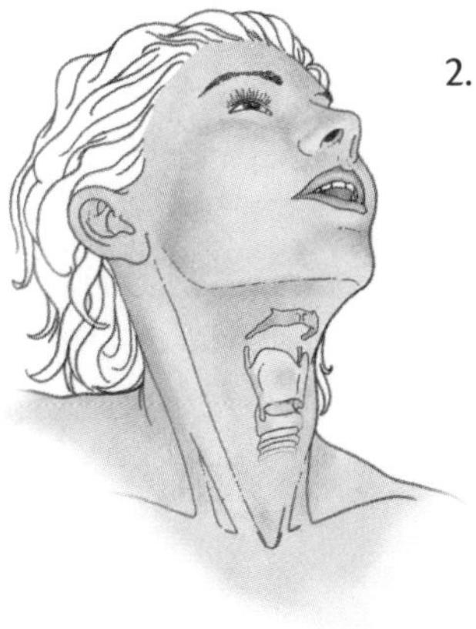

2. From there, an organ starts the air vibrating and produces the initial sound; that's the **larynx.**

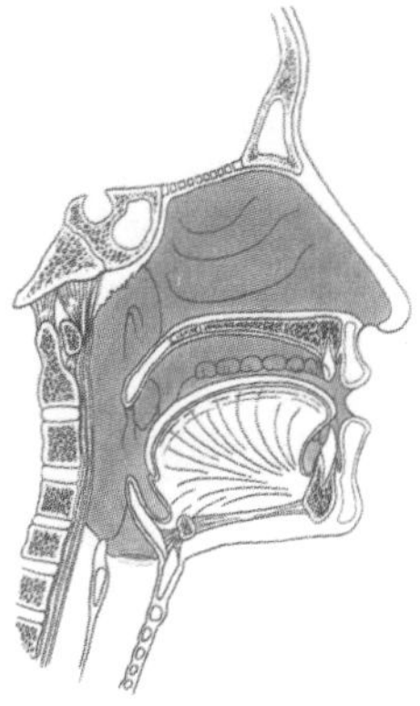

3. **Resonators** then filter and enrich this first sound; the regions of the **pharynx** and **mouth** are responsible for this.

A Couple of Points Worth Noting

- **There is no "vocal apparatus" specifically dedicated to the voice.** The various parts of the vocal apparatus all have other functions in the body. For example, we use the lungs to breathe, the mouth to eat, and the nose to smell. However, a deficiency in any one of these areas affects the voice.
- **The voice is an "event" that can't be broken down into its parts.** When we use our voice, we involve the entire body. Our internal balance and our relationship to the outside world are reflected in the voice. Our intention and our audience influence our voice. Dividing the vocal apparatus into three parts, as we've done above, is purely for the ease of physical description and isn't intended to imply that we can segment the vocal act itself. However, when we are working on our voice, this division can serve to identify areas we intend to develop.

The Moving Body (Including Static Posture), Breathing Body, and Vocal Body

If we're talking about the *vocal apparatus*, we have to include the entire body. When we talk about singers or speakers, we say that their body is their "instrument," and we tell them to sing "with their entire body."

With this in mind, understand that we're not looking at the vocal tract as a separate entity; we're looking at how it interacts with the body as a whole. To facilitate understanding of this concept, we describe three "bodies" that coexist in the voice and how they interact:

- the moving body, including static posture
- the breathing body
- the vocal body

These are "functional" bodies, which is to say that we define them by function rather than by distinct anatomical characteristics. The "anatomy" of each is intertwined with the "anatomy" of the others; however, recognizing them individually will allow us to better understand how they interact to create the vocal experience.

The Moving Body

When we talk about the moving body, we're talking about "large" movements. These would include movements such as walking, climbing stairs, dancing, lifting, pulling, throwing objects, and so on. These "full-body" movements can be seen as involving the entire body, but we assign them, for the most part, to the skeleton, muscles, and joints.

The moving body can be linked to the breathing and vocal bodies. Sometimes it is involved to a very limited degree; there are vocal techniques, in fact, that completely inhibit movement of the body. On the other hand, sometimes the moving body is very involved, whether by responding to the actions of the vocal or breathing body or by initiating the action.

Jogging, Yoga, or Pilates to Improve the Voice?

To produce certain sounds, it's necessary to be able to control the position of the body. So good physical coordination is often important for good sound production, and this coordination can be improved by regular exercise.

However, controlling the body's position is not required for all sounds. Sometimes we may be asked to let go of the body entirely, such as in certain emotional release techniques, rebirthing, primal scream, and types of theater. Activity improves the general circulation, and that includes circulation in the region of the larynx. With better circulation, this area is usually better prepared to produce any type of sound.

Static Posture

When the "moving" body is not in motion but is, instead, motionless in a vertical position, we call this *static posture* (this is not the same as, for example, a yoga posture, which is often not vertical).

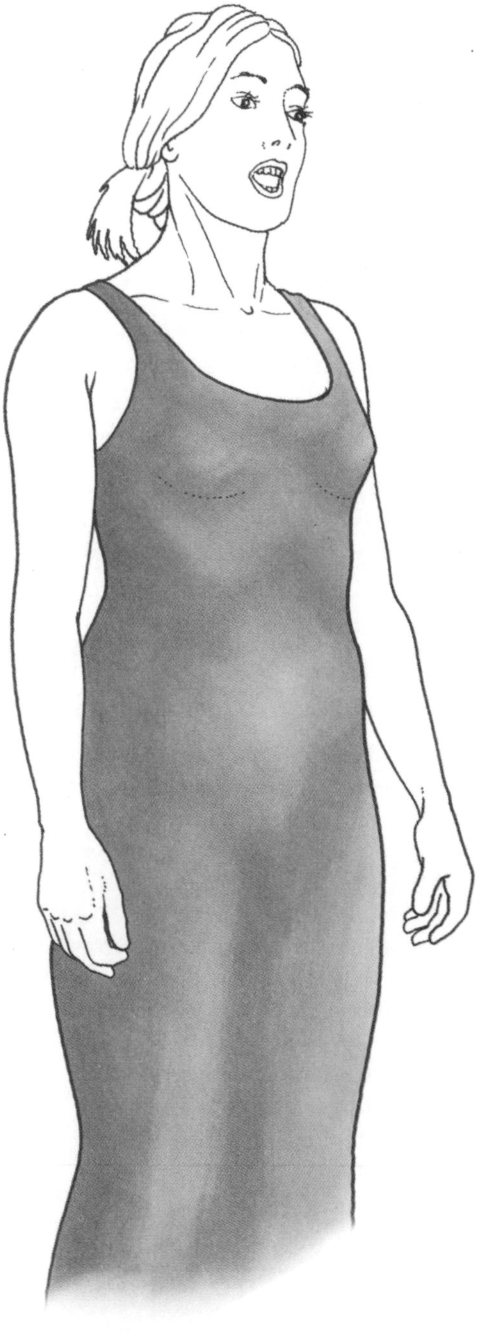

In many situations when we are using the voice, we are immobile and in this more-or-less static vertical position. In most choirs and vocal groups, for example, singers stand, their bodies barely moving. This is also true in solo vocal work or recitation where gestures are minimal.

The Bony Framework

In a standing position, the body relies, more or less, on the weight-bearing bones to support the body systems that surround them.

The Viscera

The viscera don't hold themselves in the thorax and abdomen; they are suspended in fascial envelopes, which connect them to the bones and muscles that surround and support them. In this way the viscera influence the alignment of the skeleton.

The Postural Muscles

A large part of our musculature is dedicated to keeping the body aligned and upright. Collectively, these muscles are called the **postural muscles**. Without them we'd fall over. Of course, when we move these same muscles may be involved in the execution of our movement. Saying that muscles are postural only means that posture is their primary function; it is not necessarily their only function.

Various Postural Positions

Standing posture varies from person to person depending on the combined forces acting on the body. These forces are influenced by body proportion, habits, and various muscle recruitment patterns. The study of these variations is beyond the scope of this book; however, what is important to remember is that, especially in a standing position, certain alignments are advantageous to efficient vocal production and others are disadvantageous.

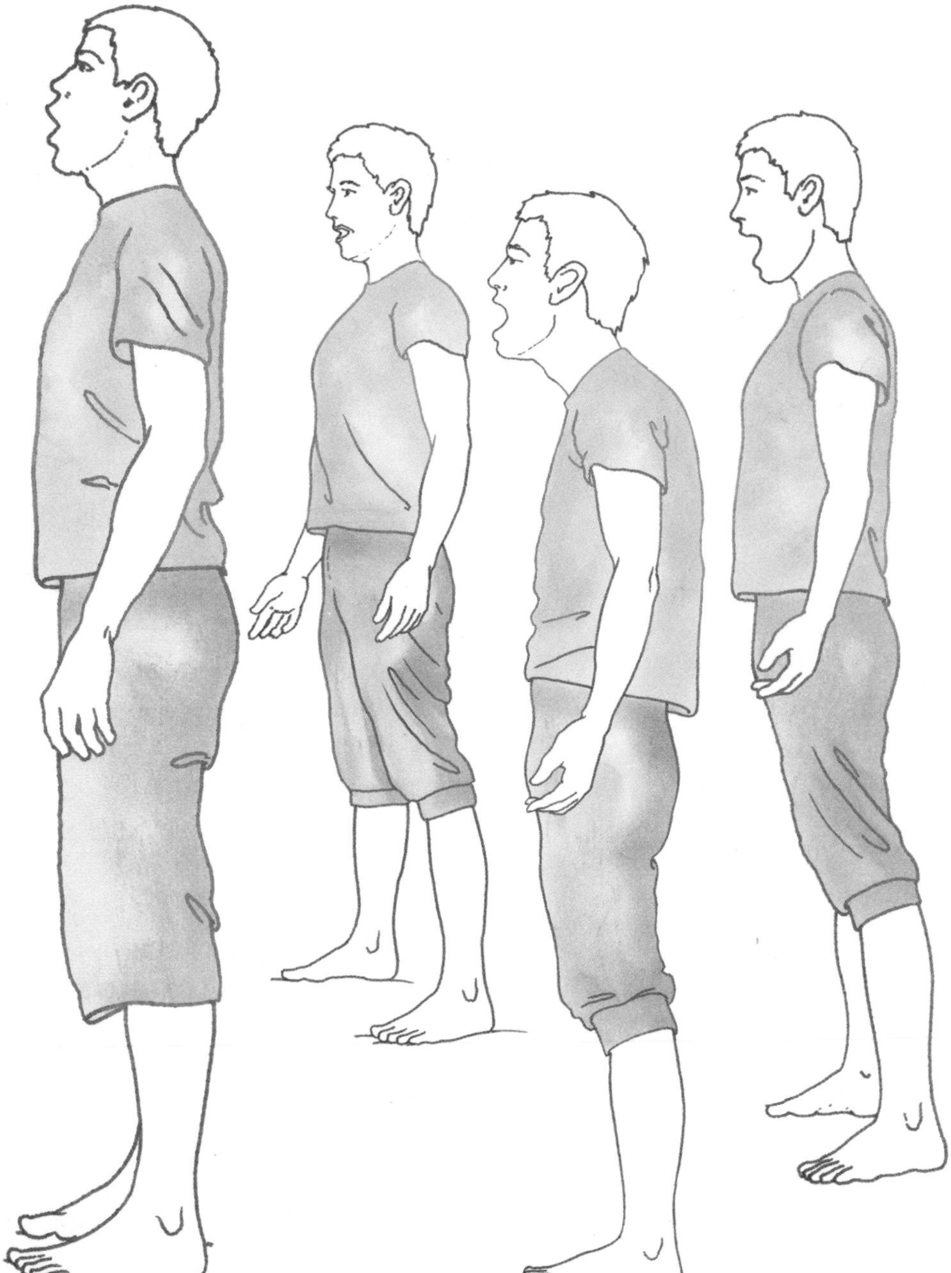

The Breathing (or "Pneumatic") Body

In this work we'll call the parts of the body that are linked to respiration the *breathing body*.

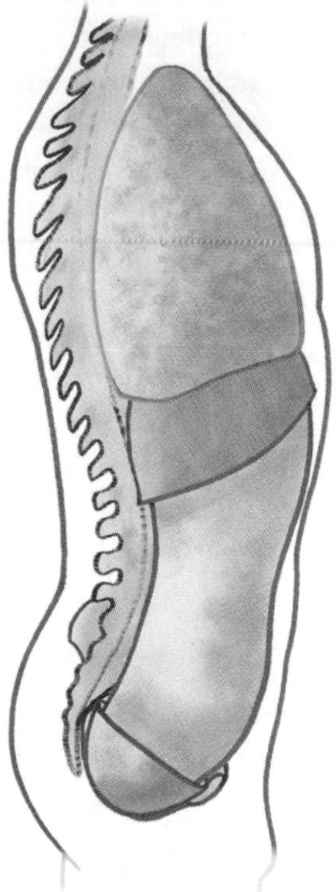

They include, of course, all of the visceral organs of respiration: the lungs and the upper and lower airways. We also include the structures that are linked to the movement of these organs: the diaphragm and abdominal cavity, the cervical spine, almost all of the bones of the skull, and the muscles that act on these parts of the body.

The breathing body moves spontaneously during respiration. It "opens" during inhalation and "closes" during exhalation. In the case of apnea, any movement spontaneously comes to a halt.

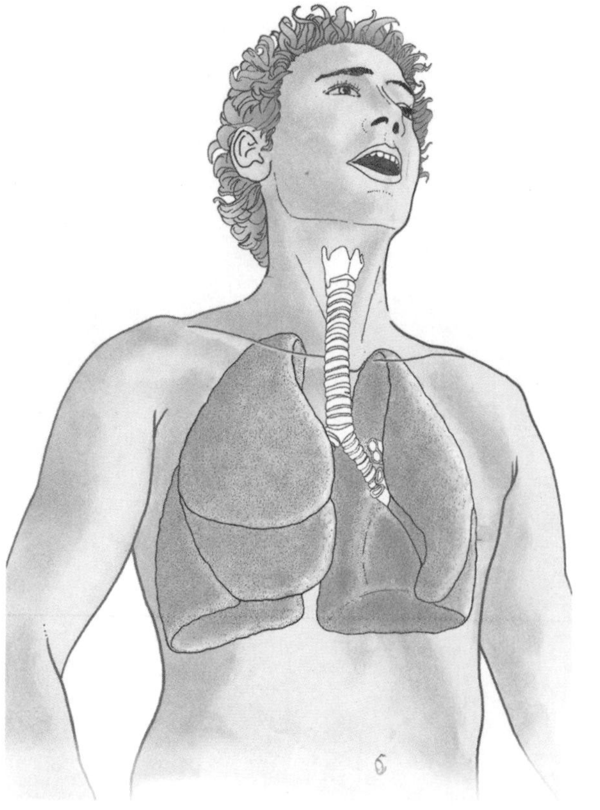

That said, the breathing body can move in opposition to these natural and automatic movement patterns. For example, it can move during apnea, and it can "open" during exhalation. In this event, there is either a change in the normal movement patterns or a change in the internal pressure, either of which can have an effect on the voice.

So Do I Need to Learn to Breathe?

Everyone already knows how to breathe to survive. When we talk about breathing effectively for using the voice most efficiently, it's not a question of taking in more air. In fact, we need very little air when singing. It's a question of better managing the air, especially the air pressure. (When we talk about taking deep breaths before singing, it's more about relaxing than increasing capacity.)

The Vocal Body or Vocal Apparatus

This is the part of the body related to phonation ("the phenomena which contribute to the production of the voice and articulated speech"*).

This includes, of course, the entire respiratory tract just discussed, but it's in action most often during the exhalation.

Certain vocal functions take place primarily in parts of the vocal apparatus that is described on page 17. Those functions are:

- vibration of the exhaled air and formation of the first sounds (the vocal cords)
- resonance of the sound that has been produced
- articulation of the sound

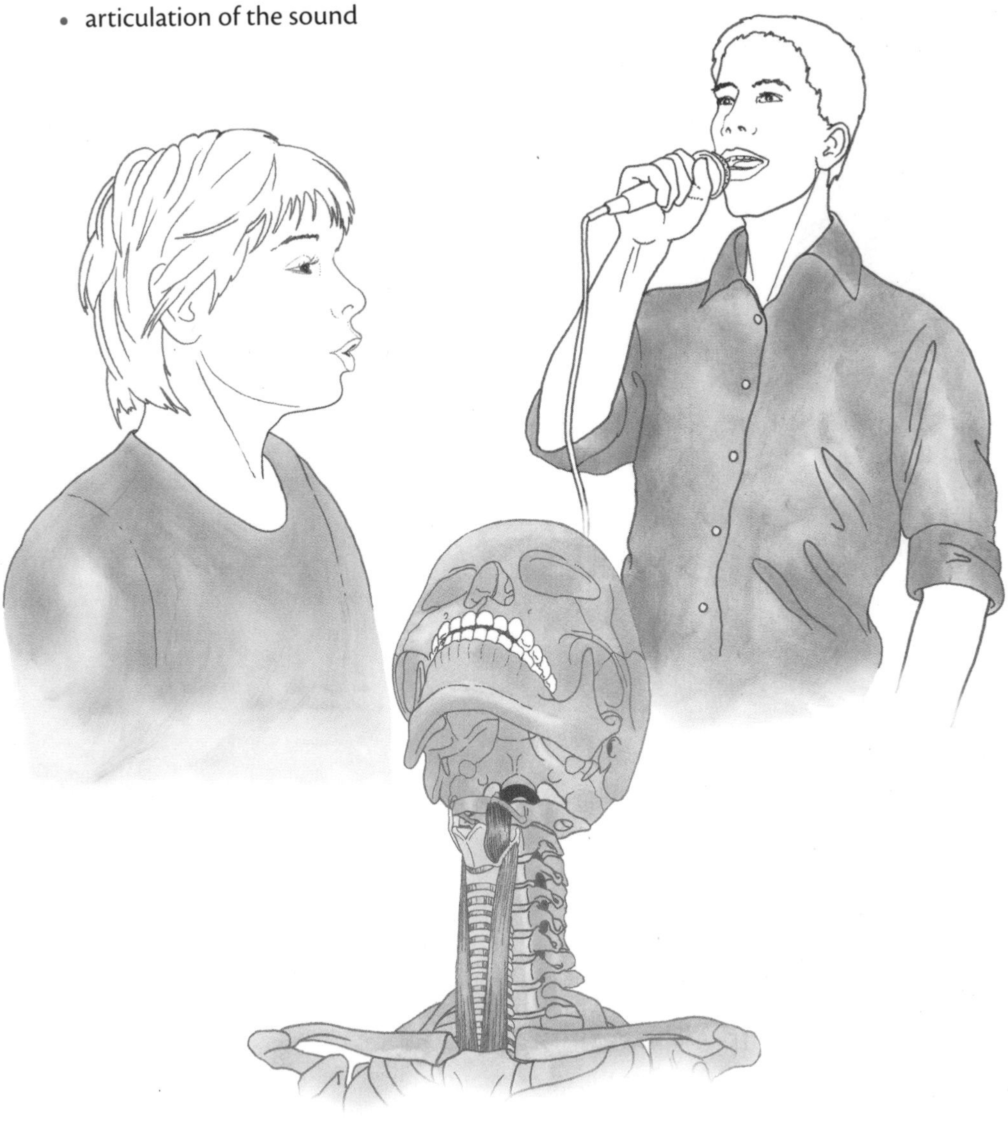

*Source: *Le nouveau petit Robert* [a classic French dictionary—Ed.]

The Moving, Breathing, and Vocal Bodies Can Interact

The moving body can influence the breathing body . . .

by the form it takes:

For example, lifting the arms results in opening the ribs, which facilitates inhalation, even a costal inhalation . . .

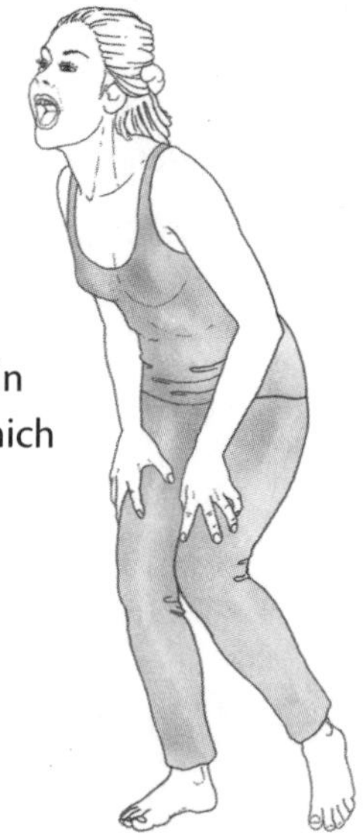

and rounding the spine forward in flexion tends to drop the ribs, which encourages exhalation, et cetera.

by the rhythm it sets:

Running or swinging the arms will change the rhythm of the breath through respiratory and cardiovascular adaptations. The force of the body against the ground, or the fact that swinging the arms shakes the trunk, will also cause a change in respiration.

The moving body can be influenced by the respiratory body . . .

For example, exhalation tends to pull the shoulders and arms forward, while a strong inhalation tends to straighten the spine.

The vocal body can be influenced by the breathing or moving body . . .

For example, if you flex your thigh rapidly toward your chest, this will most likely round the lower part of the spine and cause you to exhale. This action can contribute to the production of a sound.

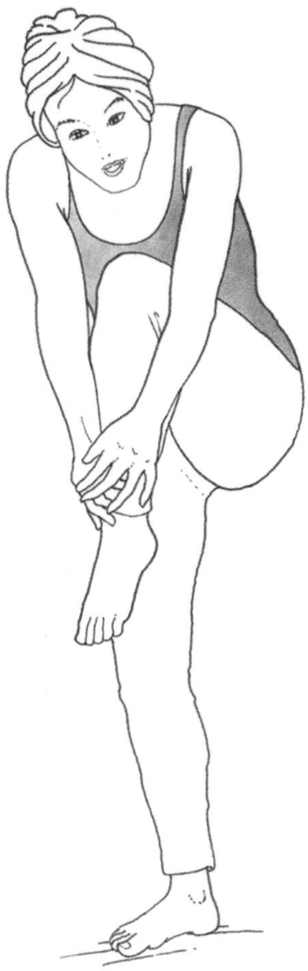

The vocal body can be influenced by the breathing body . . .

For example, dropping the sternum encourages an exhalation, which can help produce a sound.

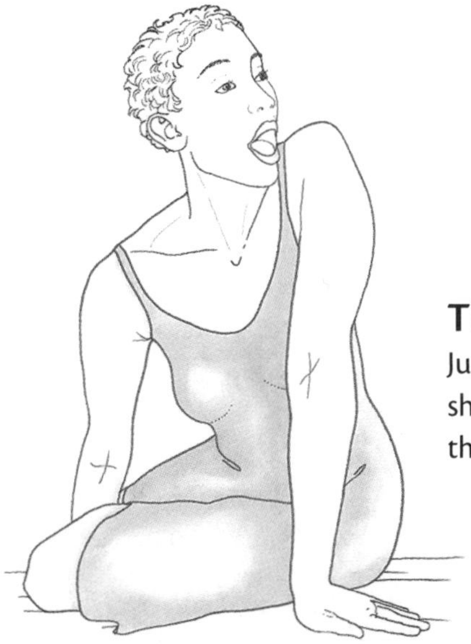

The moving body can influence the vocal body . . .

Just turning or tilting the head, or turning, lowering, or raising the shoulders, can change the position of the larynx and therefore change the voice.

2
The Skeleton of the Voice

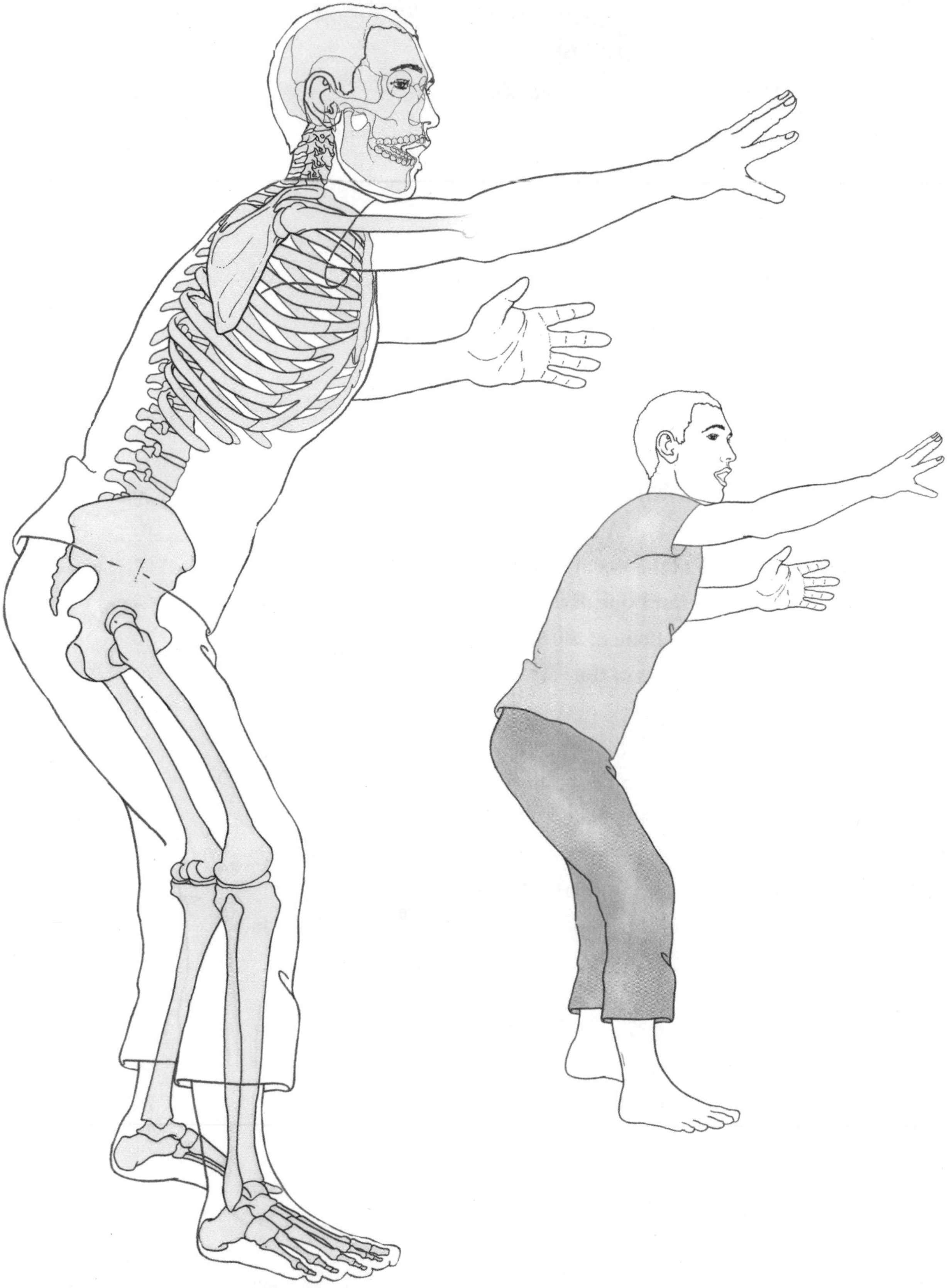

The voice is an aerial act derived from respiratory and vocal structures that are housed in a bony framework. The rigidity of this framework allows us to define a precise form for respiratory movements and some vocal movements as well.

Though we speak of a bony framework, among the numerous bones and cartilaginous structures that make up this framework, many are connected by joints that allow for a certain amount of mobility.

In this chapter we'll look at the bones that are linked to the voice:

- Those concerned with respiration, or the generator; these are the bones in the middle region of the body: the spine, pelvis, rib cage, skull.

- Those involved with the larynx; these are the cervical vertebrae, the base of the skull, the mandibles, the sternum, and the shoulder girdle.

- Those involved in resonance and articulation; these are the upper cervical vertebrae and the skull.

- Those involved in the standing posture; these are the load-bearing bones that support the weight of the body—the bones of the lower limbs, pelvis, and spine.

The following pages describe the skeleton as it relates to the voice. For a more detailed look at anatomy, see *Anatomy of Movement* and *Anatomy of Breathing,* by Blandine Calais-Germain.

The Spine: The Link between the Trunk, Neck, and Head

The spine is composed of several regions made up of vertebrae of different types.

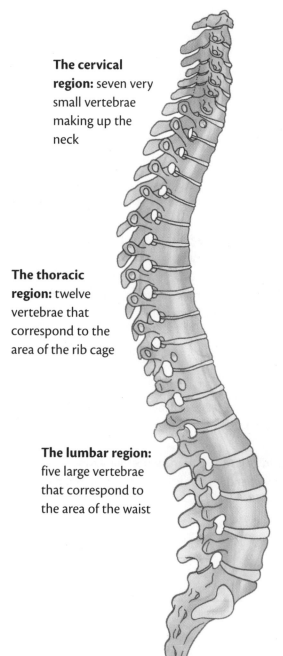

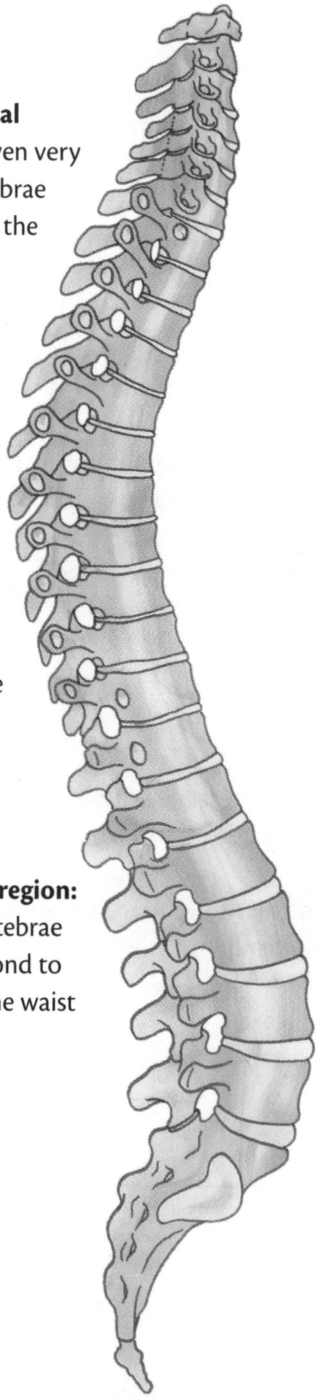

The cervical region: seven very small vertebrae making up the neck

The thoracic region: twelve vertebrae that correspond to the area of the rib cage

The lumbar region: five large vertebrae that correspond to the area of the waist

The region of the sacrum and coccyx: the lowest vertebrae, which are fused and make up a part of the pelvis

The spine doesn't serve the voice directly but acts as a "tutor" for the vocal/respiratory/moving bodies. Moreover, it's an adaptable tutor. In terms of the vocal body, it's the link between the generator, the area of the larynx, and the resonators. It's described here from top to bottom.

The spine is made up of thirty-two vertebrae, which are separated/united by articulations.

At the top the spine is connected to the skull at the occipital bone, and at the bottom it's connected to the pelvis.

When we view the spine in profile, we can see its *curves:*

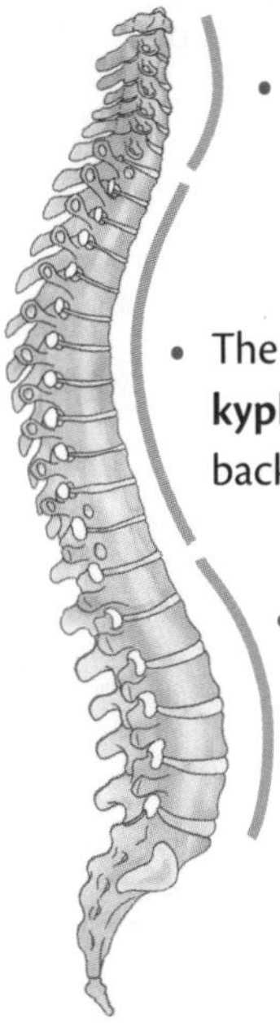

- The cervical vertebrae are in **lordosis**, a concave curve to the back.
- The thoracic vertebrae are in **kyphosis,** a convex curve to the back.
- The lumbar spine is in **lordosis,** a concave curve to the back.

These are the normal curves, and they are necessary for proper functioning. Sometimes, however, they can be

- accentuated (hyperlordosis, hyperkyphosis)
- flattened or inverted ("inversion of the curves")

When we look at a group of people singing in a standing position, we see a variety of different spinal curves. These curves are the result of each person's general posture combined with how he uses his body when he sings.

In this book we will see how sometimes these curves should be controlled during the vocal act, and sometimes they shouldn't be.

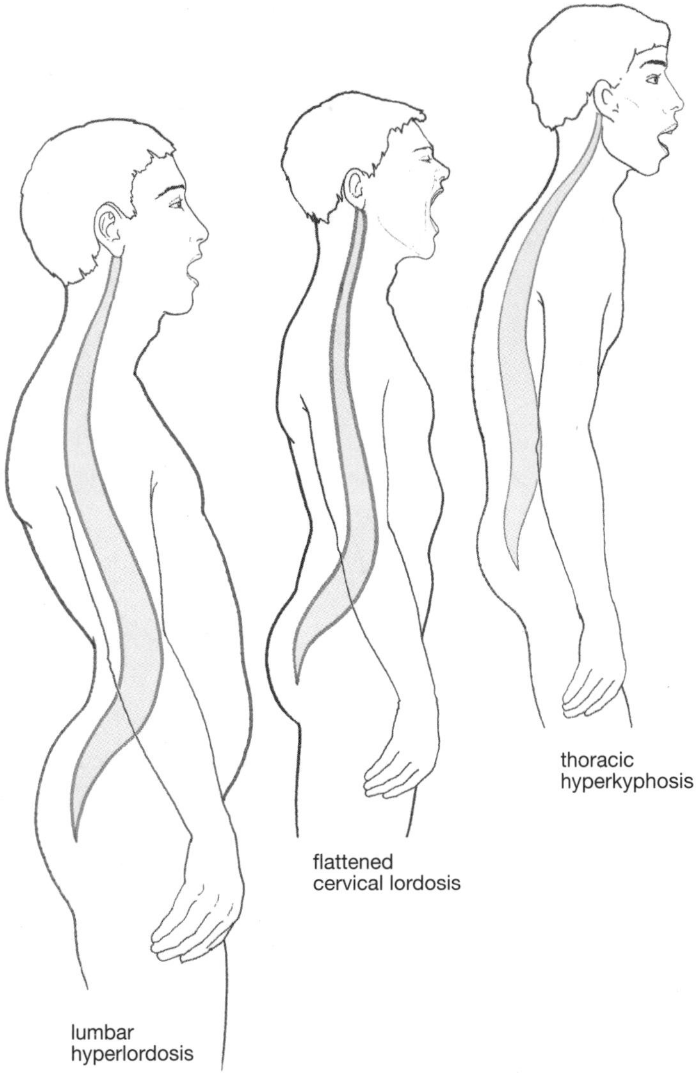

The Vertebrae

The vertebrae are the smallest bony units of the spine. Each of these small bones has two major components: the vertebral arch and the vertebral body.

The **vertebral body** is the anterior component. Cylindrical and compact, it serves as the weight bearer.

The **vertebral arch** is the more complex posterior part. It features several projections that serve as attachment points for muscles and ligaments. There are three kinds of projections:

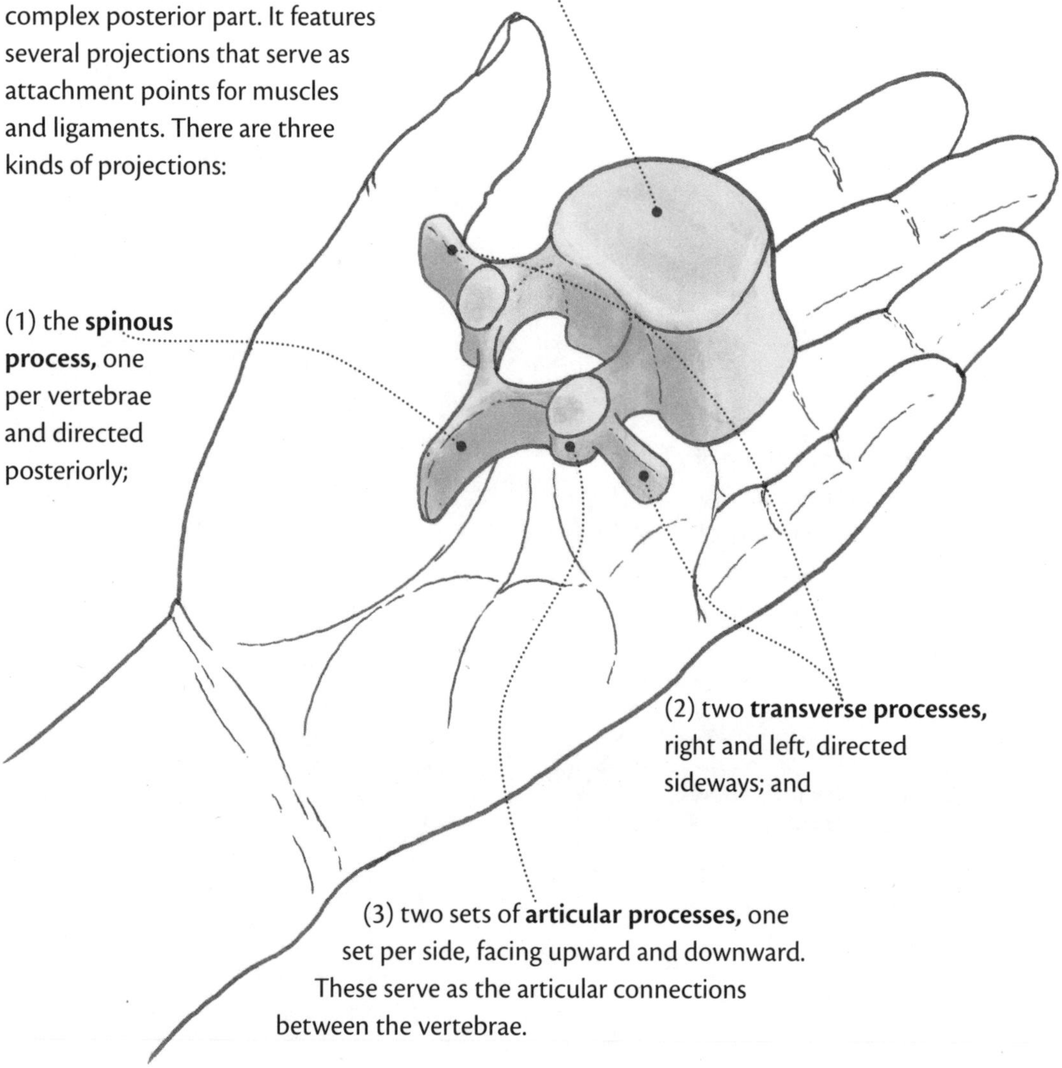

(1) the **spinous process,** one per vertebrae and directed posteriorly;

(2) two **transverse processes,** right and left, directed sideways; and

(3) two sets of **articular processes,** one set per side, facing upward and downward. These serve as the articular connections between the vertebrae.

The Vertebrae Articulate with Each Other

Between two vertebrae we find three points of articulation as shown below:

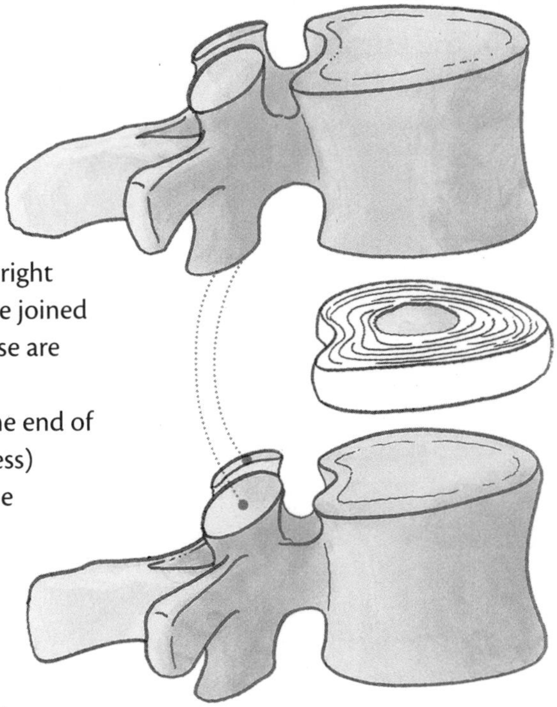

At the posterior, on both the right and left sides, the vertebrae are joined at the articular processes. These are mobile joints with:

- **articular cartilage** (at the end of the corresponding process)
- a fibrous sleeve called the **capsule**
- **synovial fluid**
- **ligaments***

At the anterior, an intervertabral disk (between the two vertebrae) serves as the third articulation.

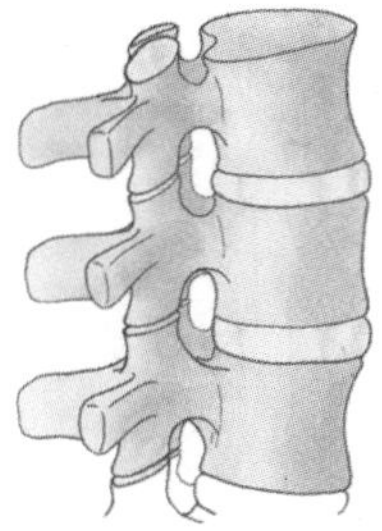

three joined vertebrae

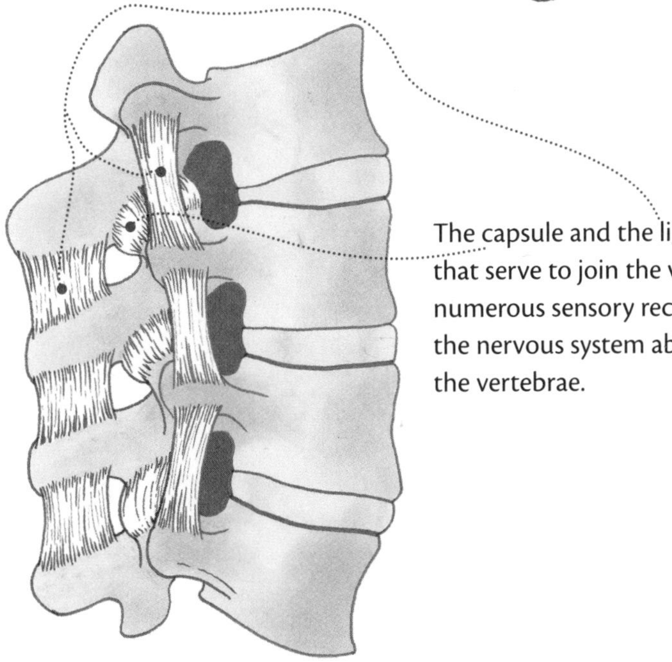

The capsule and the ligaments aren't solely structures that serve to join the vertebrae; they are also filled with numerous sensory receptors that relay information to the nervous system about pain, as well as the position of the vertebrae.

Effect on the Voice

Exercise can increase the mobility of the spine and the articulations of the vertebrae. While this mobility does not in and of itself modify the voice, it can make the body more receptive to modification in posture and breath.

*More details on ligaments can be found in *Anatomy of Movement*, pages 38–39.

The Lumbar Spine: Waist, Abdominal Cavity

This is the part of the spine between the pelvis and the ribs.

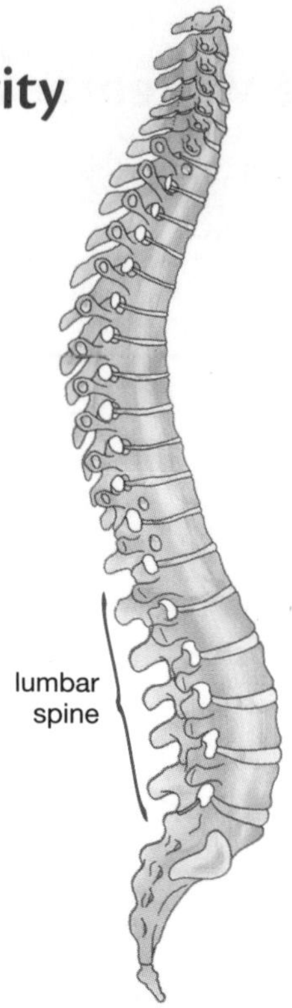

Characteristics of the Lumbar Vertebrae

The lumbar vertebral bodies are large and tall because this is the area that receives the most *load*.

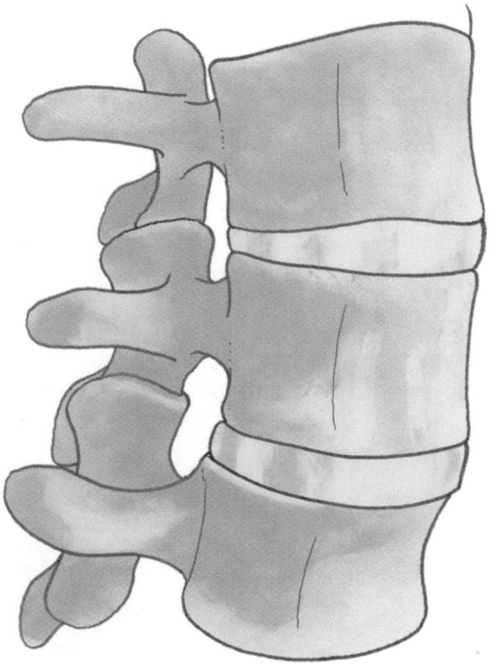

The intervertebral disks are thick.

The ends of the articular processes on both the right and left sides are covered with cartilage:

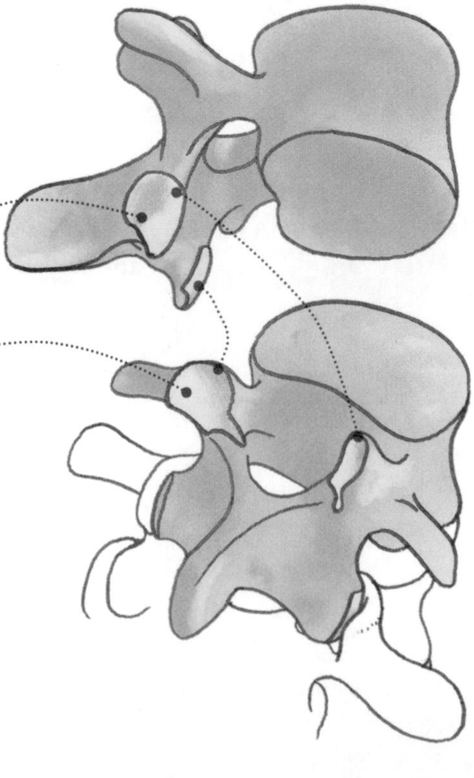

- The inferior (bottom) articular process is cylindrical and full.
- The superior (upper) articular process is cylindrical and hollow.

At each vertebral level, these articular surfaces match up and allow movement between the lumbar vertebrae.

Principal Movements of the Lumbar Spine

Rotation is impossible in this area, but apart from that . . .

- At its top, the lumbar spine is mobile in just about every direction. When we move the rib cage in one direction or another during singing or speaking, this movement usually comes from the upper lumbar spine or the lower thoracic spine.
- At its lower end, extension is the dominant direction of movement.

The Lumbar Spine When the Body Is Upright

The degree of lumbar lordosis is related to the position of the pelvis and the rib cage.

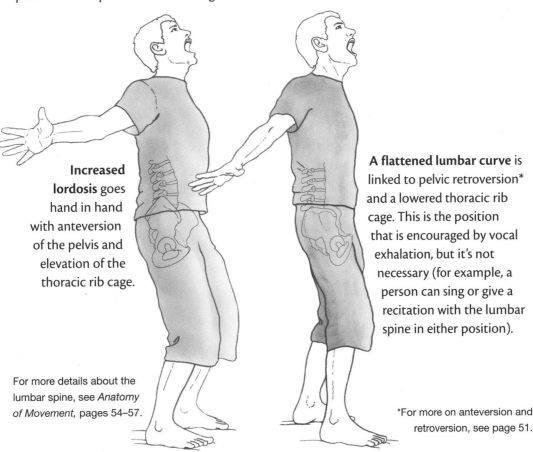

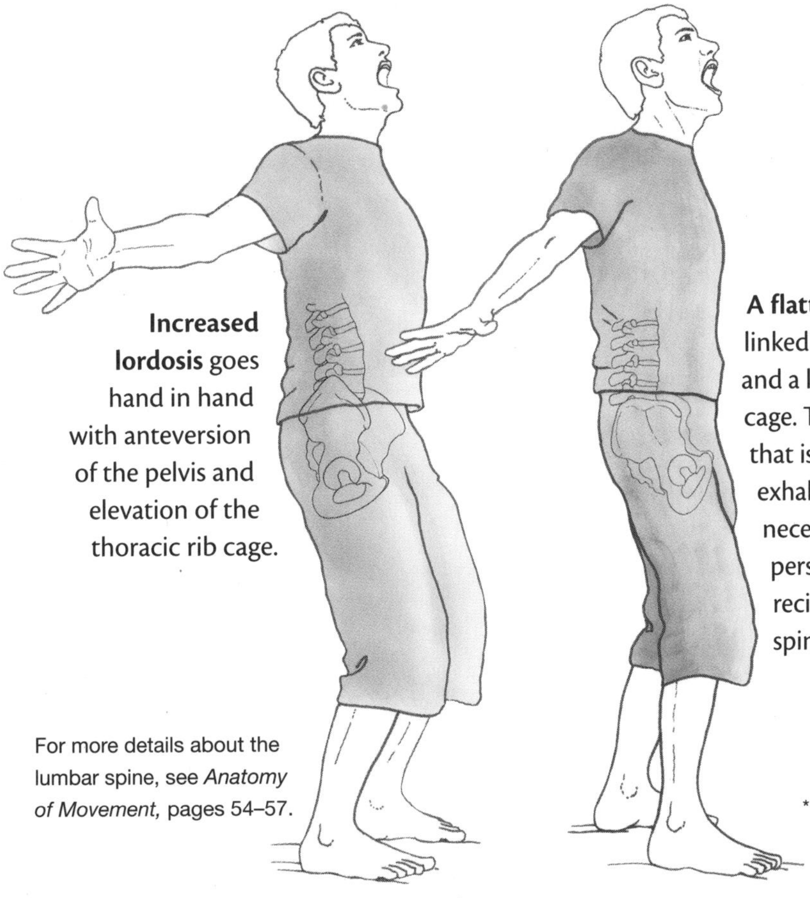

Increased lordosis goes hand in hand with anteversion of the pelvis and elevation of the thoracic rib cage.

A flattened lumbar curve is linked to pelvic retroversion* and a lowered thoracic rib cage. This is the position that is encouraged by vocal exhalation, but it's not necessary (for example, a person can sing or give a recitation with the lumbar spine in either position).

For more details about the lumbar spine, see *Anatomy of Movement,* pages 54–57.

*For more on anteversion and retroversion, see page 51.

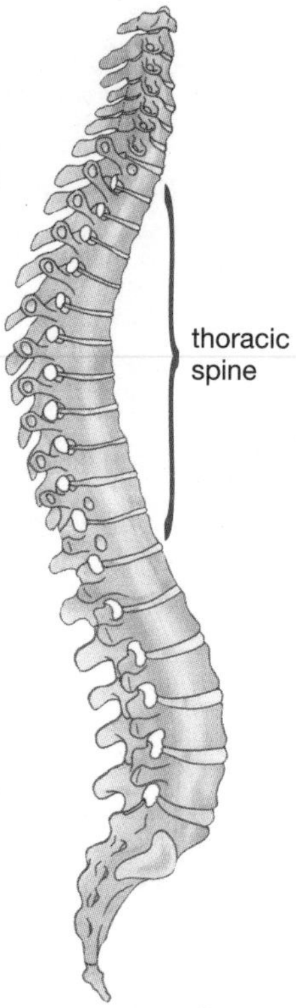

The Thoracic Spine: Back, Thoracic Rib Cage

This is the part of the spine that corresponds to the thoracic rib cage, which we cover in detail on pages 58–60.

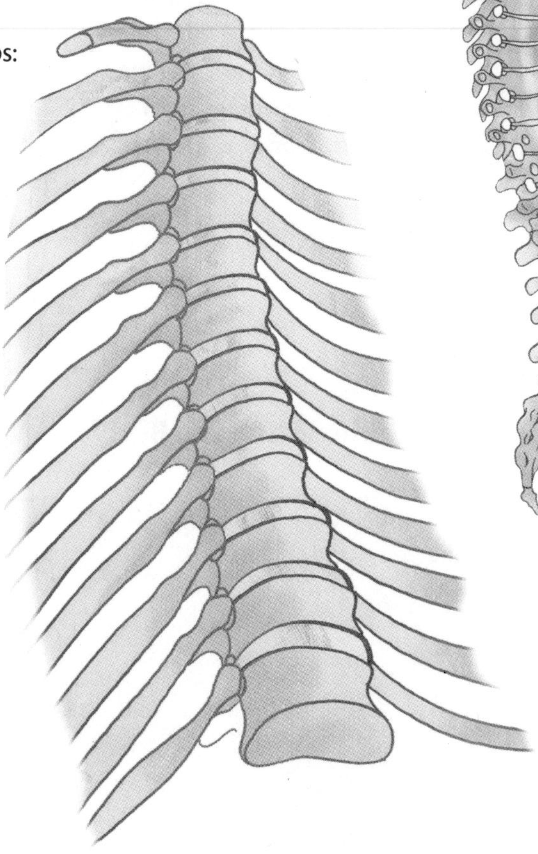

This region articulates with the ribs: between the thoracic spine and the ribs, there are numerous—no fewer than forty-eight—costovertebral joints. They tend to be small. Because of this connection, there is always an interplay between the mobility/stability of the thoracic rib cage and this region of the spine.

The Thoracic Spine When the Body Is Upright

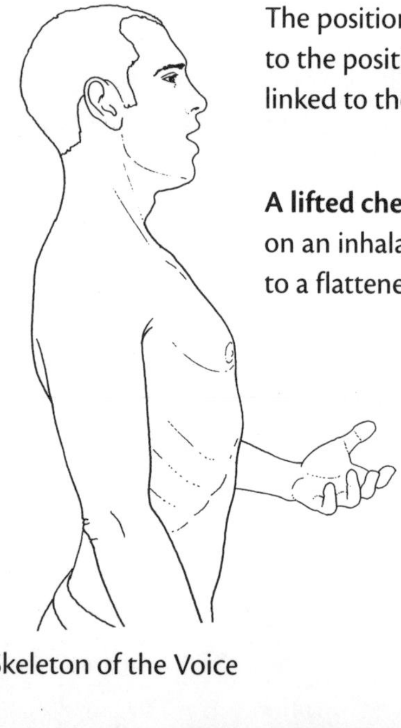

The position of the thoracic spine is related to the position of the rib cage, which is itself linked to the breath.

A lifted chest (as is common on an inhalation) corresponds to a flattened thoracic curve.

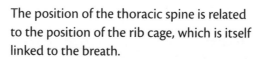

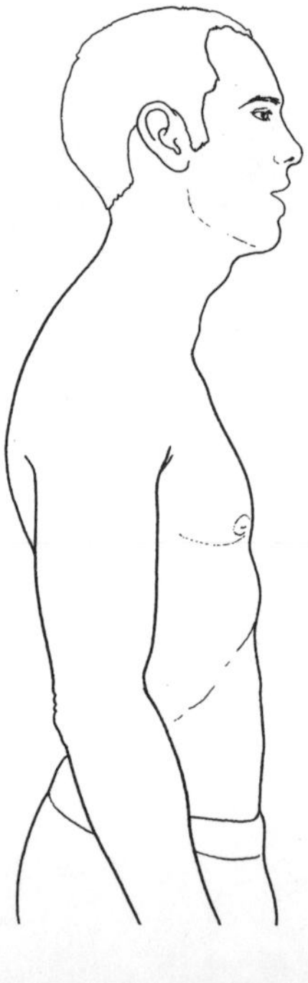

A dropped chest (as in an exhalation) often corresponds to increased thoracic kyphosis.

The Cervical Spine

The neck is made up of seven cervical vertebrae on two levels.

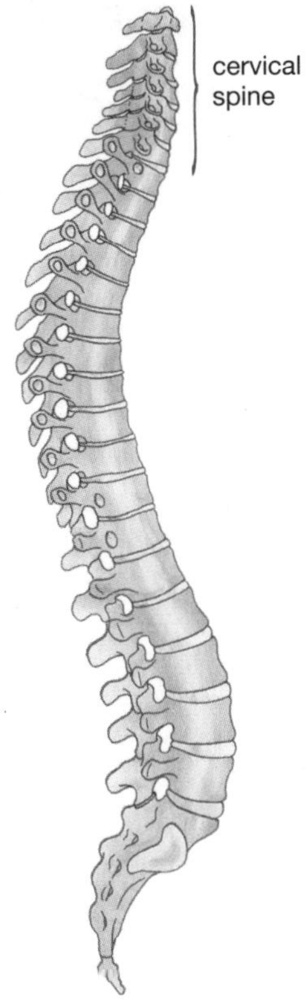

The upper level, also known as the **suboccipital region,** is composed of the **atlas** and **axis,** along with the occiput.

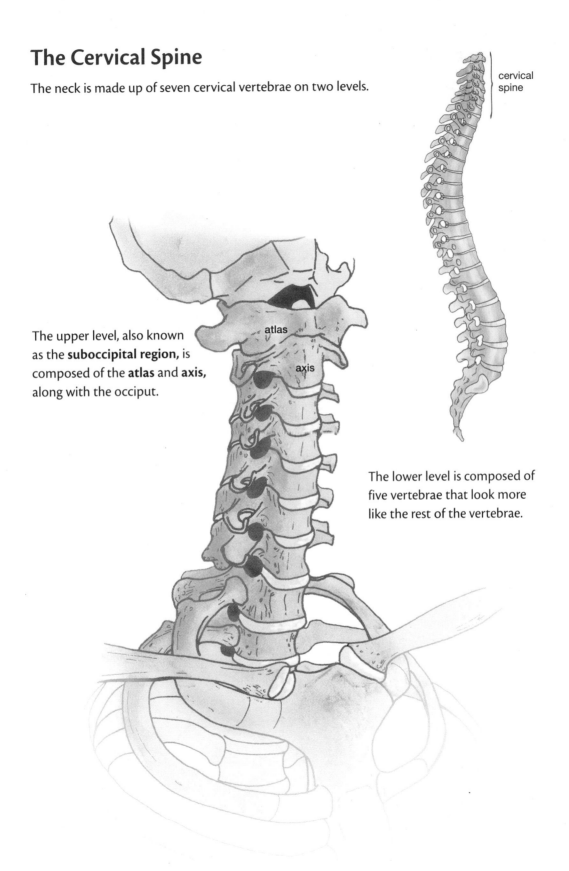

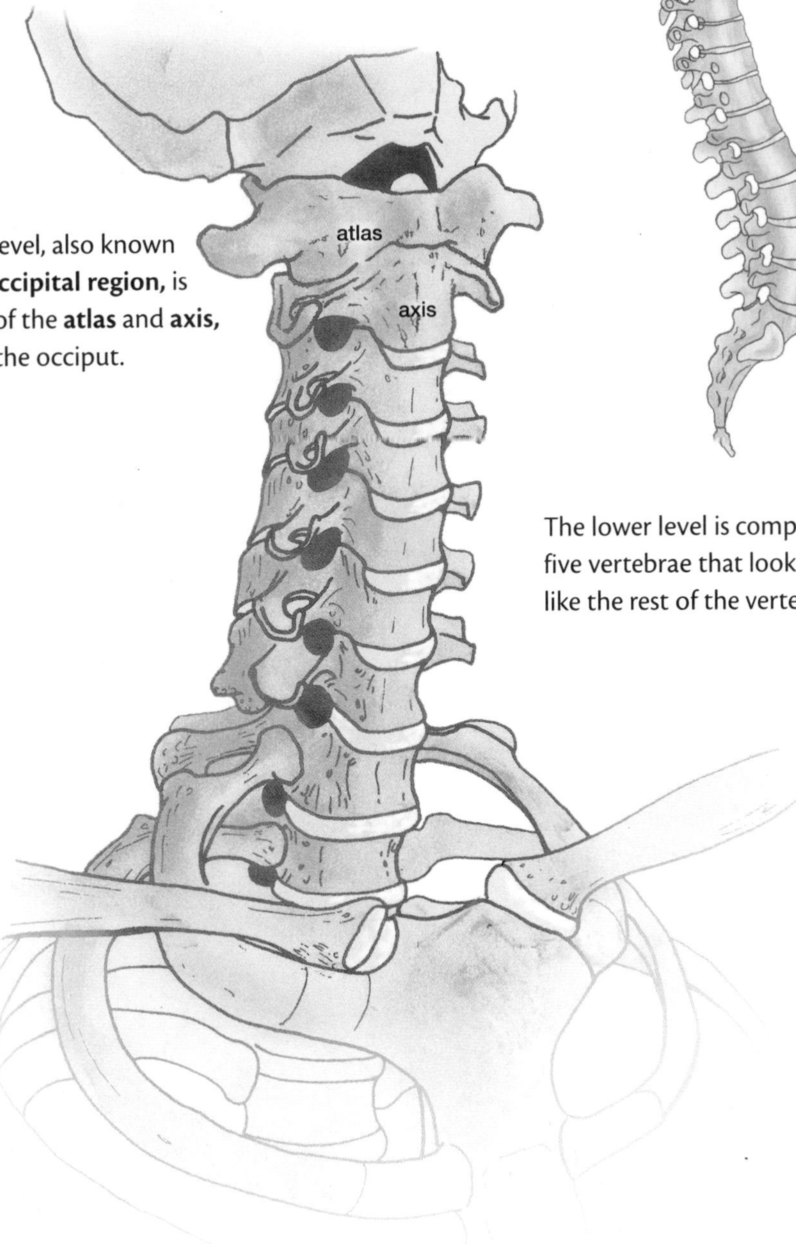

The lower level is composed of five vertebrae that look more like the rest of the vertebrae.

The Cervical Vertebrae from C3 to C7

The Shape of the Cervical Vertebrae

The **cervical vertebral body** is very small. On each side of the vertebral body we find the **transverse processes,** which are:

- grooved, to allow for passage of the cervical nerves; and
- channeled, to allow for passage of the vertebral artery, a small artery that supplies blood to part of the brain.

The **intervertebral disk** is thick, which makes it conducive to cervical mobility.

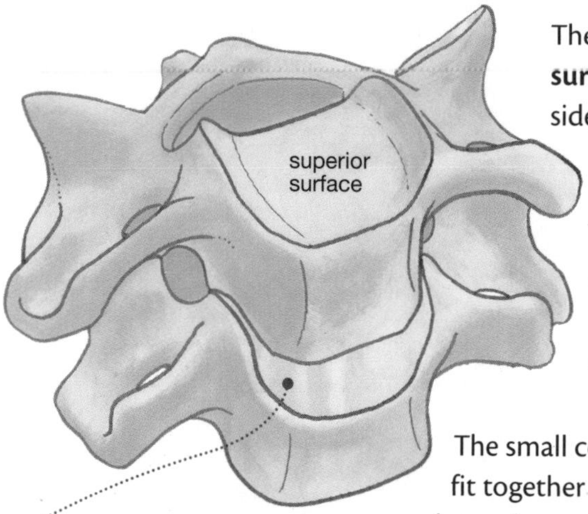

Here, two vertebrae are aligned one on top of the other and viewed from above.

The **upper (superior) surface** is raised on the sides.

The **bottom (inferior) surface,** on the contrary, is beveled on both sides.

The small cervical vertebrae fit together more snugly than other vertebrae, and this contributes to the stability and alignment of the neck.

At the back, the articular surfaces of the transverse processes are covered with cartilage:

- One set of articular surfaces is **inferior,** aimed down and forward.
- The other set is **superior,** aimed upward and back.

At the back, at the joint between each pair of cervical vertebrae, the inferior articular surface of the vertebra above articulates with the superior articular surface of the vertebra below.

Here, the cervical vertebra is lifted and seen from underneath.

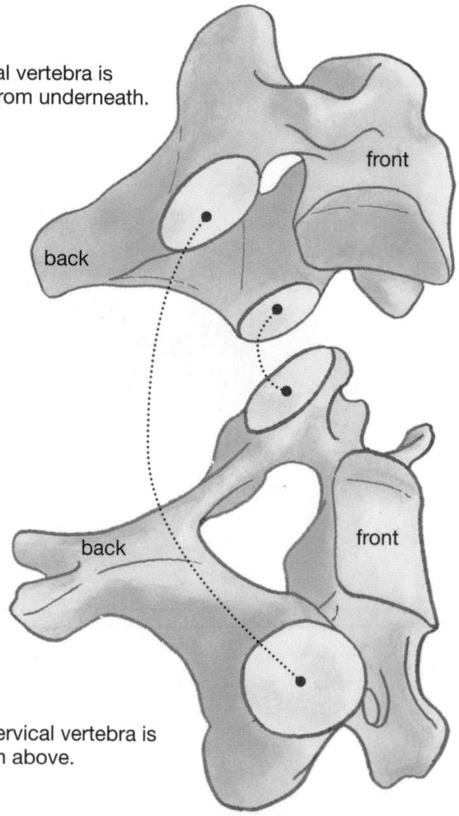

Here, the cervical vertebra is viewed from above.

How the Cervical Vertebrae Articulate with Each Other

The movements of the cervical vertebrae are ample and are distributed as follows:

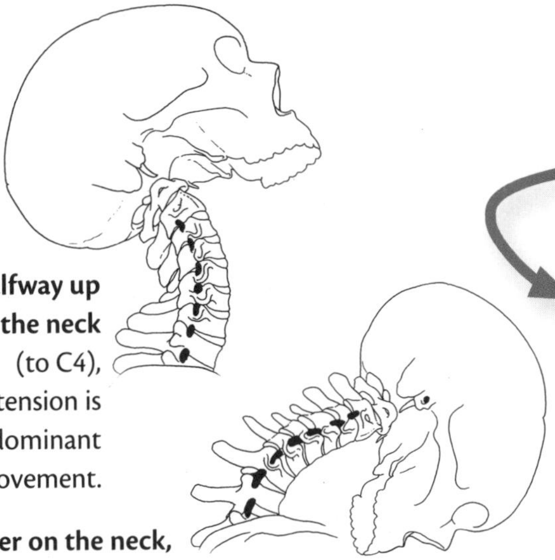

Halfway up the neck (to C4), extension is the dominant movement.

Lower on the neck, it's flexion.

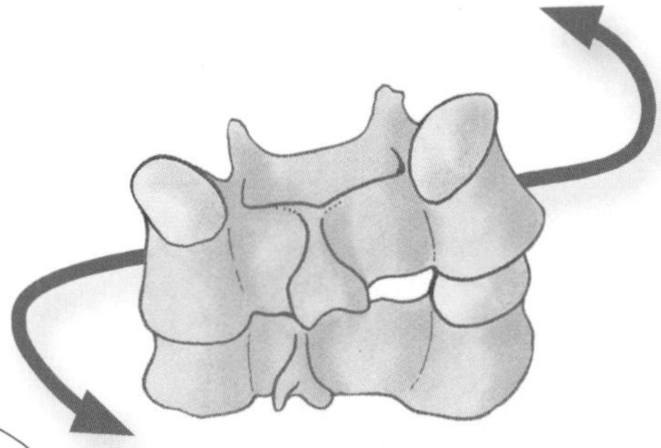

Along the entire length of the neck, lateral movement (lateral inclination) is combined with rotation to the same side, because of the shape of the posterior articular surfaces.

Mobilizing the Cervicals, Risky Movements

The mini-joints between the cervicals can become sore, and this can commonly lead to stiffness. It's important to understand how to remobilize the neck safely.

When you're trying to stretch the neck, it's important to support the weight of the head. If you don't, the muscles of the neck can hypercontract, which can compress the neck's joints and cartilage.

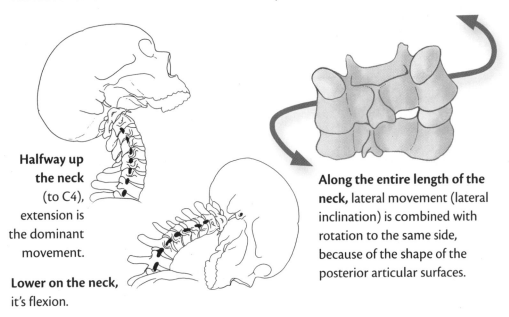

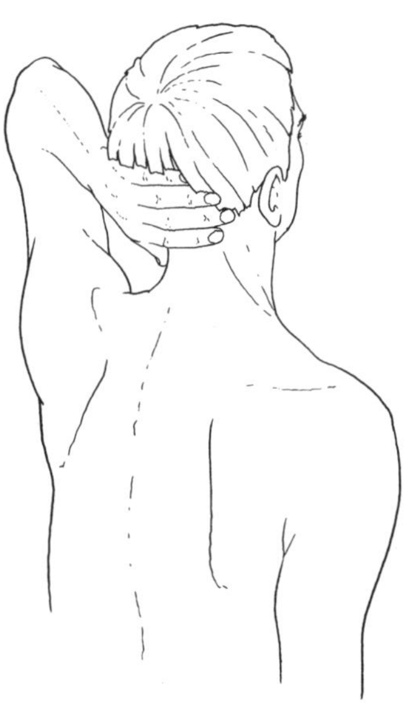

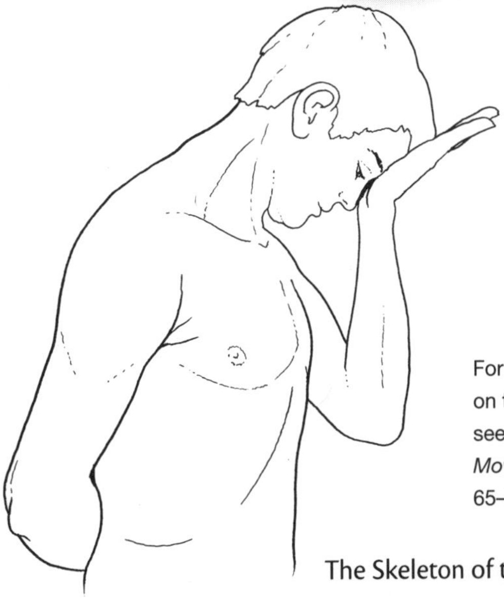

For more details on the cervicals, see *Anatomy of Movement,* pages 65–67.

The Atlas

This is the first cervical vertebra (C1), and the first of all the vertebrae starting from the top of the body. It's found under the occiput and on top of the axis.

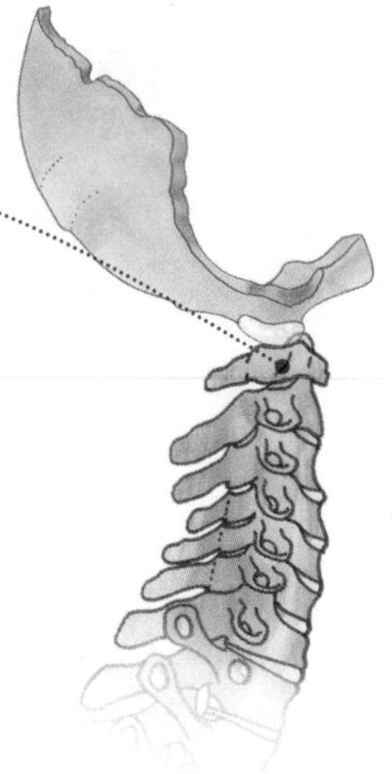

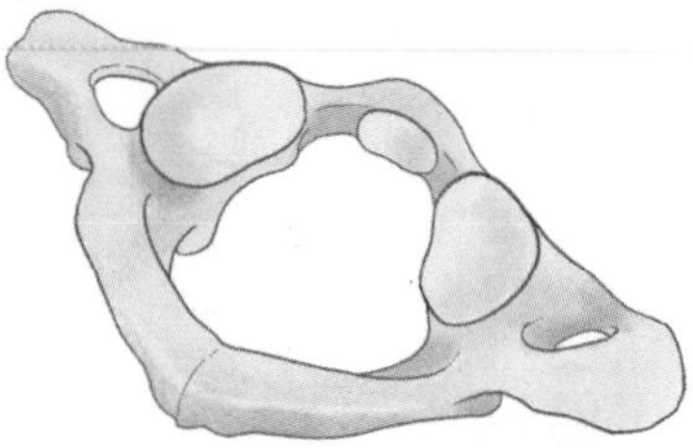

It is ring-shaped and has neither a vertebral body nor spinous processes.

On its superior (upper) surface, the atlas has two thickened areas, called **lateral masses.** The top of each lateral mass is oval in shape, concave, and covered with cartilage. These masses are part of the **occipito-atloid articulations,** where the atlas and the occiput meet.

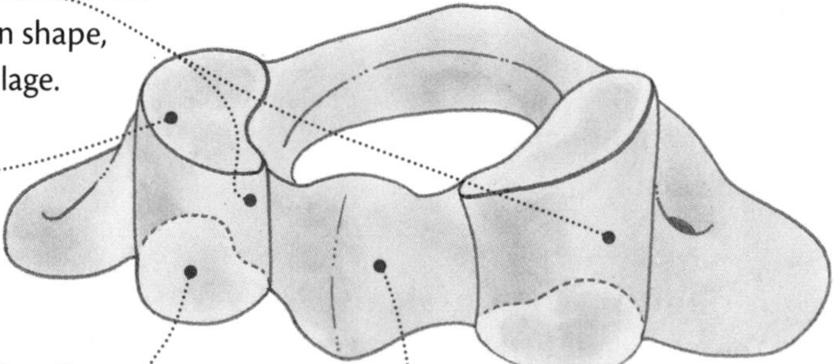

On the inferior surface, we find again two oval cartilage-covered surfaces. These articulate with the axis.

Between the lateral masses are two bony arches:

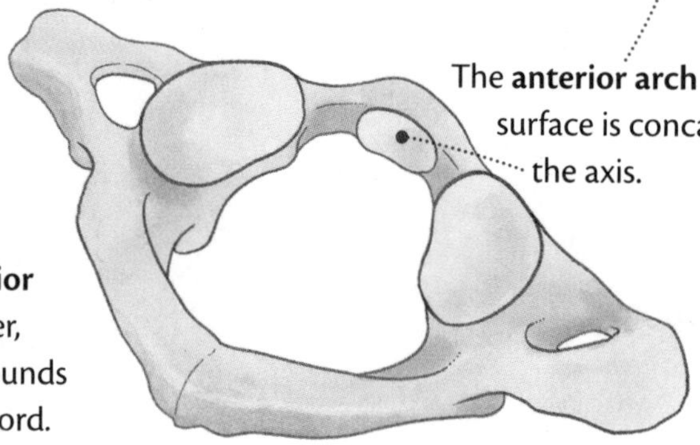

The **anterior arch** is small. Its posterior surface is concave and articulates with the axis.

The **posterior arch** is larger, and it surrounds the spinal cord.

Palpate the Atlas

The transverse processes of the atlas are more or less palpable about 1 centimeter below the earlobe.

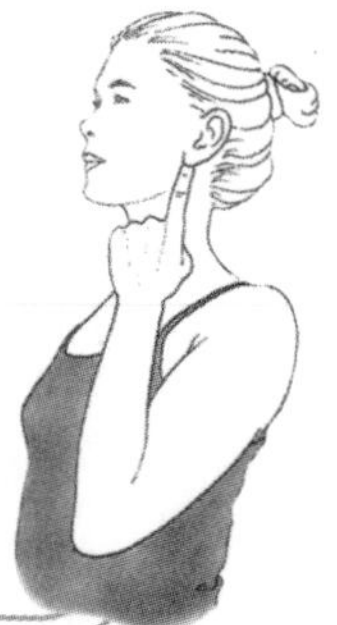

The Axis

This is the second vertebra from the top (C2). Just like the atlas, it doesn't have the "normal" vertebral structure.

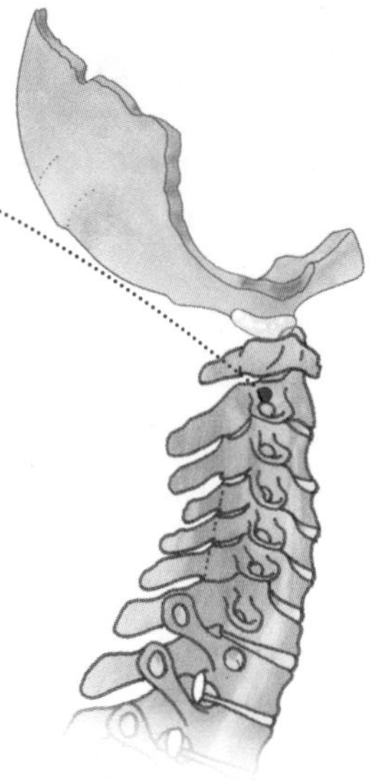

At the front, the vertebral body of the axis has a toothlike projection called the **odontoid process** (or **dens**).

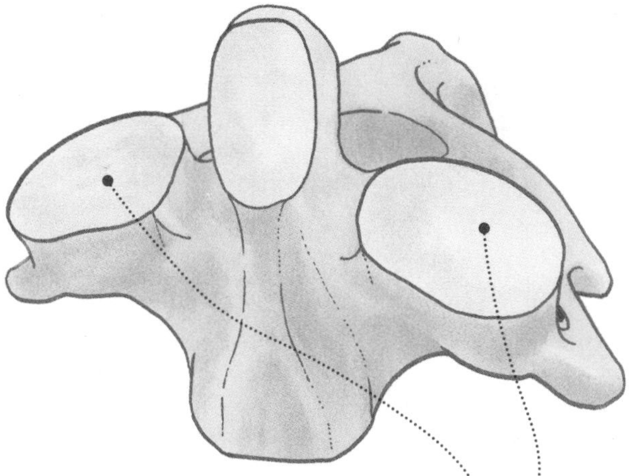

This projection is covered with cartilage on its anterior and posterior surfaces.

The axis has two articular surfaces on its superior surface at the sides of the vertebral body. They are oval and convex, and they articulate with the underside of the lateral masses of the atlas.

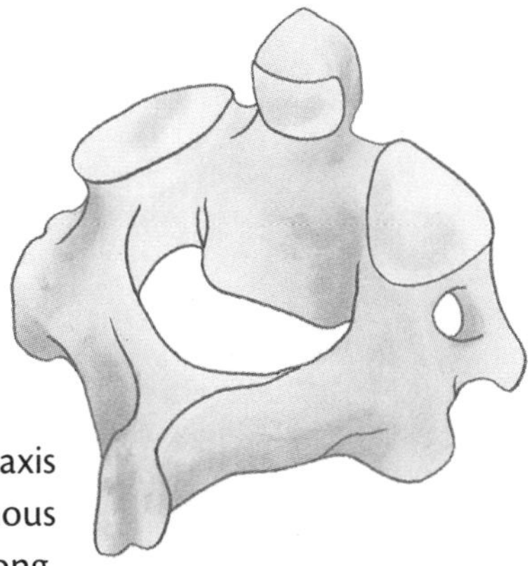

The transverse processes of the axis are short, in contrast to the spinous process, which is long.

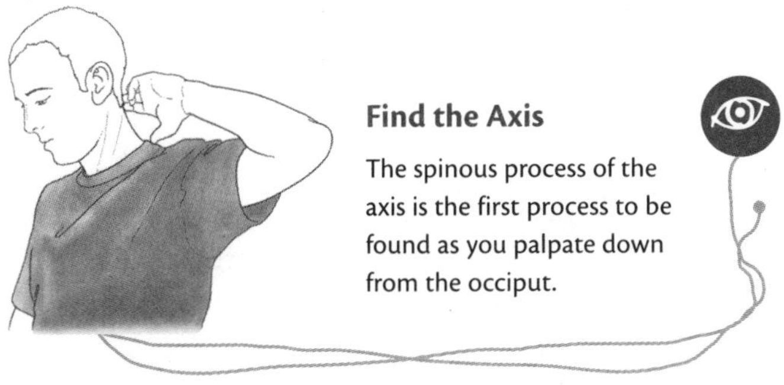

Find the Axis

The spinous process of the axis is the first process to be found as you palpate down from the occiput.

The Head and the Atlas

How the Head Articulates with the Atlas

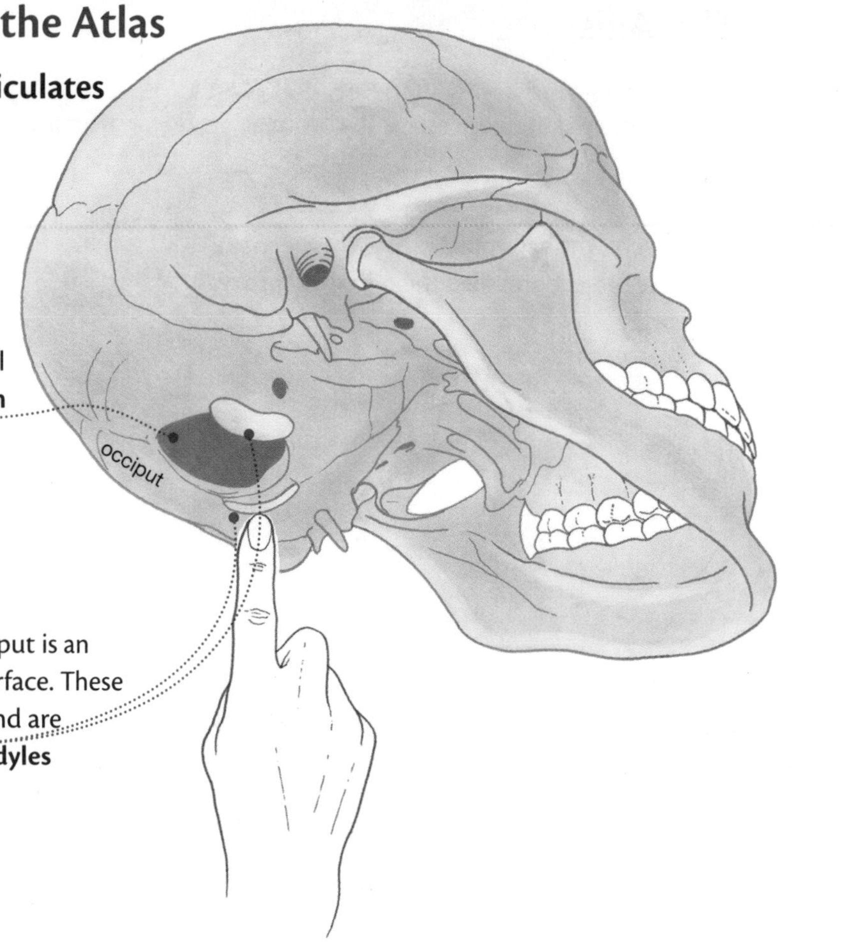

At the base of the occiput is an opening through which the spinal cord passes: the **foramen magnum** (see page 68).

On each side of the occiput is an oval-shaped articular surface. These are convex downward and are called the **occipital condyles** (see page 68)

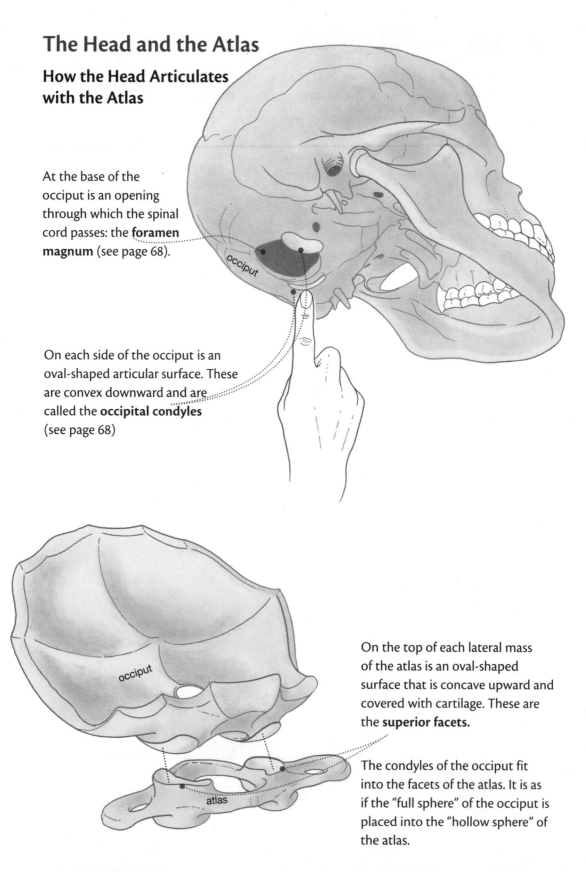

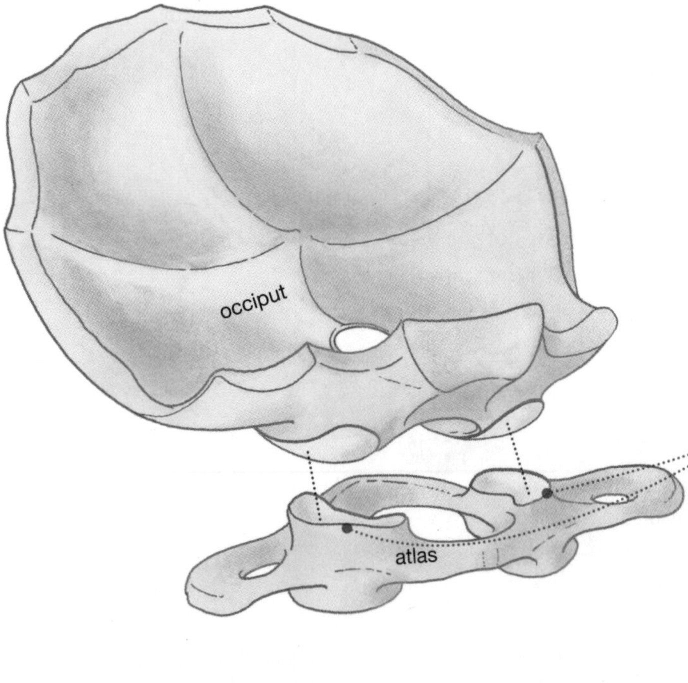

On the top of each lateral mass of the atlas is an oval-shaped surface that is concave upward and covered with cartilage. These are the **superior facets.**

The condyles of the occiput fit into the facets of the atlas. It is as if the "full sphere" of the occiput is placed into the "hollow sphere" of the atlas.

The Position of the Head on the Atlas Can Alter the Voice

This double articulation of the head and atlas would allow for movement in all directions. However, between the two bones are powerful ligaments that inhibit almost all movements except flexion and extension. It's at this junction that we evaluate the head position during singing or recitation.

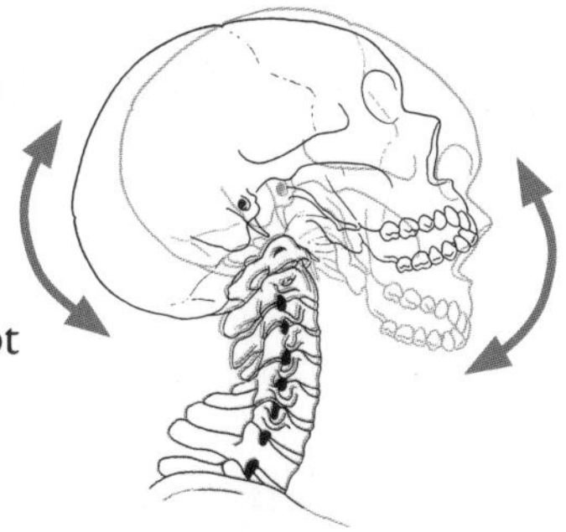

The occiput-atlas joint is where we nod the head when saying "yes." While the movement is flexion/extension, it is possible to move the head slightly to the side or in rotation as well.

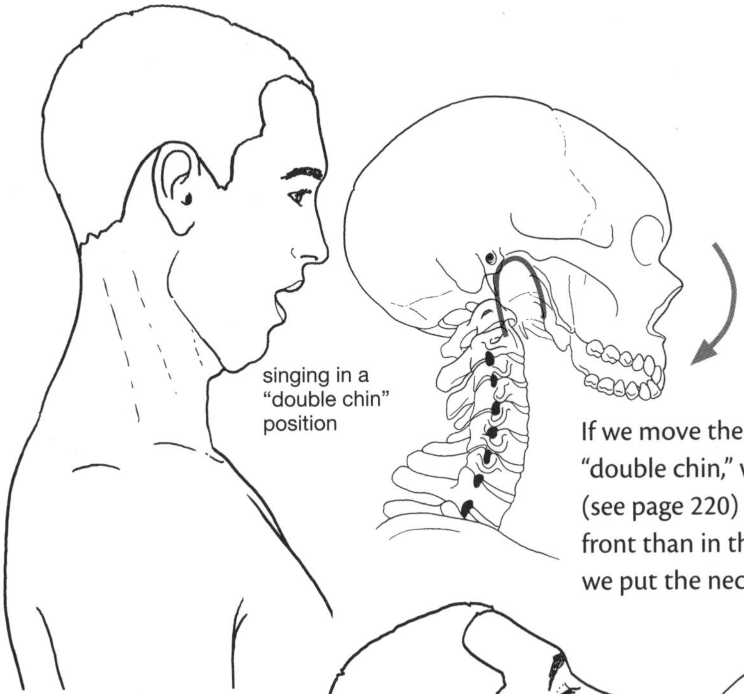

If we move the head in this plane and make a "double chin," we find that the rhinopharynx (see page 220) distance is narrower in the front than in the back, compared to when we put the neck in "extension."

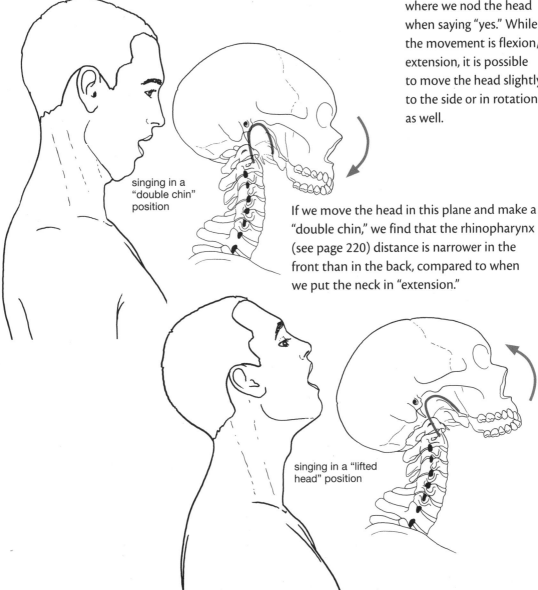

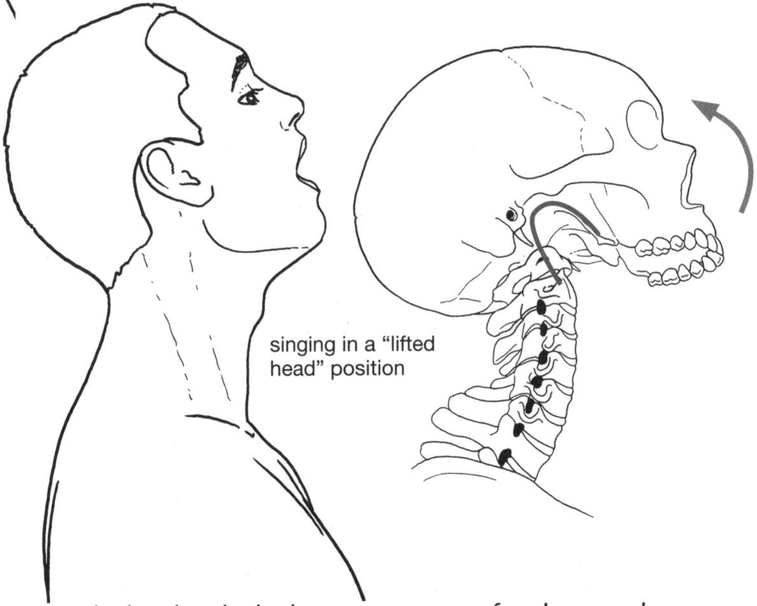

Thus the articulation between the head and atlas has consequences for pharyngeal resonance, as well as tension in the muscles of the soft palate (see pages 240–43) and the position of the mandible (see page 233) and the tongue (see pages 262–63).

How the Atlas Articulates with the Axis

There are four articulations that connect the atlas and axis.

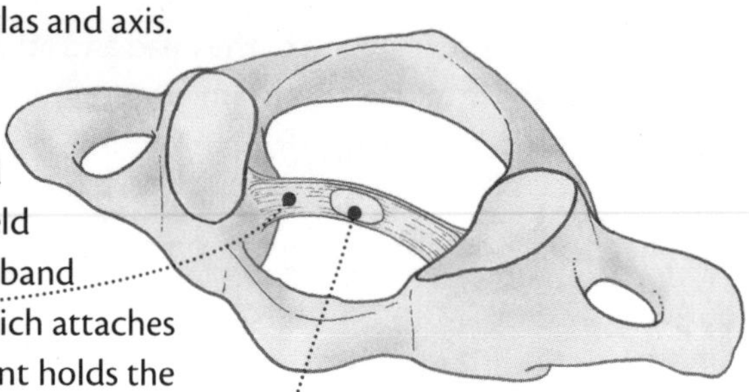

Two of these articulations support the odontoid apophysis of the axis, which is lodged in the anterior arch of the atlas. To begin, it's held in position at the front of the atlas by a fibrous band called the **transverse ligament of the atlas,** which attaches to the interior of the lateral masses. The ligament holds the odontoid apophysis in front, leaving room for the spinal cord to run through the rear or posterior arch of the atlas.

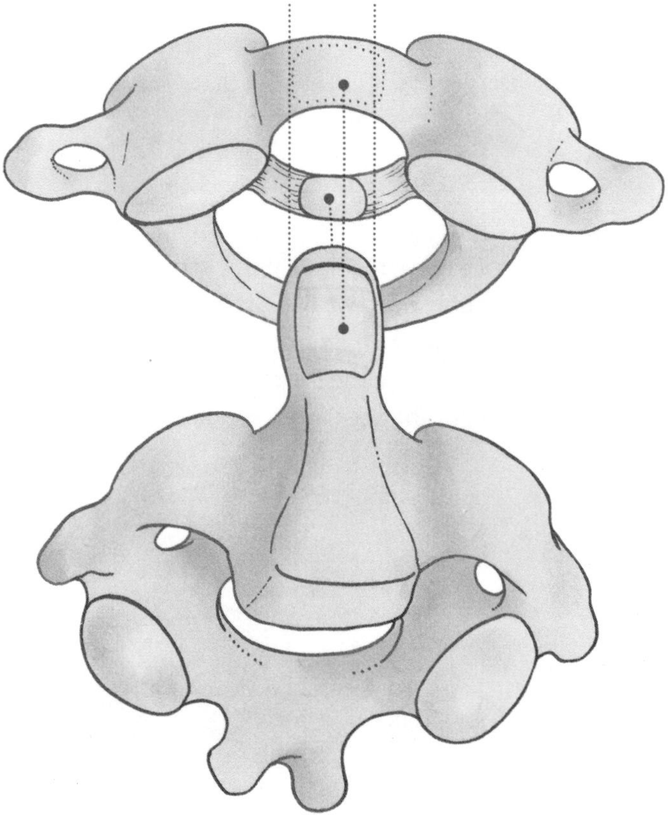

The posterior surface of the atlas's anterior arch is lined with a concave **cartilaginous facet,** which articulates with the convex cartilaginous facet at the front of the axis's odontoid apophysis.

The back of the odontoid apophysis is lined with a convex cartilaginous facet that articulates with the anterior surface of the transverse ligament, which supports a concave cartilaginous facet.

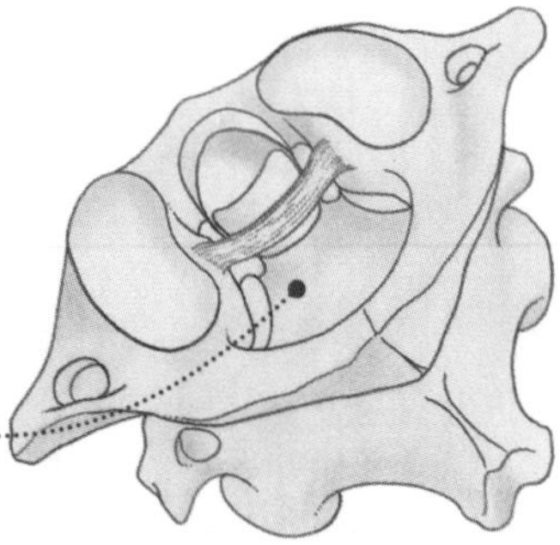

With these two articulations, the odontoid apophysis maintains its flexibility, while at the same time it is supported and it is prevented from imposing upon the **posterior arch space** of the atlas, which is occupied by the spinal cord.

Another two symmetrical articulations hold the atlas in position on the axis. Each lateral mass of the atlas articulates with a lateral surface on the top of the axis. The surfaces of the two bones are covered with thick cartilage that gives them convexity. Therefore the joints are prevented from nesting and, as a result, are highly mobile: axis-atlas movements are particularly easy.

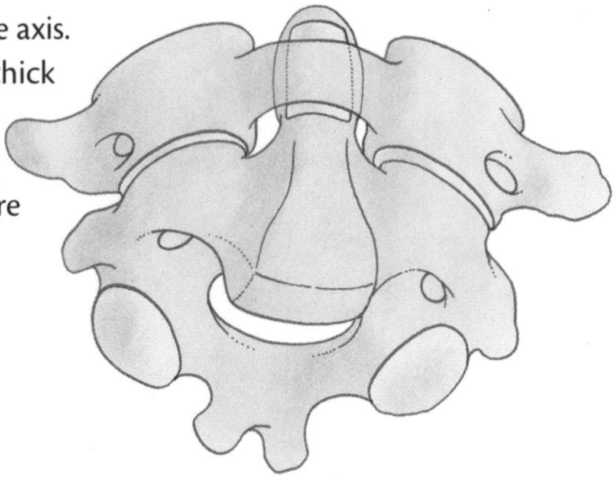

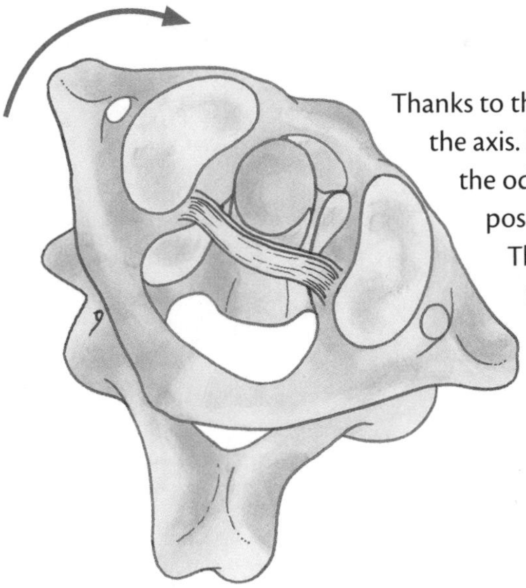

Thanks to these four joints, the atlas can turn around the axis. However, the rotation is not strictly about the odontoid process, which would push the posterior arch of the atlas into the spinal canal. The rotation of the atlas is combined with translation (the atlas sliding sideways). This movement is made possible by the deformability of the transverse ligament.

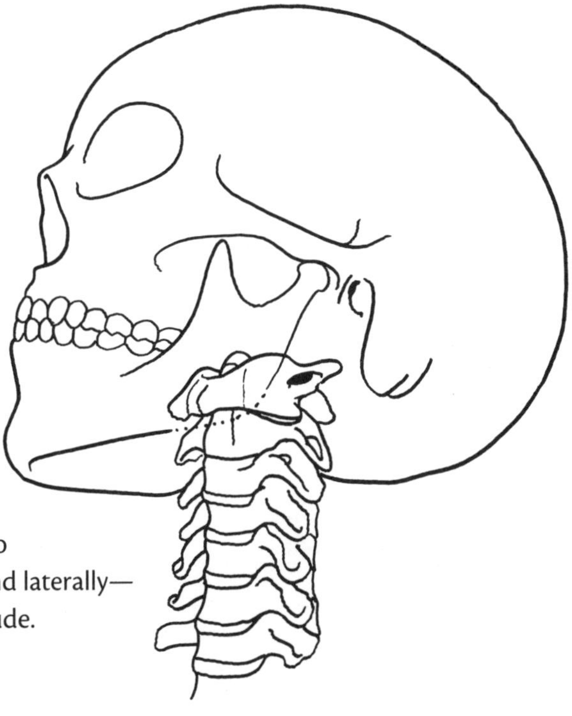

The fact that the atlas turns on the axis makes rotation of the head possible. This creates asymmetrical contraction/relaxation of the muscles that attach under the skull, in particular those that descend in the direction of the tongue and the larynx, and therefore it has an impact on the voice.

The arrangement of the atlas-axis joints also allows movements to the front and back and laterally—although these are much smaller in amplitude.

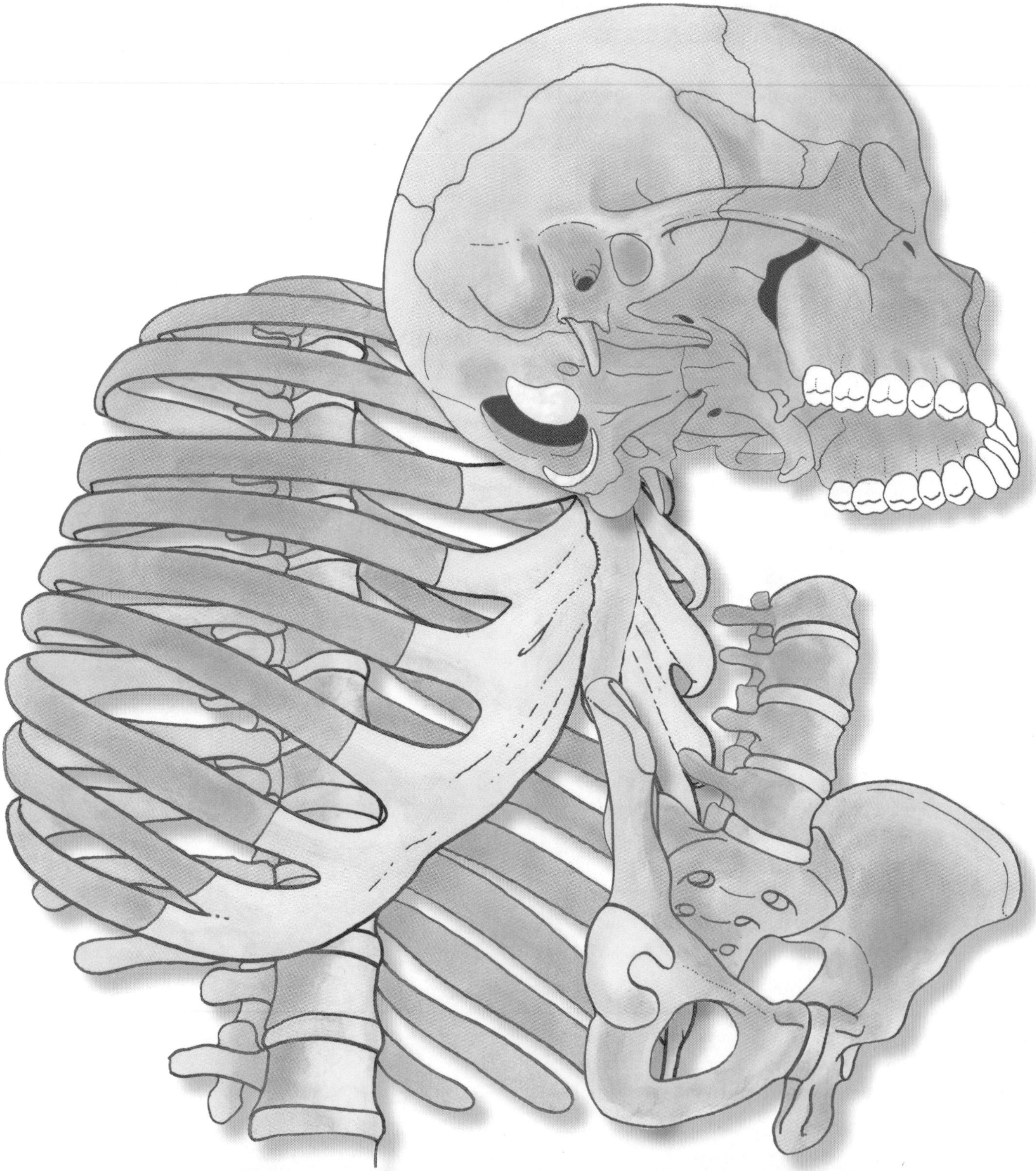

The Three Major "Blocks"—Pelvis, Rib Cage, Head

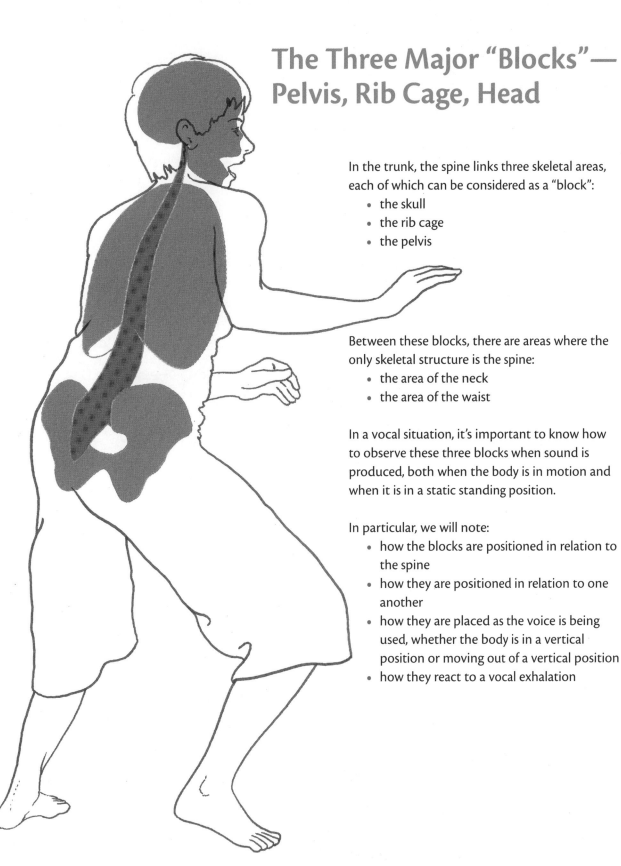

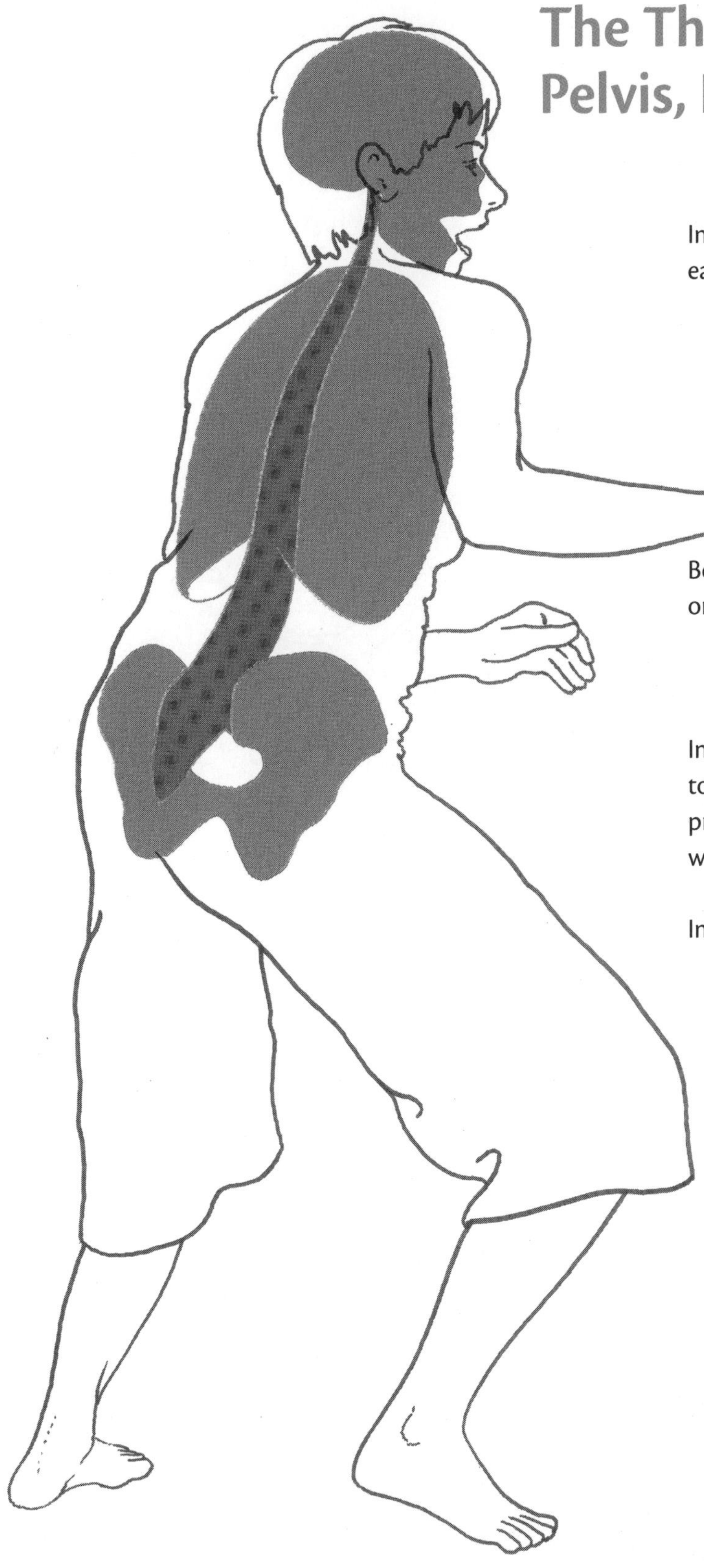

In the trunk, the spine links three skeletal areas, each of which can be considered as a "block":

- the skull
- the rib cage
- the pelvis

Between these blocks, there are areas where the only skeletal structure is the spine:

- the area of the neck
- the area of the waist

In a vocal situation, it's important to know how to observe these three blocks when sound is produced, both when the body is in motion and when it is in a static standing position.

In particular, we will note:

- how the blocks are positioned in relation to the spine
- how they are positioned in relation to one another
- how they are placed as the voice is being used, whether the body is in a vertical position or moving out of a vertical position
- how they react to a vocal exhalation

The First Major Block: The Pelvis

The skeletal structure at the base of the trunk, the pelvis, is an irregular bony ring composed of four bones:

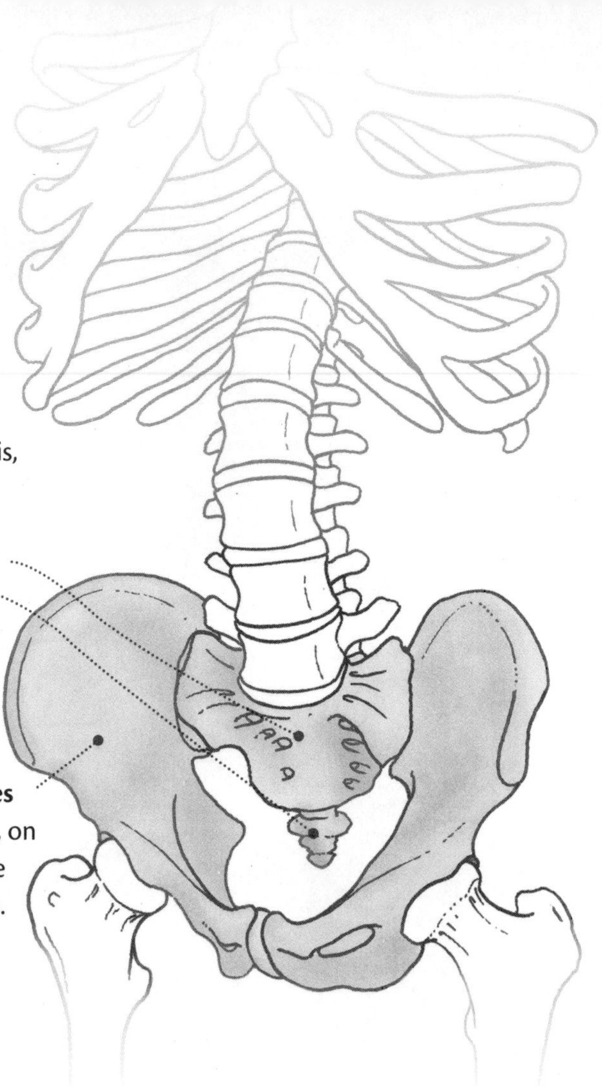

The **sacrum** and the **coccyx,** situated in the back, make up part of the spine.

The two **innominate bones** (also called the **coxal bones**), on the front and sides, are like extensions of the legs.

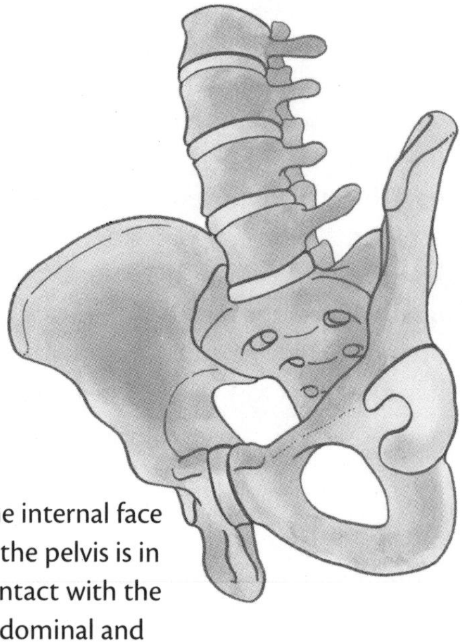

The external face of the pelvis is where we find the hip joint and the upper part of the thigh.

The internal face of the pelvis is in contact with the abdominal and pelvic viscera.

The internal pelvis is clearly divided into two parts:

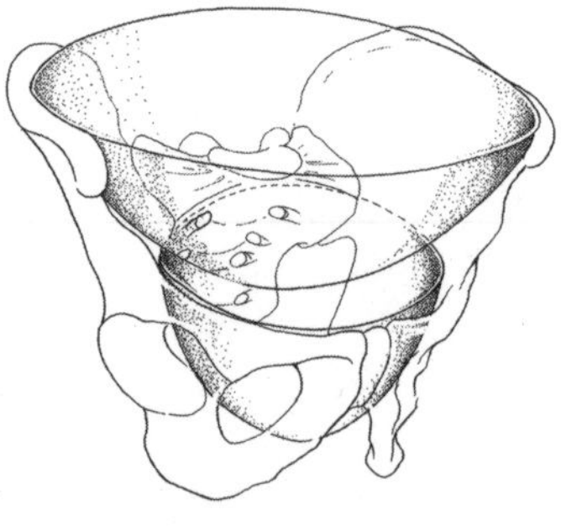

The **greater pelvis,** at the top, holds the abdominal viscera.

The **lesser pelvis,** below, holds the pelvic viscera.

Principal Landmarks (Palpable and Nonpalpable) of the Pelvis

The following are the principal landmarks of the pelvis:

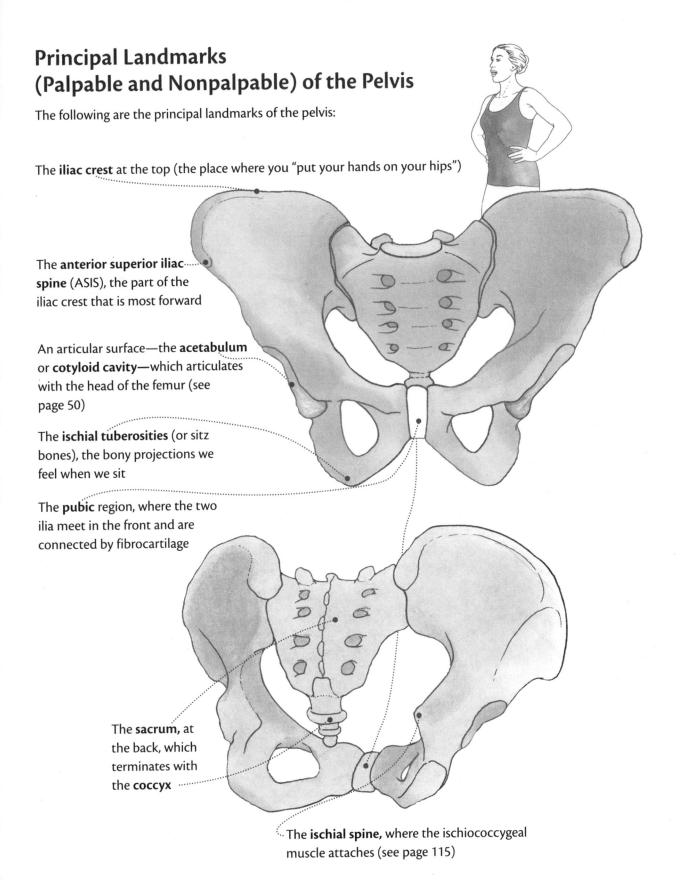

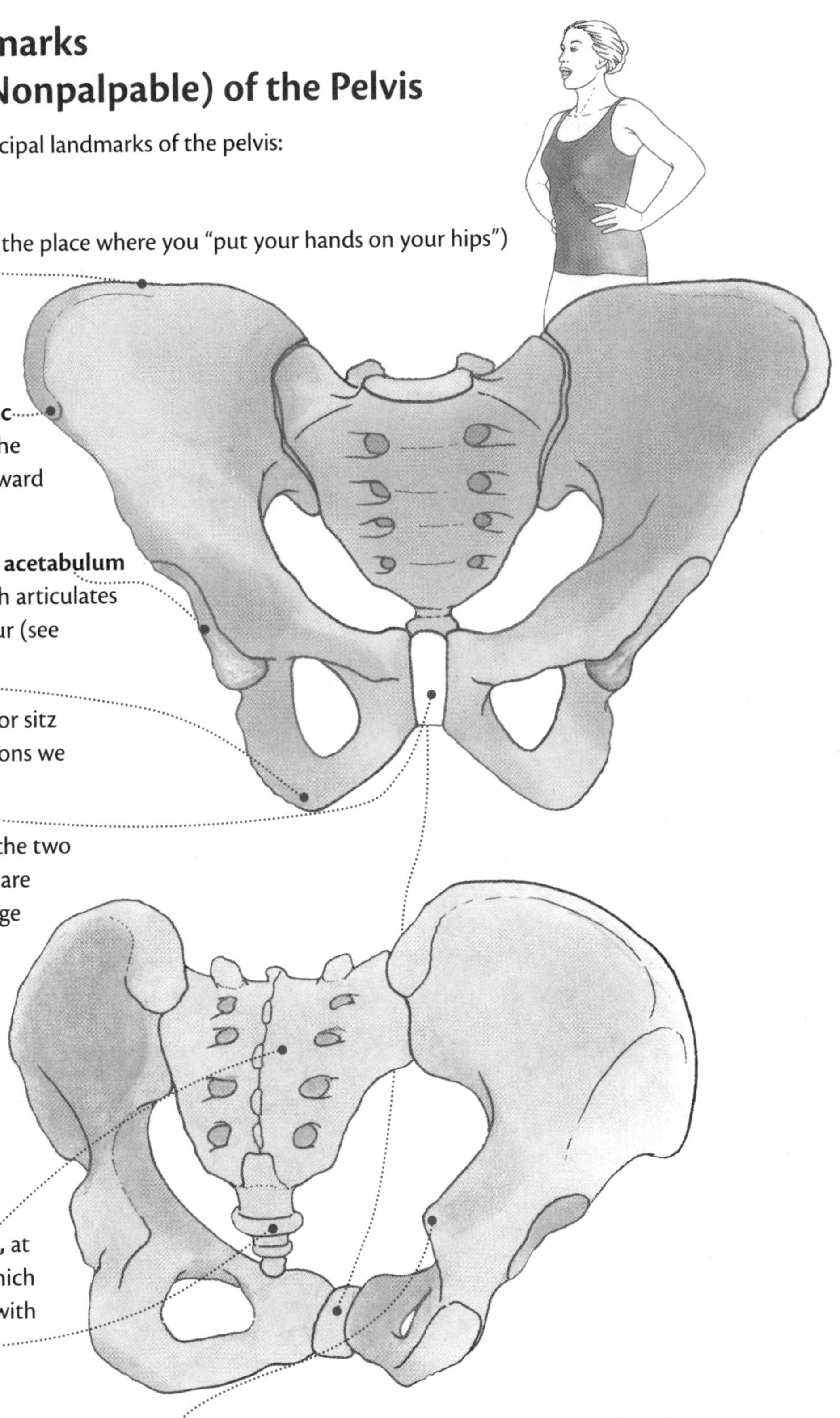

The **iliac crest** at the top (the place where you "put your hands on your hips")

The **anterior superior iliac spine** (ASIS), the part of the iliac crest that is most forward

An articular surface—the **acetabulum** or **cotyloid cavity**—which articulates with the head of the femur (see page 50)

The **ischial tuberosities** (or sitz bones), the bony projections we feel when we sit

The **pubic** region, where the two ilia meet in the front and are connected by fibrocartilage

The **sacrum,** at the back, which terminates with the **coccyx**

The **ischial spine,** where the ischiococcygeal muscle attaches (see page 115)

The Femurs and the Hip Joints

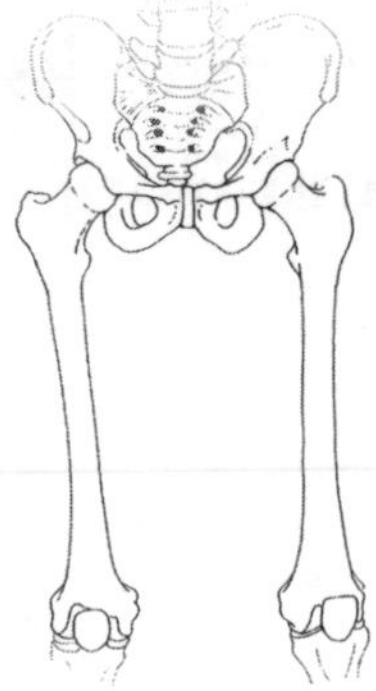

The pelvis articulates with the two **femurs,** or thigh bones. Together they form the **coxofemoral** joint, commonly known as the hip joint. Why is it important to understand this area of the anatomy? In a nutshell, when we work with the voice, the position of the pelvis has a profound effect on the two large "blocks" above it, and the pelvic position is determined at the hip joint.

The Hip Joint

This is one of the largest joints of the body. It unites the **acetabulum** (cotyloid cavity), situated on the external surface of the pelvis, and the **femur head,** the uppermost part of the femur, that is itself supported by the **femur neck.**

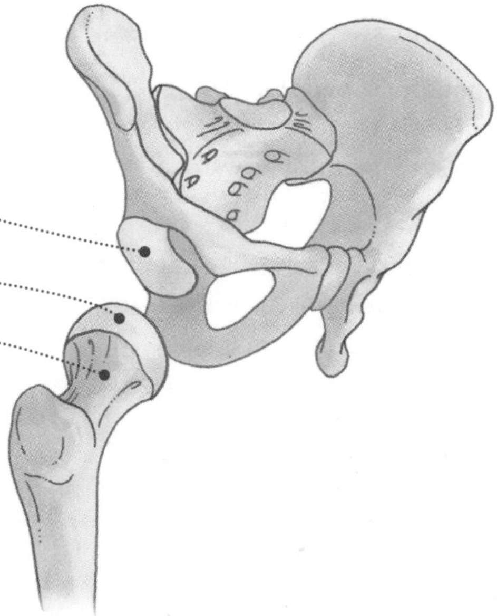

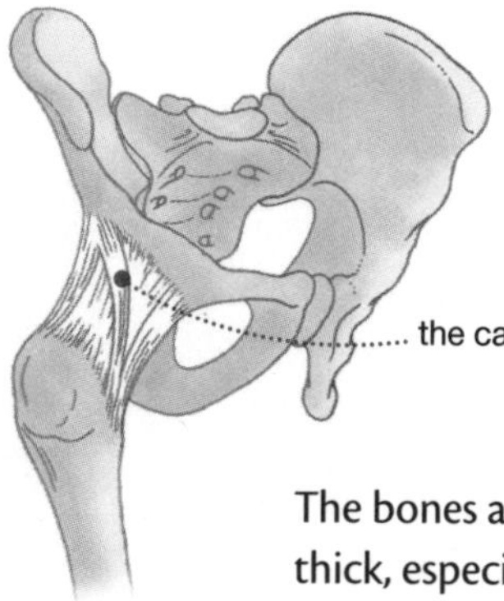

The bones are held together by a fibrous sleeve called the **capsule**. This capsule is thick, especially in the front, where it is supported by three ligaments that form a "Z."

Coxofemoral Movements

The hip unites a full sphere (the femur head) with a hollowed sphere (the acetabulum). This form makes movement in all directions possible. The femur can move . . .

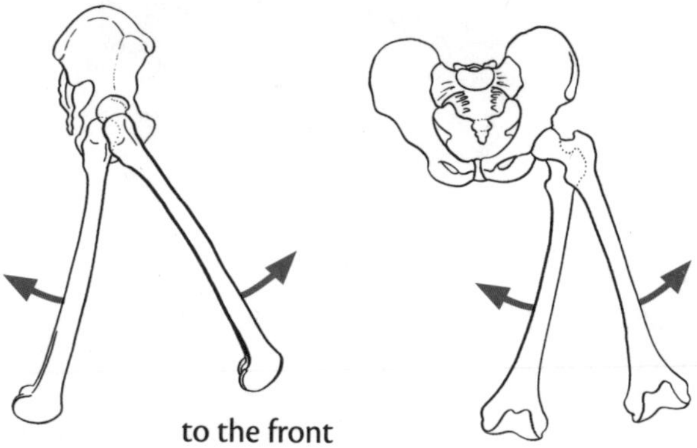

in rotation (not shown)

The "Tilting" and "Positioning" of the Pelvis

The movements that are most relevant here are those that the pelvis can make around the femur heads. We define these movements by the position of the anterior superior iliac spine, or ASIS. (In this book we are only going to look at the pelvis as it rocks to the front and back, since these are the movements most commonly employed when using the voice.)

In anteversion
the ASIS tips forward,
the ischial tuberosities
(sitz bones) tip backward.

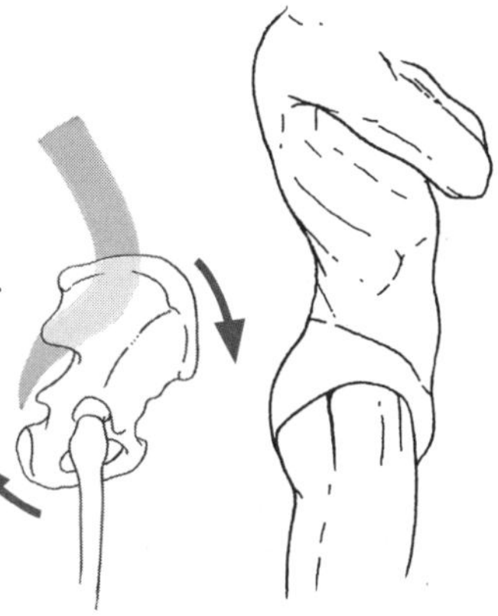

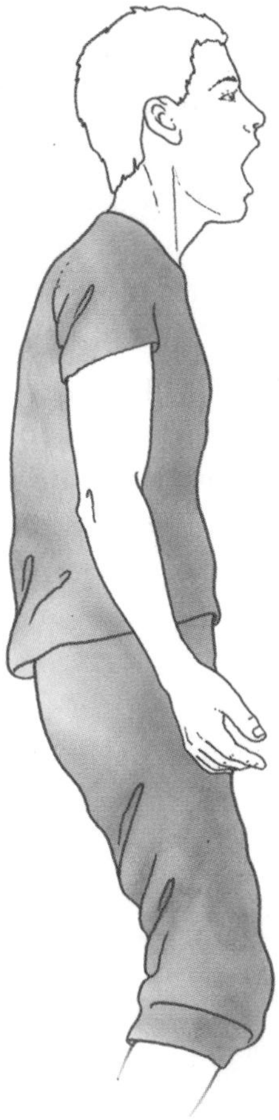

In retroversion
the ASIS tips backward,
the ischial tuberosities
tip forward.

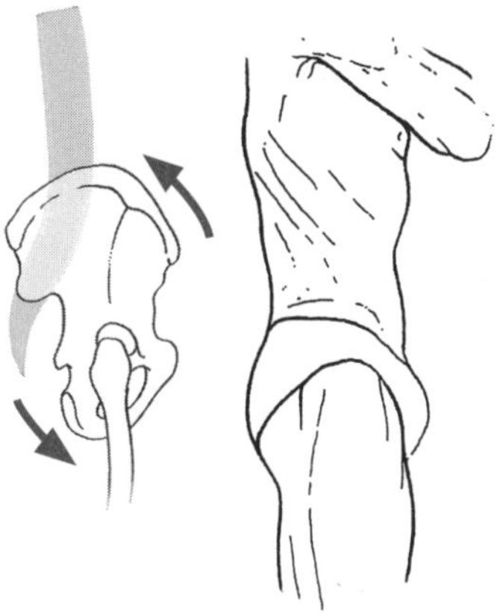

Almost all of the abdominal muscles retrovert the pelvis. They also encourage exhalation. This is why it's easy to see the association between retroversion and the voiced exhalation. However, we can produce sounds in anteversion as well.

The bones of the lower limbs (the feet, knees, and so on) won't be covered in this book; they are put into play during the vocal act, but not in a specific way.

More details on movement of the hips and pelvis can be found in *Anatomy of Movement*, pages 194–99.

The Second Major Block: The Rib Cage—the Transformable Block

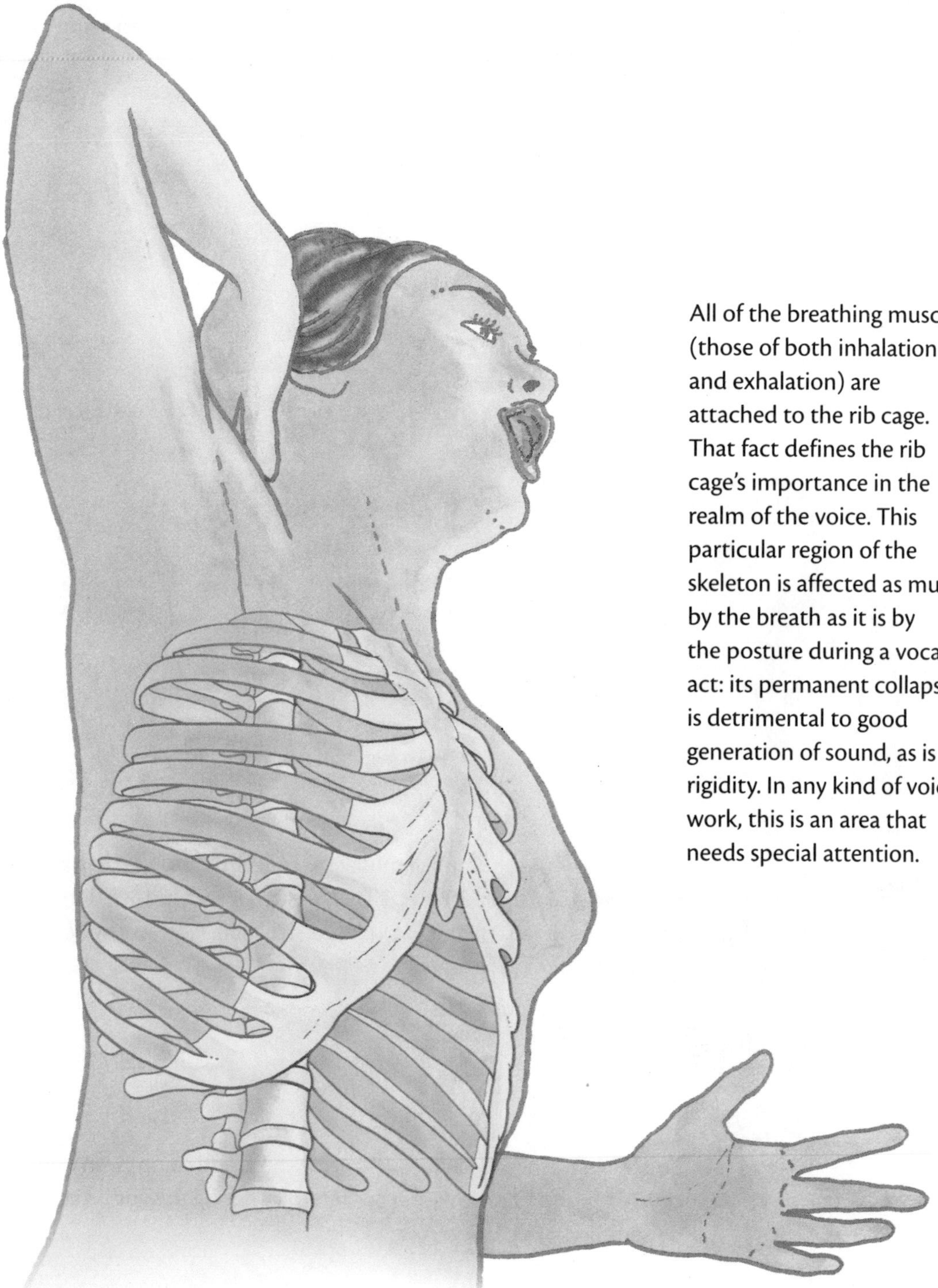

All of the breathing muscles (those of both inhalation and exhalation) are attached to the rib cage. That fact defines the rib cage's importance in the realm of the voice. This particular region of the skeleton is affected as much by the breath as it is by the posture during a vocal act: its permanent collapse is detrimental to good generation of sound, as is its rigidity. In any kind of voice work, this is an area that needs special attention.

The rib cage is an ensemble of many united elements. In terms of its osteoarticular components, it's the most complex area of the body. It's made up of:

- the **thoracic spine** (in back)
- twelve pairs of **ribs,** with costal cartilage
- the **sternum** (in front)

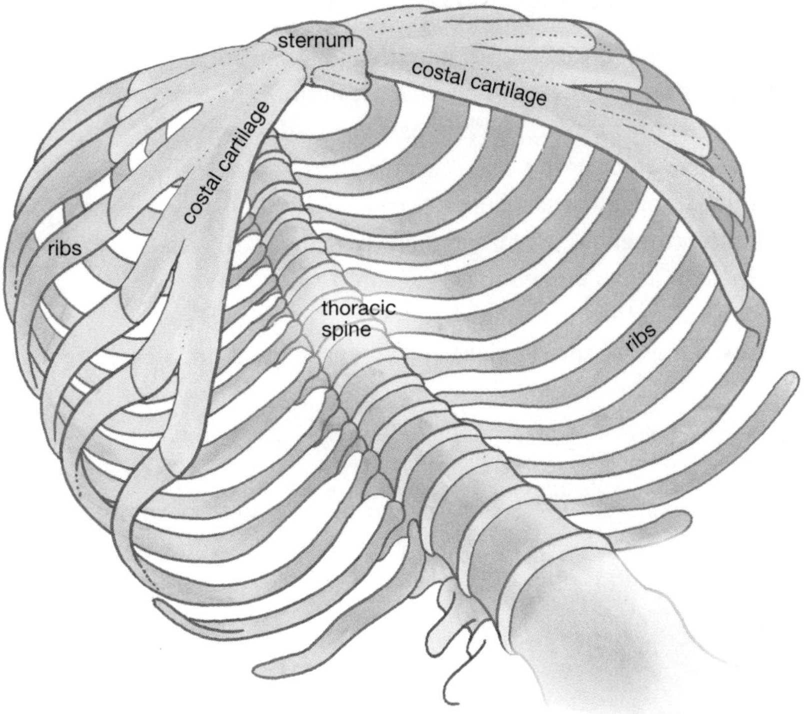

In the following pages, we will look at these elements as well as the joints that link them.

The **costal arch** at each vertebral level is made up of:

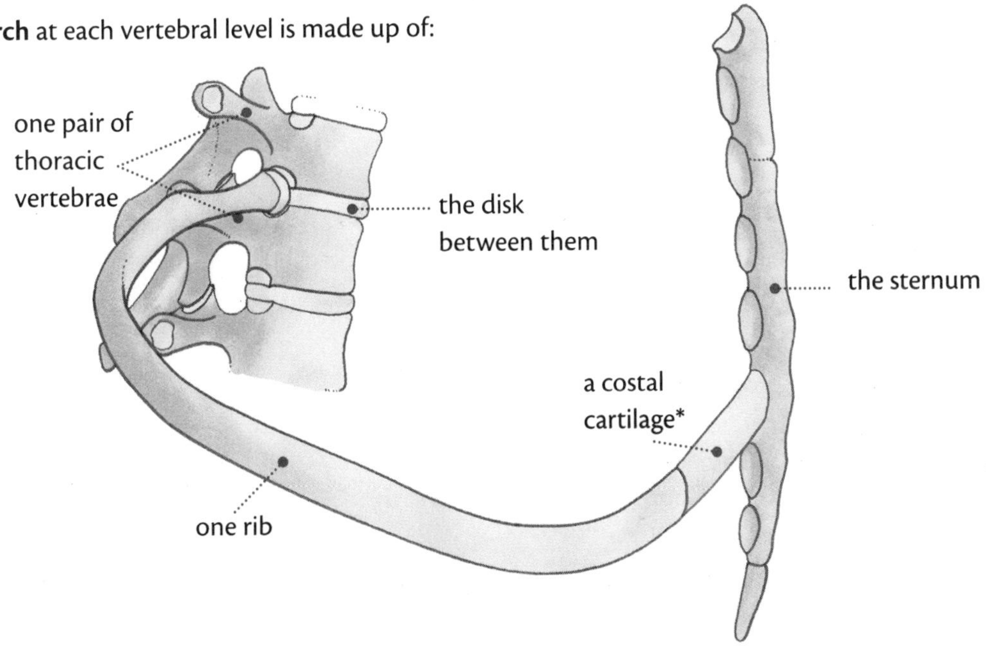

*Except for the eleventh and twelfth ribs, which lack cartilage.

The Ribs: The Only Flexible Bones

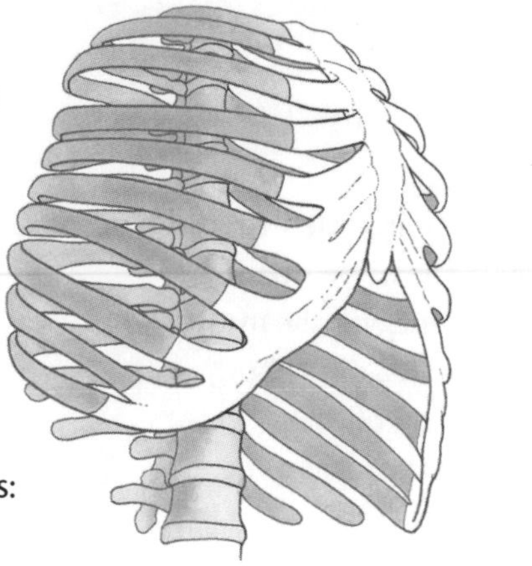

Each rib is at once flat and curved. Among the bones of the body, these are the only ones that are flexible; they are, in fact, rather elastic. This is in part due to the form of the ribs, which have a triple curvature: they encircle the thorax, and at the same time, (viewed in profile), they descend, then rise again, and then twist in on themselves. Each rib has an inside surface (oriented toward the viscera) and an external surface (oriented toward the costal muscles and the skin), and are made up of three parts:

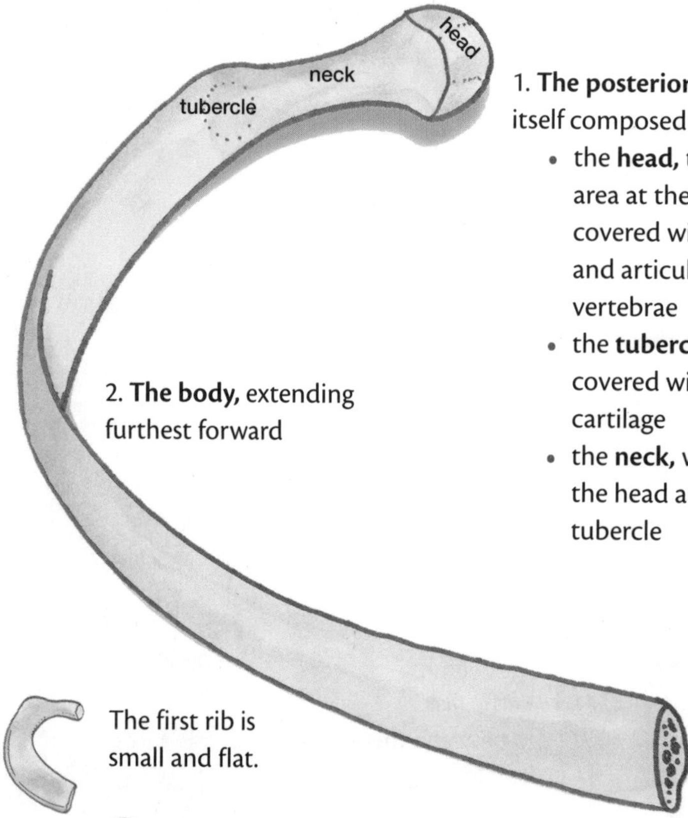

1. **The posterior part,** which is itself composed of three parts:
   - the **head,** the bulging area at the end that is covered with cartilage and articulates with the vertebrae
   - the **tubercle,** which is covered with articular cartilage
   - the **neck,** which joins the head and the tubercle

2. **The body,** extending furthest forward

3. **The anterior end,** which meets the costal cartilage

The first rib is small and flat.

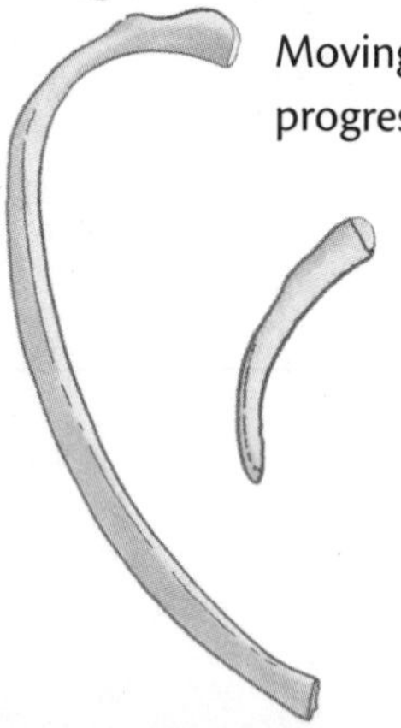

Moving down, the ribs that follow are progressively longer and longer.

The eleventh and twelfth ribs are short, particularly the twelfth.

The ribs may follow each other, but no two are alike!

Maintain Flexibility

The flexibility of the ribs is maintained by movements of the rib cage independent of breathing.

The First Rib: An Important Area to Observe

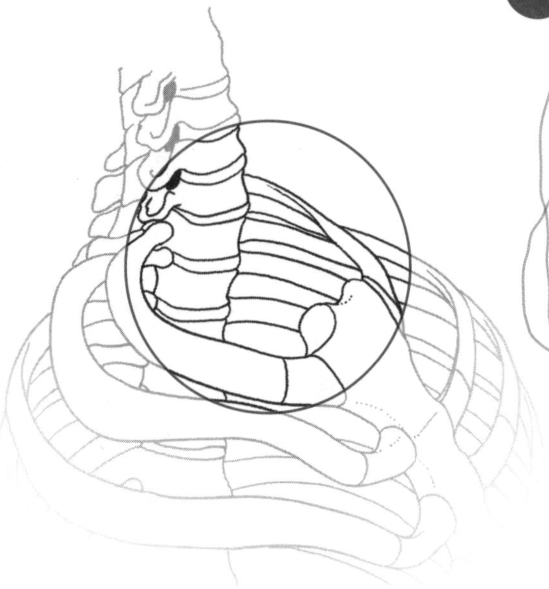

Palpation

To find the "circle" of the first two ribs, follow the line formed by the collar of a T-shirt. Or place your fingers around the base of the neck, your thumbs at the sternum and second fingers touching the most prominent cervical vertebra (C7). The circle formed by the fingers, or the neck of the T-shirt, marks the dimensions and orientation of the first rib.

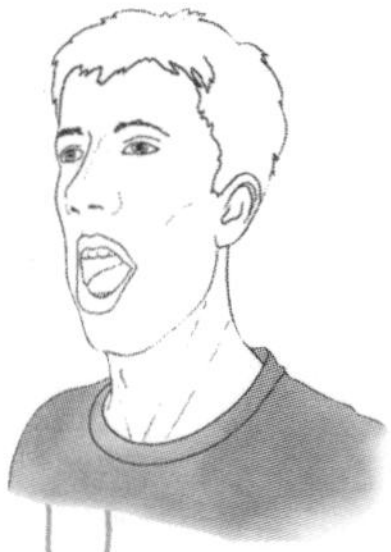

It's important to look at the orientation of the circle formed by the two first ribs (right and left). When we watch someone in a standing position as they sing or breathe and the circle of the two first ribs is rather horizontal (sternum lifted), it's an indication of an open rib cage and a flattened thoracic spine. If the circle is more vertical, it indicates the opposite—that the rib cage is dropped and the thoracic spine is in kyphosis (see page 36).

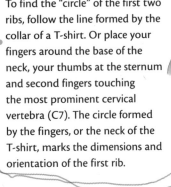

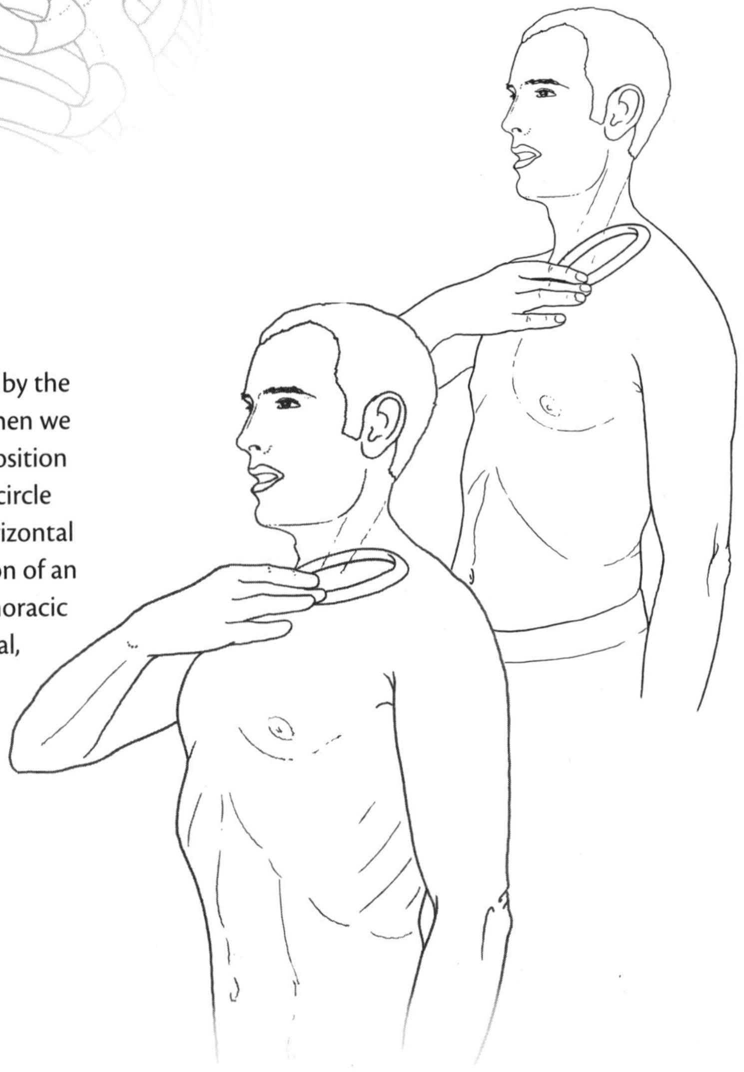

The Costal Cartilages

In the front, the **costal cartilages** link the ribs to the sternum. They get progressively longer as we descend down the length of the sternum.

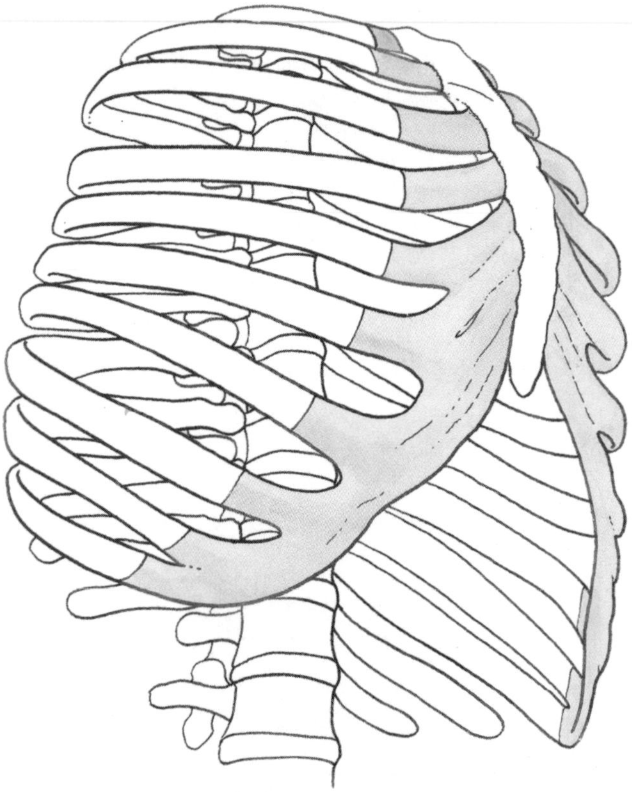

From top to bottom:

- The first seven are relatively short and link directly to the sternum.
- The following three (eighth, ninth, and tenth) merge with the seventh costal cartilage.
- The last two ribs don't have cartilage linking them to the sternum. These are the "floating ribs."

Being more flexible than the ribs themselves, the costal cartilages increase the overall flexibility of the rib cage. In terms of mobility, the rib cage can be broken down into three regions:

From the first to the seventh ribs: The ribs are not long and the cartilage is short; at this level there is reduced mobility.

From the eighth to tenth ribs: The ribs are longer and the cartilage is longer; at this level mobility is greatly increased, especially laterally.

At the eleventh and twelfth ribs: The ribs here lack cartilage; this level is very mobile.

How the Costal Cartilages Connect to the Ribs

Every costal cartilage has an oval-shaped external (lateral) extremity.

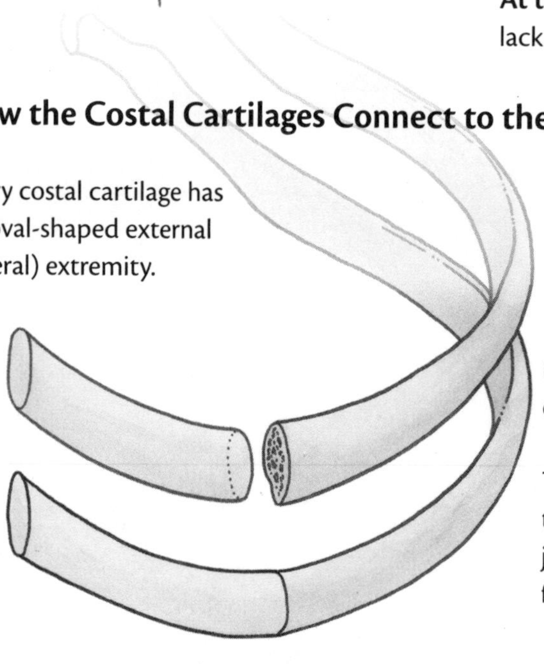

Every rib has an oval-shaped anterior extremity.

The rib and the cartilage are joined at these surfaces. These joints are simple junctions without a capsule or synovial fluid.

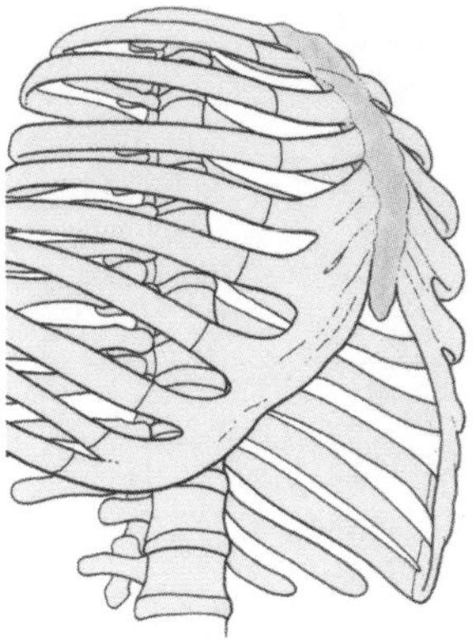

The Sternum

The sternum is a flat bone that resembles a sword. It's located at the front and center of the rib cage. Its inner surface faces the heart and pericardium. Its outer surface is just under the skin.

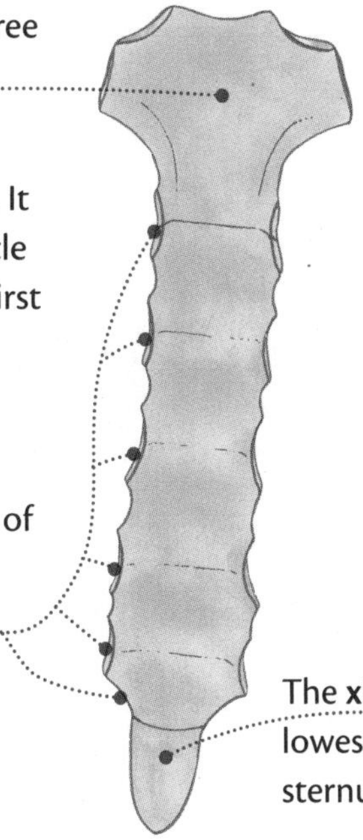

We can divide the sternum into three distinct areas:

The **manubrium,** at the top, resembles the handle of the sword. It articulates at its top with the clavicle and at its lateral borders with the first two costal cartilages.

The **body** is the largest area. Its lateral borders have notches that correspond to the costal cartilages of ribs 2 through 7.

The **xiphoid process** is the lowest segment, where the sternum terminates.

How the Costal Cartilages Join the Sternum

On the lateral borders of the sternum, the bone has hollowed areas that form small oval surfaces.

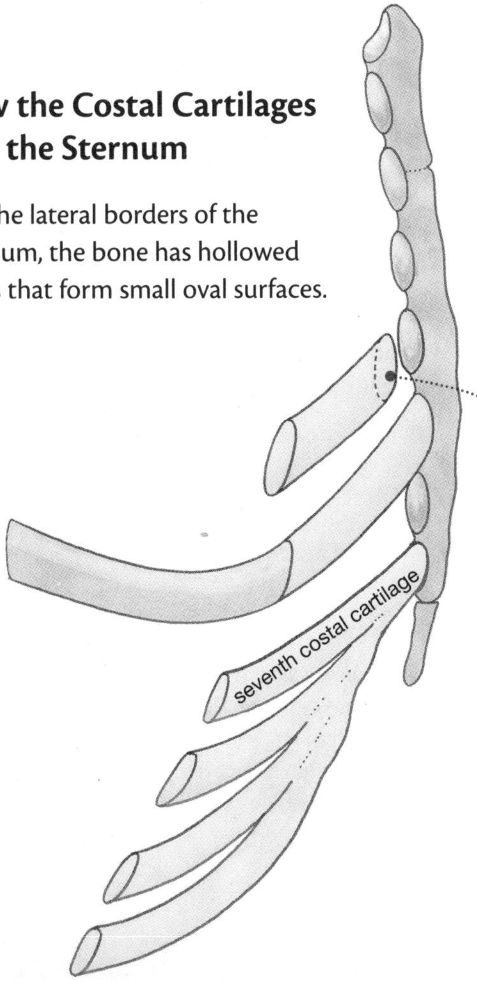

The inside end (medial) of every costal cartilage (from the first through the seventh ribs) terminates in a slightly protruding oval where it joins with the sternum. These cartilage-sternum joints, like the cartilage-rib joints, are simple junctions, with no capsule or synovial fluid.

The eighth, ninth, and tenth costal cartilages join up with the seventh costal cartilage; they don't articulate directly with the sternum.

The Thoracic Spine

This is the part of the spine that is associated with the rib cage, with various localized characteristics. In general, this part of the spine is not extremely mobile, except for the lowest vertebrae.

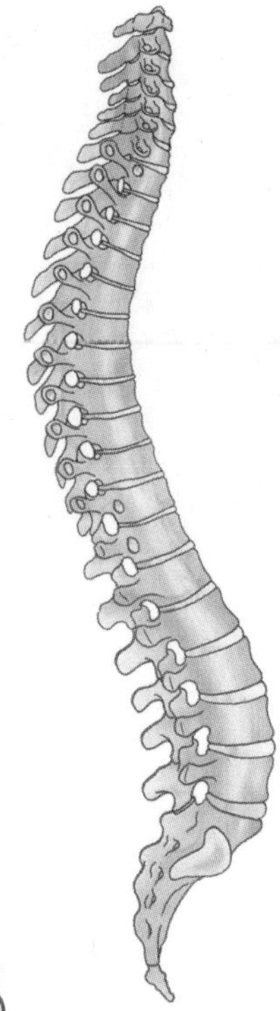

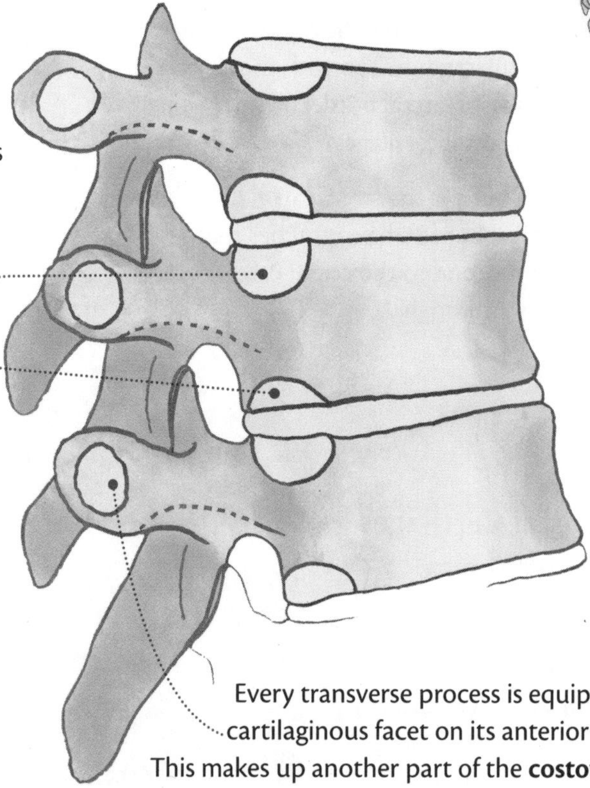

Each vertebral body has articular facets toward the back of its lateral surface: one at the top and one at the bottom. Along with the intervertebral disks, these form the part of the **costovertebral joint** that aligns with the head of a rib (see the facing page).

Every transverse process is equipped with a cartilaginous facet on its anterior surface. This makes up another part of the **costovertebral joint;** this one joins the vertebra to the rib's tubercle (see the facing page).

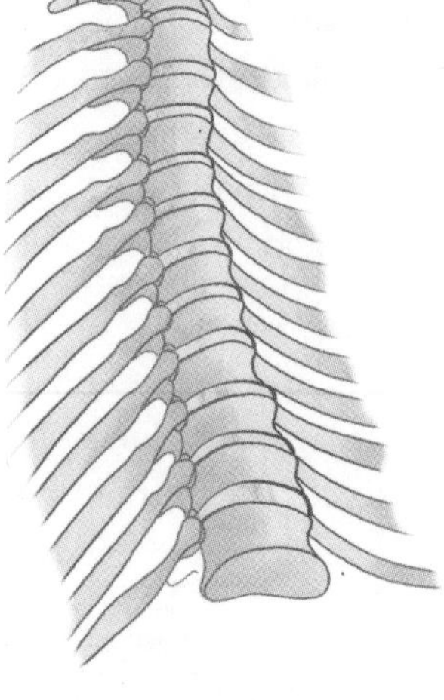

"Breathing into the Back"

Mobilization of all of these articulating facets is the key not only to thoracic mobility but to efficient respiration as well. This mobilization is important when you are learning effective "posterior breathing" or "back breathing."

The Costovertebral Joints

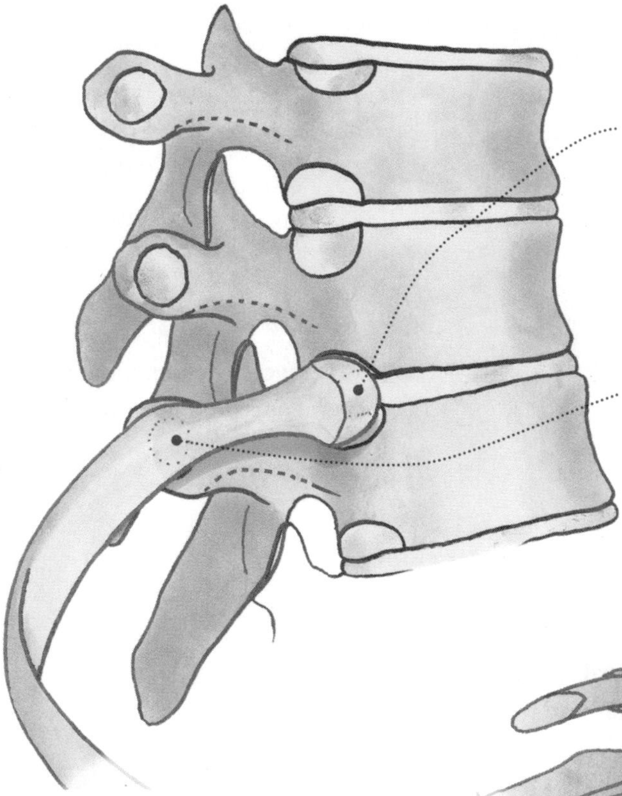

At every vertebral level of the thoracic spine, we find two different joints:

One joint unites the head of the rib with an intervertebral disk and the two articulating facets that are situated on the bodies of the vertebrae above and below.

Another joint unites the tubercle of the rib with the facet located on the vertebra's transverse process.

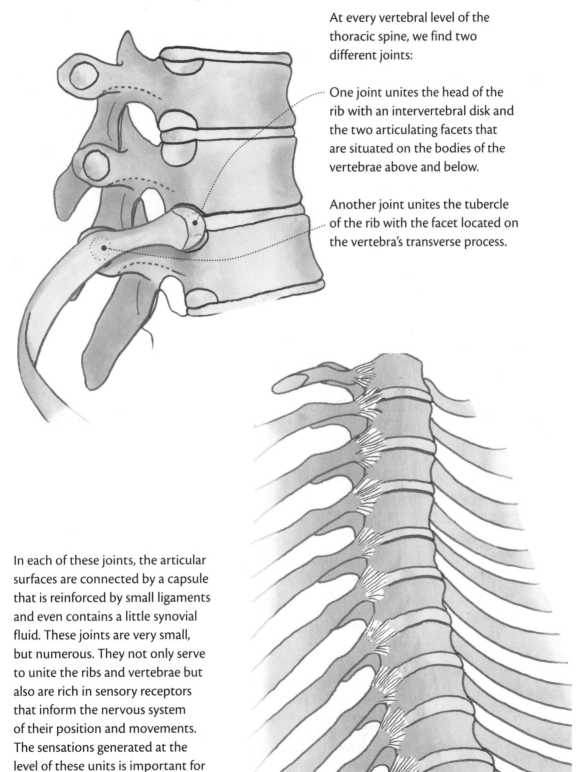

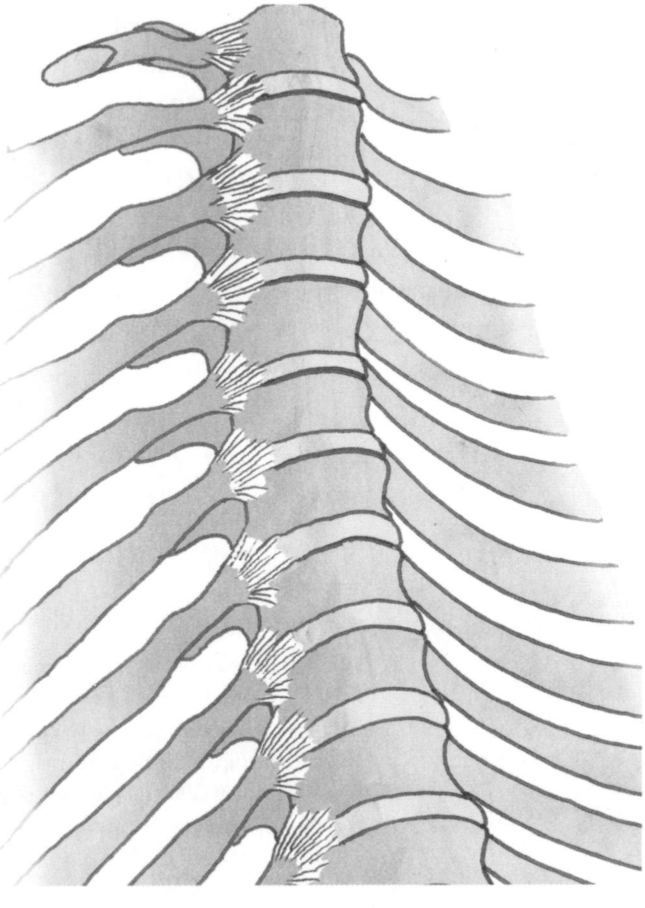

In each of these joints, the articular surfaces are connected by a capsule that is reinforced by small ligaments and even contains a little synovial fluid. These joints are very small, but numerous. They not only serve to unite the ribs and vertebrae but also are rich in sensory receptors that inform the nervous system of their position and movements. The sensations generated at the level of these units is important for proprioception.

Variations of the Costovertebral Axes

The axis formed between the two costovertebral joints varies at each vertebral level, which means that the types of movements the ribs can make at each level varies as well.

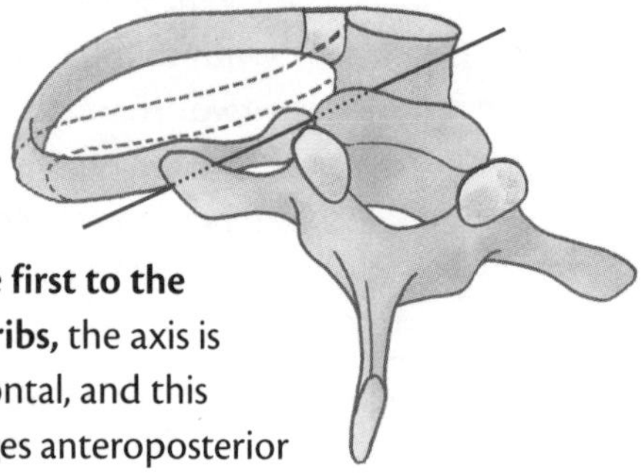

From the first to the seventh ribs, the axis is rather frontal, and this encourages anteroposterior movement.

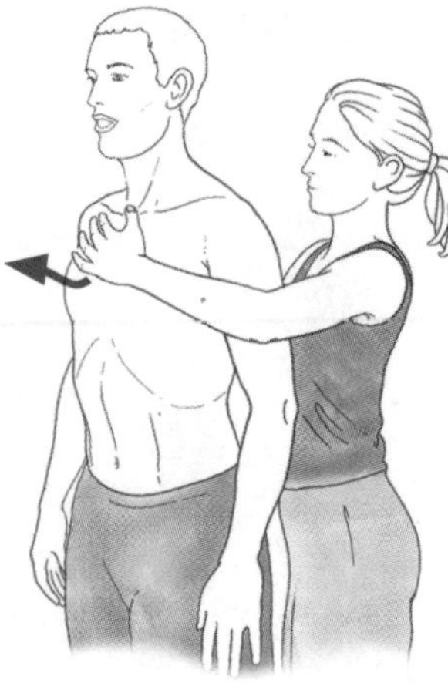

In other words, in the upper chest movement is predominantly to the front and back.

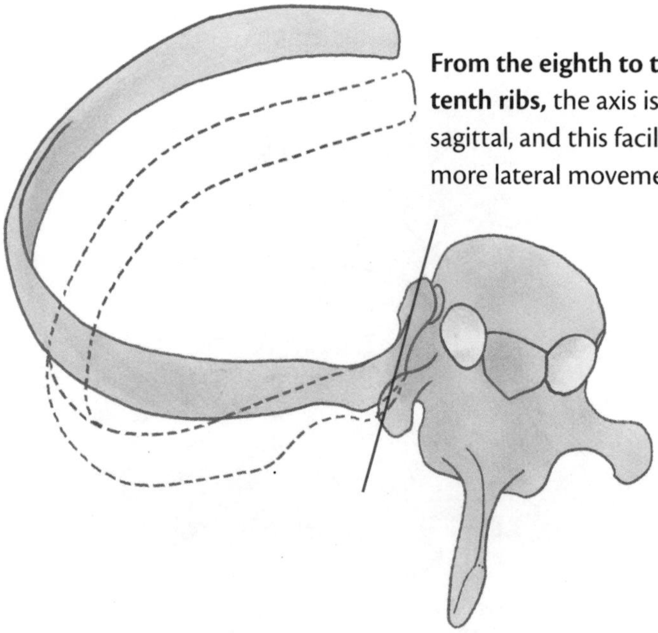

From the eighth to the tenth ribs, the axis is more sagittal, and this facilitates more lateral movement.

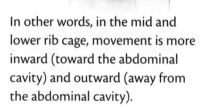

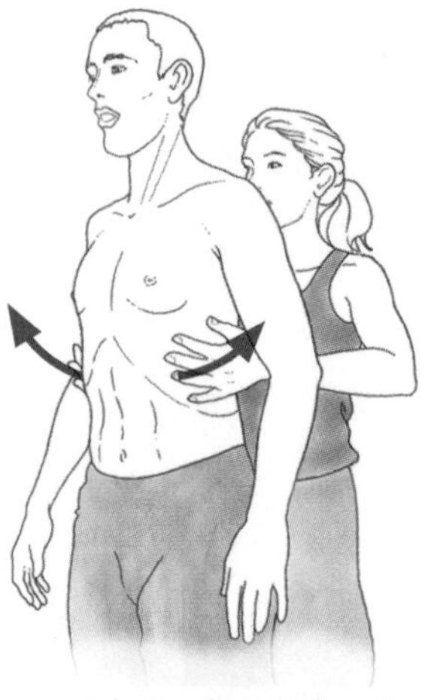

In other words, in the mid and lower rib cage, movement is more inward (toward the abdominal cavity) and outward (away from the abdominal cavity).

The Important Directions . . .

Knowing how the rib cage moves most efficiently can be helpful when you're guiding someone during vocal work, both in terms of the words you use to cue the person (for example, "expand the top of the rib cage" may not be effective) and in terms of manual cuing (application of your hands) to solicit a specific respiratory movement.

The Two Kinds of Rib Cage Movements

Independent of the costovertebral axes, the ribs and the rib cage move in two major directions, and people generally favor one direction or the other.

Movement in the Sagittal Plane ("Like a Pump Handle")

The ribs can follow the movement of the sternum forward or backward, which increases the sagittal diameter of the thorax when the sternum rises and diminishes it when the sternum drops.

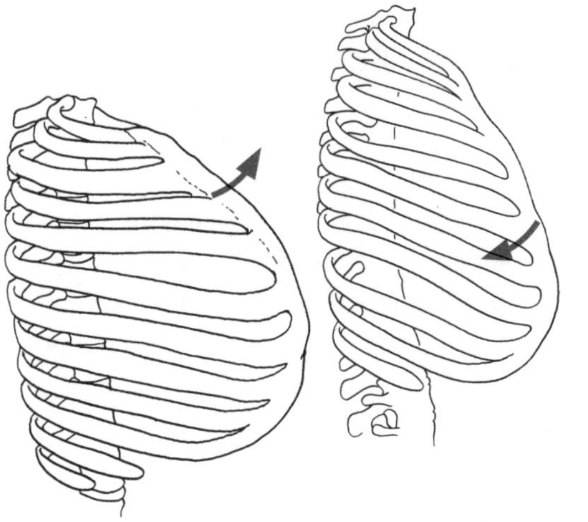

Note that movement of the sternum itself can be differentiated:

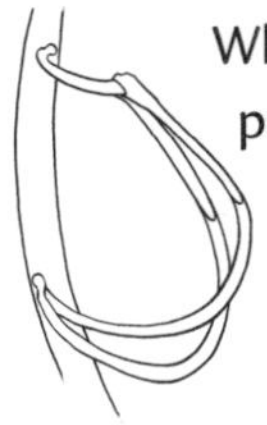

When it's predominantly the lower part of the sternum that lifts and moves forward, the movement is mostly in the lower ribs.

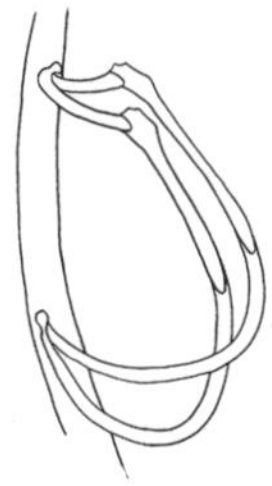

When the whole sternum moves, all of the ribs move.

Observe the Sternum

It's important to observe and guide the sternum in vocal work. Its orientation is a result of forces from both inside and outside of the thoracic rib cage, in particular the tonus of the inspiratory and expiratory muscles.

Movement in the Frontal Plane ("Like a Bucket Handle")

The ribs can also move more laterally, which increases the lateral diameter of the chest when the sternum rises and decreases it when it drops.

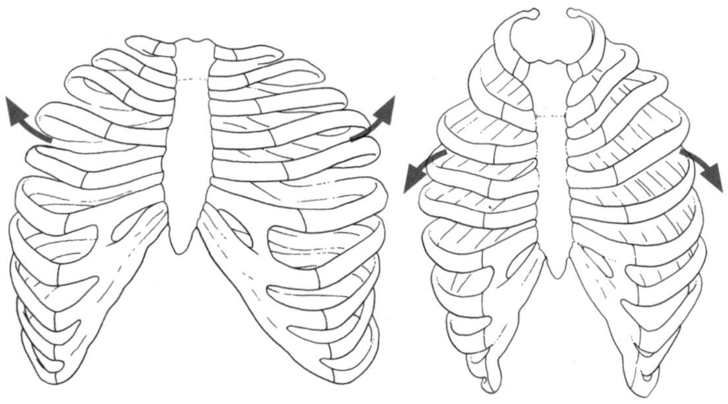

The Shoulder Girdle

The shoulder girdle is made up of the clavicle and shoulder blades, with the sternum in the middle. The shoulder girdle is the bony and articular structure that allows the arms to attach to the trunk. We will cover it briefly here.

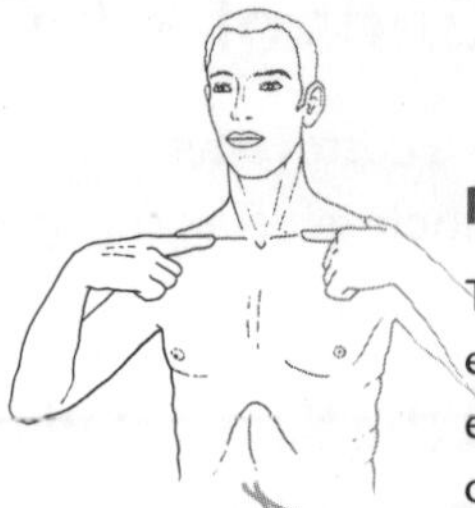

Palpation

The clavicle can be easily palpated on either side of the base of the neck

The Clavicle

This is a bony rod located between the scapula and the sternum. It features two surfaces (superior and inferior), two borders (anterior and posterior), and two ends. At both ends we find articular cartilaginous surfaces.

At its outer extremity the clavicle's articular surface aligns with that of the acromion (the outer end of the scapula), forming the **acromioclavicular joint.**

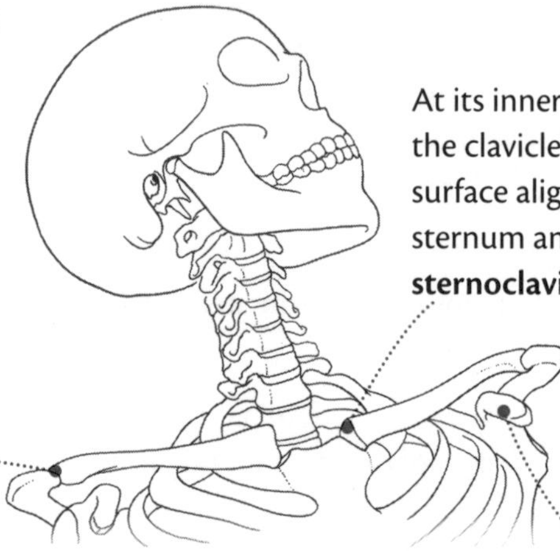

At its inner extremity, the clavicle's articular surface aligns with the sternum and forms the **sternoclavicular joint.**

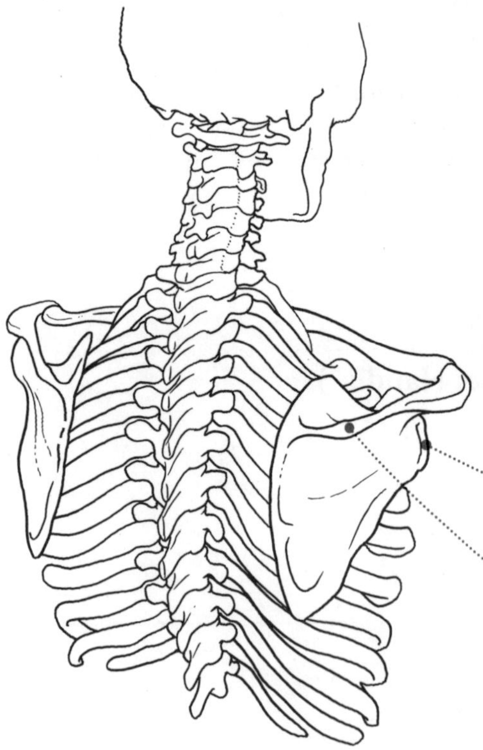

The Scapula

Commonly called the shoulder blade, this is a flat, triangular bone positioned on the posterolateral aspect of the thorax. We describe it as having two surfaces (anterior and posterior), three borders (superior, internal or medial, and external or lateral), and three angles (superior, external, and inferior).

When discussing the voice, we can point to some important landmarks:

- The **scapular notch** is located on the superior border, just medial to the base of the **coracoid process.**
- On the external angle, the **glenoid cavity** is a cartilaginous articulating surface that is part of the shoulder joint.
- On its external surface, the projecting blade called the **spine of the scapula** crosses obliquely and terminates at the **acromion.**
- On the anterior tip of the acromion, there is an articular surface.

More details on the clavicle and shoulder blade can be found in *Anatomy of Movement*, pages 110–15.

Movements of the Shoulder Girdle

The arrangement of the clavicle and shoulder blades allows for a variety of movements:

abduction
adduction

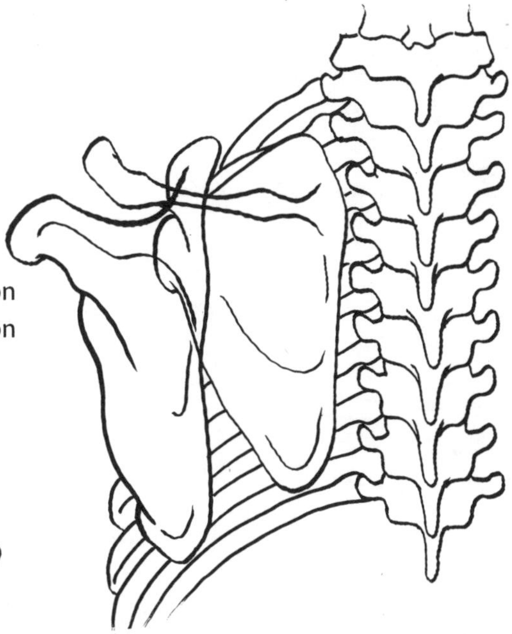

elevation
depression

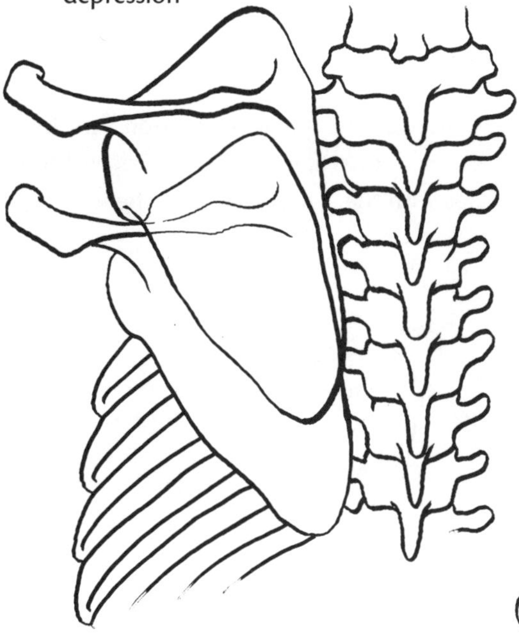

upward rotation
downward rotation

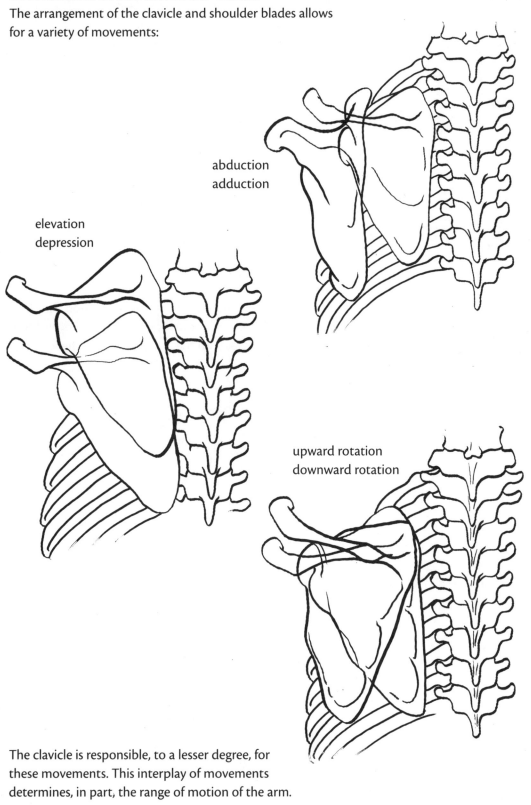

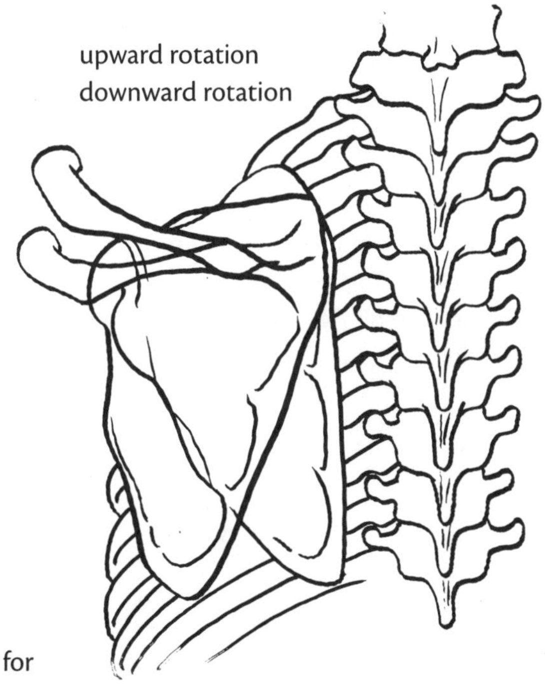

The clavicle is responsible, to a lesser degree, for these movements. This interplay of movements determines, in part, the range of motion of the arm.

The Arms and Shoulders

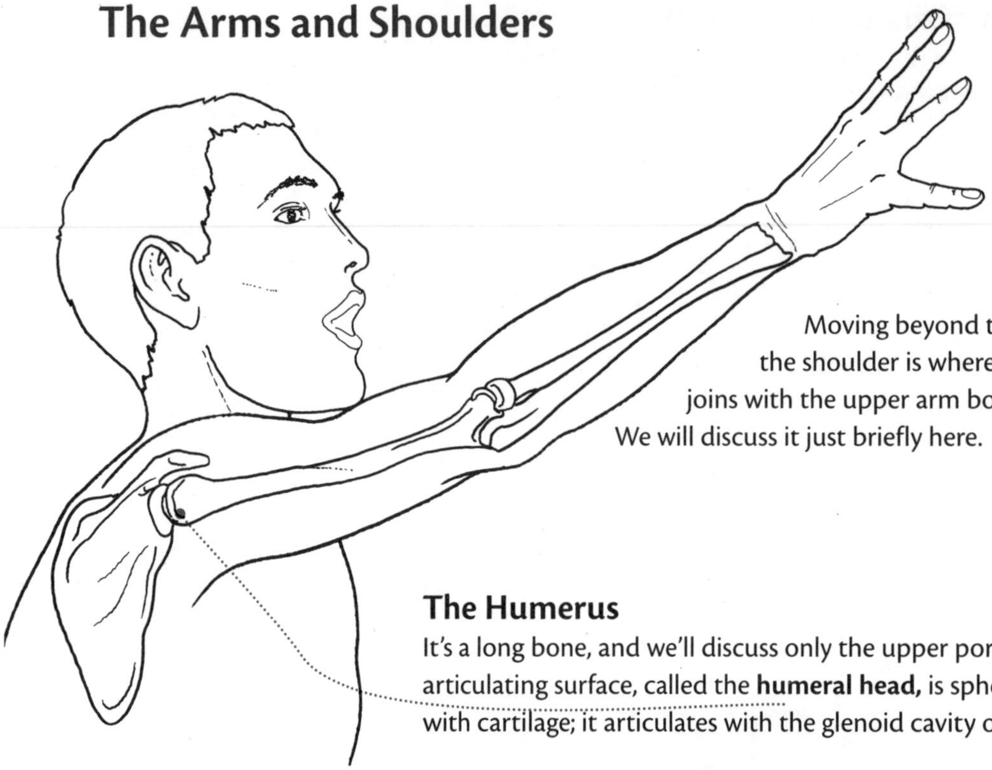

Moving beyond the shoulder girdle, the shoulder is where the shoulder blade joins with the upper arm bone: the humerus. We will discuss it just briefly here.

The Humerus

It's a long bone, and we'll discuss only the upper portion here. Its articulating surface, called the **humeral head,** is spherical and covered with cartilage; it articulates with the glenoid cavity of the scapula.

Shoulder Movements

The ball-and-socket form of the joint allows for movement in all directions. That movement can be amplified by the mobility of the shoulder girdle. Arm movements often move the rib cage as well, making them important in facilitating respiration and vocal preparation.

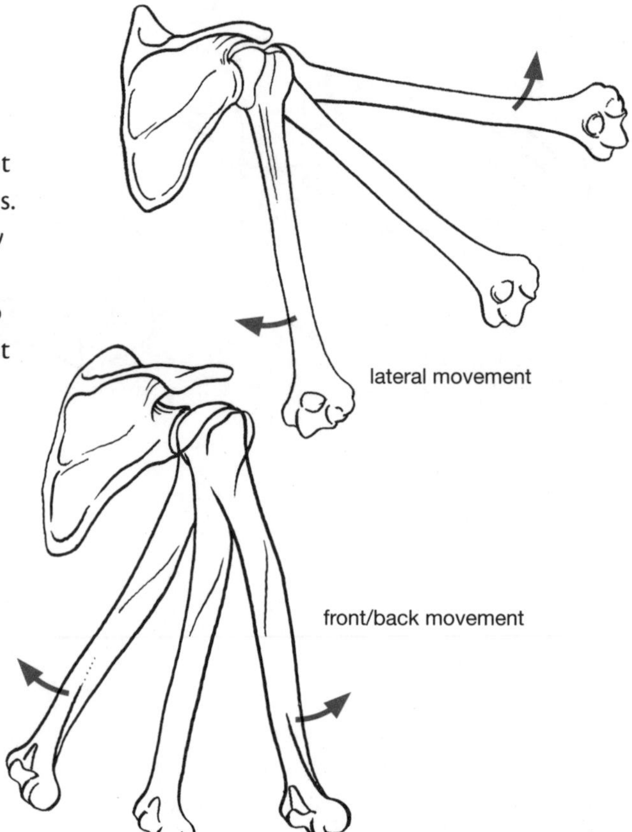

More details on the shoulders and arms can be found in *Anatomy of Movement,* pages 116–18.

Arms and the Voice

Singing with the arms just hanging at your sides, as is the practice in many choirs and in some classical singing techniques, doesn't help keep your chest wide or lift your ribs. This is even more the case if you're forced to hold sheet music in your hands, which tends to pull the head and spine forward.

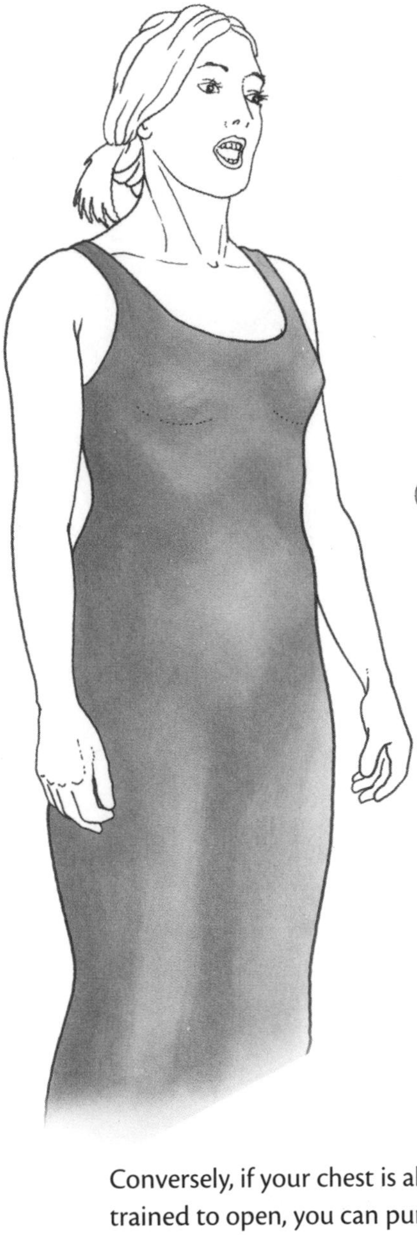

When doing vocal work, it's sometimes interesting to place your hands on your hips or head, behind your pelvis, or on the back of a chair or on a piano, to relax the thorax.

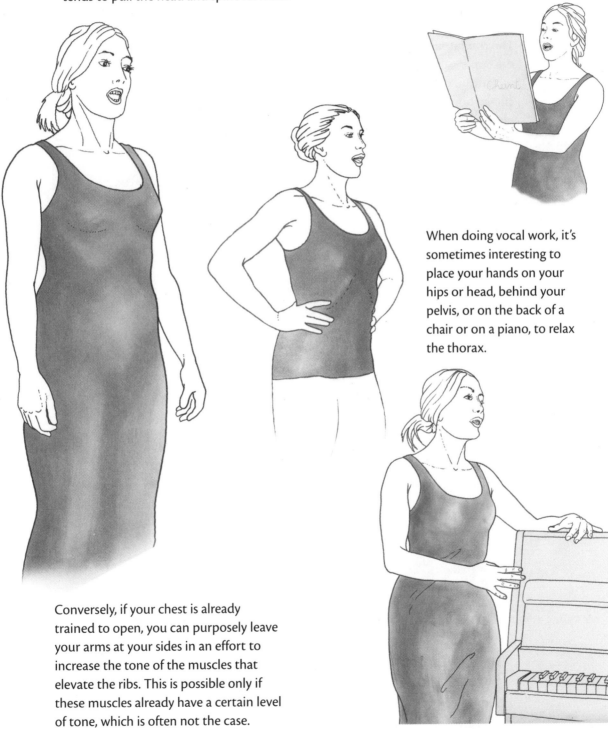

Conversely, if your chest is already trained to open, you can purposely leave your arms at your sides in an effort to increase the tone of the muscles that elevate the ribs. This is possible only if these muscles already have a certain level of tone, which is often not the case.

The Third Major Block: The Head—the Vocal Skull

The skull, the bony skeleton of the head, is composed of three major parts:

The cranial vault—the upper part of the skull—is made up of flat and curved bones that house the brain and the cerebellum.

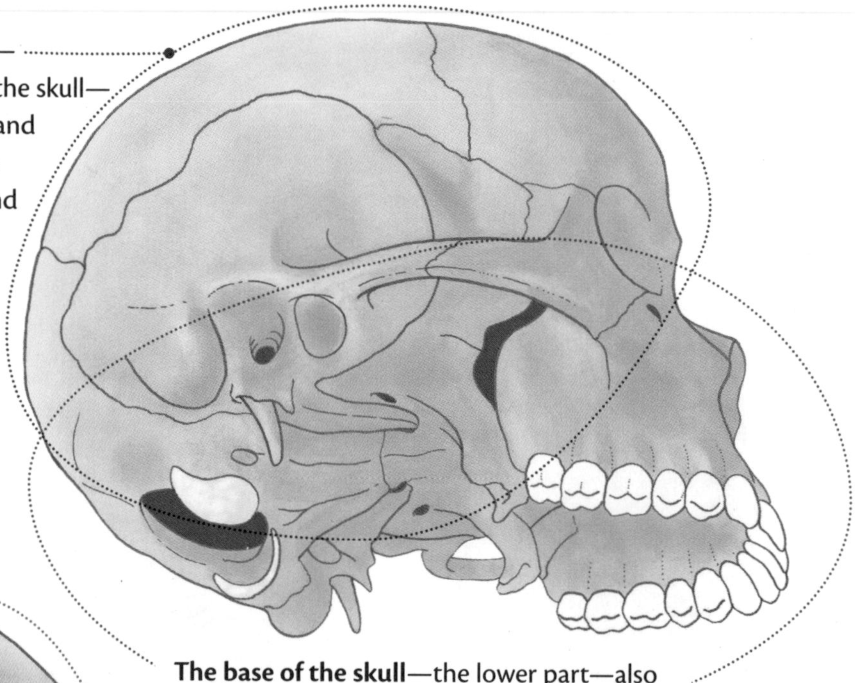

The base of the skull—the lower part—also houses the brain and is composed of thicker areas of bone.

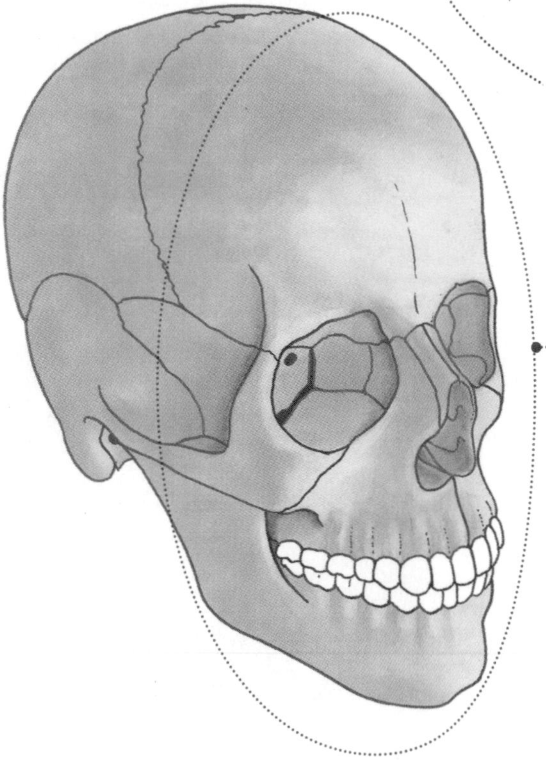

The facial skull—the anterior part—is composed of solid and complex bones that correspond to the face.

All of the bones of the skull have a connection to the vocal apparatus, whether by way of muscles, ligaments, or mucous membranes. This means that vocal work involves the whole skull. In the following pages, we will look at each bone with an emphasis on its link to the vocal body.

The Base of the Vocal Skull

If we look at the skull from below, with the mandible (lower jaw) removed, we see a continuous structure made up of bones that fit together like a puzzle. This is the **base of the skull,** and it's composed of the following:

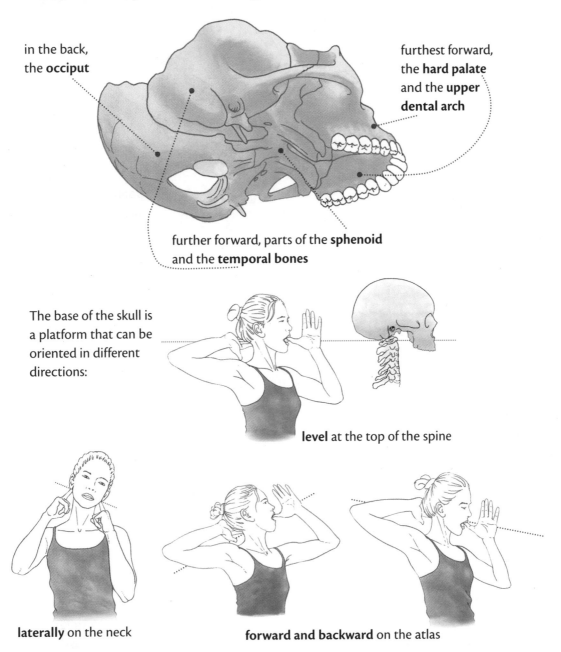

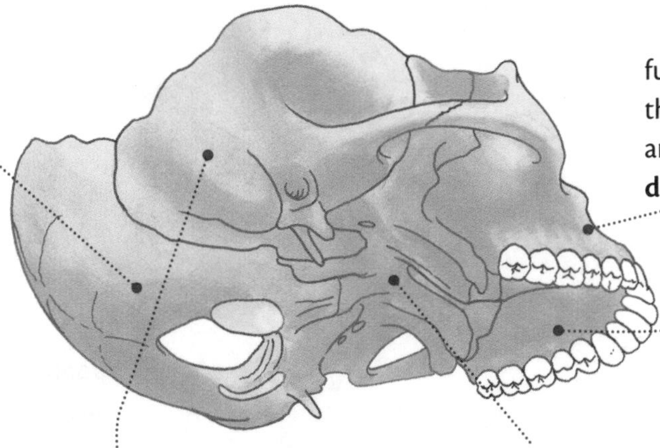
in the back, the **occiput**
further forward, parts of the **sphenoid** and the **temporal bones**
furthest forward, the **hard palate** and the **upper dental arch**

The base of the skull is a platform that can be oriented in different directions:

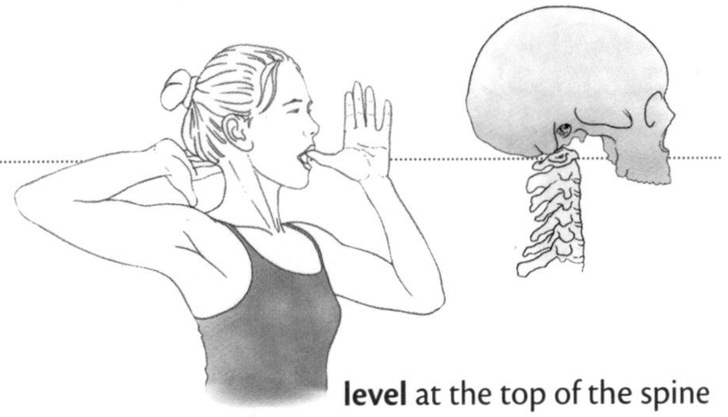

level at the top of the spine

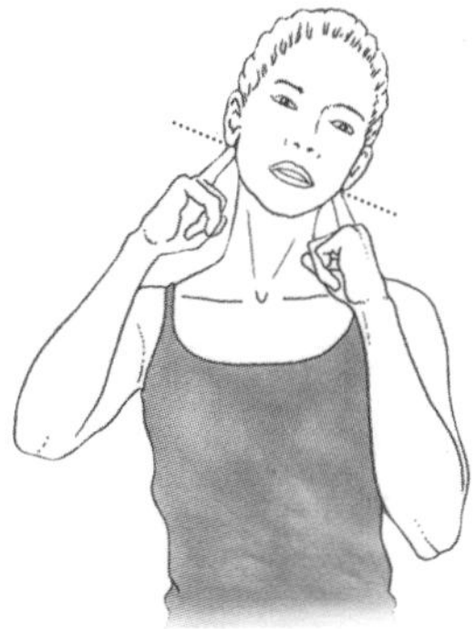

laterally on the neck

forward and backward on the atlas

This base of the skull is stabilized and mobilized by the muscles of the neck. Components of the vocal tract are suspended from the base of the skull (see page 204), and its orientation is important because it influences the tension and relaxation of these components.

The Posterior Bone of the Skull: The Occiput

The occiput is a flat bone that is composed of three parts: the **basion,** the **squama,** and the **condyles.**

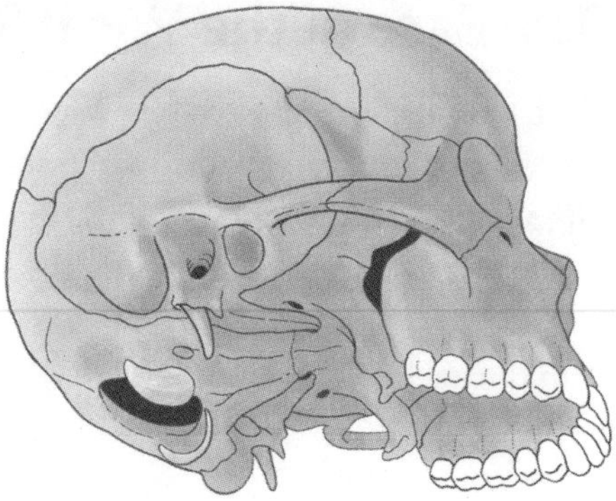

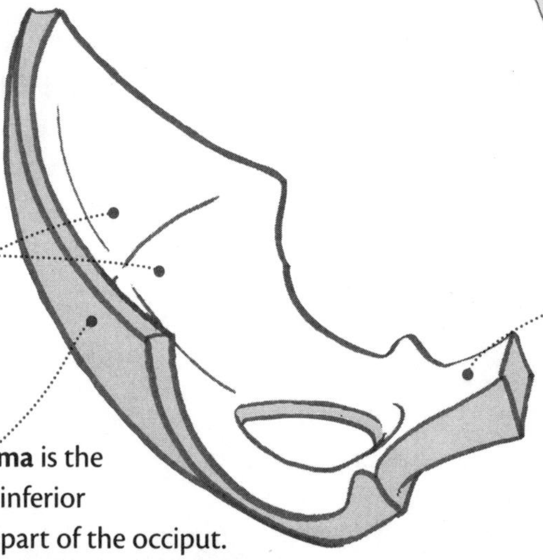

The **squama** is the flattened inferior posterior part of the occiput. Its internal surface, which is deeply concave toward the top and front, forms the back of the cranium and houses the cerebellum and brain; we won't get into its details here.

The external surface is convex and has a palpable projection called the **external occipital protuberance.** Parallel ridges run on each side: these are the **superior and inferior nuchal lines,** on which attach the posterior muscles of the neck.

The **basion** or basilar apophysis is the wedge-shaped anterior portion of the occiput that articulates with the body of the sphenoid. Together, their inferior surfaces make up the roof of the nasopharynx. On the inferior surface of the basion, we find a projection: the **pharyngeal tubercle.**

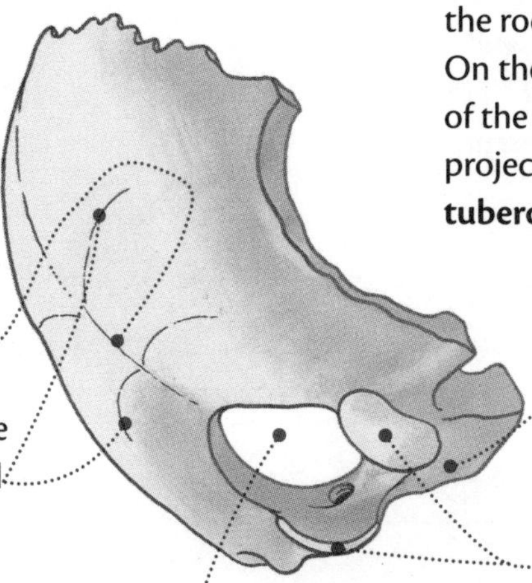

Between the basion and the squama we find an opening called the **foramen magnum** through which the spinal cord passes. On each side of this opening are articular surfaces: the **occipital condyles** (see page 42).

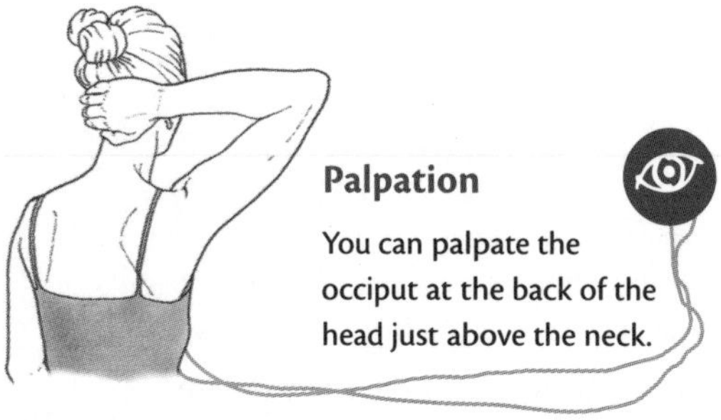

Palpation

You can palpate the occiput at the back of the head just above the neck.

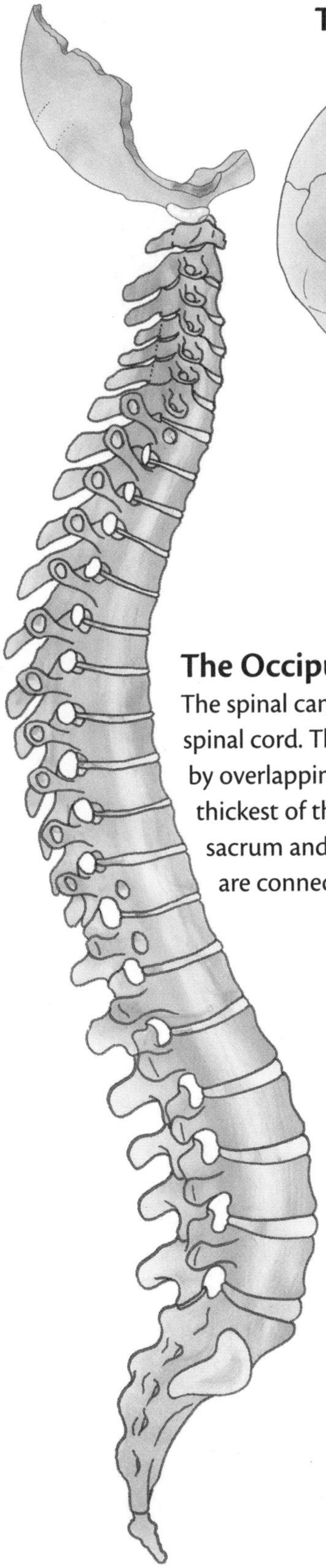

The Occiput and the Sphenoid: Micromovements

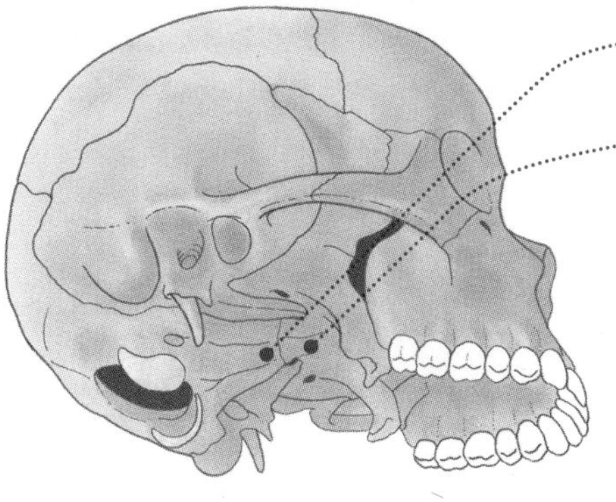

The *front of the occipital basion* articulates with the *back of the sphenoid body*. This is the **sphenobasilar symphysis** (or sphenobasilar synchondrosis), a joint that allows the skull to have micromovements of flexion or extension. These movements aren't possible unless the occiput is relatively free on the spine.

The Occiput and the Sacrum: A Flexible Connection

The spinal canal runs up through the vertebrae of the spine and houses the spinal cord. Throughout its length, the spinal cord (like the brain) is enveloped by overlapping layers of membranes called the **meninges**. The outermost and thickest of these membranes, the dura mater, attaches at the bottom to the sacrum and at the top to the foramen magnum. The sacrum and the occiput are connected by this membrane.

Microflexibility for Vocal Fluidity

In vocal work, we avoid stiffening the spine (which can cause overcontraction of the postural muscles) to allow for the microflexibility between the sacrum and the occiput—except at specific times when we need extreme vocal power to "belt it out."

The Central Bone of the Skull: The Sphenoid

The sphenoid bone resembles a butterfly or a bat. It is structured around a central part, or **body,** that is cubical in shape and hollow, and from which originate three kinds of symmetrical projections:

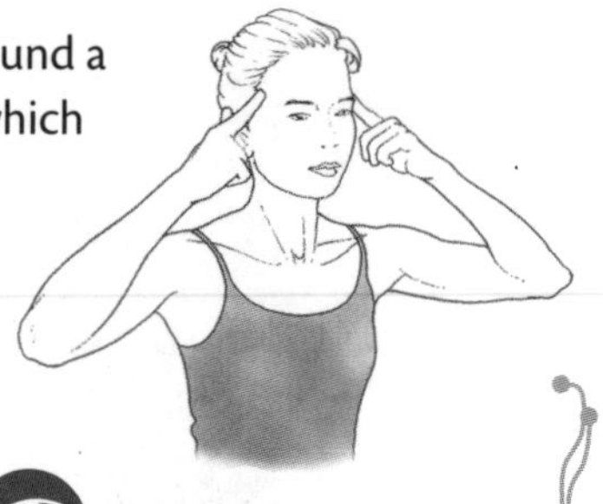

Palpation

You can palpate the sphenoid at the temples.

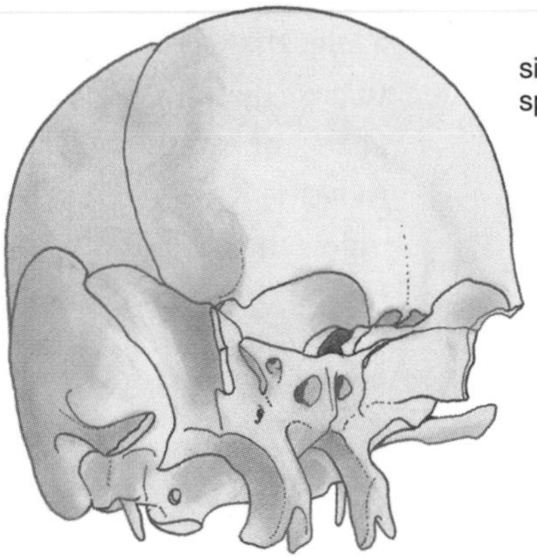
side view of
splenoid

1. The **lesser wings** arise from the superior surface of the sphenoid body and are located in the cranial cavity.

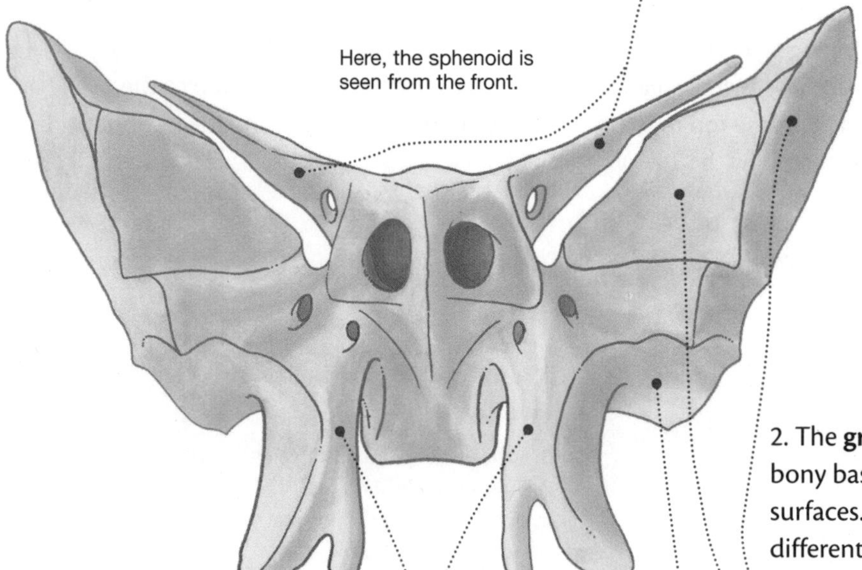
Here, the sphenoid is
seen from the front.

2. The **greater wings** emerge from a bony base on the sphenoid's lateral surfaces. Three facets, with three different orientations, arise from this base:
   - The **temporal facet** makes up an area of the temple.
   - The **orbital facet** makes up a part of the eye socket.
   - The **maxillary facet** constitutes a part of the pterygopalatine fossa (see page 230).

3. The **pterygoid processes** extend downward; they are the most important prominences in regard to the voice, forming part of the skeletal structure of the vocal tract.

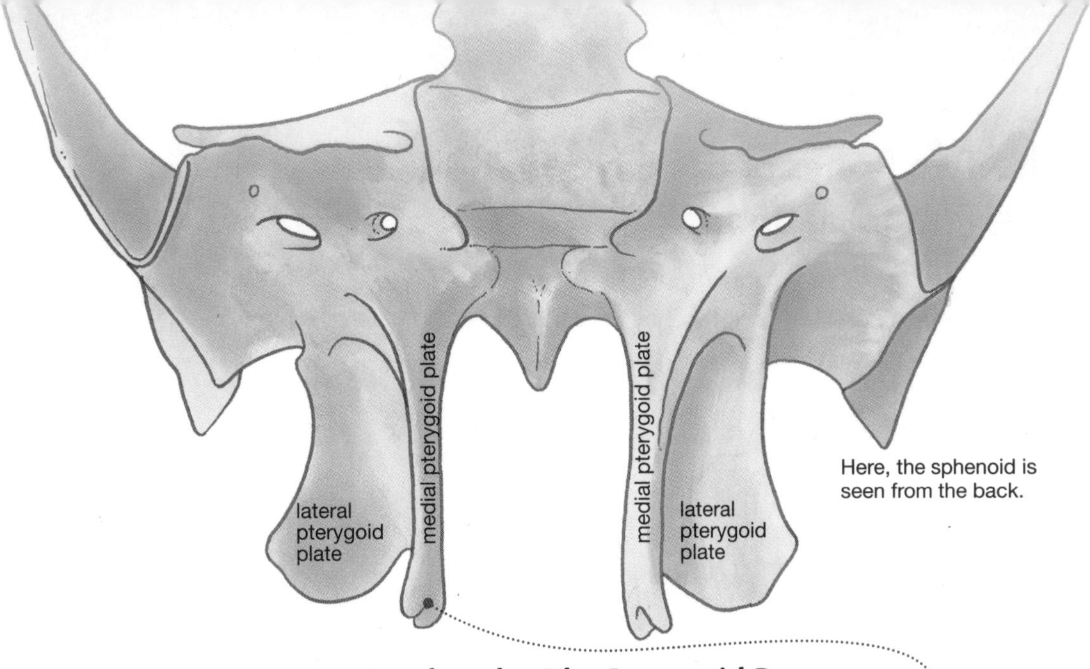

Important Landmarks: The Pterygoid Processes

The two pterygoid processes extend from the inferior surface of the sphenoid body and hang downward (if the sphenoid is seen as a bat, the pterygoid processes are its feet). Each process has a medial plate (or lamina) and a lateral plate. The bottom tip of the medial pterygoid lamina forms a hook, or **hamulus,** through which glides the tendon of the muscle that tenses the soft palate (see page 242).

These plates form the lateral boundary of the choanae (see page 279).

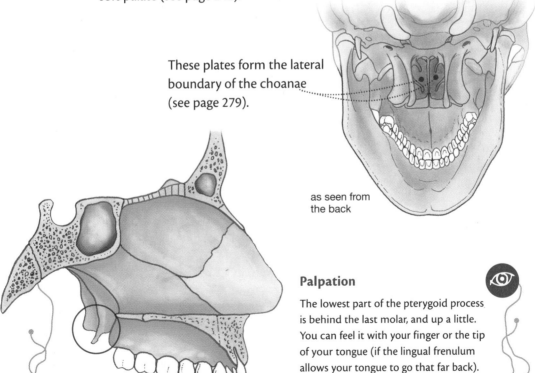

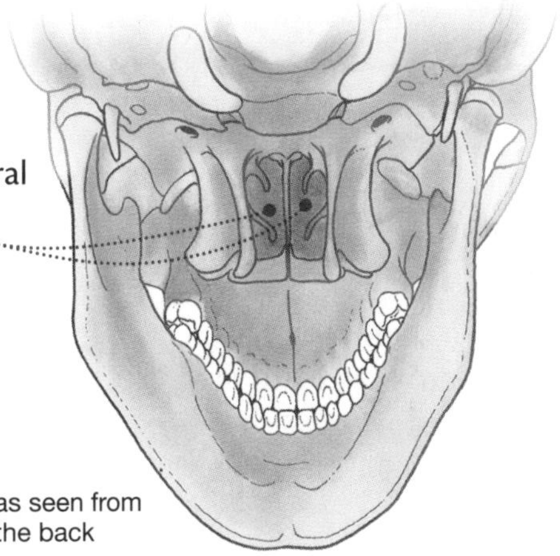

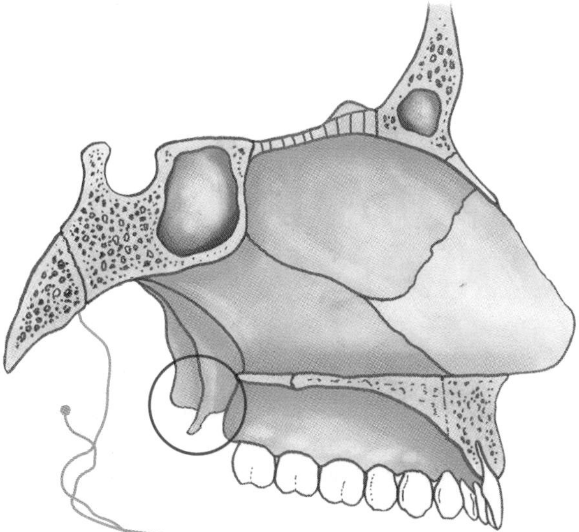

Palpation

The lowest part of the pterygoid process is behind the last molar, and up a little. You can feel it with your finger or the tip of your tongue (if the lingual frenulum allows your tongue to go that far back).

The Two Bones of the Ears: The Temporal Bones

The temporal bones are located more or less where headphones would cover the ears. They are composed of three major parts: the **squamous, tympanic,** and **petromastoid**.

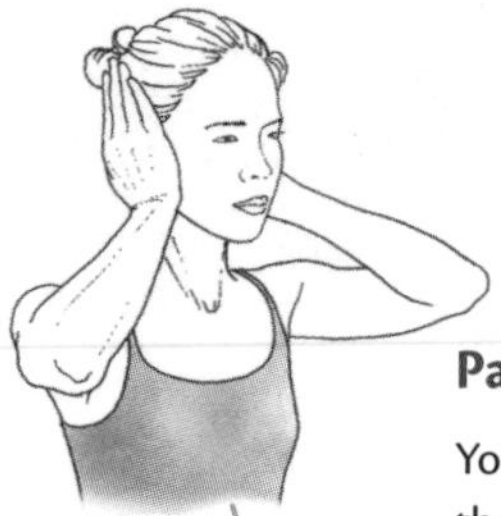

Palpation

You can palpate the temporal bones around the ears.

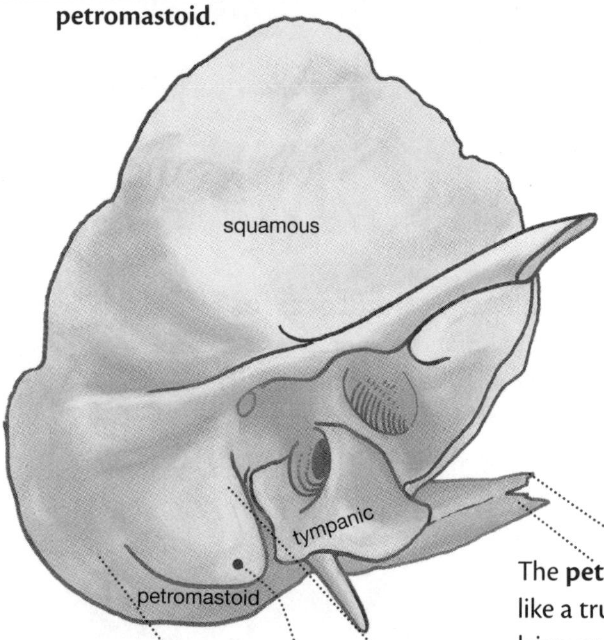

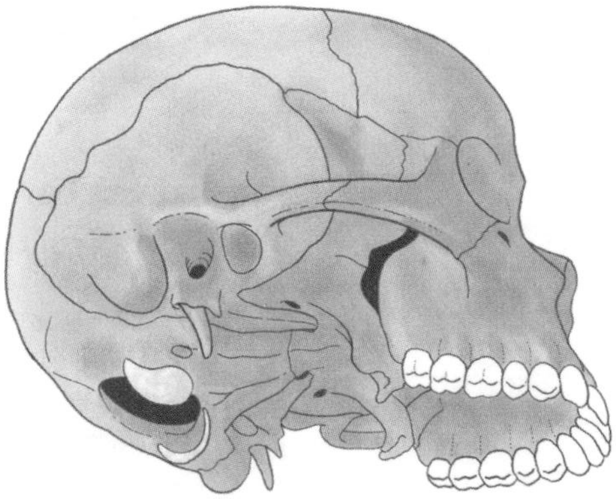

The **petromastoid** is shaped like a truncated pyramid lying on one side.

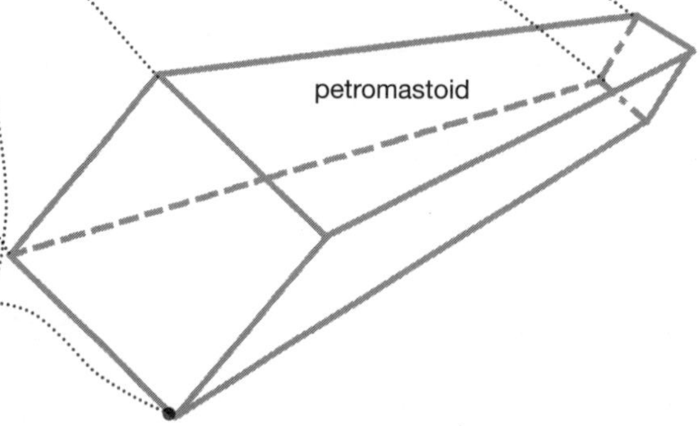

The summit of the pyramid is situated near the foramen magnum (see page 68).

The base of the pyramid is massive, and projecting downward at its bottom is the **mastoid process.**

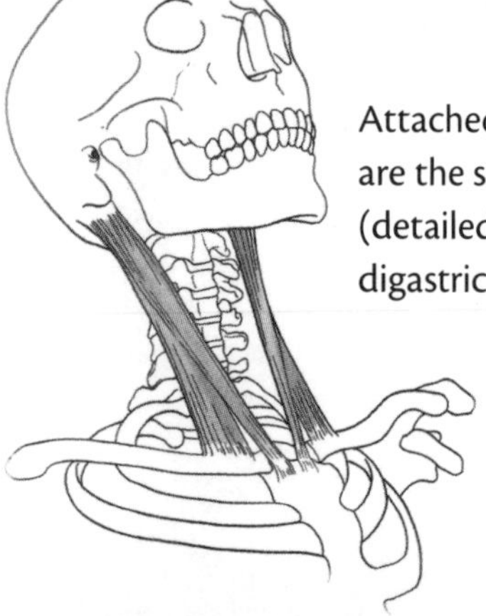

Attached to the mastoid process are the sternocleidomastoid muscle (detailed on pages 129 and 216) and digastric muscle (see page 188).

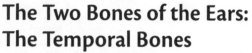

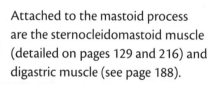

Palpation

You can palpate the mastoid process behind the ear.

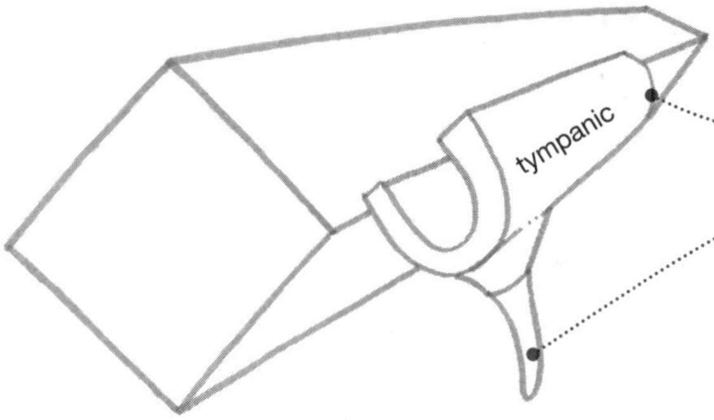

The **tympanic** part of the temporal bone is gutter-shaped and joins the front edge of the petromastoid pyramid. The outer side forms the bony part of the ear opening. Two projections arise from the tympanic:

- the **torus tubarius,** which is internal and forms a part of the Eustachian tubes (see page 283)
- the **styloid process** (see page 189), which points downward, and from which the muscles and ligaments of the larynx, the pharynx, and the tongue (see page 255) are suspended

The **squamous** is a bony circular plate. The anterior inferior quarter is folded inward. It is part of the cranial vault. About midway to the top is a projection that is directed forward. This is the **zygomatic process,** which forms, with the malar and maxilla bones, the zygomatic arch (see page 228). The lower edge of the arch has a recess and a projection: the **glenoid fossa** and the **temporal condyle,** which make up part of the temporomandibular joint (see page 82).

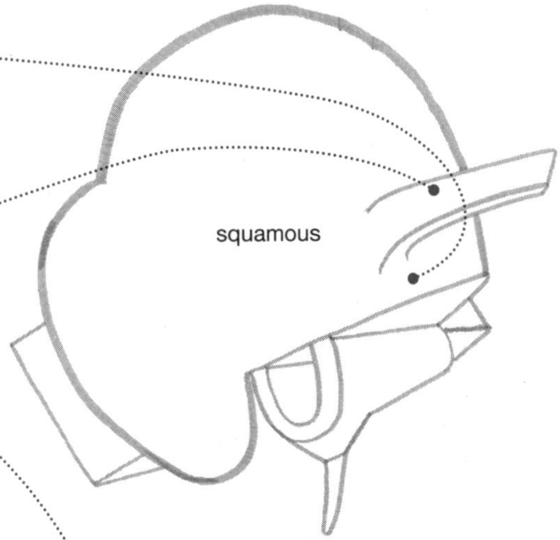

The Malar Bone

This is commonly referred to as the cheekbone. It is joined with the superior maxilla, the frontal bone, and the temporal bone.

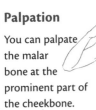

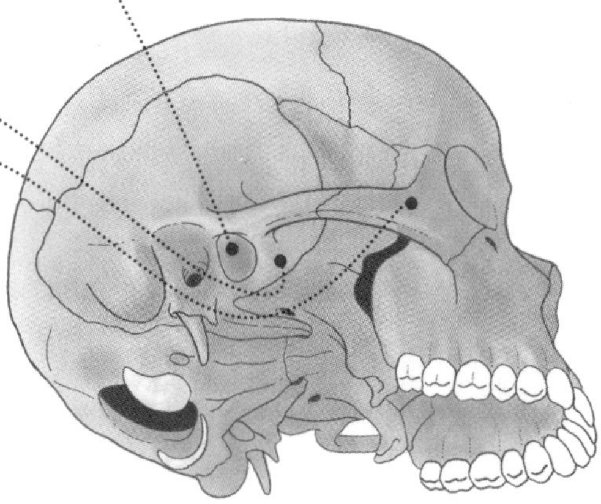

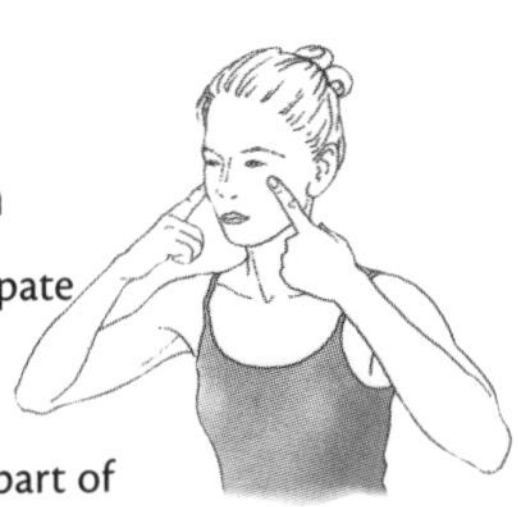

Palpation

You can palpate the malar bone at the prominent part of the cheekbone.

The Bones That Make Up the Nose

The Uppermost Bone of the Face: The Frontal Bone

This flat, dome-shaped bone makes up part of the cranial vault and the face. It has two parts: the squama frontalis and the pars orbitalis.

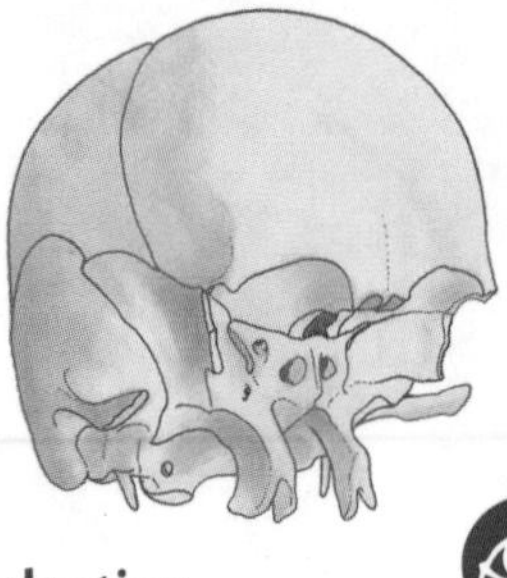

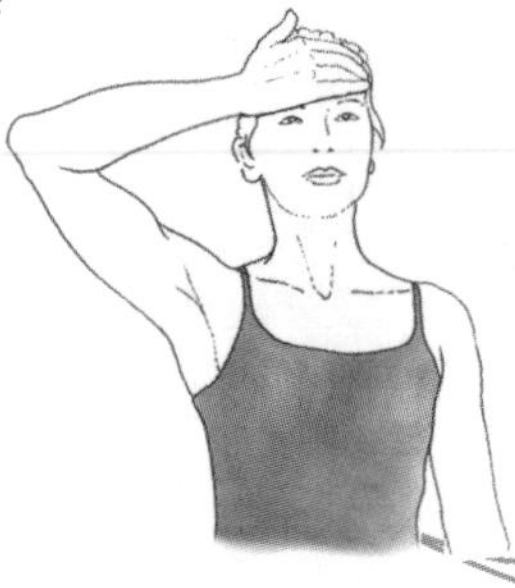

Palpation

You can palpate the frontal bone from the forehead to the top of the head.

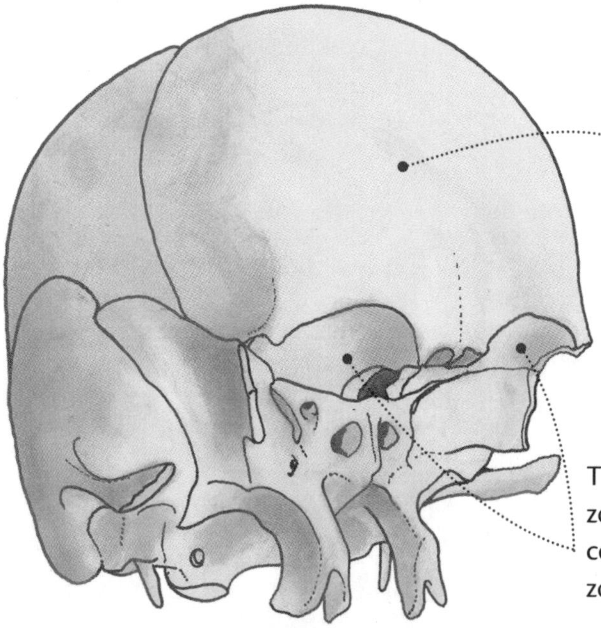

The **squama frontalis** runs from the top to the front of the skull. Above the orbital arches it features two projections: the supraorbital ridges, or brow ridges. These ridges are connected by the smooth prominence called the **glabella**.

The **pars orbitalis** is made up of three zones: two symmetrical zones that constitute the roof of the *eye sockets* and a zone in the middle.

To better show the details of the frontal bone here, the rest of the facial skeleton has been omitted.

That middle zone (in purple) of the pars orbitalis articulates with the ethmoid bone (in green), the lacrimal bone (in orange), the nasal bone (in blue), and the frontal process (in yellow). These bones constitute the skeleton of the nose and create a wall between the two orbits.

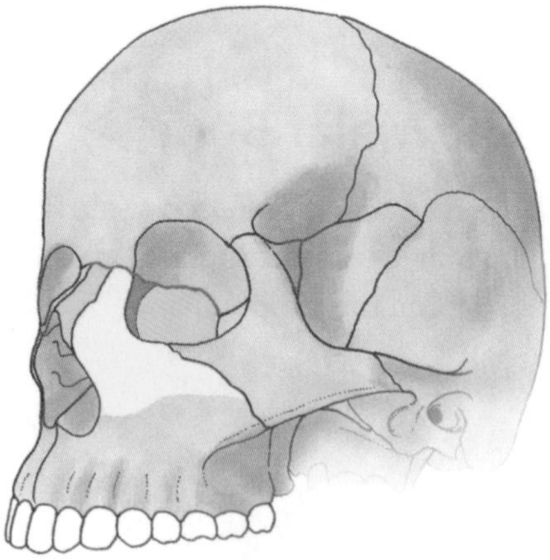

Here, the rest of the facial skeleton is in place.

Details

Behind each supraorbital ridge is a hollow cavity: the frontal sinus.

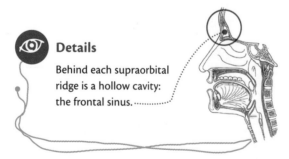

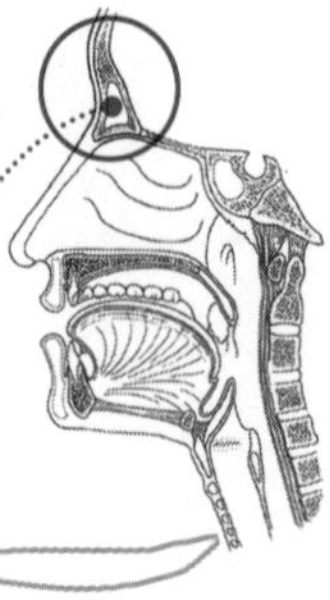

The Ethmoid: The Bone at the Roof of the Nasal Cavity

The ethmoid is a very complex bone, with extremely thin walls, located under the central area of the frontal bone. It is made up of:

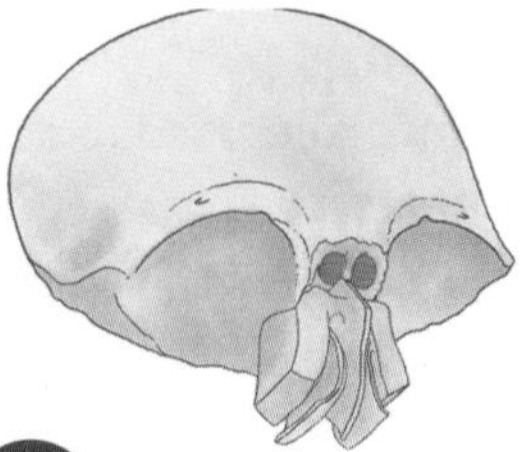

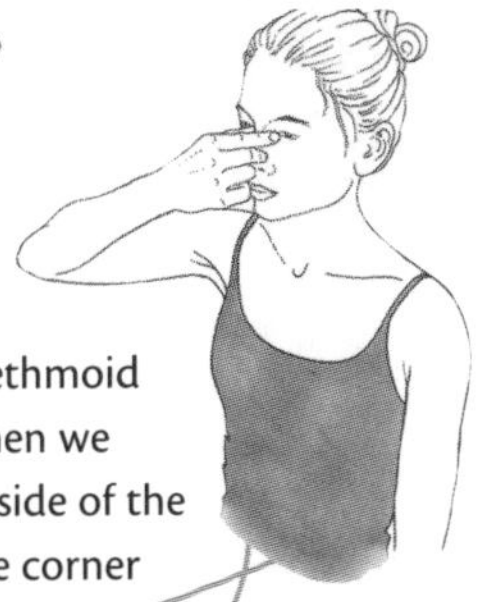

Palpation

We can envision the ethmoid behind the fingers when we place them on either side of the top of the nose, at the corner of the eyes.

- the **perpendicular plate,** in the middle, which forms part of the septum
- the **cribriform plate,** perforated with many holes, which forms the roof of the nasal cavity (see page 278) and through which pass the ends of the olfactory nerve
- on each side, the **ethmoidal labyrinth,** which is partitioned into tiny alveoli or air cells: the **ethmoidal sinuses** (see page 280)

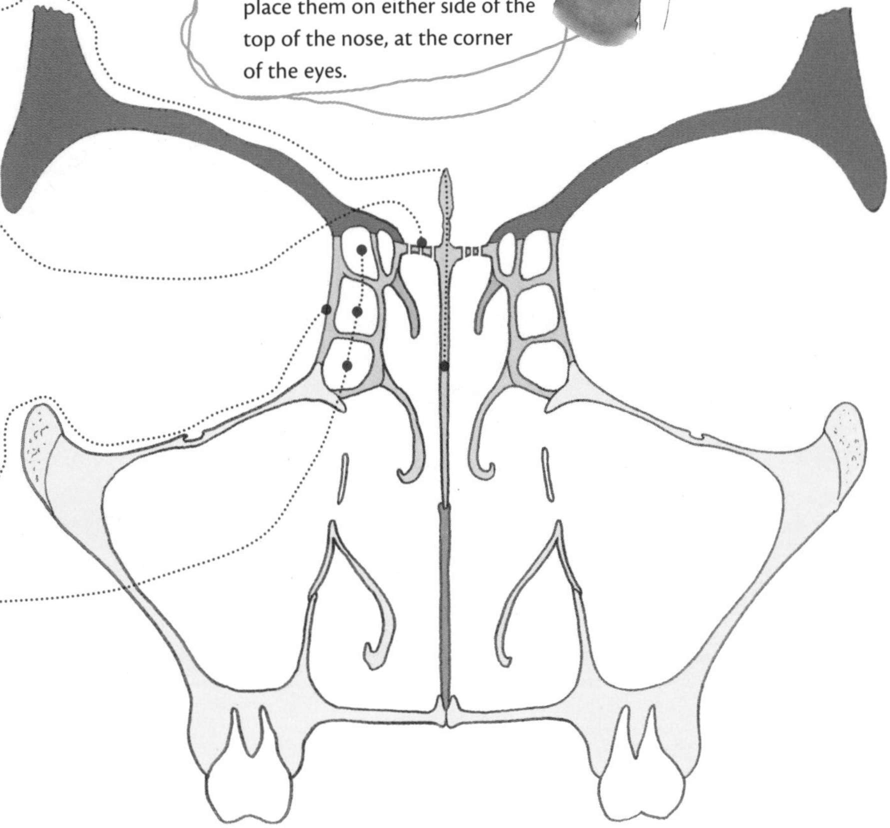

This frontal section of the facial bones shows the ethmoid (green), frontal bone (purple), nasal conchae (blue), maxilla (yellow), and vomer (orange), which is described on the following page.

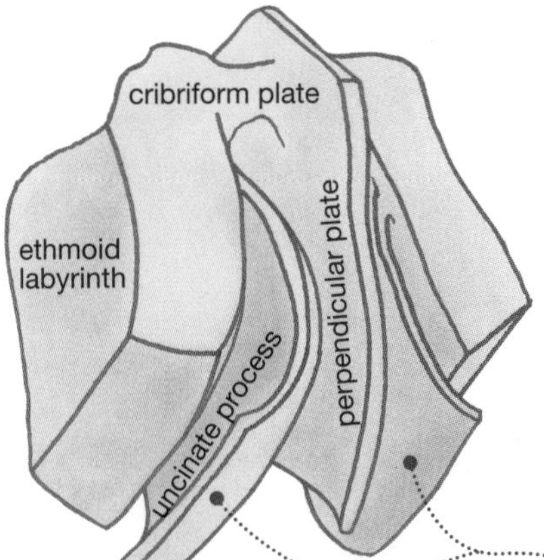

Two curving lamina project down from the ethmoid labyrinth. These are the **uncinate processes,** and they form part of the medial wall of the sinuses.

Image

To get a better idea of the form, imagine that you're looking at the rear of a bicycle. The back wheel is the perpendicular plate, the luggage rack is the cribriform plate, and the two bags on either side are the ethmoid labyrinths.

The Vomer: The "Spur" Bone

This is a thin, flat bone, sometimes compared to a spur, that makes up part of the "puzzle" that is the nasal septum. It's situated under the perpendicular plate of the ethmoid and in front of the sphenoid.

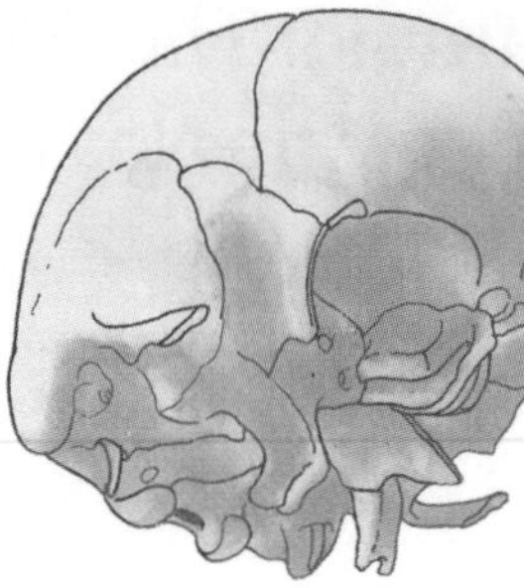

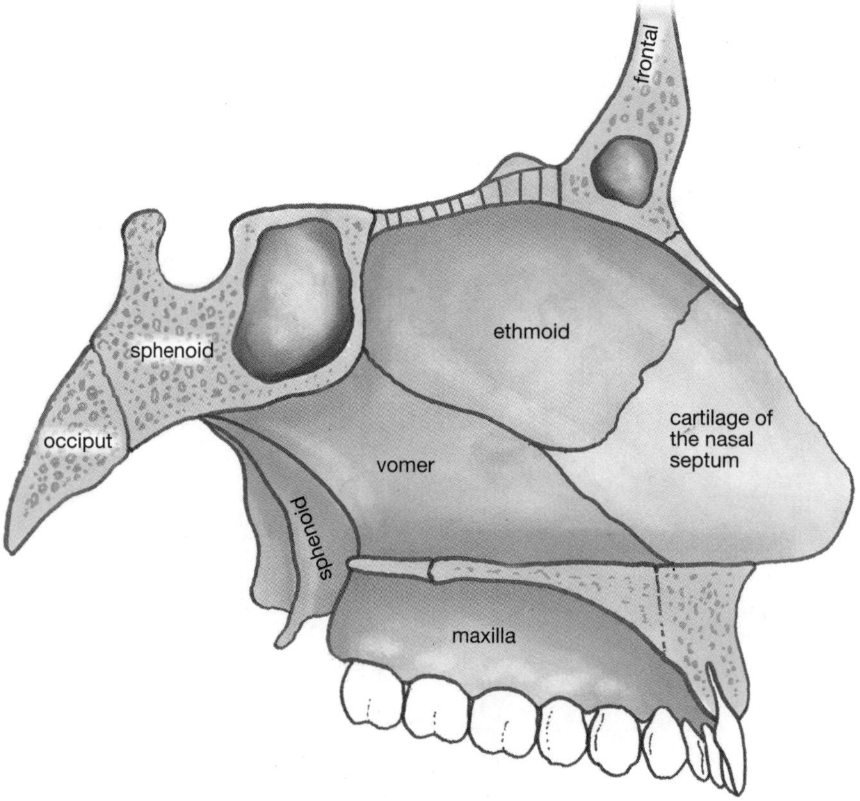

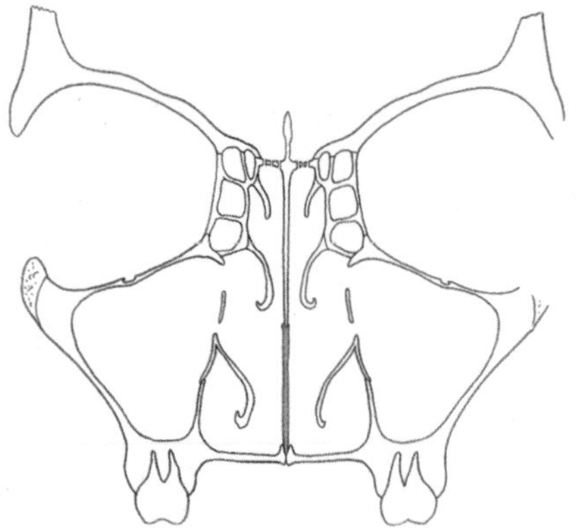

This frontal view shows the vomer (in orange) in relation to the other bones of the face: the conchae, maxilla, ethmoid, and frontal bones.

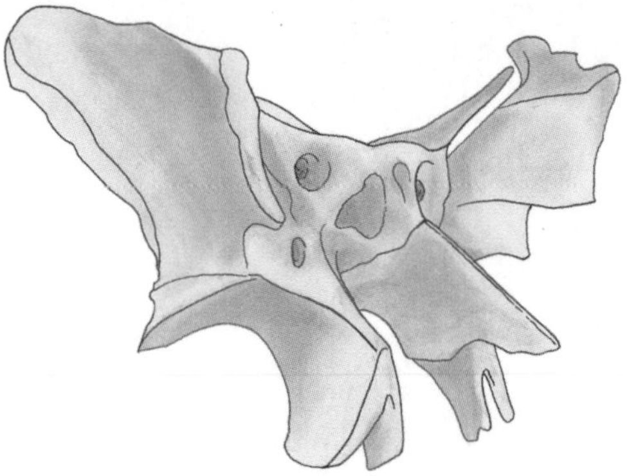

The vomer extends the body of the sphenoid to the front.

The Conchae

In each nasal cavity we find small bones: the **nasal conchae,** or turbinates. These long, curved bones are similar in shape to the lip of a conch shell.

They attach horizontally to the lateral face of the bones of the nasal cavity and curve medially and downward into the nasal cavity.

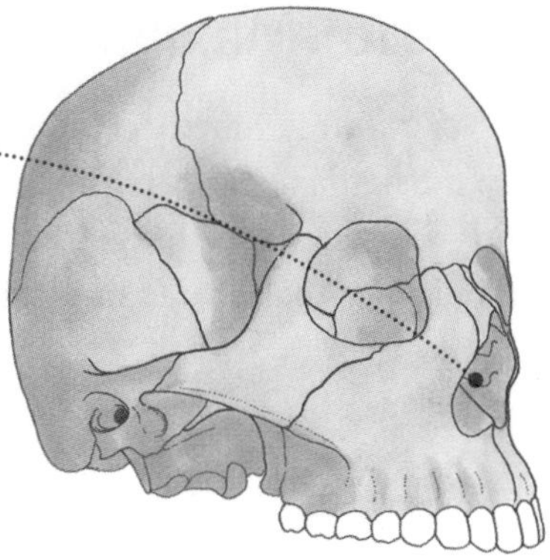

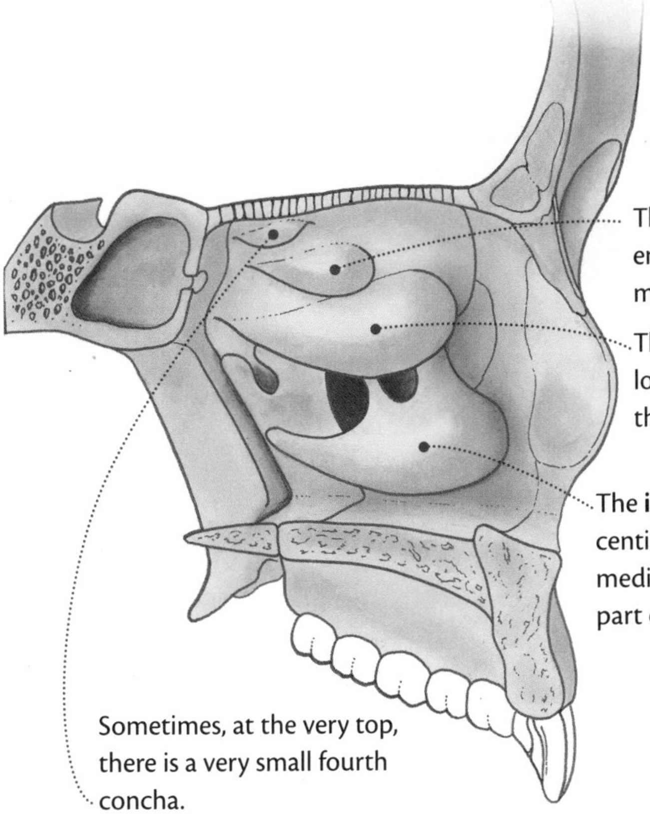

The **superior concha** is the smallest and is embedded in the medial surface of the lateral mass of the ethmoid.

The **middle concha** is about 7 centimeters long and also arises from the ethmoid, below the superior concha.

The **inferior concha** is the longest (at about 10 centimeters in length) and is attached to the medial surface of the superior maxilla. It closes a part of the maxillary sinus (see page 280).

Sometimes, at the very top, there is a very small fourth concha.

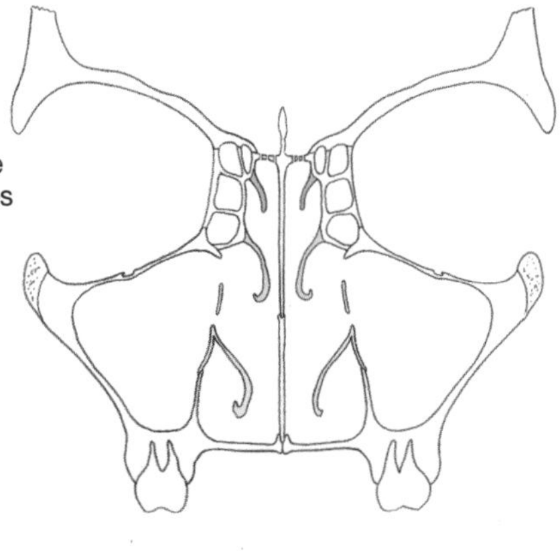

This frontal view shows the conchae (in blue) in relation to the other bones of the face: the vomer, maxilla, ethmoid, and frontal bones.

The Superior Maxilla

The two halves of this symmetrical bone form part of the hard palate, the nasal fossa (cavity), and the cheeks. Each half of the superior maxilla has a base, three surfaces, and a summit.

Palpation

You can palpate the maxilla on each side of the nostrils, above the upper teeth.

The **superior surface** forms part of the floor of the orbital socket. It extends upward by way of a projection: the **frontal process**.

The **base** forms part of the lateral wall of the **nasal fossa** and part of the hard palate. It is bordered by the **dental arch**.

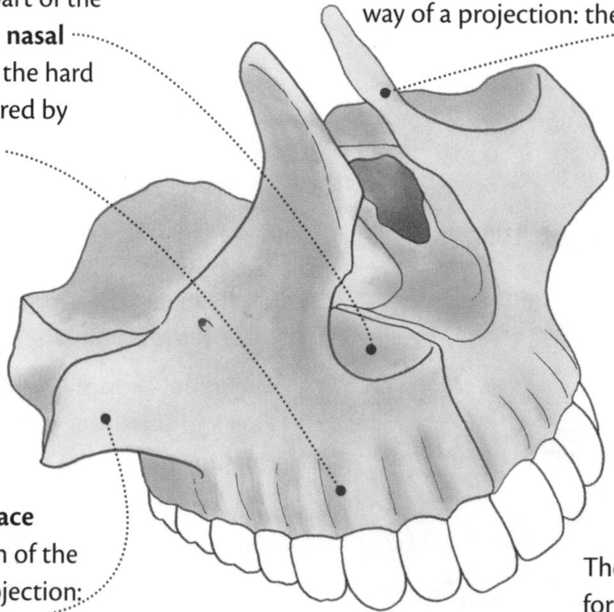

The **anterior surface** forms the skeleton of the cheek. It has a projection: the **malar process**, which joins it to the malar bone.

The **posterior surface** forms part of the **pterygopalatine fossa** (see page 230).

The maxilla is home to the large **maxillary sinus**, which lies above the molars and communicates with the nasal cavity.

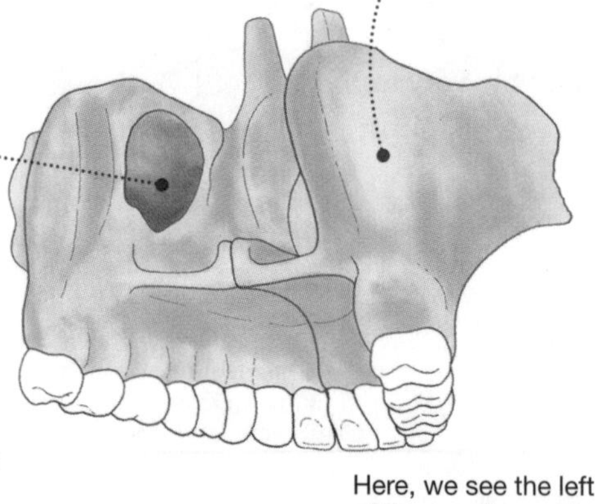

Here, we see the left side of the maxilla, as viewed from behind.

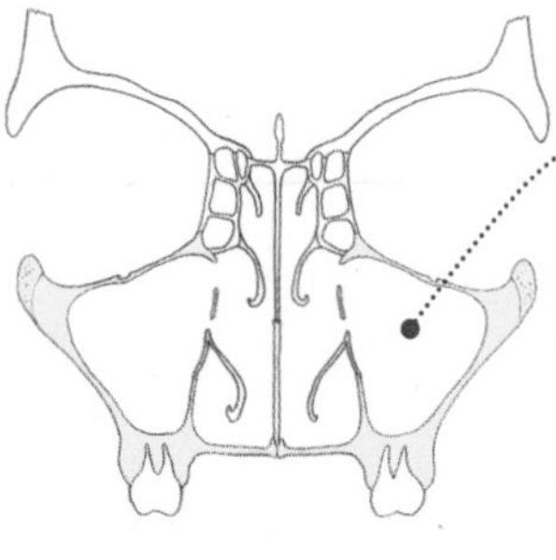

This frontal view shows the superior maxilla (in yellow) in relation to the other bones of the face: the conchae, ethmoid, vomer, and frontal bones

The Bones of the Palate and Nasal Fossa: The Palatines

These small bones are located behind the maxilla on each side of the nasal cavity, and they contribute to the structure of the palate, the nasal fossa (nasal cavity), and, at the top, the eye sockets. Seen from the side, each palatine bone is shaped like a capital L, comprising perpendicular and horizontal plates.

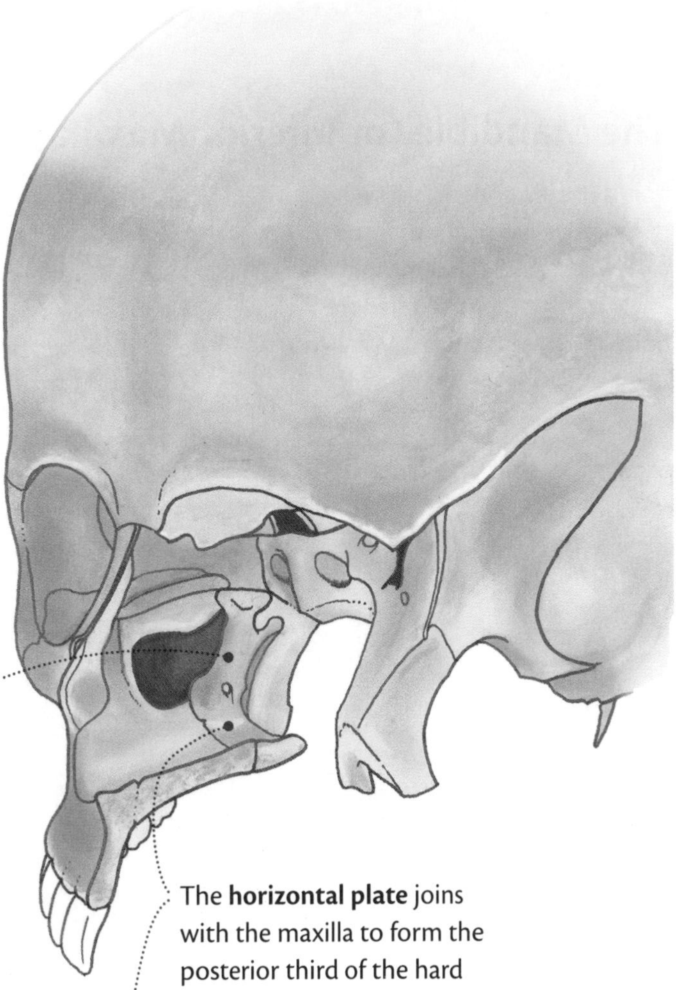

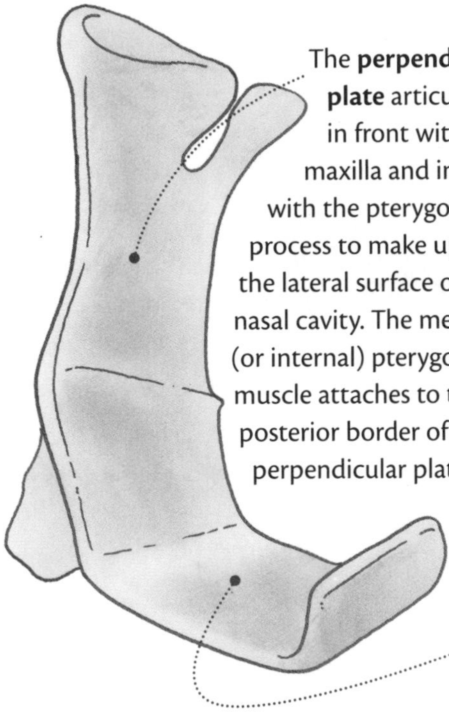

The **perpendicular plate** articulates in front with the maxilla and in back with the pterygoid process to make up the lateral surface of the nasal cavity. The medial (or internal) pterygoid muscle attaches to the posterior border of this perpendicular plate.

The **horizontal plate** joins with the maxilla to form the posterior third of the hard palate. At its medial border it joins its symmetrical partner coming from the other side.

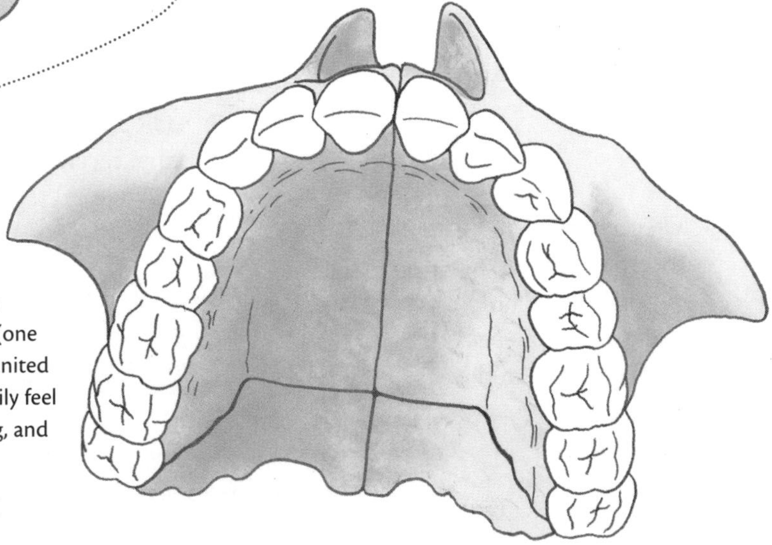

The Bony Palate: A Four-Bone Puzzle

The bony palate is formed in front by the horizontal halves of the superior maxilla (one on the right, one on the left), which are united in the middle by a suture that we can easily feel with the tongue through the palate lining, and in back by the two palatine bones.

The Mandible (or Inferior Maxilla)

This is the jawbone. It has two symmetrical vertical parts, one on each side, called the **rami,** and a lower horizontal part, called the **body**.

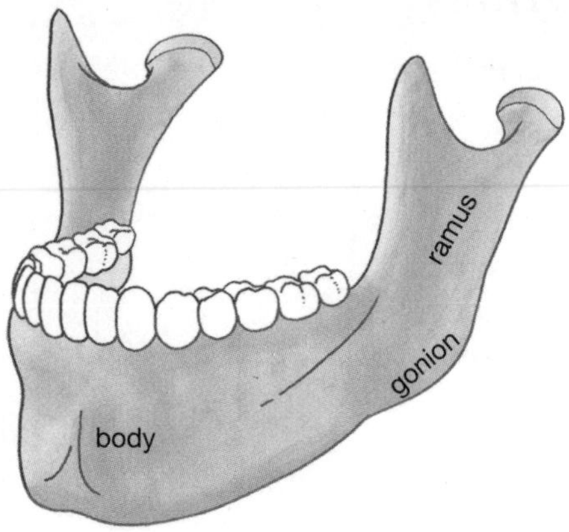

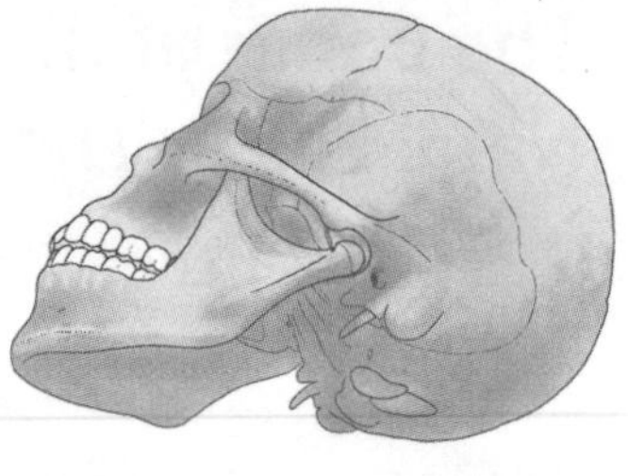

The junction of the ramus and the body forms an angle called the **gonion**.

The Body

The body is shaped like a thick, curved blade with a concave internal surface and a convex external surface. We divide it into two parts:

- the **alveolar part,** at the top
- the **base of the mandible,** at the bottom

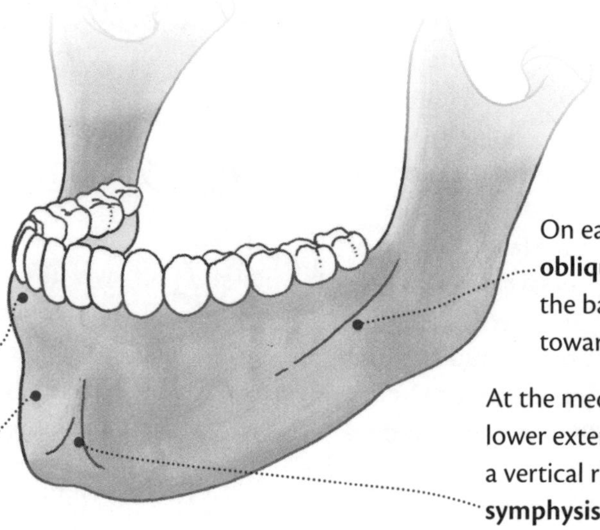

On each side, the **external oblique ridge** rises toward the back and continues toward the ramus.

At the medial line on the lower external surface, we find a vertical ridge known as the **symphysis menti**.

The upper border of the mandibular body is filled with cavities, known as the **dental alveoli,*** that receive the dental roots.

On each side, the **internal oblique ridge** (or **mylohyoid line**) rises toward the back and continues to the ramus.

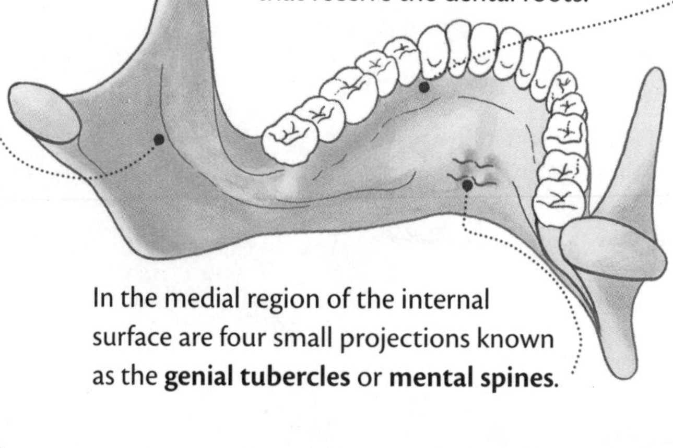

In the medial region of the internal surface are four small projections known as the **genial tubercles** or **mental spines.**

*By extension, we call the lower part of the teeth the **alveolar edge** because of their proximity to the alveolar ridge.

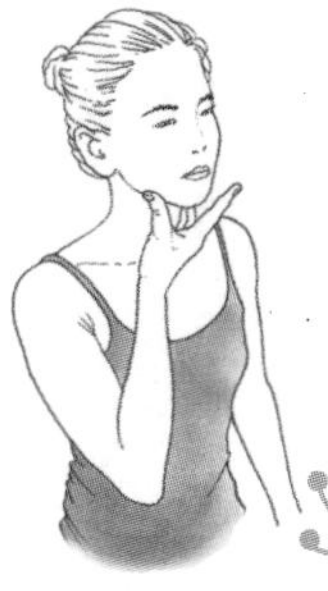

Palpation

You can palpate the inferior maxilla on each side of the chin.

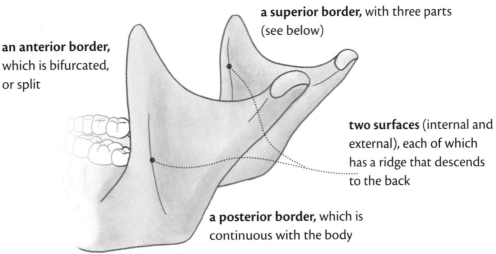

The Rami

The rami have:

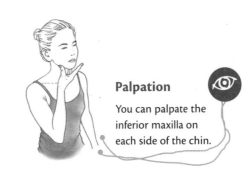

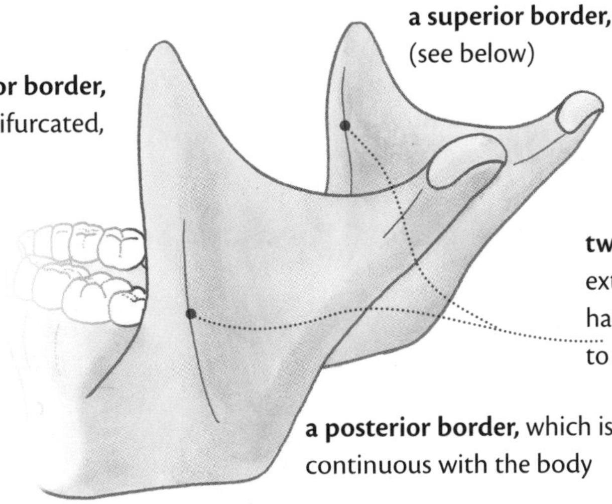

an anterior border, which is bifurcated, or split

a superior border, with three parts (see below)

two surfaces (internal and external), each of which has a ridge that descends to the back

a posterior border, which is continuous with the body

The three parts of the superior border are:

the **mandibular notch,** a thin, concave area between the condyloid process and the coronoid process

the **condyloid process,** which is oval-shaped and has an articular cartilaginous surface

the **coronoid process,** a thin projection that points upward

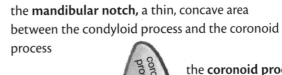

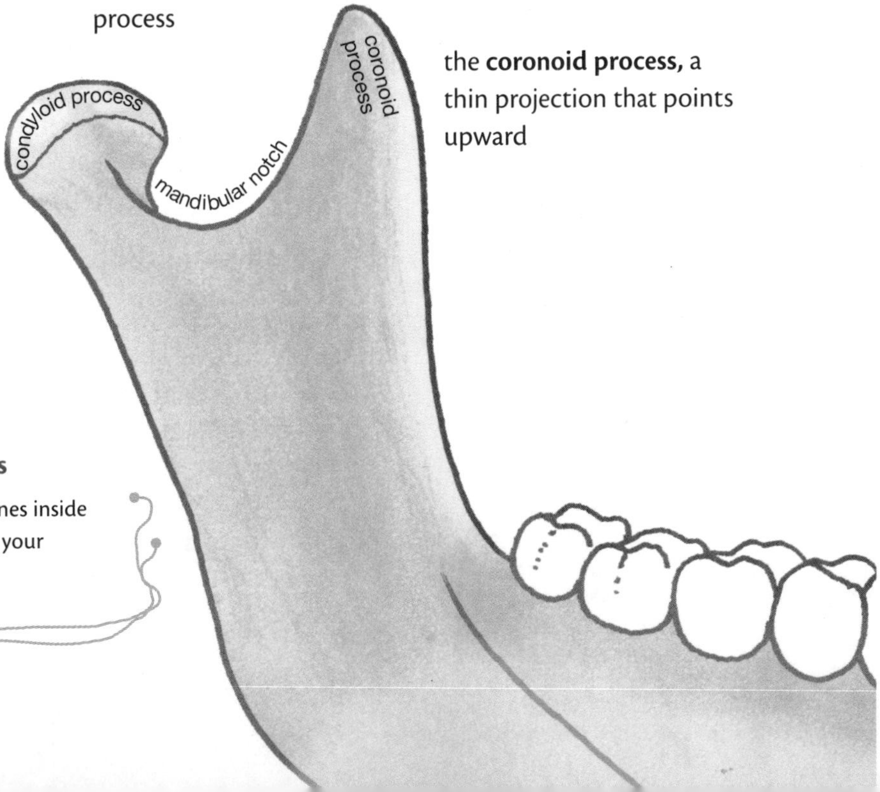

Feel the Mental Spines

You can feel the mental spines inside your mouth with the tip of your tongue.

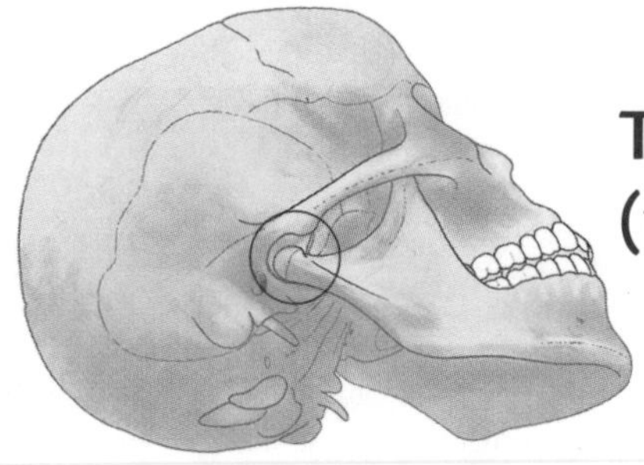

The Joints of the Mandible (the Temporomandibular Joint or TMJ)

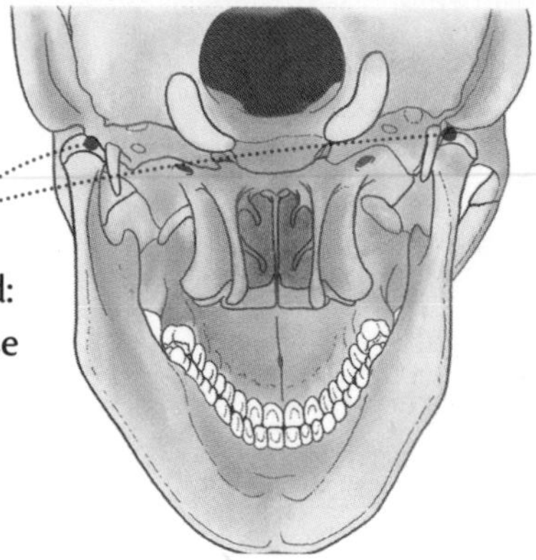

There are two symmetrical joints for the mandible. *The two joints always function together.* On each side we find:

- the temporal bone, which is part of the lateral base of the skull
- the inferior maxilla, which is known by several names: the jaw, the mandible, the maxilla

On the Temporal Bone: The Glenoid Fossa and the Temporal Condyle

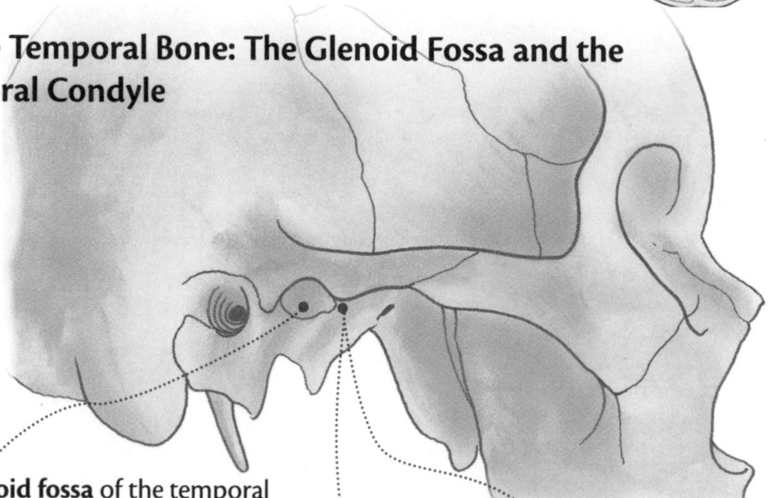

The **glenoid fossa** of the temporal bone is a concave surface that faces downward, situated behind the condyle and in front of the opening of the ear. It comes in contact with the mandibular condyle when the mandible is moved forcefully backward.

The **temporal condyle,** whose surface is in the form of a partial cylinder, and convex toward the bottom, is situated on the inferior surface of the zygomatic process (see page 73).

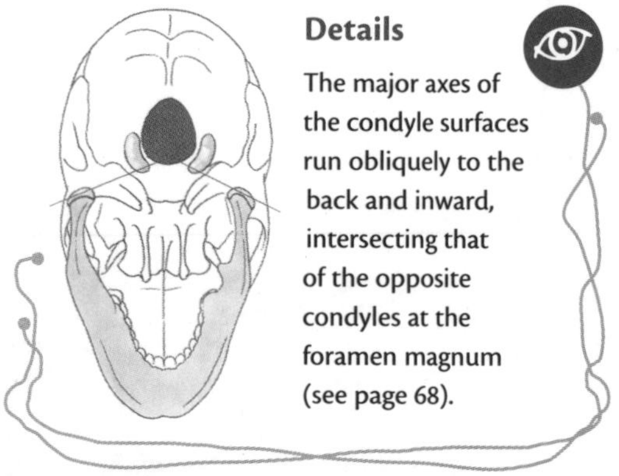

Details

The major axes of the condyle surfaces run obliquely to the back and inward, intersecting that of the opposite condyles at the foramen magnum (see page 68).

On the Mandible: The Condyloid Process

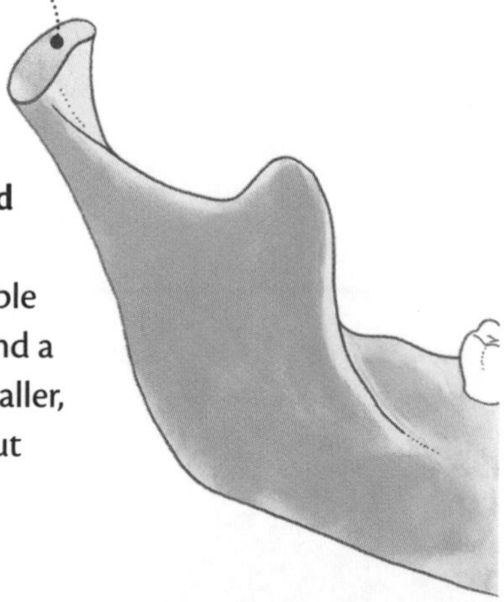

The **mandibular condyloid process** is made up of an anterior part, which is sizable and covered in cartilage, and a posterior part, which is smaller, almost vertical, and without cartilage.

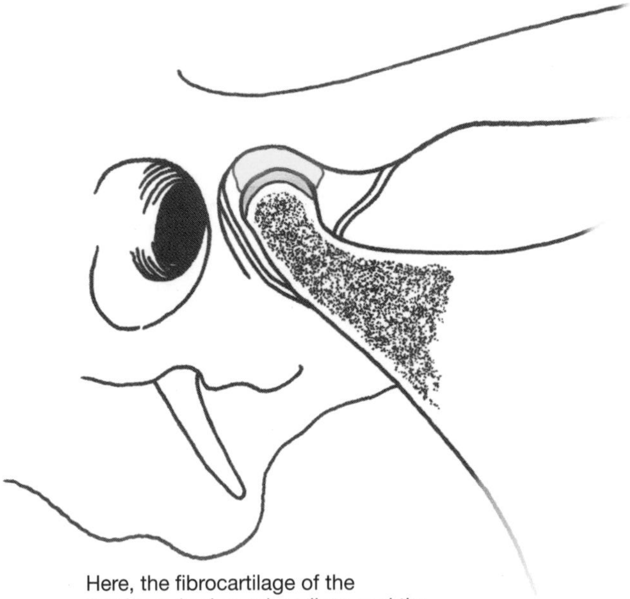

Here, the fibrocartilage of the meniscus is shown in yellow, and the cartilage of the condyle in blue.

Between the Surfaces: The Articular Meniscus

This is a fibrocartilaginous disk. Its superior surface is convex in the back, where it matches up with the glenoid fossa of the temporal bone, and concave in the front, where it meets the condyloid process. Both the temporal and the mandibular condyles are convex, so they don't fit together. The biconcave form of the meniscus allows the two condyles to be joined.

The meniscus is loosely attached to the mandibular condyle by fibrous rings that leave it some ability to move.

The Means of Union

Cartilage surfaces are surrounded by a fibrous capsule that is thin and loose. The meniscus divides the capsule, forming two capsular enclosures: a top and a bottom. Each is lined with a synovial membrane that produces synovial fluid.

The capsule is reinforced by ligaments—in particular, the lateral ligaments:

- the **internal lateral ligament** (not shown)
- the **external lateral ligament,** shown here in two parts: the *posterior* and the *anterior*

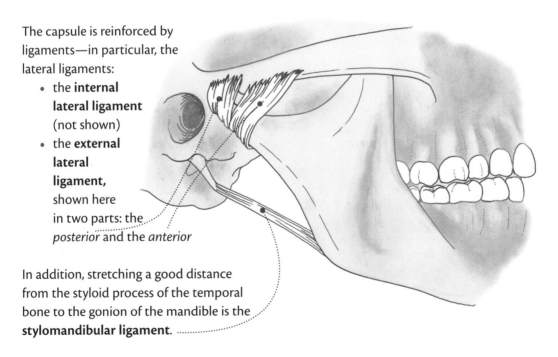

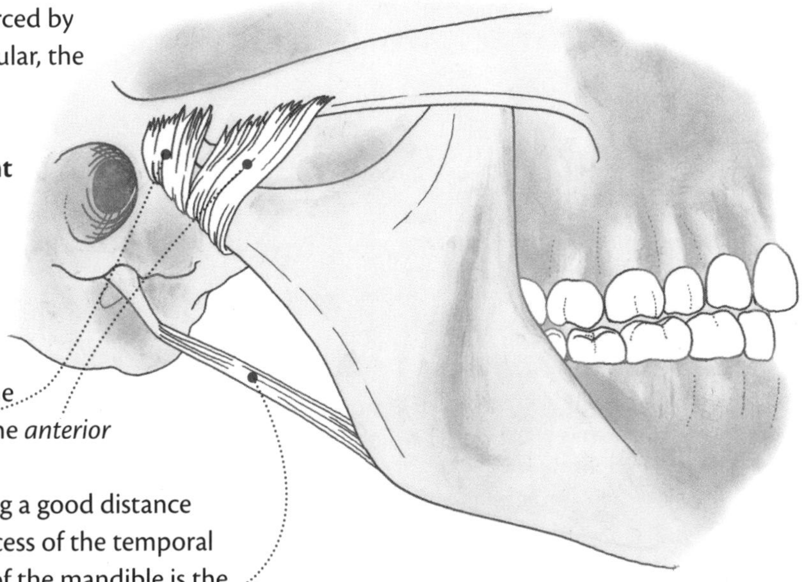

In addition, stretching a good distance from the styloid process of the temporal bone to the gonion of the mandible is the **stylomandibular ligament**.

The Movements of the TMJ

The TMJ allows many movements that change the shape of the mouth and, therefore, affect the voice.

Opening and Closing the Jaw

When the mandible is dropped, the mouth opens. When the mandible is elevated, the mouth closes (closing not shown).

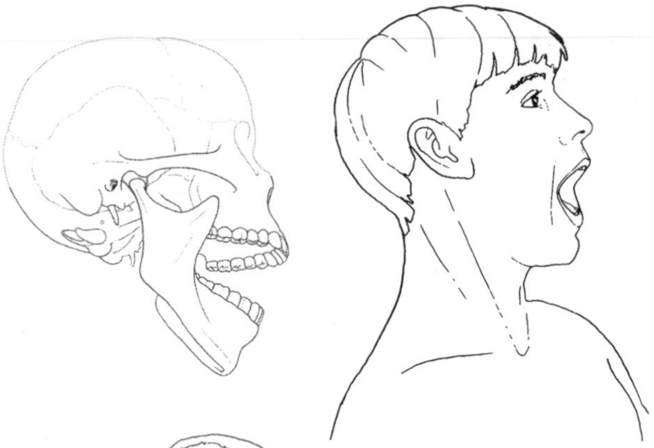

If the mouth is opened *passively* (as in "your jaw drops"), the condyloid process of the mandible rolls in the glenoid fossa of the temporal bone. If the mouth is opened *actively* (as in "open your mouth really wide"), the condyloid process does not roll in the glenoid fossa; rather, it is displaced slightly forward in relation to the temporal condyle. You can feel this action if you place your fingertips on each condyle (about 1 centimeter in front of the ear hole).

Head or Jaw

The movements described above are most often made with the head fixed and the jaw moving. That said, there is usually slight movement of the head as well. It can be interesting to experiment with the movement, mobilizing the TMJ by fixing the jaw and making the movement with the head.

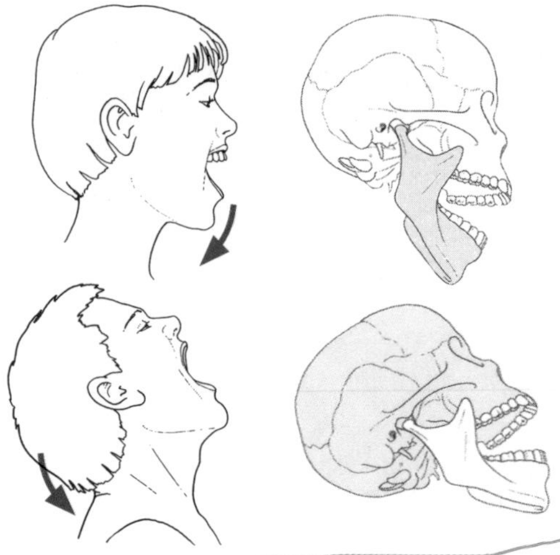

Movements in the Sagittal Plane

These movements are effectuated by the gliding of temporal condyle on the meniscus. The mandible can slide . . .

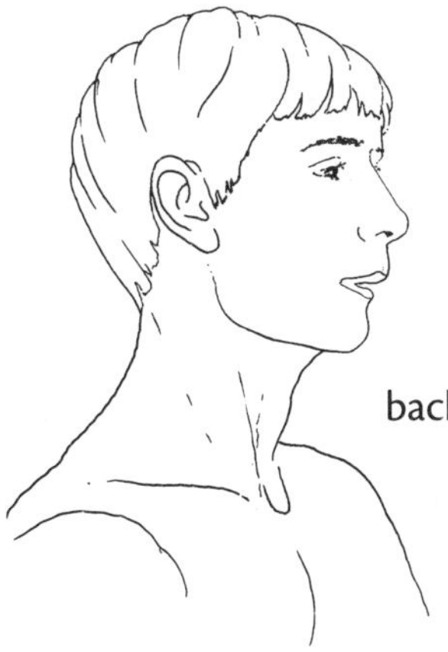

backward; this is **retrusion**

forward; this is **protrusion**

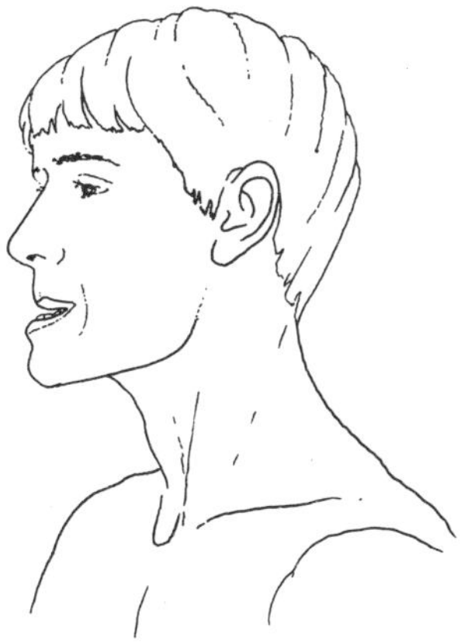

The Mandible and Carriage of the Head

When you are relaxed and lying on your back, your jaw lowers (the mouth opens), and at the same time it moves backward a bit (retrusion). These movements are passive and happen minimally whenever the head is in extension on the atlas (even when the extension is slight). When the head is flexed on the atlas, the opposite happens—the movements are reversed. We can thus see that the position of the head influences the position of the mandible. Sometimes the amplitude of the movement is minimal, but in vocal work this is an important variable, since each change in the position of the mandible in turn influences the movement of the tongue and everything that is attached to it.

Movements in the Lateral Plane

These movements carry the maxilla laterally (from side to side). This is *laterotrusion*. One condyle rotates on itself, while the other goes into protrusion.

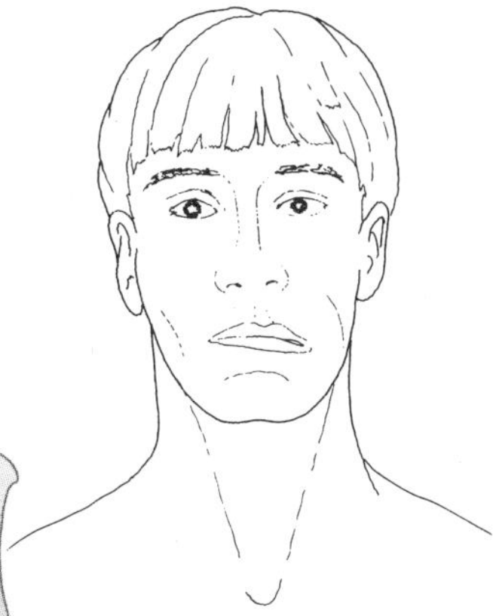

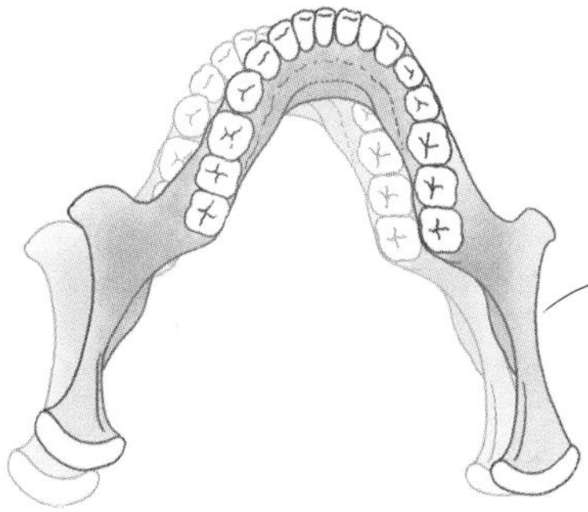

The Dental Arches and Teeth

We'll touch on this part of the mouth here specifically because the precise placement of the tongue in relation to the teeth plays a role in the accuracy of articulation.

The Upper (Superior) Dental Arch

The superior dental arch is formed by the inferior borders of the right and left superior maxillae, which are joined together to form a single curve. The bone is thickened at this joint and supports an alignment of **dental alveoli** (cavities).

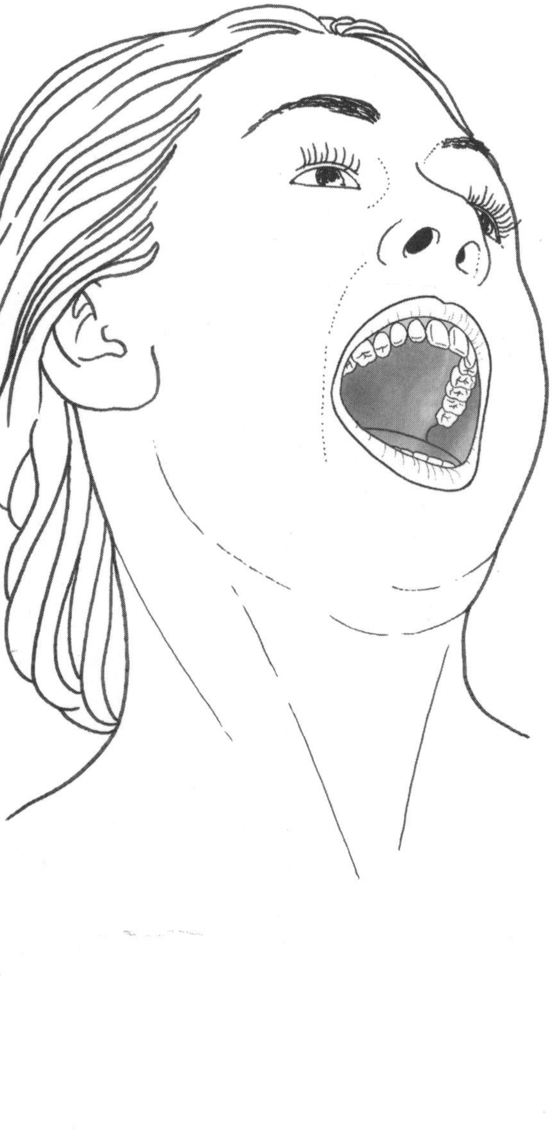

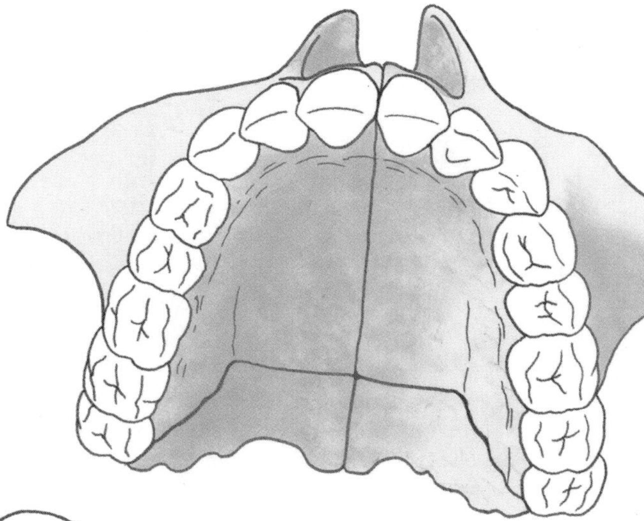

The Lower (Inferior) Dental Arch

The inferior dental arch is formed by the upper part of the mandible body. Here, too, the bone has an alignment of dental alveoli.

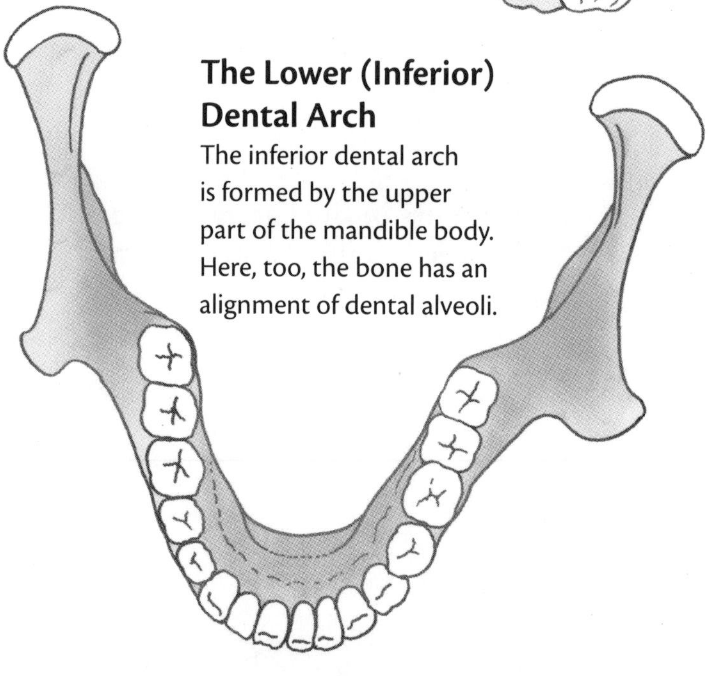

The Teeth

A tooth can be considered an organ composed of a hard crown and one or several roots that are implanted in the dental alveoli. The tooth is essentially formed of ivory. Externally, it is covered with enamel at the crown, which is very hard, and cementum at the root. Internally, the tooth is composed of soft tissue: the dental pulp, which is innervated and vascularized.

There are four types of adult teeth:

- flattened **incisors** at the front
- **canines,** which are pointed
- **premolars** and **molars,** with large and irregular crowns

The teeth are fixed to the maxilla in two ways:

- by being embedded: each tooth corresponds precisely to an alveolar cavity.
- by a periodontal ligament: at the root level, it unites the alveolar wall to the cementum.

The Tongue and Teeth

Other than when the tongue is involved with chewing or the articulation of specific sounds, it should rest against the hard palate. The tip of the tongue should be placed against the **incisive papilla,** a small projecting fold of mucous membrane situated behind and above the alveoli of the incisors, or front teeth.

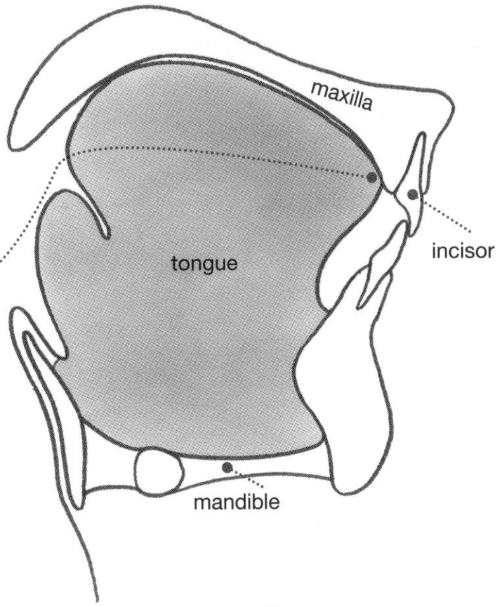

The tongue shouldn't push against the back of the teeth. Even when swallowing, it shouldn't push beyond the incisive papilla. If it pushes against the teeth, dental alignment is compromised.

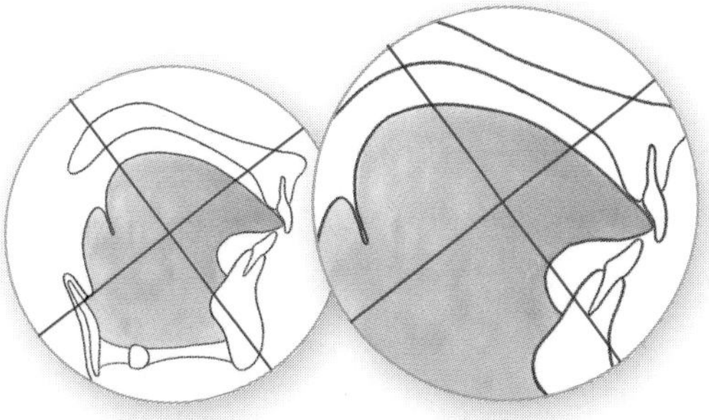

The Hyoid Bone

Located just above the larynx, the hyoid bone is small (20 to 25 mm wide, 30 mm long) but very important for the voice, as many ligaments and muscles of phonation attach to it.

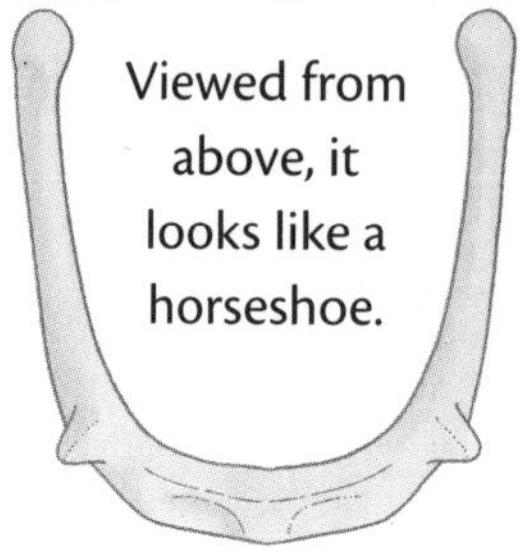

Viewed from above, it looks like a horseshoe.

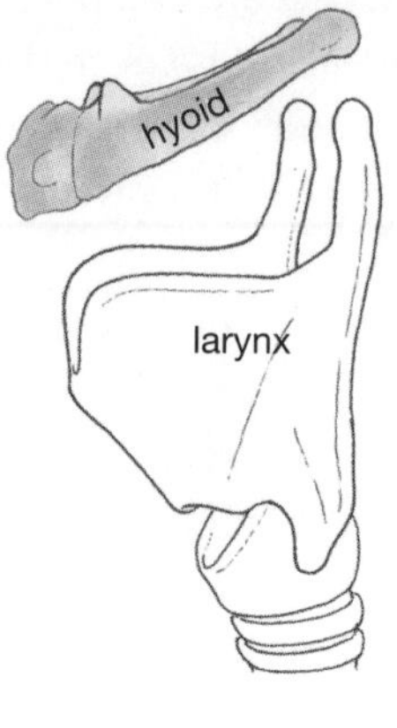

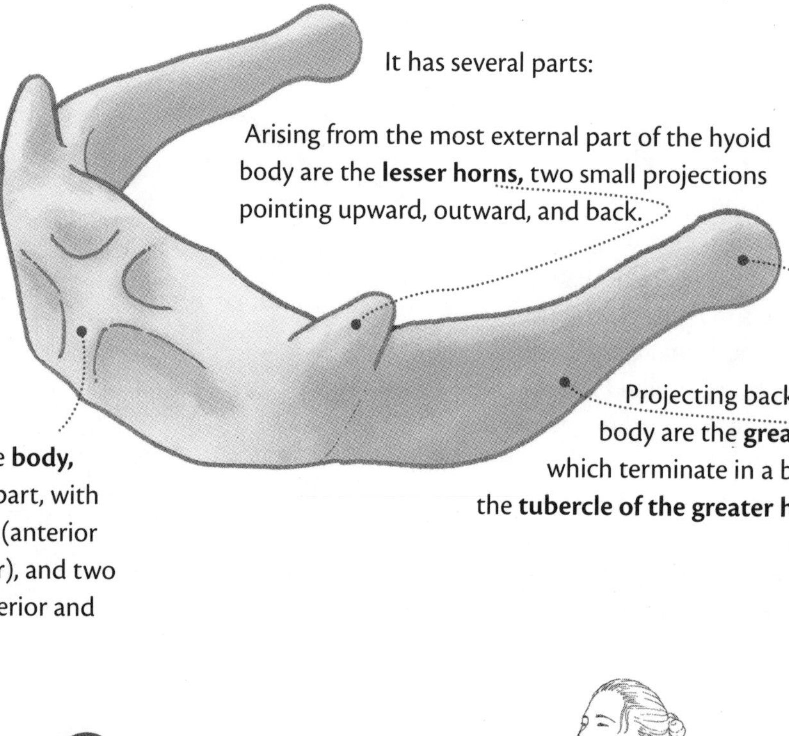

It has several parts:

Arising from the most external part of the hyoid body are the **lesser horns,** two small projections pointing upward, outward, and back.

Projecting back from the body are the **greater horns,** which terminate in a bulge called the **tubercle of the greater horn**.

In front is the **body,** the thickest part, with two surfaces (anterior and posterior), and two borders (superior and inferior).

Palpation

You can find the hyoid at the junction of the anterior surface of the throat (vertical) and the underside of the chin (horizontal). Palpate it using your thumb and the index finger, just above the thyroid cartilage (see page 146).

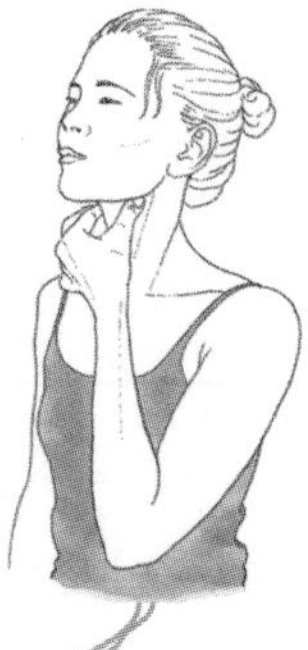

The hyoid bone doesn't articulate with any other bone in the body. It is suspended from the skull and mandible by muscles and ligaments, in particular:

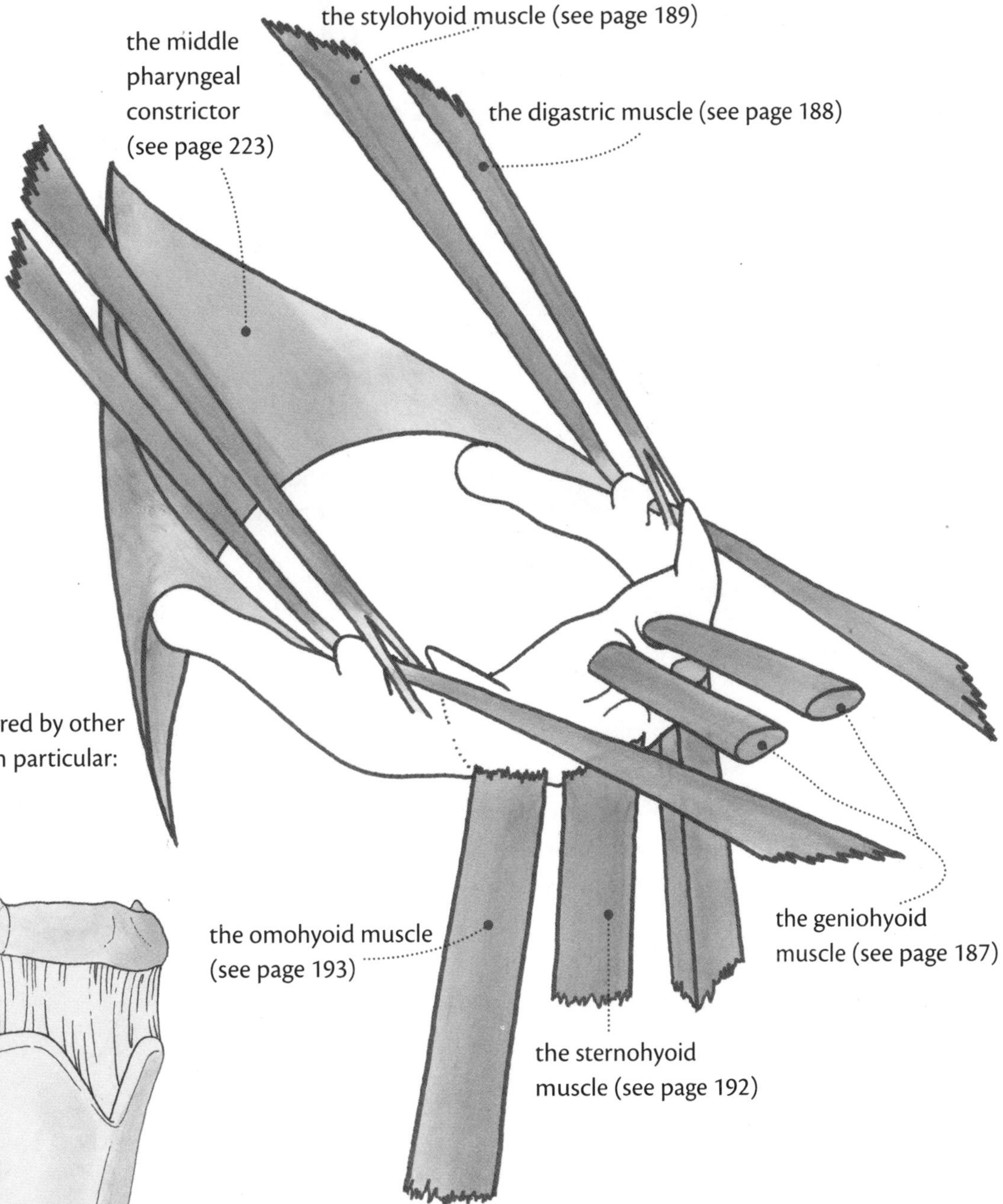

It is anchored by other muscles, in particular:

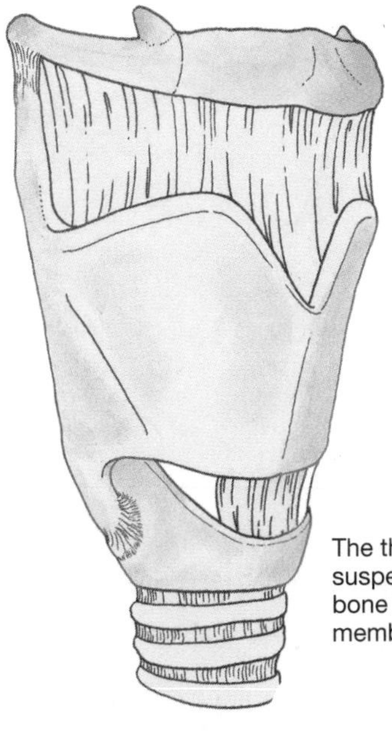

The thyroid cartilage is suspended from the hyoid bone by the thyrohyoid membrane.

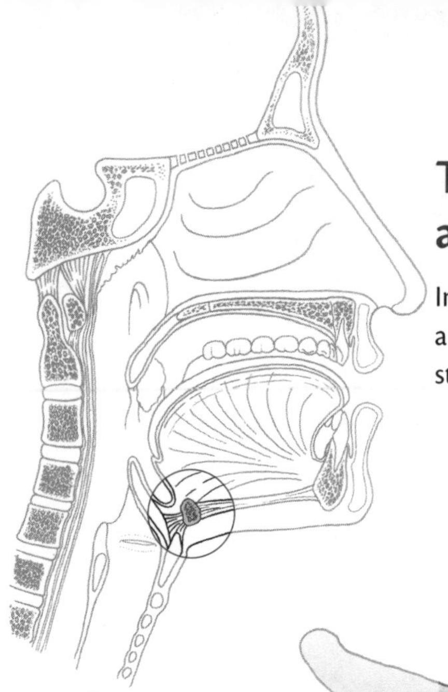

The Hyoid: Its Environment and Roles

In a cross-sectional view, the hyoid appears as a small bony point that is connected to many structures.

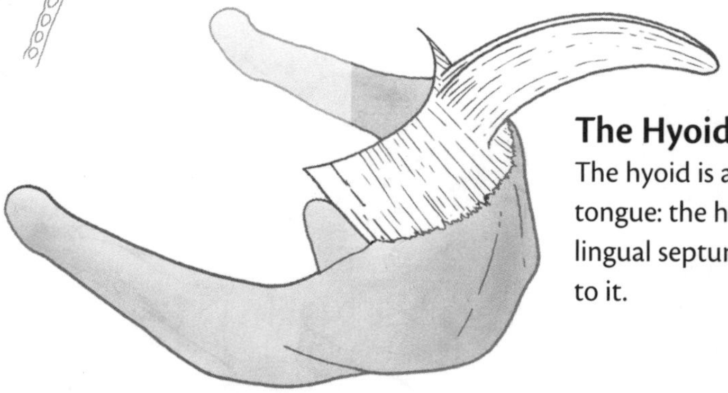

The Hyoid and Its Environment

The hyoid is a base for the structures of the tongue: the hyoglossal membrane and the lingual septum (see page 250) are attached to it.

It is an anchor for the floor of the mouth: the mylohyoid and geniohyoid muscles attach to its anterior part (the body).

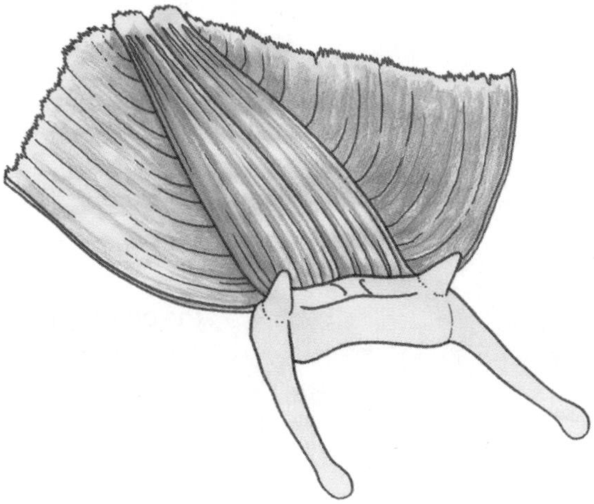

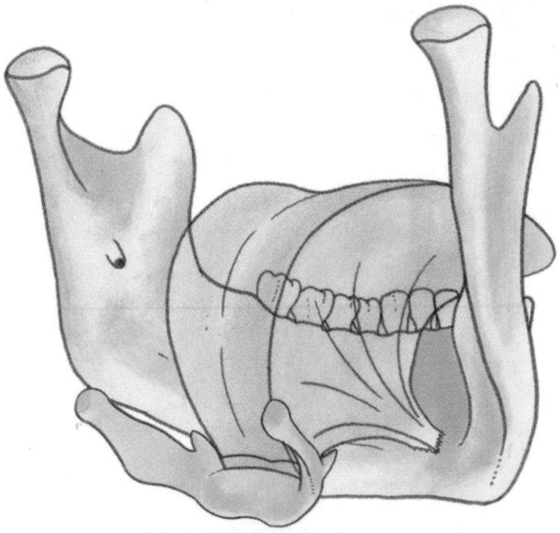

The hyoid body also serves as an attachment point for the genioglossus (see page 252), the muscle that makes up the bulk of the tongue.

The hyoid also serves as one of the anterior attachment points for the pharynx: the middle pharyngeal constrictor (see page 223) inserts on the greater horn.

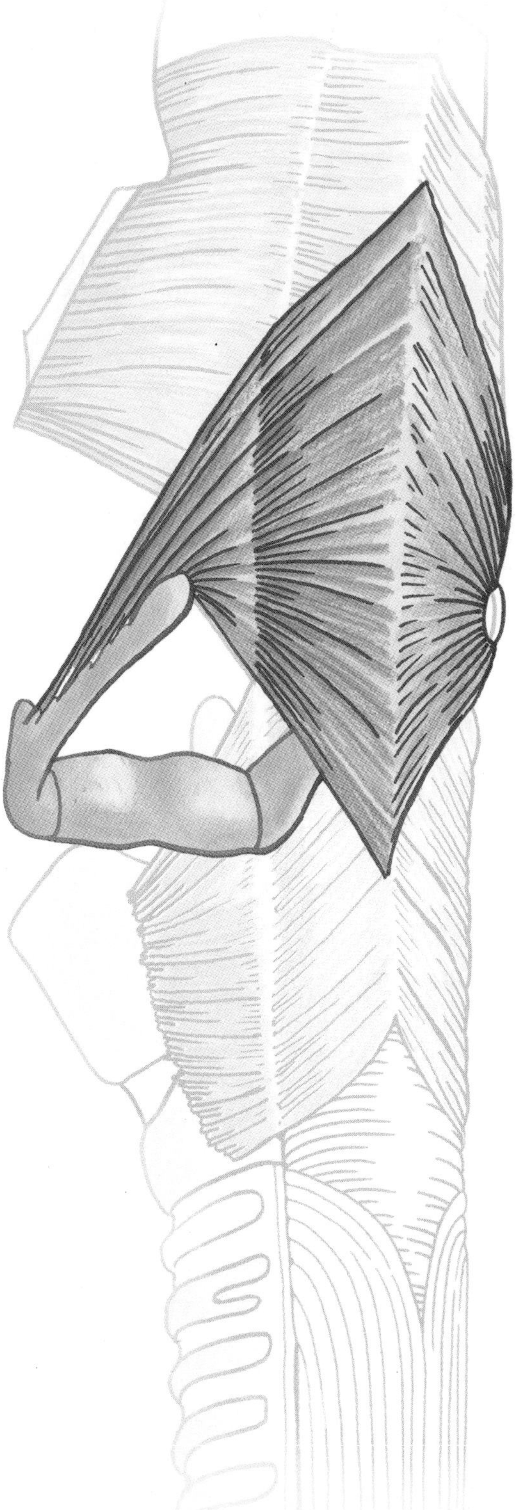

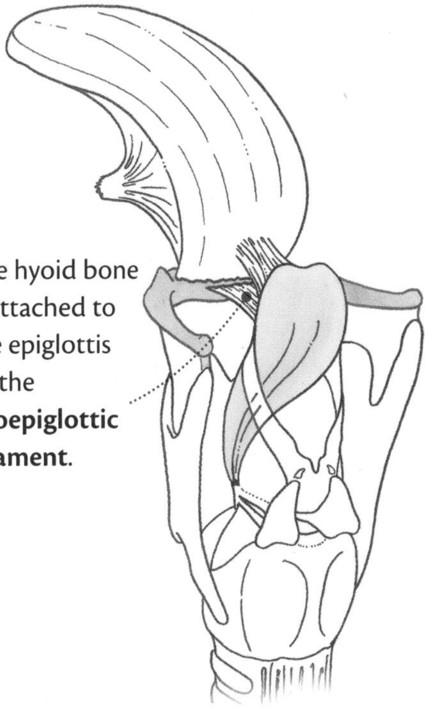

The hyoid bone is attached to the epiglottis by the **hyoepiglottic ligament**.

The Hyoid Plays Numerous Roles

In regard to respiration, the hyoid is like a "miniature horseshoe" that is located below and behind the "horseshoe" that forms the jaw. Here it keeps the tube of the larynx (voice box) away from the pharynx and helps keep it open.

In regard to the voice, in babies the hyoid bone is quite high in the pharynx. Around the age of two, the hyoid descends into the neck, which increases pharyngolaryngeal resonation. This allows for a second resonator and makes the formation of vowels possible (see page 203).

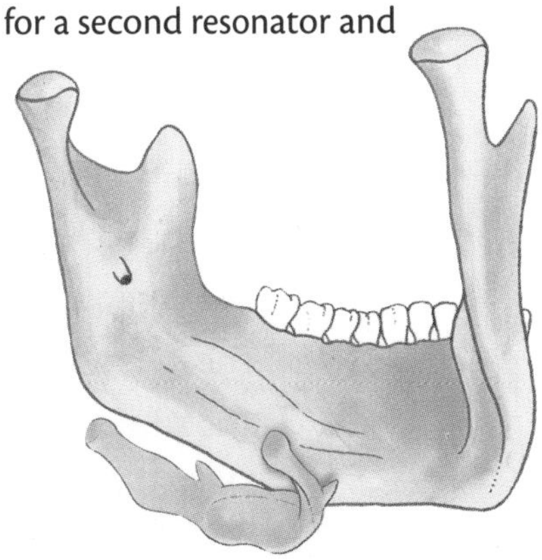

3
The Generator

Introduction

In order for sound to be generated, air under pressure (for more on pressure, see page 288) has to arrive at the vocal cords (see page 152) from below. This air is contained in the lungs (just the lungs, not the stomach), the bronchi, and the trachea.

The buildup of air just below the vocal cords creates what is called *subglottal pressure*, and it is the means by which sound is born. (More detail on the three levels of the pharynx, including the subglottis, can be found beginning on page 174.)

In the vocal apparatus, the **generator** is the ensemble of anatomical parts that allows this subglottal pressure to be produced.

This chapter, which is dedicated to the generator, will look at all of the elements that produce this pressure and measure it out. The generator includes, of course, the lungs, but also the entire trunk and the part of the neck that lies below the glottis.

Details

The pressurized air is used for vibrating the vocal cords, which creates a sound wave. (We'll discuss this phenomenon in the chapter on the larynx; see page 154.)

In general, the mechanism of the generator can be described as a two-step process.

Step 1. Air fills the lungs and the glottis closes (see page 177).

Step 2. As lung volume shrinks, air tries to escape but can't because the glottis is closed. Consequently, the pressure increases (illustrated here in a darker shade), building up under the glottis.

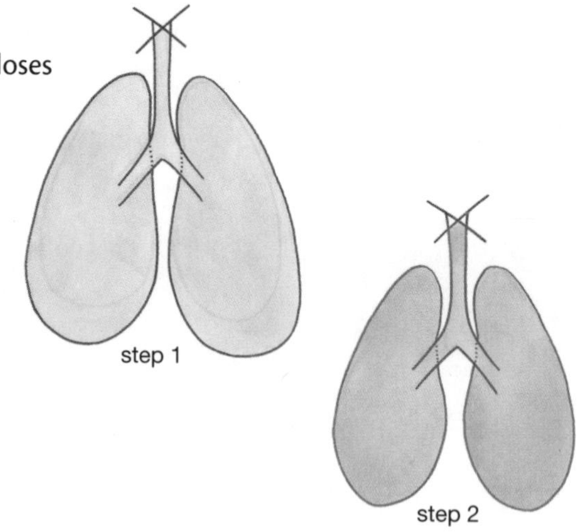

We will see that the pressure can be measured out based on the needs of the voice, in particular by the interplay of the expiratory muscles (page 106), but also by the inspiratory muscles (page 118), which act on the volume of air in the lungs.

The Two Cavities

The generator is based in two "box-like" anatomical regions: the thoracic cavity and the abdominal cavity. In a person standing upright, in a vertical position, these cavities are layered.

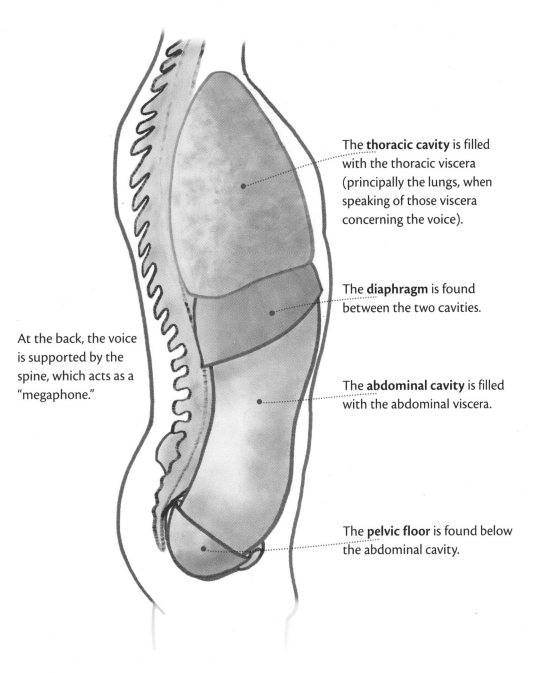

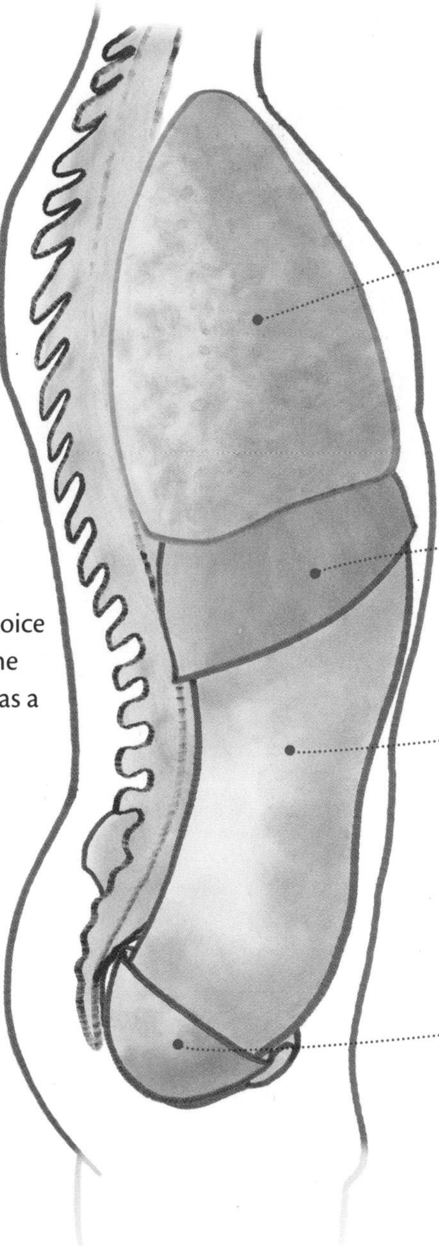
The **thoracic cavity** is filled with the thoracic viscera (principally the lungs, when speaking of those viscera concerning the voice).
The **diaphragm** is found between the two cavities.
At the back, the voice is supported by the spine, which acts as a "megaphone."
The **abdominal cavity** is filled with the abdominal viscera.
The **pelvic floor** is found below the abdominal cavity.

Note: These structures are described in detail in *Anatomy of Breathing* and are presented here with a focus on those that principally concern the voice.

The Thoracic Cavity

This houses all of the viscera that are located in the chest above the diaphragm, as well as the tissues that surround them. We can describe the thoracic cavity in terms of a container and its contents.

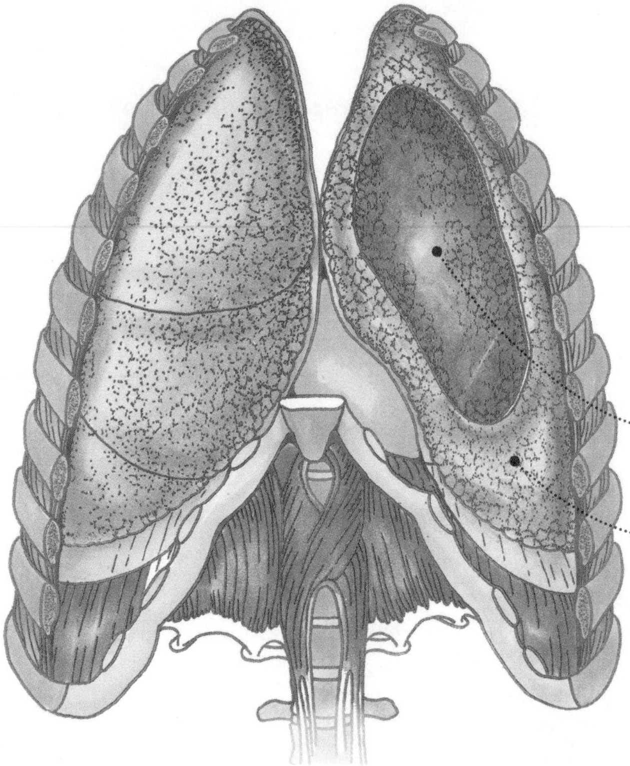

Here, the pleura is removed and we see the exposed lung.
Here, the pleura covers the lung.

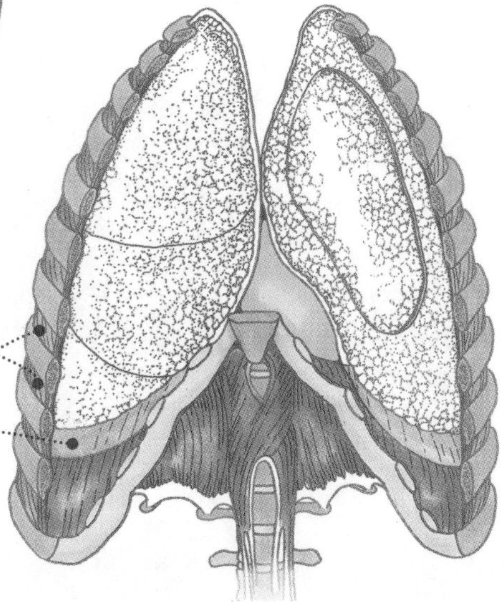

The Container

The **thoracic rib cage**
The **intercostal muscles**
The **diaphragm**

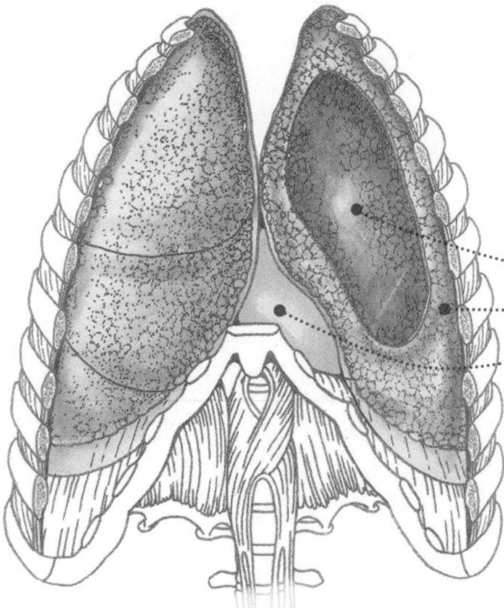

The Contents

The **lungs**
The **pleurae,** enveloping the lungs
The **heart** (enveloped in the pericardium)
The **mediastinum** (not shown)
The **bronchi** and the **trachea** up to the glottis (not shown)

Mechanical Properties of the Thoracic Cavity

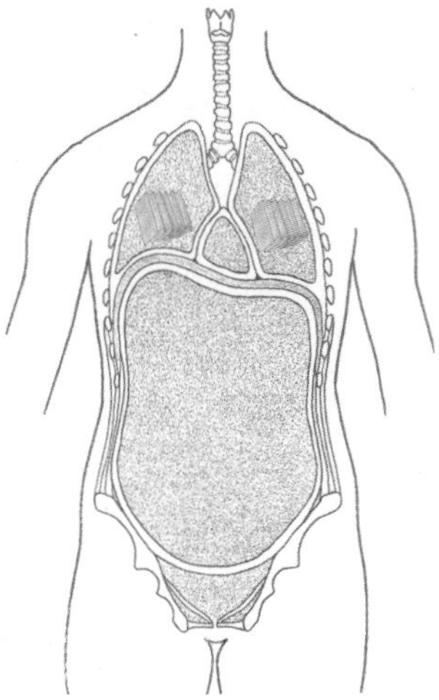

It's in the thoracic cavity that subglottal pressure, which serves to form the voice, is created. To understand the way in which the pressure is put into play, imagine a box filled with air. The volume of air can be regulated, a little like an accordion, by using a valve (the glottis) which permits, or not, the exchange of air with the exterior.

When we reduce the volume of the box . . .

With the valve (glottis) open, there is a flux of air outward; this is exhalation.

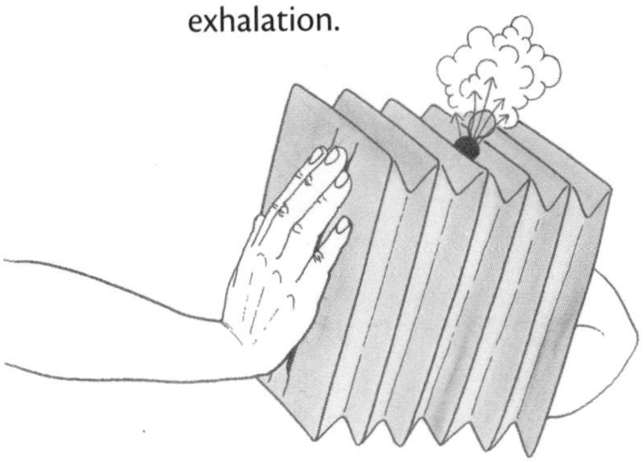

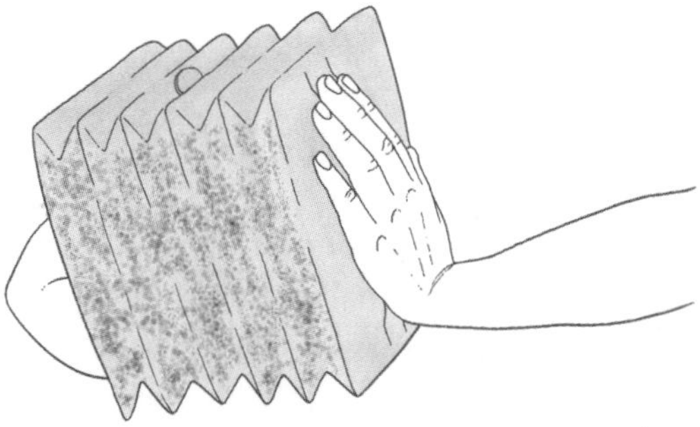

With the valve (glottis) closed, there is an increase in subglottal pressure.

This action is often effectuated by the expiratory muscles, but not always.

When we increase the volume of the box . . .

With the valve (glottis) open, there is a flux of air inward; this is inhalation.

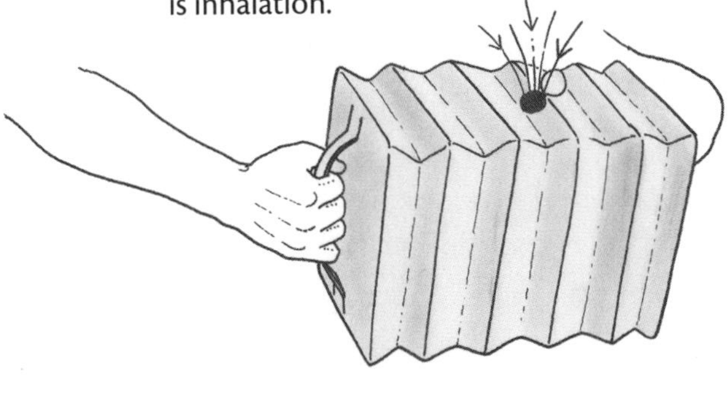

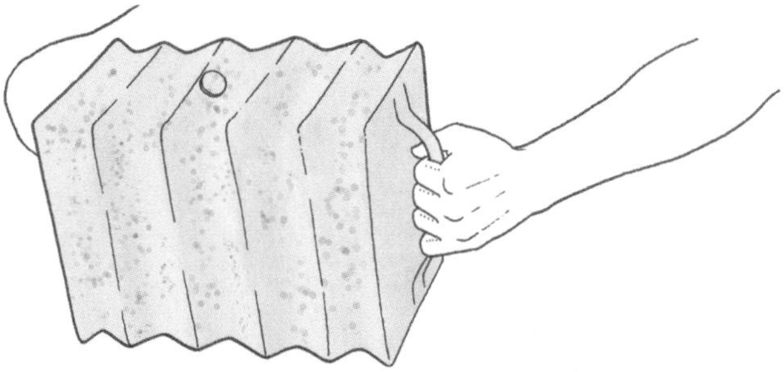

With the valve (glottis) closed, there is a reduction of subglottal pressure.

This action is often effectuated by the inspiratory muscles, but not always.

The Abdominal Cavity

The abdominal cavity encloses all of the viscera found under the diaphragm, as well as the tissues that surround them. We can describe it in terms of a container and its contents.

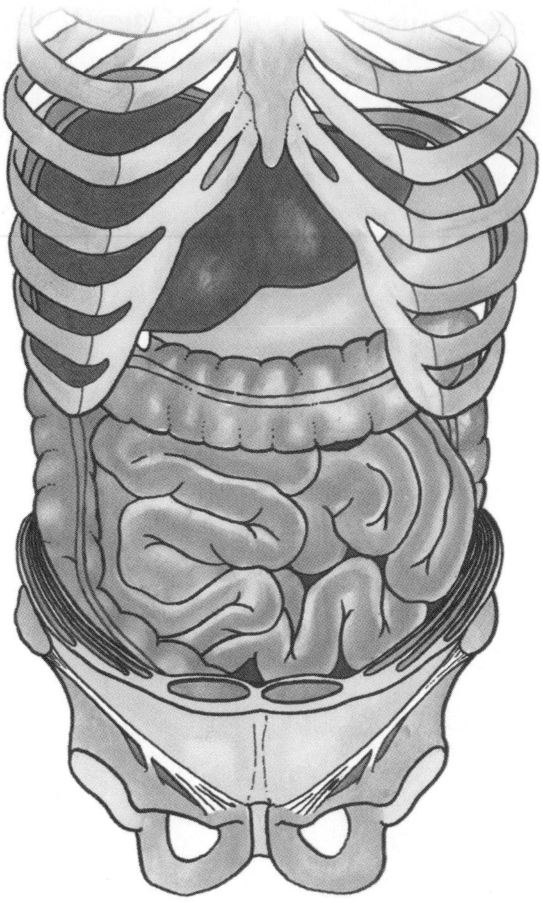

The Container

The **pelvis**
The **pelvic floor** (not shown)
The **abdominal muscles**
The **lumbar spine** (not visible)
The lower part of the **thoracic rib cage**
The **diaphragm**

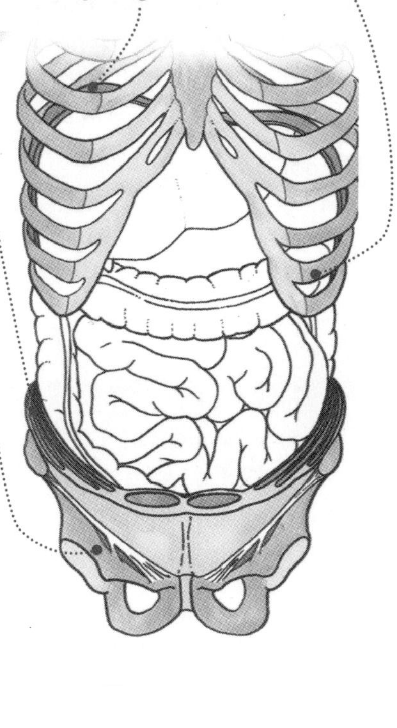

The Contents

The **abdominal and pelvic viscera**

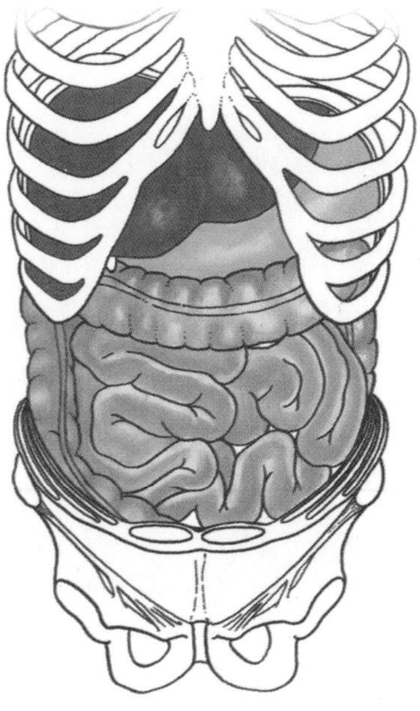

Mechanical Properties of the Abdominal Cavity

In contrast to the thoracic cavity, which can be filled with air, the viscera of the abdominal cavity can be considered a liquid mass, represented here by a water bottle. The visceral mass has two properties:

- It is deformable.
- It is incompressible.

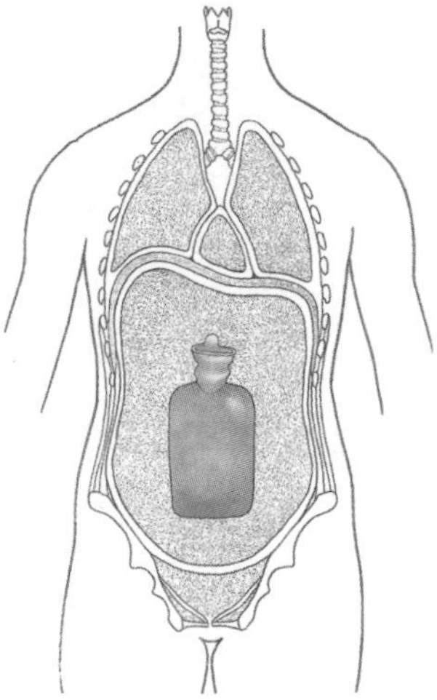

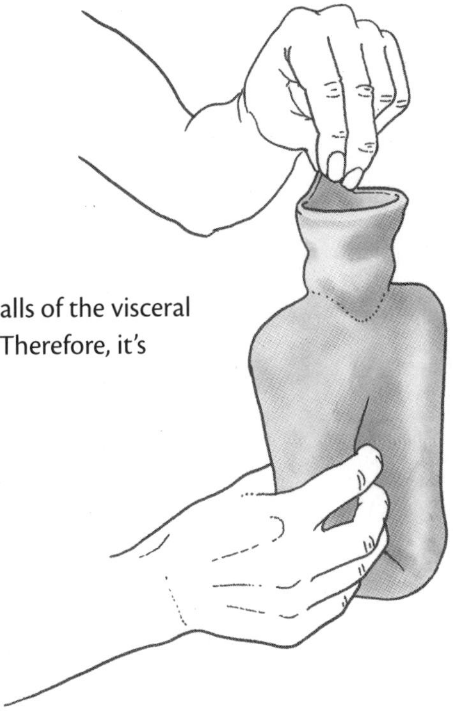

Deformable

If we push or pull on the walls of the visceral mass, its form will change. Therefore, it's *deformable.*

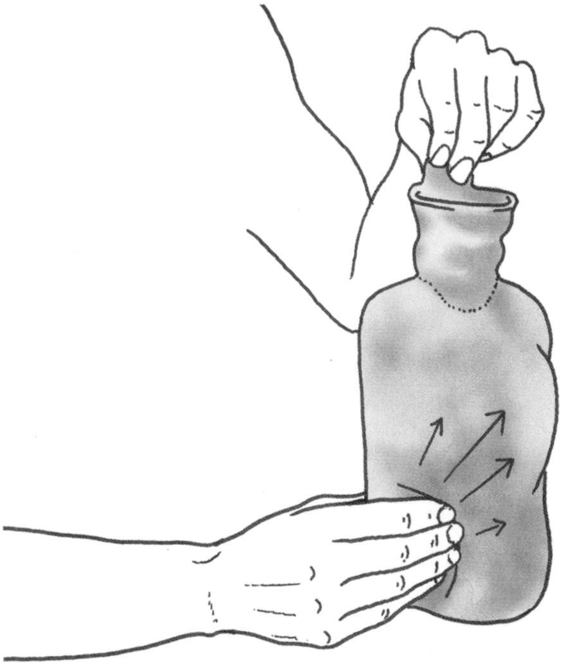

Incompressible

On the contrary, its volume is always the same. If one area is deformed, a balancing deformation will take place in another area; that is, if we push or pull on one area, it will deform elsewhere. So, we say that it's *incompressible.*

Deformation of the visceral mass often results from the action of the muscles surrounding it (the diaphragm, the abdominals, the pelvic floor). However, it can also be due to other forces, specifically weight.

The Diaphragm: The "Double-Sided Tape" between the Two Cavities

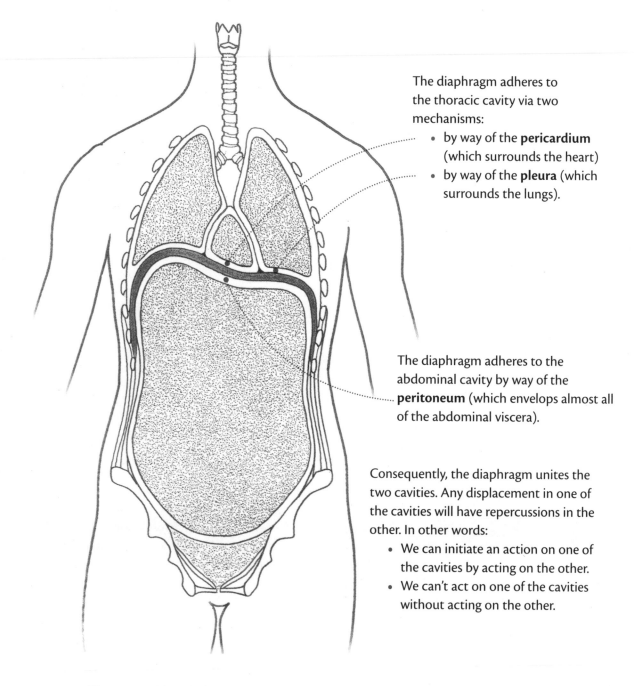

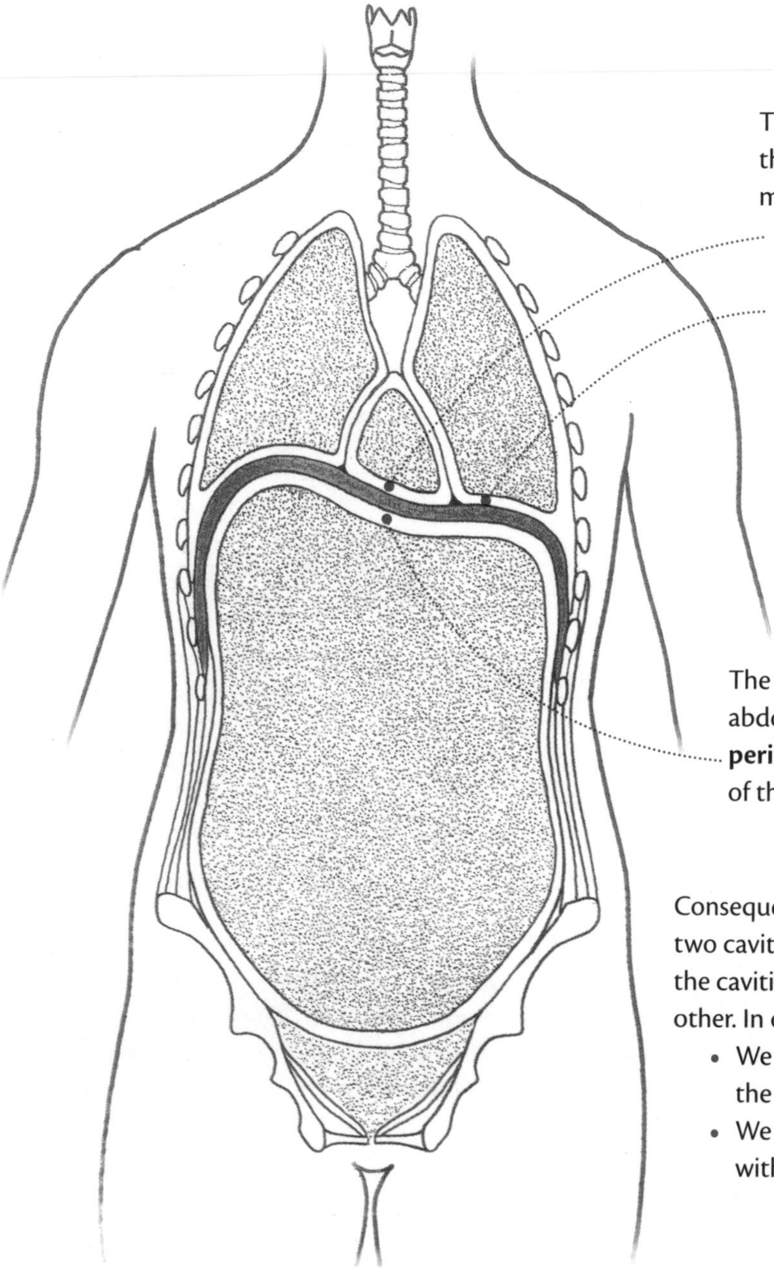

The diaphragm adheres to the thoracic cavity via two mechanisms:

- by way of the **pericardium** (which surrounds the heart)
- by way of the **pleura** (which surrounds the lungs).

The diaphragm adheres to the abdominal cavity by way of the **peritoneum** (which envelops almost all of the abdominal viscera).

Consequently, the diaphragm unites the two cavities. Any displacement in one of the cavities will have repercussions in the other. In other words:

- We can initiate an action on one of the cavities by acting on the other.
- We can't act on one of the cavities without acting on the other.

The two cavities are always connected, which means that their mechanical properties interact. This interaction is one key to the generation of the voice; it explains, for example, how we can create subglottal pressure with displacement of the abdomen, and even the pelvic floor.

The Breathing Body and Control of Expiratory Air Pressure

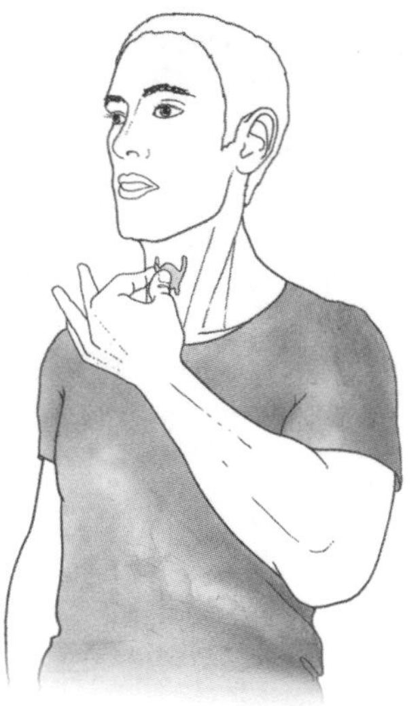

We can control expiratory air pressure with the laryngeal sphincter, which can narrow and tighten like a bottleneck (see page 170).

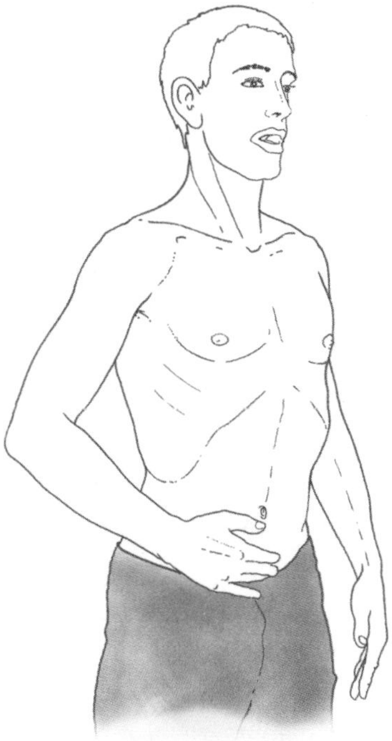

We can also control expiratory air pressure by playing with the propelling force that can be generated in the thorax and abdomen. Our expiratory power can be . . .

augmented, most often by contracting the expiratory muscles (principally the abdominal and pelvic floor muscles), or

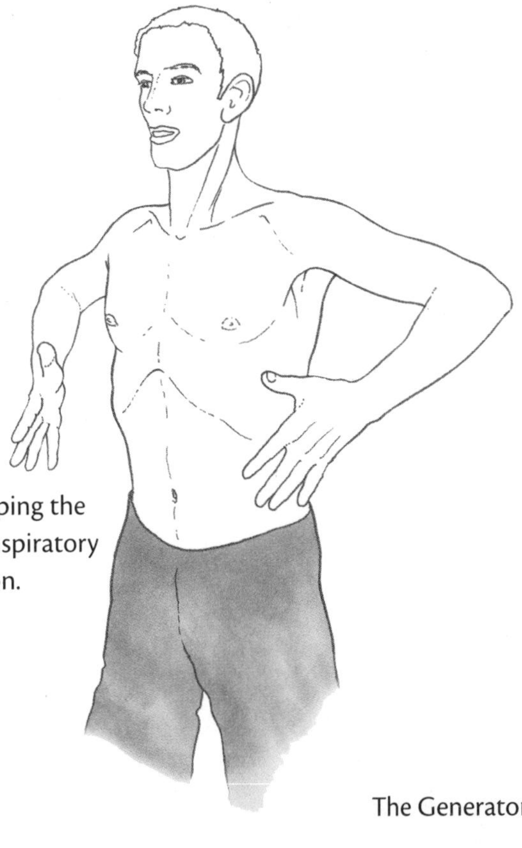

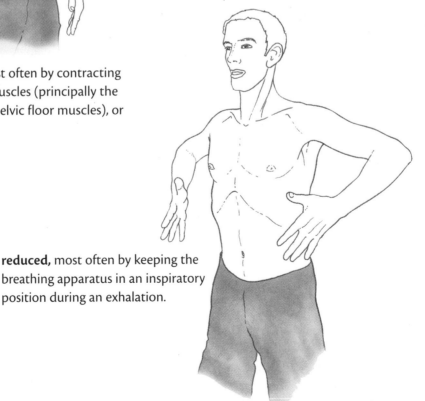

reduced, most often by keeping the breathing apparatus in an inspiratory position during an exhalation.

The Organs of Respiration and the Surrounding Area

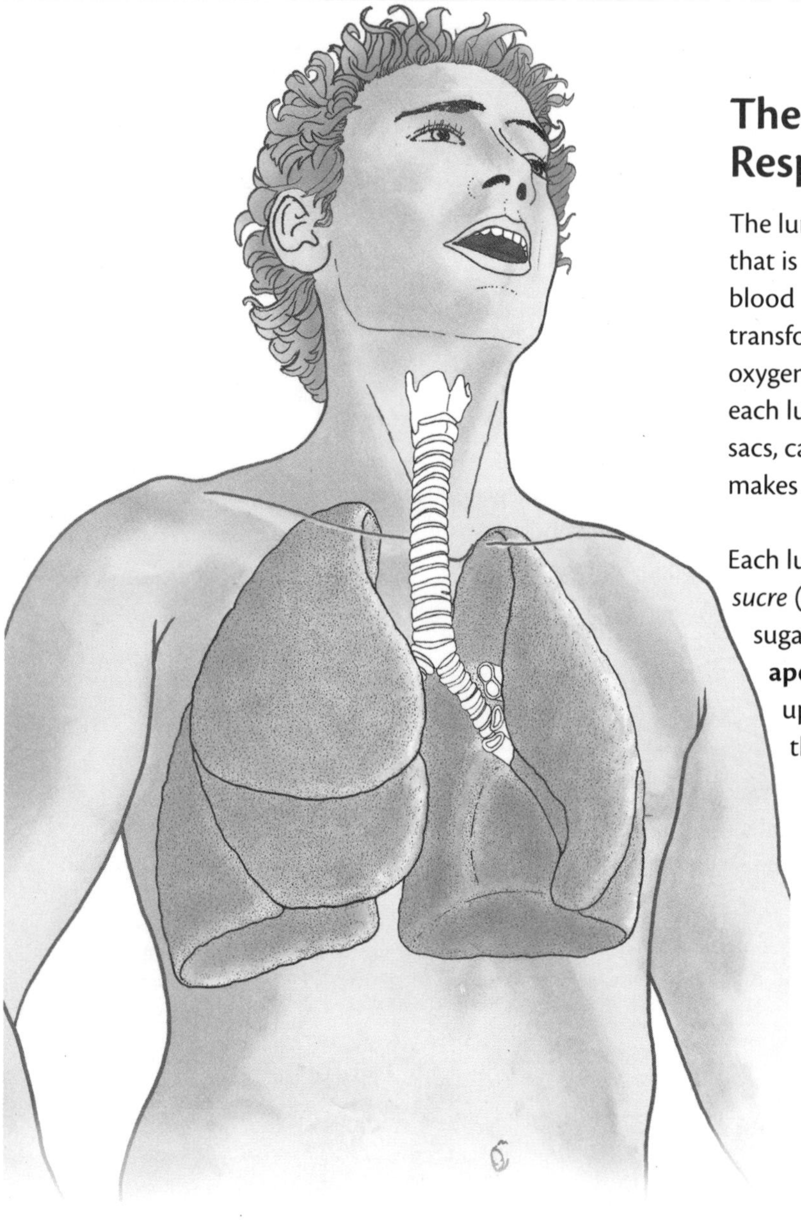

The Lungs: The Organs of Respiration

The lungs are the organs of oxygenation; that is to say, here is where venous blood (rich in CO_2 and low in oxygen) is transformed into arterial blood (rich in oxygen and low in CO_2). To accomplish this, each lung contains roughly 300 million small sacs, called **alveoli,** where the air arrives and makes this gas exchange possible.

Each lung is shaped a bit like a *pain de sucre* (an elongated cone-shaped "loaf" of sugar), with a peak at the top, called the **apex,** and a broad base that is concave upward, corresponding to the shape of the diaphragm located below it.

The Lungs and the Thoracic "Container"

Each lung is wrapped in a double envelope of tissue called the **pleura,** which loosely adheres to the rib cage. At the bottom the pleura is connected to the diaphragm. Thus, any movement of the ribs and diaphragm affects the lungs and changes their shape.

More details on the organs of respiration can be found in *Anatomy of Breathing,* chapter 3.

The Elasticity of the Lungs and Voice

Between the alveoli we find conjunctive, or connective, tissue that is rich in elastin.

Attention!

The illustrations on this page show not a lung but an enlargement of a group of alveoli.

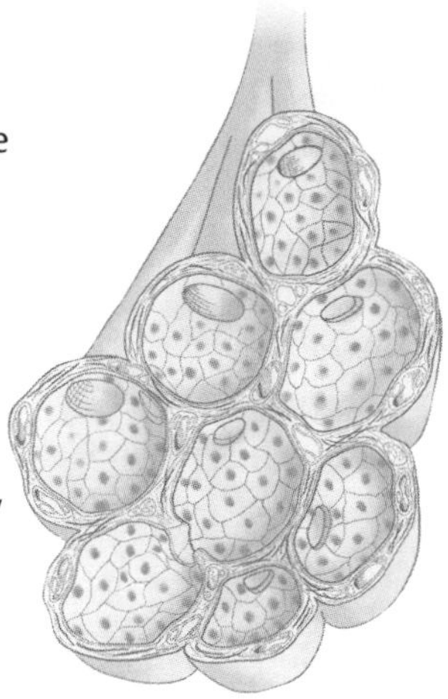

pulmonary alveoli at rest

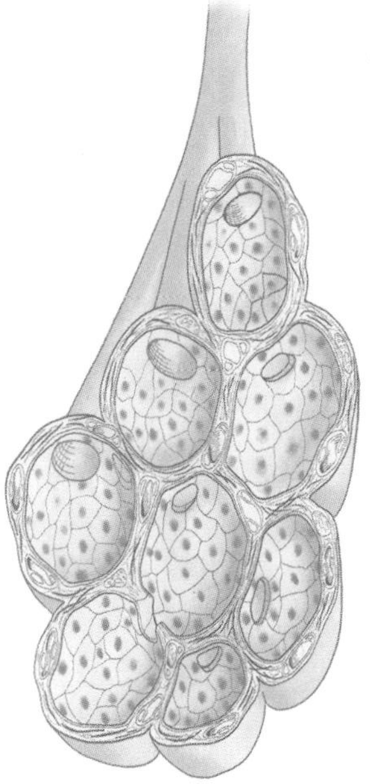

As a result, each lung as a whole has an elastic quality.

We can stretch the lungs *laterally*, which is essentially what happens when we separate the ribs.

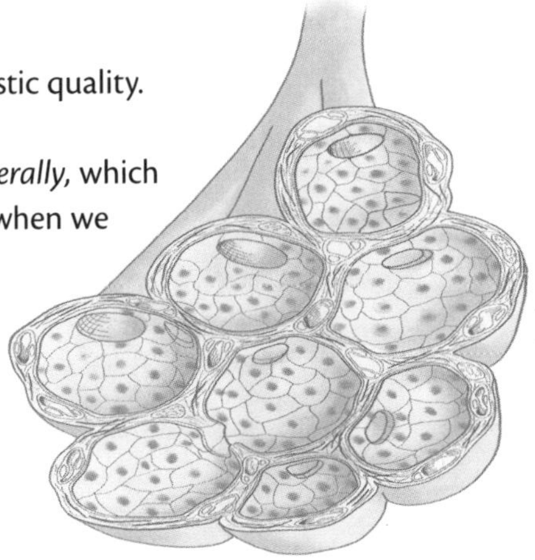

We can also stretch the lungs *vertically*, which is what happens when they are pulled by the diaphragm.

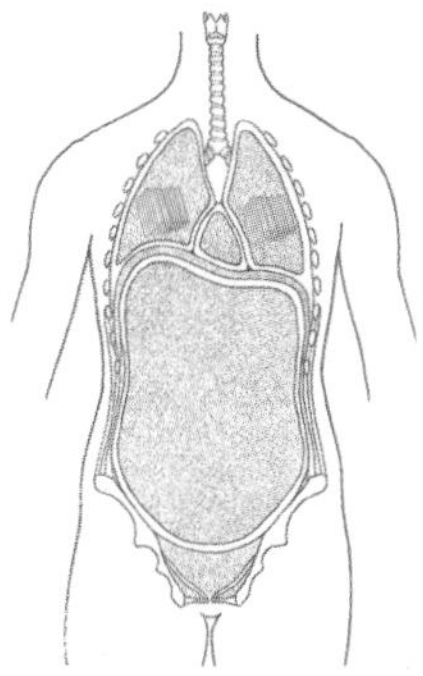

This is a fundamental element of respiratory dynamics: after an inhalation, the elastic recoil of the lungs is the force behind the exhalation for all passive breathing.

Regarding vocal dynamics, the elastic recoil can produce a weak sound of short duration—for example, when we say a few words in a low voice. It's as if we had pulled an elastic "box of air" (see page 97) taut in all directions. This elastic force becomes important when we take a deep breath and then start the exhalation that follows; at that moment, the alveoli are wide open thanks to the force of the inspiratory muscles, and "pulmonary elasticity" is very strong. Maybe too strong: sometimes we need to brake the exhalation, especially in vocal techniques where we want very even pressure on the glottis (to create an even sound). In this case, we use the inspiratory muscles to inhibit the elastic recoil.

The Bronchi and Trachea

The alveoli of the lungs are linked to the exterior of the body by a series of tubes. These conduits start out very tiny, with the **alveolar ducts,** and then increase in size, becoming the **bronchioles** and then **bronchi**, which join to form one large primary or **stem bronchus** per lung. The stem bronchi meet, between the two lungs, in a single conduit: the **trachea,** which travels through the upper thorax and the lower half of the neck to the **larynx**.

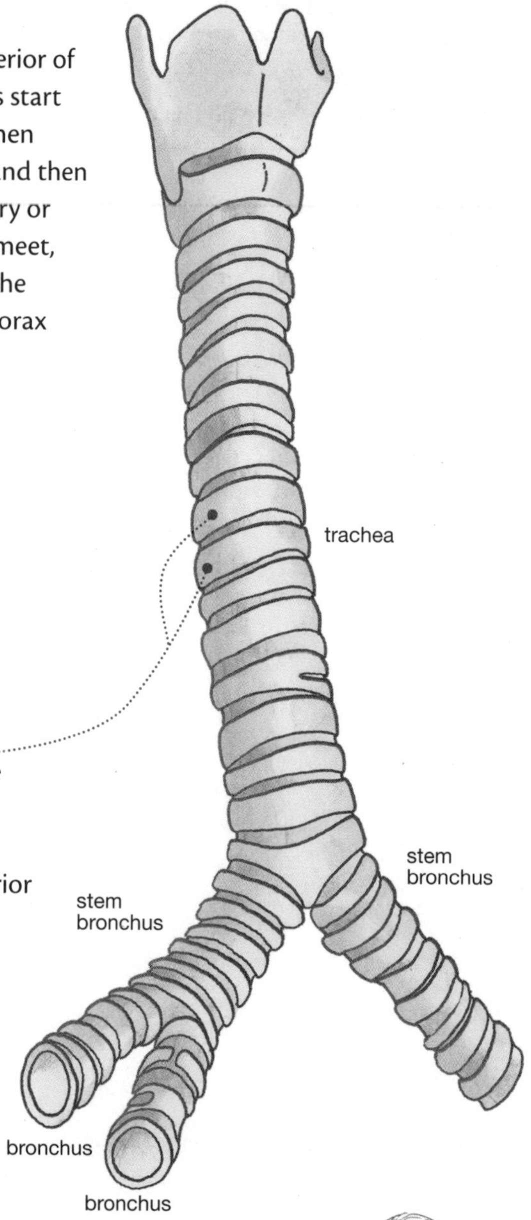

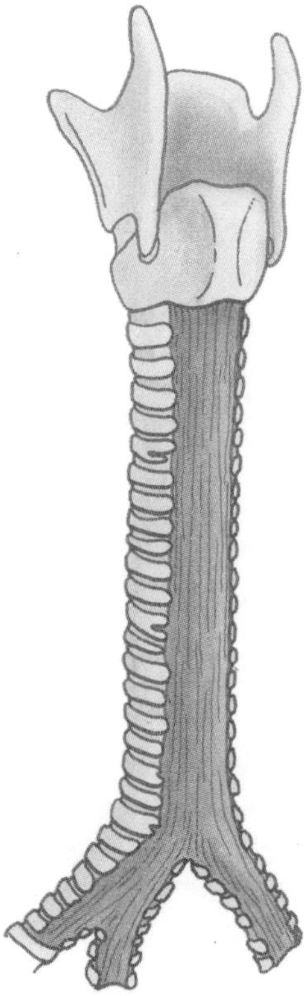

Rings of cartilage stiffen the trachea, except in the back, which is composed of a muscular wall located anterior to the esophagus.

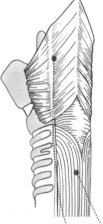

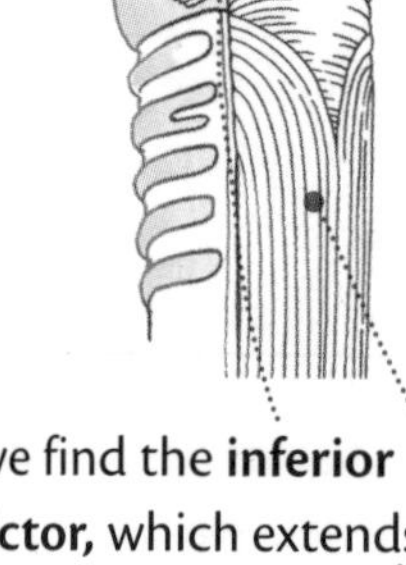

Behind the larynx we find the **inferior pharyngeal constrictor,** which extends downward with the **esophagus** behind the trachea.

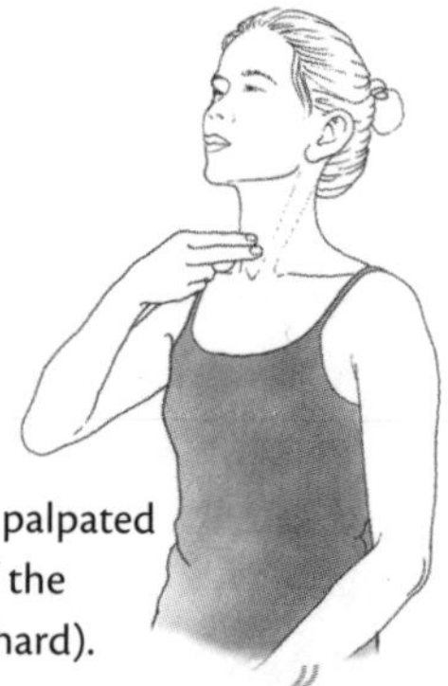

Palpation

The trachea can be palpated at the lower part of the throat (don't push hard).

The Muscles of Respiration and the Voice

Phonation requires tremendous use of the respiratory muscles. In the following pages, we will look at these muscles from the point of view of the breathing body and the vocal body.

In particular, we'll look at the **expiratory muscles** (pages 106–17):

- their respiratory role in exhalation
- their vocal role as the muscles that can increase subglottal pressure

We'll also look at the **inspiratory muscles** (pages 118–31):

- their respiratory role in inhalation
- their vocal role as the muscles that can reduce expiratory subglottal pressure

In the vocal act, these two kinds of muscles—with their opposite action—often contract simultaneously to mete out the right amount of force; this is called *synergy/antagonism*.

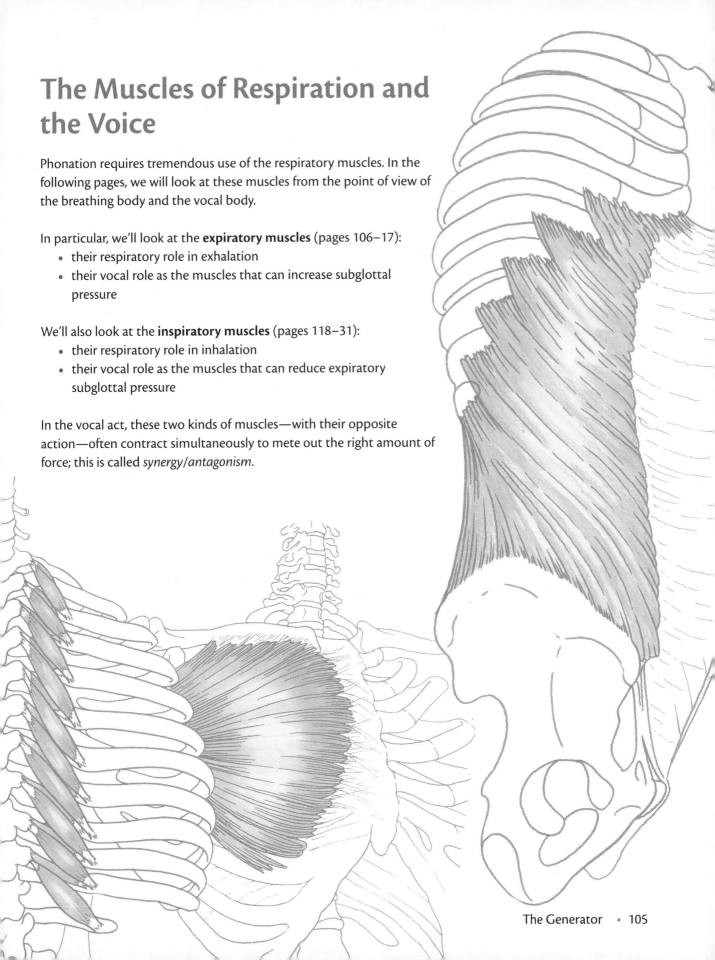

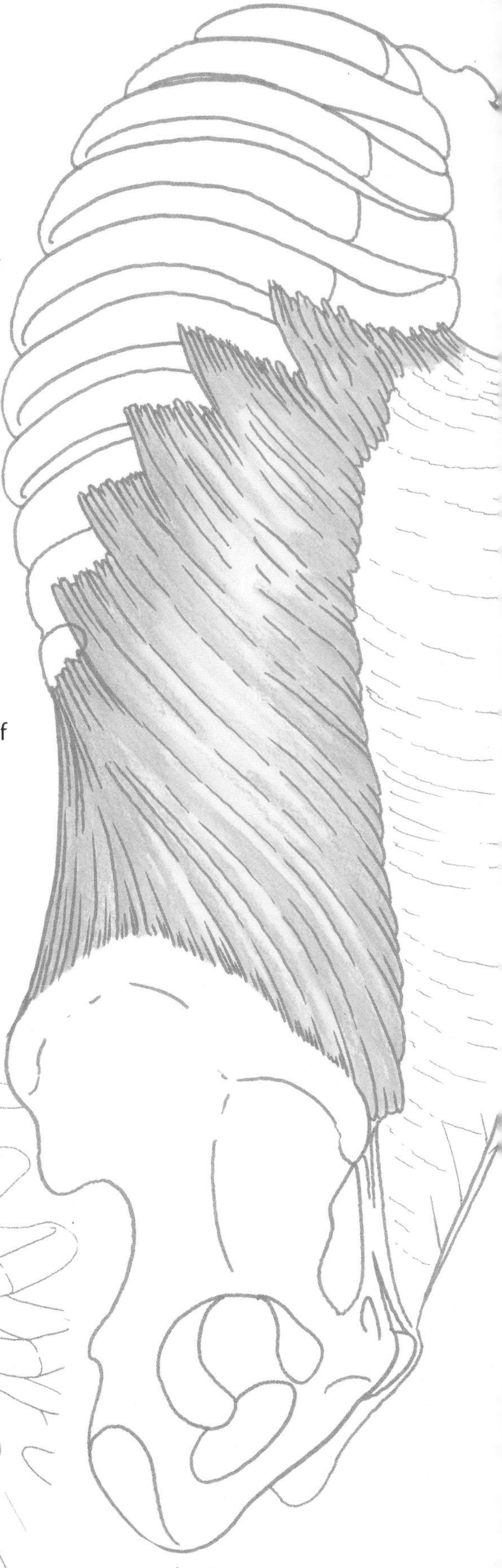

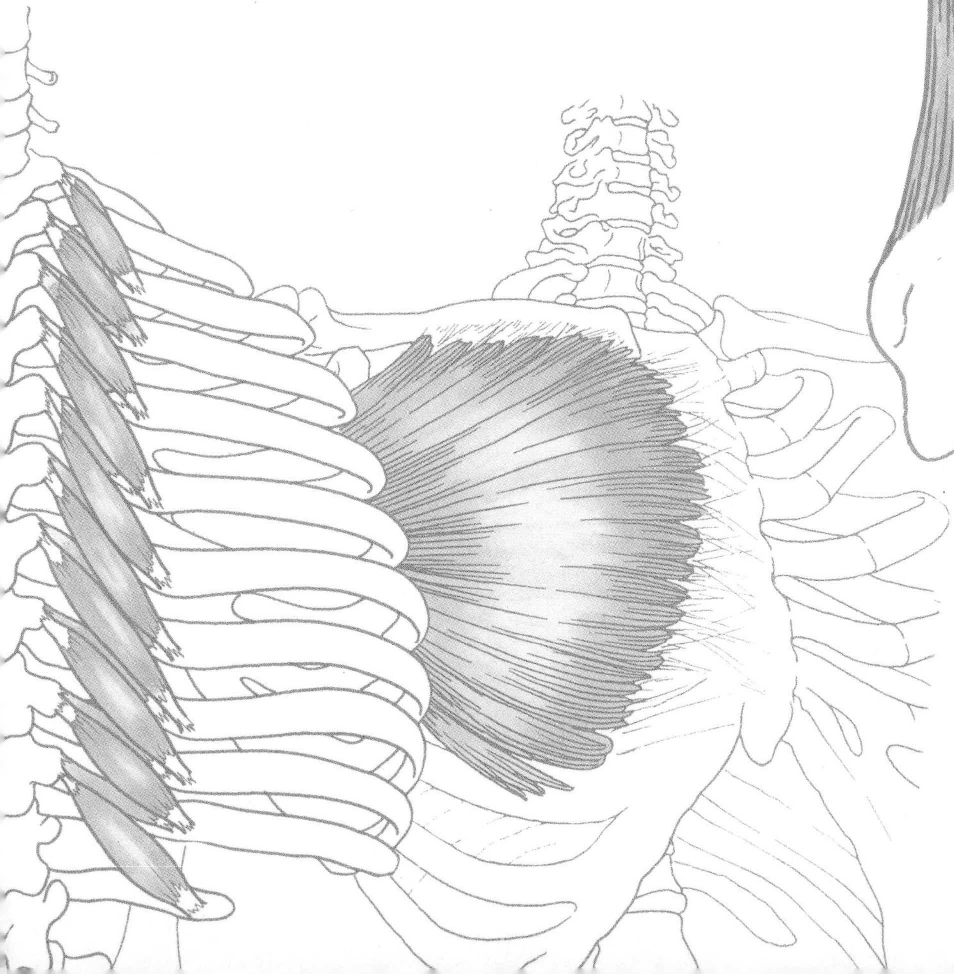

The Expiratory Muscles: The Muscles That Produce the Vocal Breath

The following pages will review the muscles that produce the exhalation and support the voice.

We should first note that the exhalation at rest—that is, passive breathing—requires no muscular action; the exhalatory pressure is solely a result of the elastic recoil of lung tissue.

In certain situations, such as when we speak in a confidential manner (such as reading in a low voice, or speaking to a listener who is in very close proximity), or when we speak in very short sentences, this pressure from the elastic recoil is sometimes sufficient for our vocalization.

So, why do we use the expiratory muscles? To exhale more completely, which can serve to:

- extend the duration of the exhalation (we are then using the expiratory reserve volume, which is very common when using the voice, especially when singing)
- augment subglottal pressure, specifically to increase our volume or to reach higher notes

The expiratory muscles are principally the **abdominals,** eight muscles in four pairs arranged on the front and sides of the abdomen.

The four sets of abdominal muscles are the primary expiratory muscles.

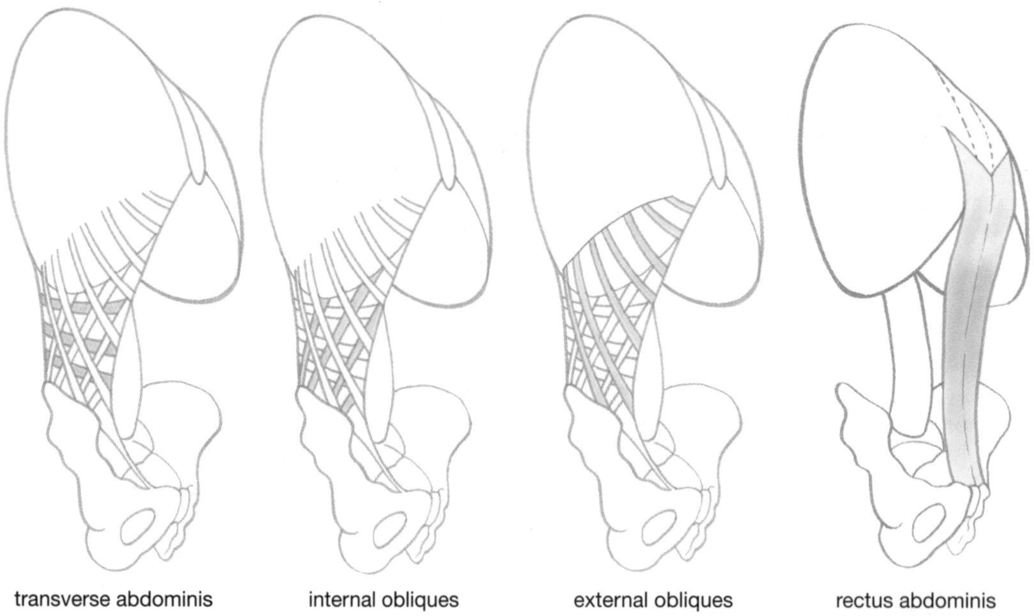

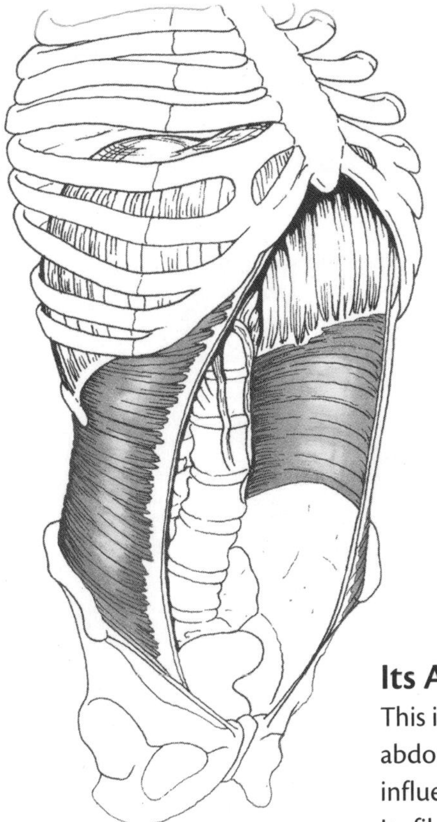

The Transverse Abdominis

Located on the sides and slightly to the front of the trunk, this muscle is the deepest of the abdominals. It arises from the lumbar vertebrae, from fibrous attachments. Its contractile fibers wrap horizontally around the waist, and its force is extended further to the front by a wide fibrous web: the **transverse aponeurosis**.

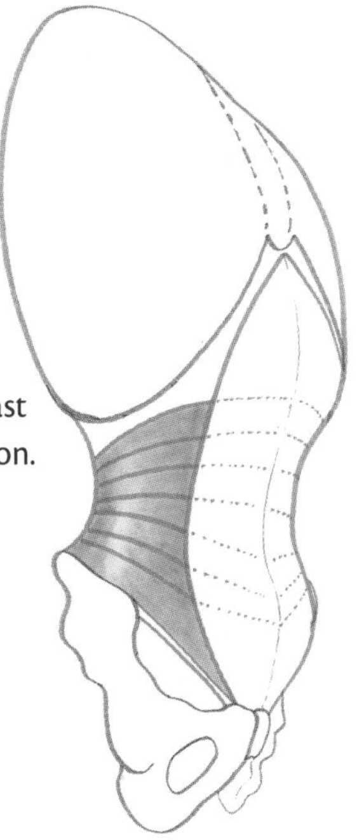

Its Action in Respiration

This is the most "visceral" of the abdominals, or at least the one that least influences the movement of the skeleton. Its fibers constrict the diameter of the waist, causing an exhalation.

Its Role in the Voice

In vocal work, the transverse abdominis can act synergistically/antagonistically with the diaphragm: it contributes to the exhalation while the diaphragm holds back the same exhalation. This action allows us to mete out subglottal pressure.

Attention!

This muscle acts, for the most part, in the area of the waist, and its activation can put downward pressure on the pelvic floor (see page 117).

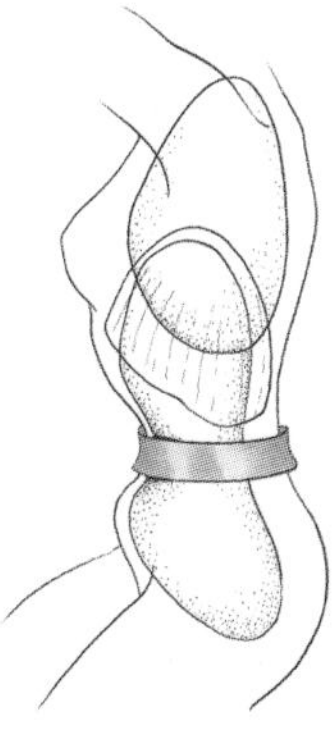

The Obliques

The oblique muscles are situated on the sides and front of the trunk. They arise from the iliac crests.

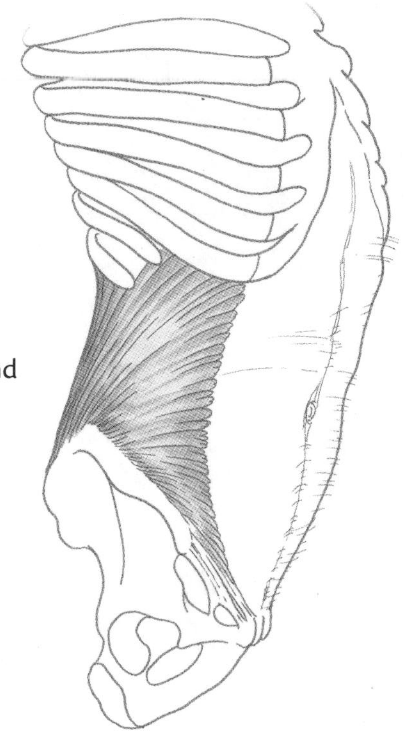

The Internal Oblique

This muscle covers the transverse abdominis and forms the middle layer of the lateral abdominal muscles. Its contractile fibers ascend anteriorly around the waist. One part ends at the inferior border of the thoracic rib cage. The other part continues anteriorly by way of its fibrous fascial sheath, the **aponeurosis of the internal oblique,** and joins with its symmetrical partner in the middle of the abdomen, where it forms the **linea alba**.

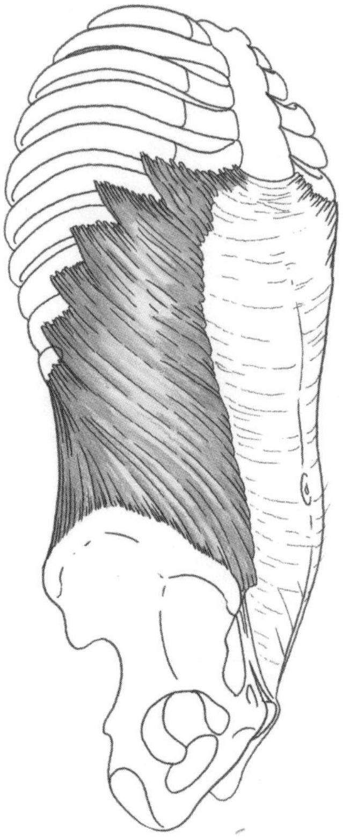

The External Obliques

This muscle covers the internal oblique and forms the outermost layer of the lateral abdominal muscles. Its contractile fibers wrap around the waist as they move upward and to the back. One part terminates on the last seven ribs. Another part continues anteriorly by way of its fibrous fascial sheath, the **aponeuroris of the external oblique,** and joins with its symmetrical partner in the middle of the abdomen, where it forms the **linea alba.**

The Action of the Obliques in Respiration

In addition to mobilizing the bones, the obliques, along with the other abdominals, lift the abdominal mass. Their fibers contract around the area of the waist, so they also are involved with the exhalation.

The Role of the Obliques in the Voice

In vocal work, the obliques can act synergistically/antagonistically with the diaphragm: they contribute to the exhalation while the diaphragm can hold back the same exhalation. This action allows us to mete out subglottal pressure.

Attention!

If the action of these muscles is not directed upward, some of the force they generate will be directed to the pelvic floor (see page 117).

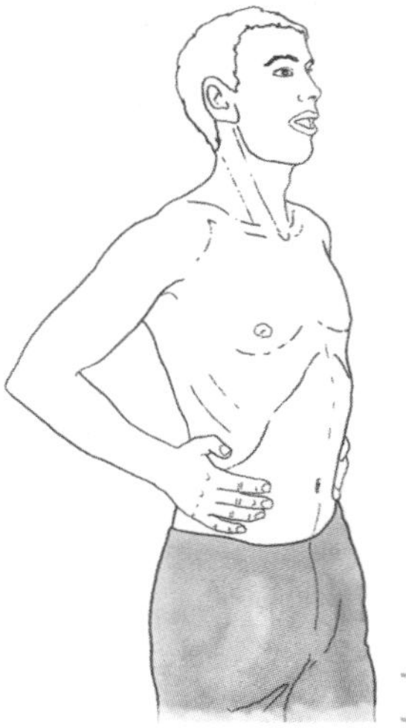

Palpate the Internal Obliques

Just like with the transverse abdominis, the internal obliques can't be palpated because they are situated under the external obliques. But you can palpate the area of their insertion, which is between the lower ribs and the iliac crest, at the sides of the waist.

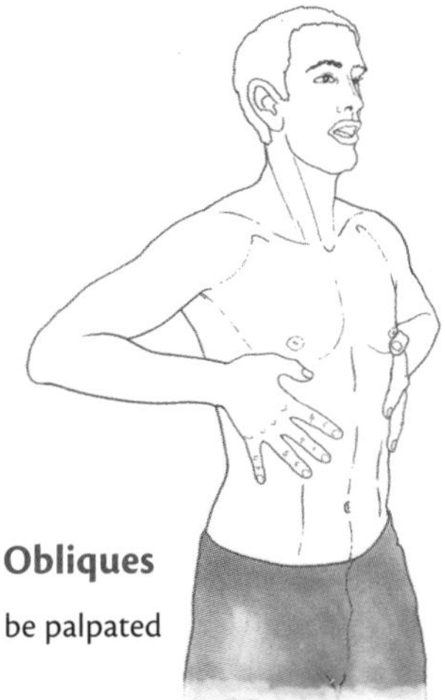

Palpate the External Obliques

The external obliques can be palpated at the sides of the waist.

More details about the action of the abdominals can be found in *No-Risk Abs*.

The Rectus Abdominis

Among the abdominal muscles, this is the only one located at the front of the trunk. It attaches to the pubis at its lower end. Its fibers ascend almost parallel to the thoracic rib cage, ending at the sternum and the costal cartilage of the fifth, sixth, and seventh ribs.

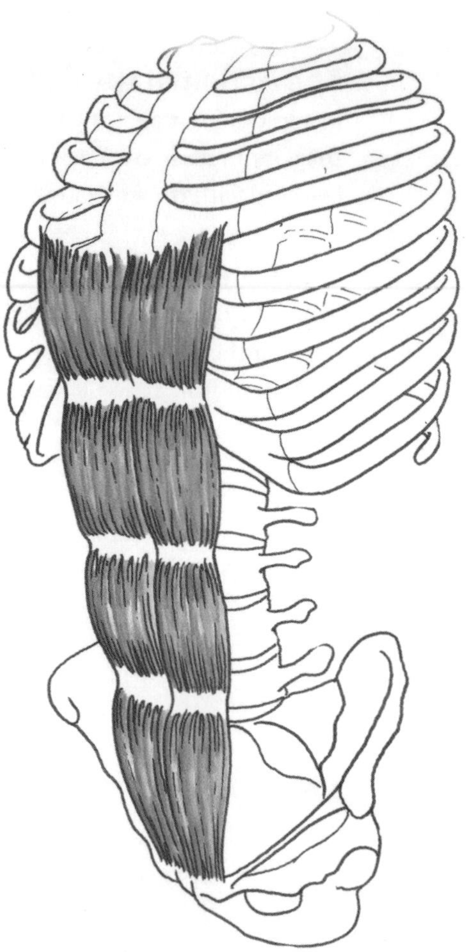

Its Action in Respiration

In addition to mobilizing the bones, the rectus abdominis, along with other abdominals, lifts the abdominal mass. The fibers tighten at the front of the belly and therefore encourage exhalation.

Its Role in the Voice

In vocal work, the rectus abdominis can act synergistically/antagonistically with the diaphragm: it contributes to the exhalation while the diaphragm holds back the same exhalation. This action allows us to mete out subglottal pressure.

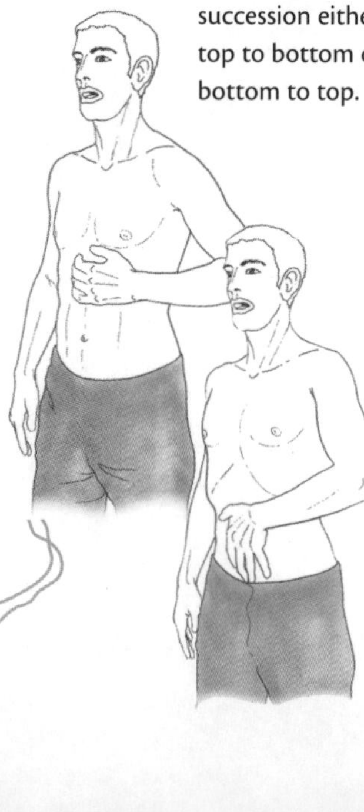

The different segments of the rectus abdominis can be contracted in succession either from top to bottom or from bottom to top.

Vocal Application

The different segments of the rectus abdominis can contract individually. Therefore, it is possible to:

- contract them successively from bottom to top to create a force directed upward, toward the larynx, which will contribute to the production of sound
- contract them successively from top to bottom, and this action can be used to control vocal intensity (but it should be noted that this will increase pressure on the pelvic floor; see page 117)

The Abdominals and Vocal Work

The Abdominals Act at Once on the Skeleton and the Viscera

The abdominals can mobilize, fix, or brake the movement or position of the skeleton, and they can as well mobilize or fix the abdominal visceral mass. These two actions are most often combined and influence each other.

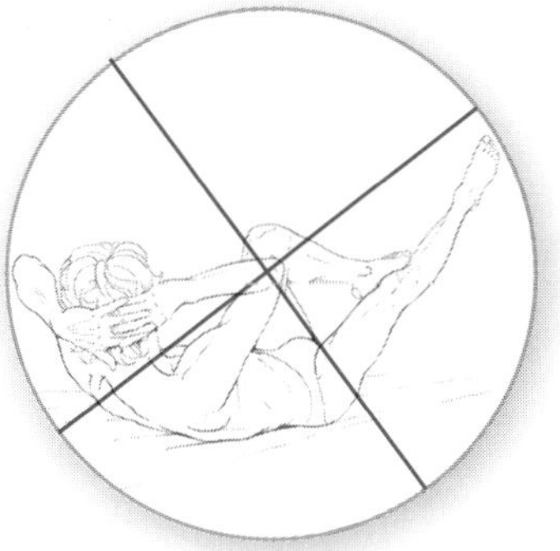

In vocal work, it's not particularly useful to train the abdominals in their skeletal function by doing push-ups or leg lifts. Instead, it's important to refine their ability to move the *viscera* in different ways, especially during the vocal exhalation.

The Abdominals Can Act on the Viscera Segmentally

We can contract different areas of the abdominals individually. This is possible because the abdominals are innervated from spinal nerves on seven different levels.

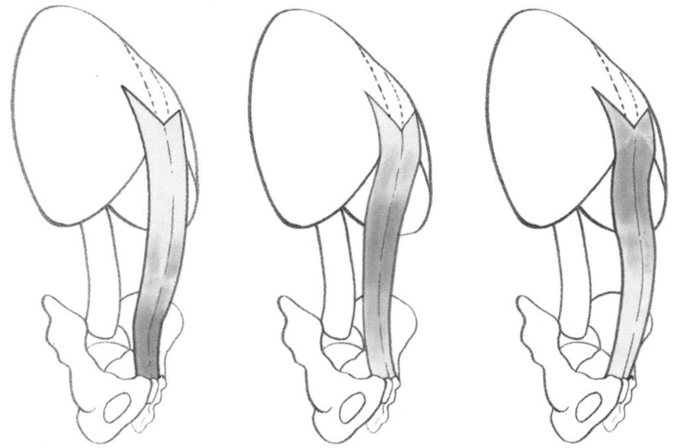

Segmentalize Abdominal Contractions

It is important to learn how to easily segmentalize abdominal contraction. Think of the belly as a keyboard. If we learn this, we can coordinate the contraction at a specific level of the abdominals with, for example, the diaphragm or the perineum, to create various expiratory outcomes.

The Action of the Abdominals on the Viscera Can Be Ascending or Descending

We can contract the abdominals successively from top to bottom or bottom to top. The direction of contraction determines the direction of force on the viscera.

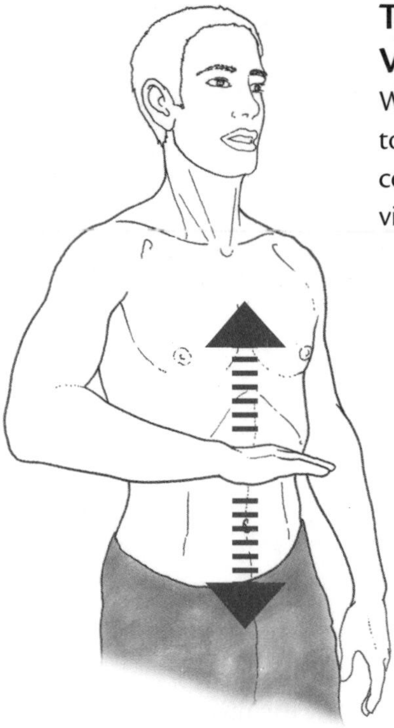

Direct the Contraction from Bottom to Top

It's important to learn how to easily direct the contraction upward. However, you should intermittently direct the contraction in the other direction to balance the trunk.

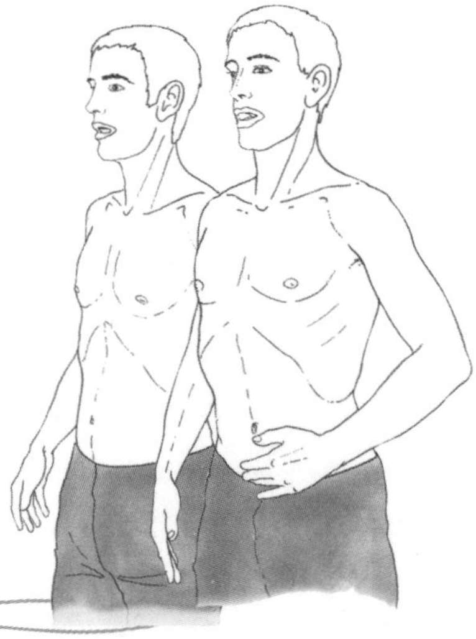

Abdominal contraction directed downward can create hernias at the level of the inferior insertions of the abdominals. The pressure can push on the lowest part of the small intestine, which can then slide between the fibers of the abdominals that are attached to the rim of the pelvis. This effect can be augmented when the abdominals are working synergistically/antagonistically with the diaphragm, which also exerts downward pressure toward the pelvis.

Confirmation

This risk of exerting excessive downward pressure confirms the importance of learning to also direct the contraction of the abdominals from the bottom upward, which builds upward force.

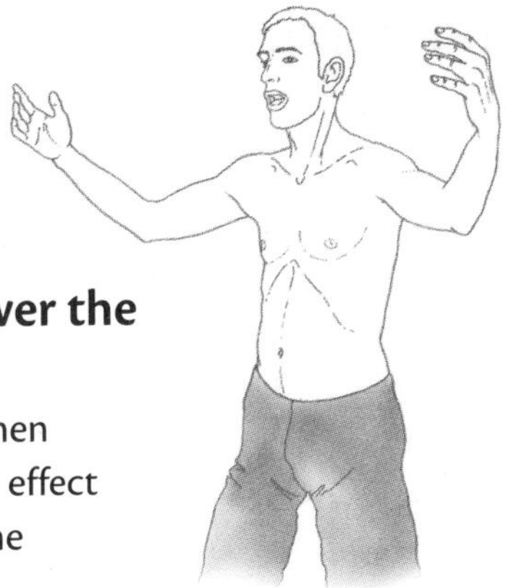

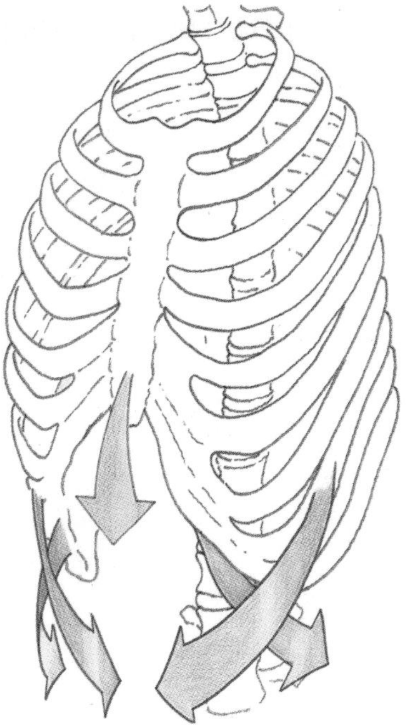

All of the Abdominal Muscles Lower the Thoracic Rib Cage

However, this is not always what we want when doing vocal work. And this will augment the effect of downward force that we mentioned on the previous page.

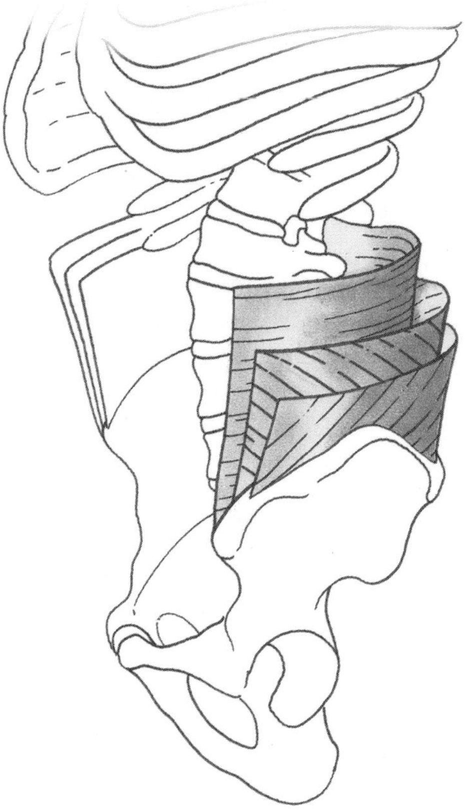

Not All of the Abdominal Muscles Are at the Front of the Trunk

Only the two recti are located in the front, which means that the majority of the abdominal muscle mass is at the sides of the trunk, in three superimposed layers on each side:

- the transverse abdominis
- the internal obliques
- the external obliques

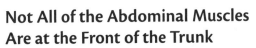

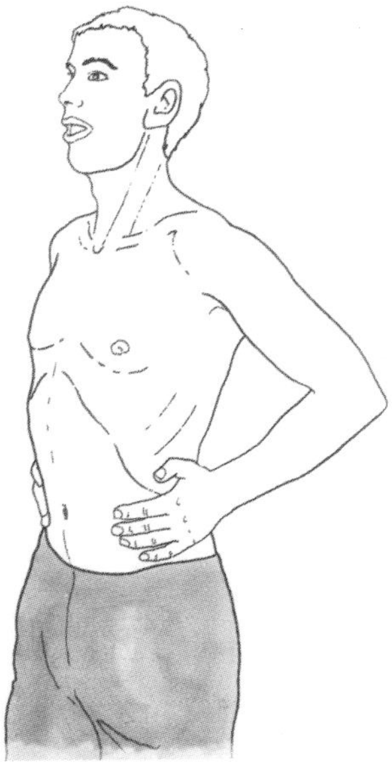

It's important to remember where the force is coming from with the contraction of the abdominals in a vocal exhalation: the action is coming as much from the sides as it is from the front.

The Perineum and Pelvic Floor

These two words are often used interchangeably, and sometimes defined a bit differently. For the purpose of this work, we will define them as follows: the perineum is the area of the body found at the base of the pelvis. In this sense, the **perineum** includes all of the structures found there, collectively: the bones, muscles, viscera of the pelvis, skin.

In this region the muscles lying in a horizontal plane are grouped under the name **pelvic floor**. The term itself evokes the support that the muscles give to the pelvic and abdominal viscera lying above them.

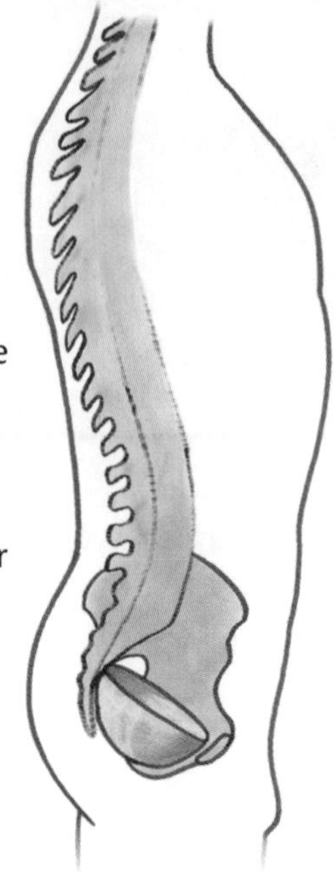

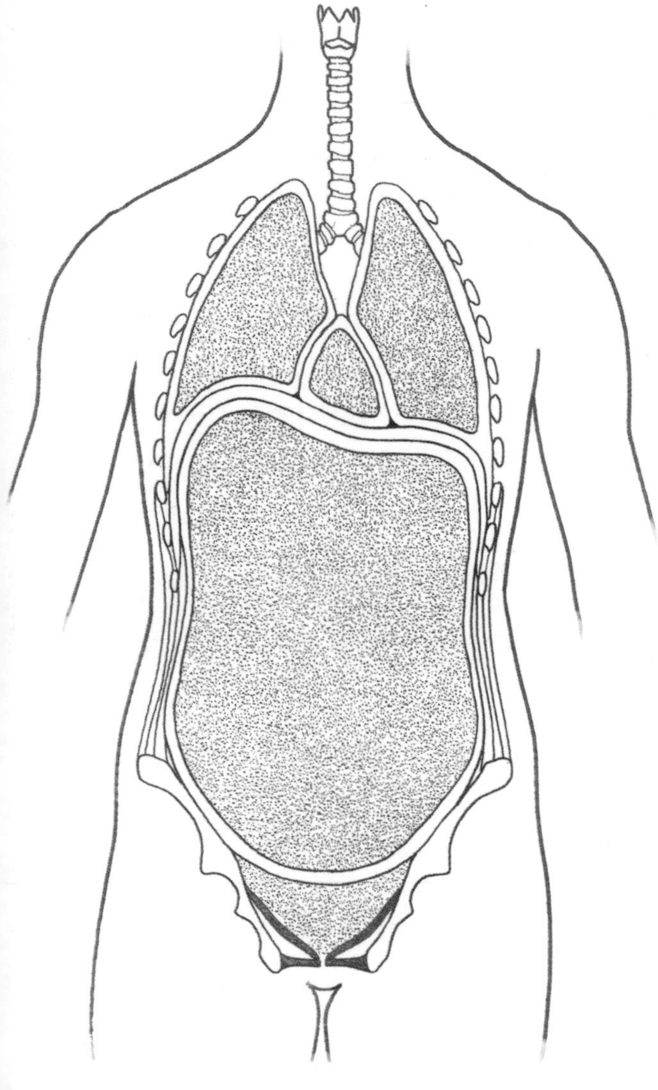

But what this "floor" must withstand, more than the weight of the viscera, is the pressure that is generated downward from the trunk by different kinds of events, such as a cough, evacuation (urination, defecation, childbirth), and sometimes . . . the voice. How could that be?

1. Generation of sound requires the production of a certain amount of pressure under the vocal cords. This pressure comes most often from the abdomen and is created by the abdominal muscles. It is rare that a contraction will send pressure only upward, and part of the pressure is directed toward the bottom of the trunk, where it is received by the pelvic floor.

2. At the top, the pressure meets a "brake," the glottis, which is practically closed during phonation. So, the pressure is again sent downward, where it's received by the pelvic floor.

This demonstrates the importance of the pelvic floor, which is at once the support for the pressure directed to it (which it must be able to receive without collapsing) and, at the same time, an active participant with the ensemble of muscles that generate pressure to create the sound—and send the pressure upward. For these tasks, the pelvic floor needs to have sufficient tone.

The Muscles of the Pelvic Floor

These muscles exist in men just as they do in women. However, in men, the pelvic floor is "closed," and the external genitalia are suspended from the pelvic floor. In women, the pelvic floor has in the anterior part a slit that is necessary for copulation and childbirth. This opening is a weak point in a woman's abdominal cavity, and it's important that the muscles bordering it are sufficiently toned to allow the pelvic floor to perform its support role.

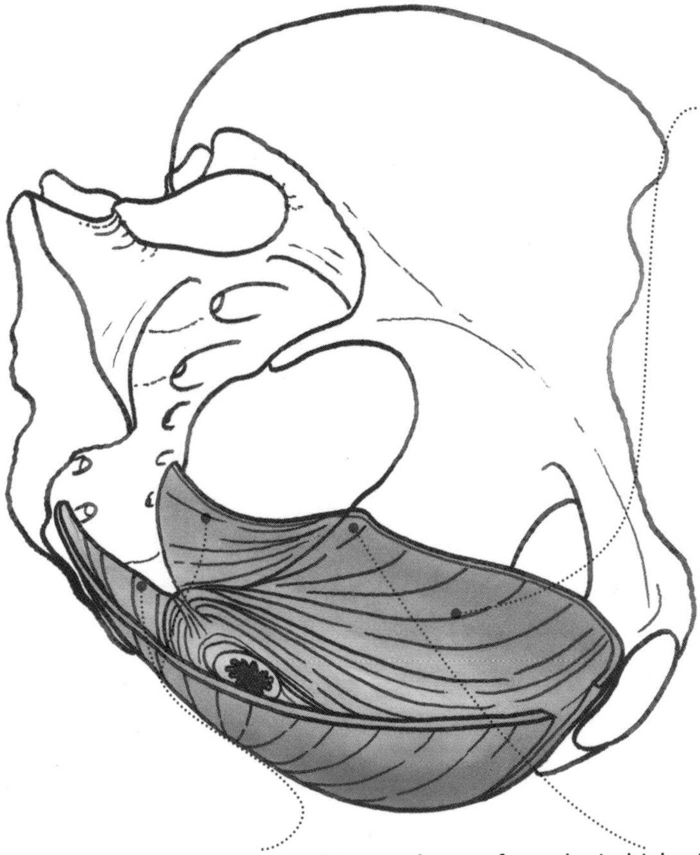

The Deep Muscles of the Pelvic Floor

The levator ani. This muscle is attached to the pubis in the front, the obturator foramen, and the raised area of the ischium. Its fibers run to the back and middle and terminate symmetrically. It forms something like a hammock, which is more or less convex downward. (This form changes depending on the pressure or load on it.)

Its function: The levator ani lifts toward the middle part of the lesser pelvis. It moves the anus forward and upward.

The ischiococcygeus. This muscle runs from the ischial spine (see page 49) to the sacrum and coccyx. Its structure completes that of the levator ani to form a hammock-shaped bowl.

Its function: The ischiococcygeus completes the action of the levator ani posteriorly. (That is, it slightly mobilizes the sacrum and the coccyx to the front, which we call *counternutation*. This action is sometimes noticeable when we do vocal work and look to consciously contract the pelvic floor.)

For more details, see *The Female Pelvis,* pages 57–77.

The Superficial Muscles of the Pelvic Floor

These muscles are very small and are located below those discussed on the previous page. When we do vocal work, for the most part they supplement the action of the deep muscles, and so we will just briefly describe them here.

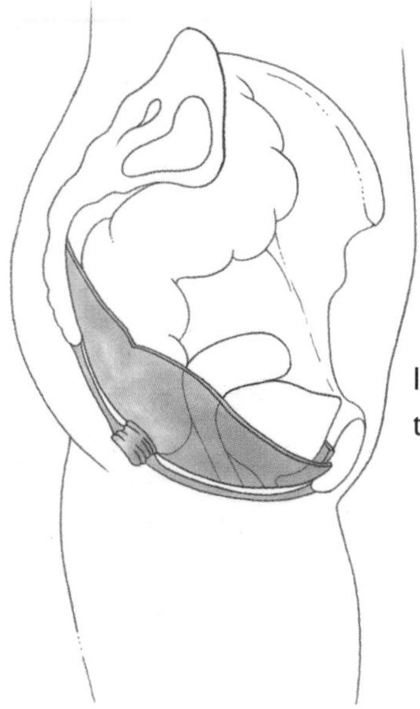

The bulbospongiosus. This muscle is located at the front of the lesser pelvis. It is connected to the anal sphincter and, further back, to the coccyx by a fibrous band.

In women, this muscle borders the vulva.

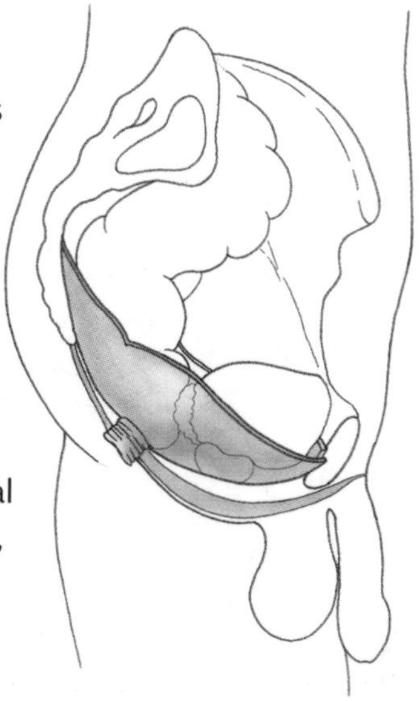

In men, this muscle is under the horizontal part of the penis, which it supports. Then, further forward, it passes above and in front of the vertical part.

The superficial transverse perineal. This muscle (not shown), called the **central tendon,** runs from the ischium to the central area of the perineum.

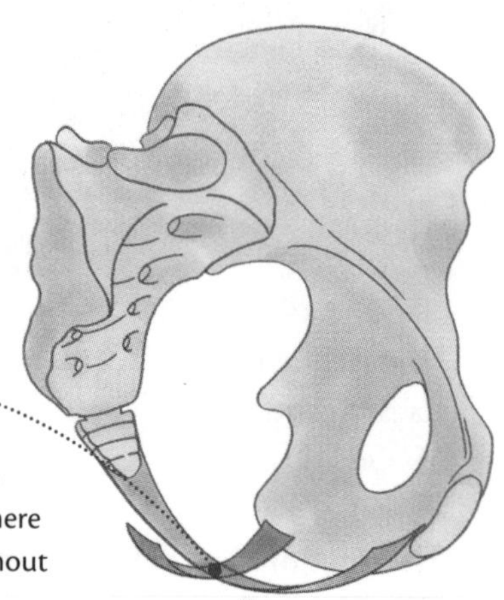

The Central Tendon of the Perineum

This area is important as it makes up both the center and the place where the muscle fibers of the deep and superficial layers intersect. It has no direct role in the voice; however, weakness here often leads to a disorganized generation of sound, which is without a "foundation." This confirms the importance of toning and coordination of the perineum for vocal work.

Read more about the female perineum in *The Female Pelvis.*

The Pelvic Floor and Pressure

Certain situations in the generation of sound modify the pressure that is put on the muscles of the pelvic floor. Below are some examples; however, the situations can be much more diverse.

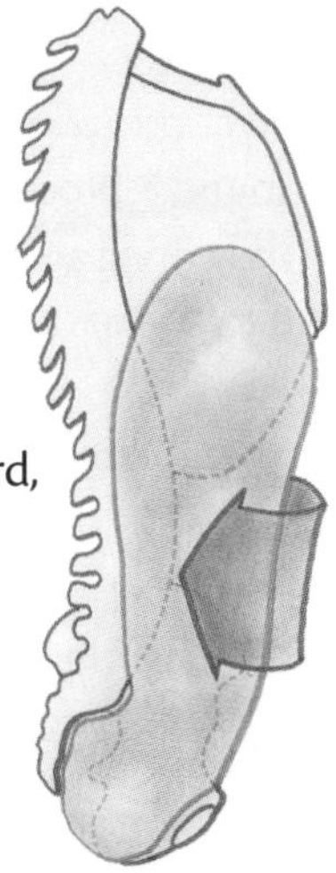

When we exhale and contract the abdominals not from bottom to top, but in the middle or upper abdomen, the abdominal mass can be sent downward, creating vertical pressure on the pelvic floor.

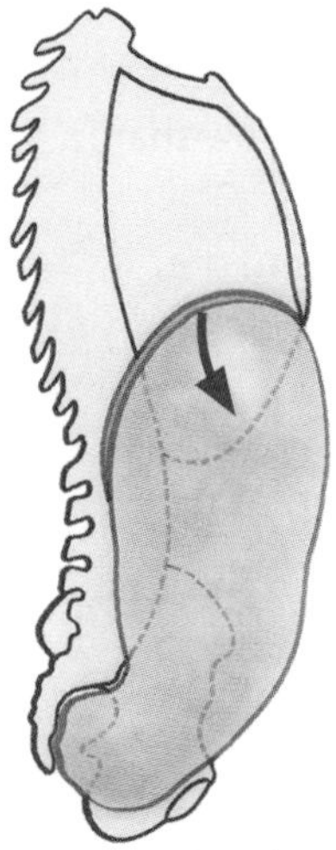

When we inhale while lowering the diaphragm, and then we exhale while keeping the diaphragm dropped, the abdominal mass is sent toward the walls of the abdomen and also toward the pelvic floor, where it creates pressure.

In certain vocal situations, we play with these two processes at the same time: when the diaphragm is kept lowered to keep the lungs open, and at the same time we create pressure by contracting the abdominals. The resulting pressure on the pelvic floor is very high. In these situations, it's important that the pelvic floor is able to contract to support the pressure.

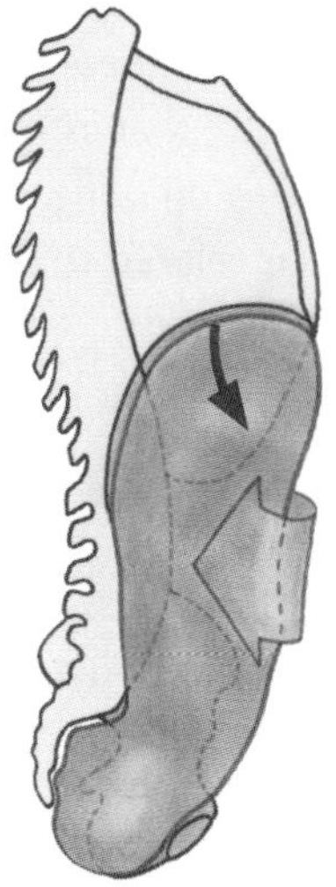

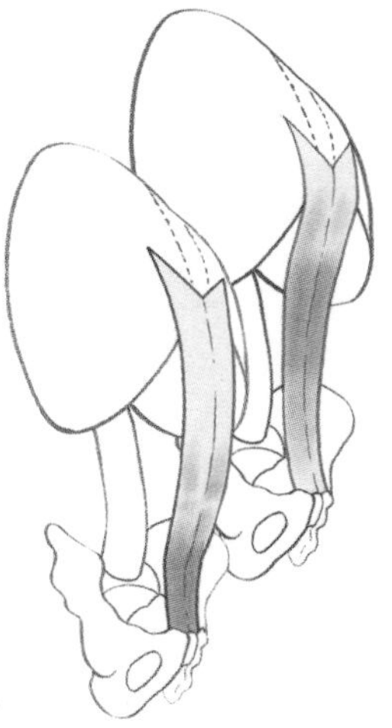

When we exhale and contract the abdominals from the bottom to the top, the abdominal mass is sent upward, and this *minimizes* pressure on the pelvic floor.

When we inhale and open the ribs or keep the ribs open during the exhalation, the abdominal mass is pulled toward the thorax, and this *minimizes* the pressure on the pelvic floor (not illustrated).

The Inspiratory Muscles

Over the following pages, we will review the muscles responsible for inhalation. Since sound is produced upon exhalation, how do these muscles serve the voice? Of course, they bring air back into the lungs. But very often they act to brake the breath, and because of this, they mete out the *subglottal airflow.*

There are two kinds of inspiratory muscles:

- The **diaphragm,** which is, of course, the most powerful; it is located below the lungs and inside the thorax.
- The **intercostals** and **accessory inspiratory muscles,** which all attach to the outside of the thoracic rib cage. These muscles can help open the alveoli; they aren't always involved in routine breathing but are recruited as needed.

In vocal preparatory work, we are interested in toning the intercostal and accessory inspiratory muscles for two reasons: they allow us to vary the inhalation, and they can keep the rib cage from collapsing, which means that the abdominals can work more efficiently on the exhalation.

We'll present all of these muscles one by one, beginning with the diaphragm. Note that in vocal work, we aim to recruit the muscles of the thoracic region rather than those of the neck and throat, which we want to be relaxed so that the muscles of the larynx are free to work more efficiently.

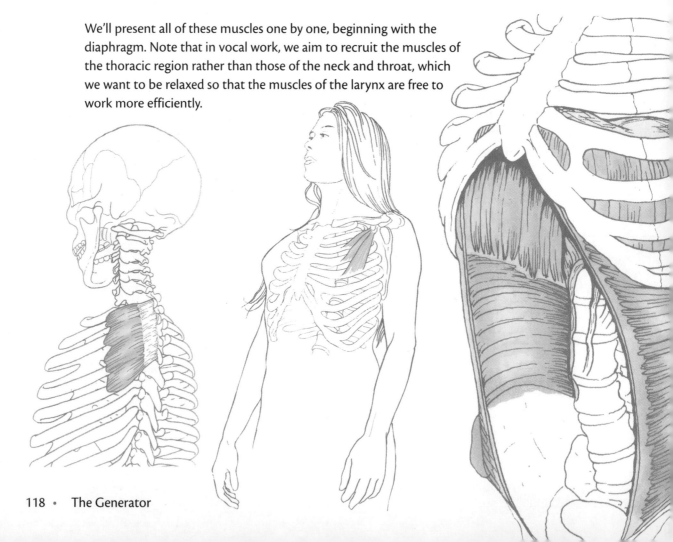

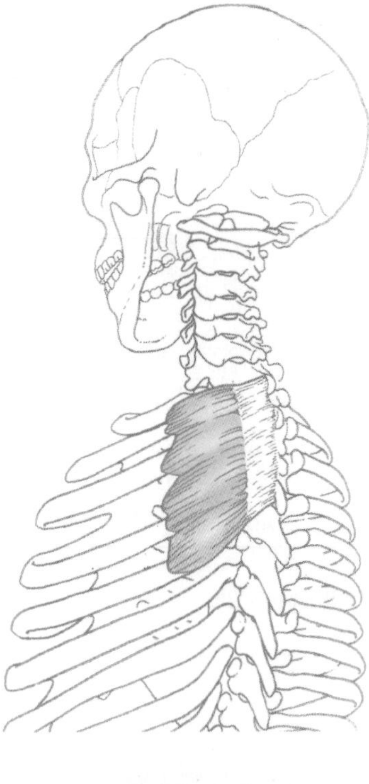

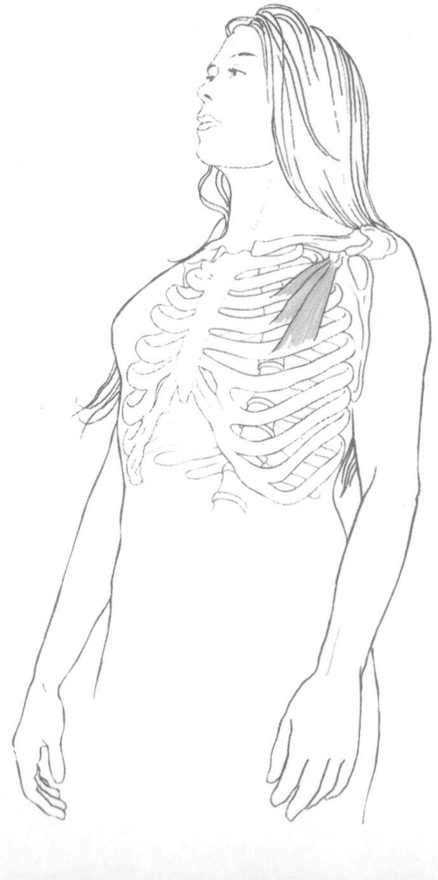

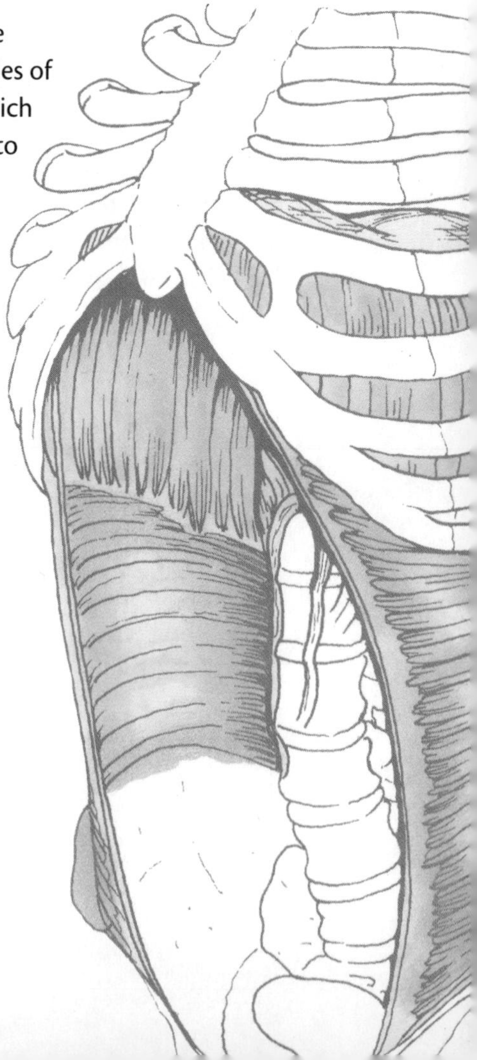

The Diaphragm

This muscle, located between the thoracic and abdominal cavities, is shaped like a dome-shaped sheet. It is composed of two parts:

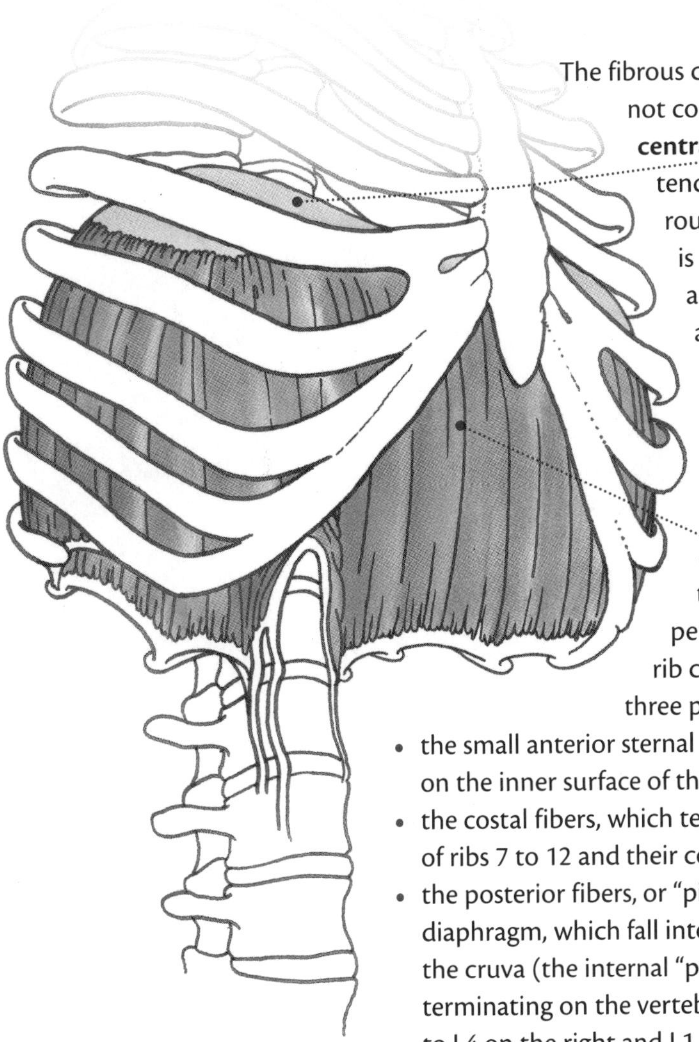

The fibrous center, which does not contract, is called the **central tendon** (centrum tendineum). It is shaped roughly like a clover, and is therefore described as having three leaflets: an anterior leaflet and two posterior leaflets, right and left.

The central tendon gives rise to **contractile fibers** that radiate out to the perimeter of the thoracic rib cage and terminate in three parts:

- the small anterior sternal fibers, which terminate on the inner surface of the xiphoid process
- the costal fibers, which terminate on the inside of ribs 7 to 12 and their costal cartilage
- the posterior fibers, or "pillars" of the diaphragm, which fall into two categories: the cruva (the internal "pillars"), terminating on the vertebral bodies of L1 to L4 on the right and L1 to L3 on the left, and the arcuate ligaments (the external "pillars"), terminating in fibrous arches

Specifics Regarding the Diaphragm

The diaphragm meets the transverse abdominis (see page 107) at the interior of the thoracic rib cage. In fact, the fibers of the two muscles mesh, which makes them easy synergists/antagonists in vocal work.

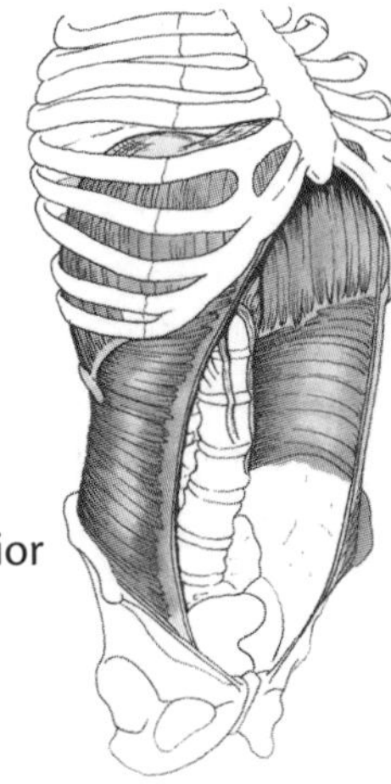

Find more details on the diaphragm in *Anatomy of Breathing,* pages 80–86 and 134–39.

Action of the Diaphragm in Respiration

The diaphragm is the principal muscle of inhalation. And it can act in two ways, often simultaneously.

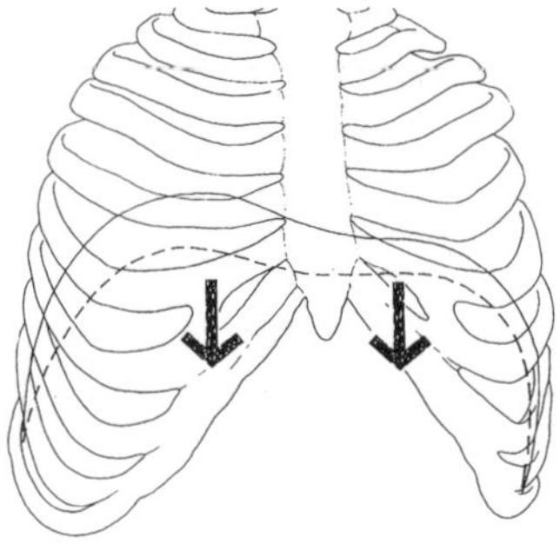

Up-and-Down Action

Its contraction lowers the central tendon. This pulls the base of the lungs downward toward the lower part of the trunk, increasing lung volume and causing inhalation. It also has the effect of pushing the abdominal viscera downward, which is often called (incorrectly) "inflating" the abdomen.

Sideways Action

The abdominals can inhibit the lowering of the central tendon. The diaphragm is then acting from a fixed center while pulling on the ribs, which it lifts, from the inside of the thorax. However, any elevation of the ribs expands the diameter of the thorax (see page 61). We say that the diaphragm "opens" the bottom of the rib cage. Again, this movement causes inhalation.

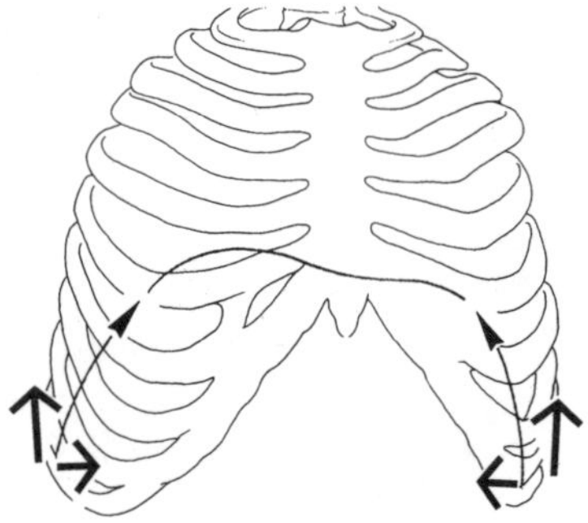

Much More Important: The Action of the Diaphragm on the Voice

Sound is made not on the inhalation but, rather, on the exhalation. The action of the diaphragm during the production of sound is, therefore, not respiratory but consists of "braking" expiratory actions. The diaphragm "brakes" the elastic recoil of the lungs. Considering the two actions of the diaphragm, this meting out of the vocal exhalation can take two forms:

- either the belly bulges when we sing; or
- the lower ribs are kept open when we sing.

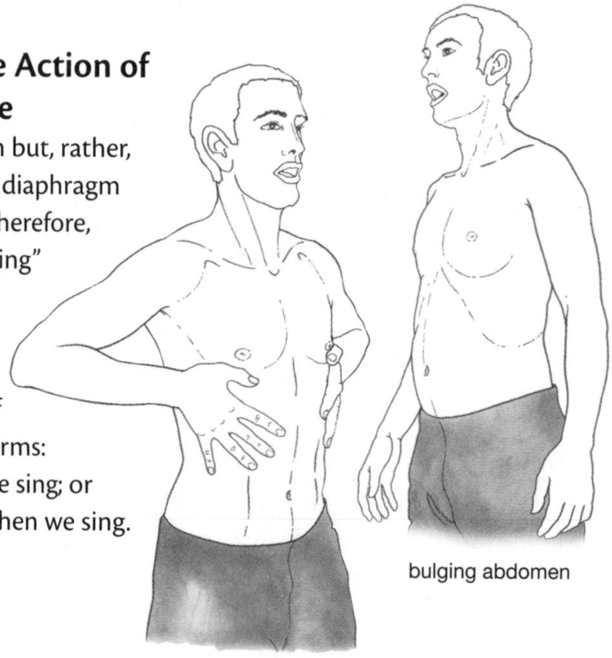

bulging abdomen

"open" ribs

The Diaphragm Is Connected to the Pharynx and Larynx

The diaphragm adheres to the base of the lungs and heart by way of the pleura and pericardium. These in turn adhere to many elements located between the lungs that, collectively, are called the **mediastinum**. In particular, we can cite the esophagus and the trachea.

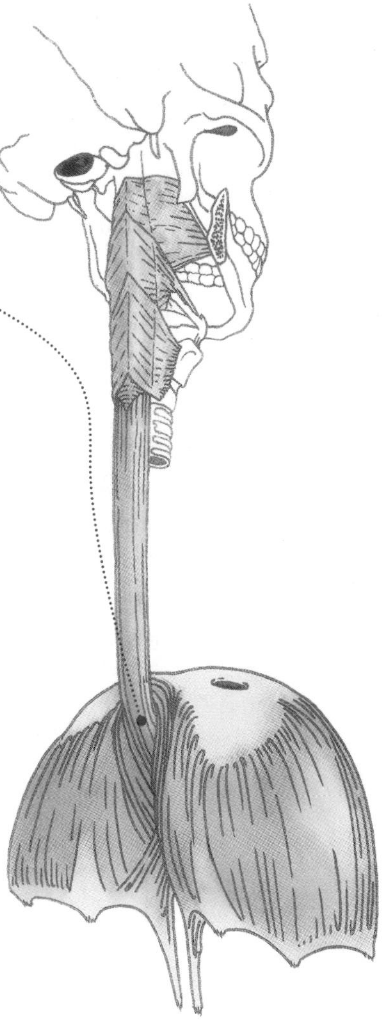

At the bottom of the chest, the esophagus crosses the diaphragm through an opening called the **esophageal hiatus** and then continues down into the abdomen and stomach. At the esophageal hiatus, the posterior fibers of the diaphragm surround the esophagus like a lasso. The esophagus is connected in part to the diaphragm by fibrous connective tissue: it therefore follows the downward movement of the diaphragm. However, the top of the esophagus inserts at the cricoid cartilage of the larynx and thyroid (see page 142), which means that a lowering of the diaphragm has repercussions higher up on the larynx and pharynx:

- The larynx lowers (see the consequences on pages 194–95).
- The pharynx lengthens (see the consequences on page 225).

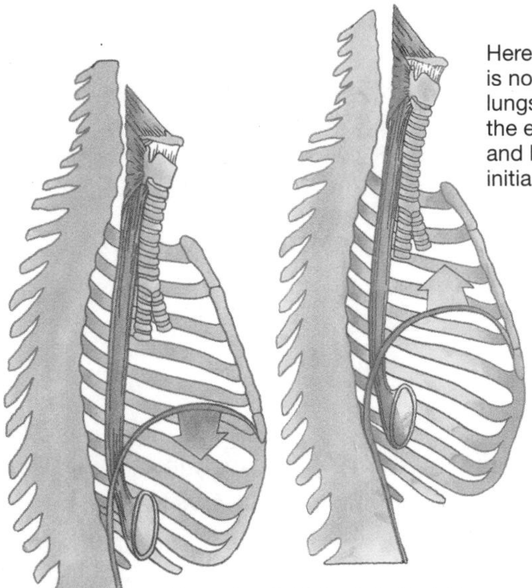

Here, the diaphragm is no longer active; the lungs come back up and the esophagus, trachea, and larynx return to their initial position.

Here, the diaphragm contracts and lowers the esophagus, trachea, and larynx.

Palpation

You can easily feel the lowering of the larynx when you put your fingers on the thyroid cartilage and yawn.

Note: The diaphragm and the larynx/pharynx can be lowered independent of an inhalation.

Releasing the Belly on the Inhalation: The "False" Diaphragmatic Breath

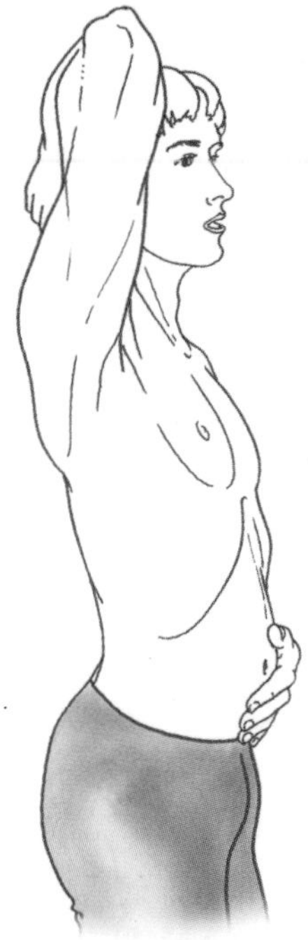

All abdominal breaths are not necessarily diaphragmatic. In some positions, the contents of the abdomen are pushed forward and/or downward mechanically. These are:

- the vertical position (standing, sitting)
- the position on all fours
- the side-lying position

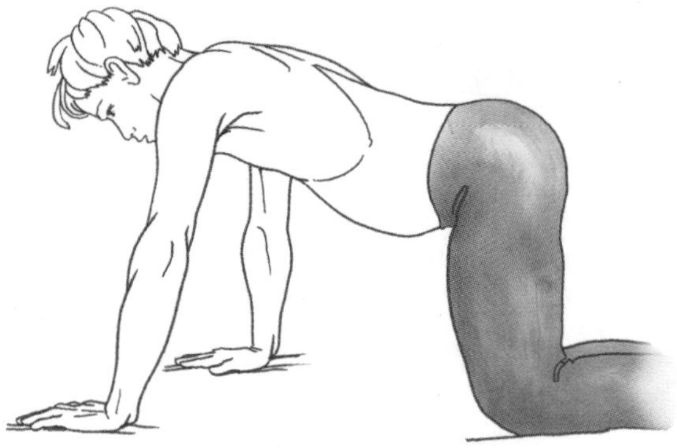

In these positions, the diaphragm does not need to be activated to push the belly out; instead, gravity causes it to bulge. In this situation, the diaphragm follows the fall of the abdomen because it adheres to it (see page 100).

Details

It should be noted that this inhalation resembles in form a diaphragmatic inhalation. But the feeling is different: we don't "push" the belly with the diaphragm; instead it is "loose"—that is to say, we "relax" the abdominals.

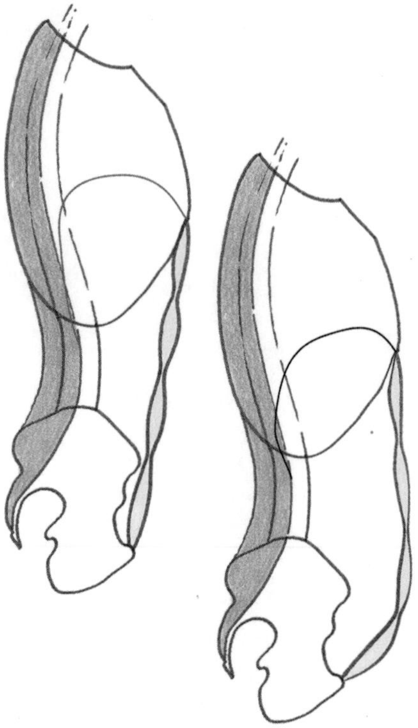

This method of inhalation is particularly interesting when we are singing in a standing position if we need to take in air between vocal sequences faster than the diaphragm works. We produce the vocal sound by progressively recontracting the abdominal muscles. Of course, this presupposes a very strong "vocal posture" (see the discussion of postural muscles on pages 132–35) that can be maintained despite the complete relaxation of the abdominals; otherwise their release will pull on the trunk and destabilize the entire body.

Misconceptions about the Diaphragm

"We cannot feel the diaphragm because it has no sensory capacity."
Not quite true. The diaphragm is innervated by the sensory networks that arise from the last six intercostal nerves and nerves that radiate from the solar plexus. You can feel, as well, the serous membranes that are against the diaphragm: the pleura, the pericardium (above), and the peritoneum (below). These membranes are richly innervated by the phrenic and intercostal nerves.

"We cannot control the diaphragm because breathing is a reflex action."
Not always true. Our everyday breath is usually a reflex action. That is to say, it's under the control of the nervous system and is unconscious (we don't really have to think about it) and involuntary (we don't decide to breathe). In this context, it is true that the action of the diaphragm is controlled mostly by reflex.

However, this reflex activity can become, intermittently, a conscious and voluntary act. That is to say, we can decide momentarily to breathe with a movement . . . to breathe more fully, or to direct our breath more to the ribs or to the belly, or even to suspend breathing (apnea), et cetera. It needs to be emphasized that we can only do this within the range of what we are capable of surviving. It's impossible to lengthen or shorten the breath more than is physiologically possible. In this context, the action of the diaphragm is controlled on a voluntary and conscious level. This conscious control of breath is what some breathing techniques and vocal methods implement, at least initially. If we repeat these actions a number of times, they no longer require conscious effort: our body learns to perform these actions unconsciously.

"All inhalations are diaphragmatic."
To qualify: Though undoubtedly the diaphragm is the main muscle of inhalation, it is easy to experiment on yourself and see if you can inhale using the other inspiratory muscles—especially if they are toned and the rib cage is mobile. And it is theoretically possible to inhale without using any muscles (as seen on the previous page).

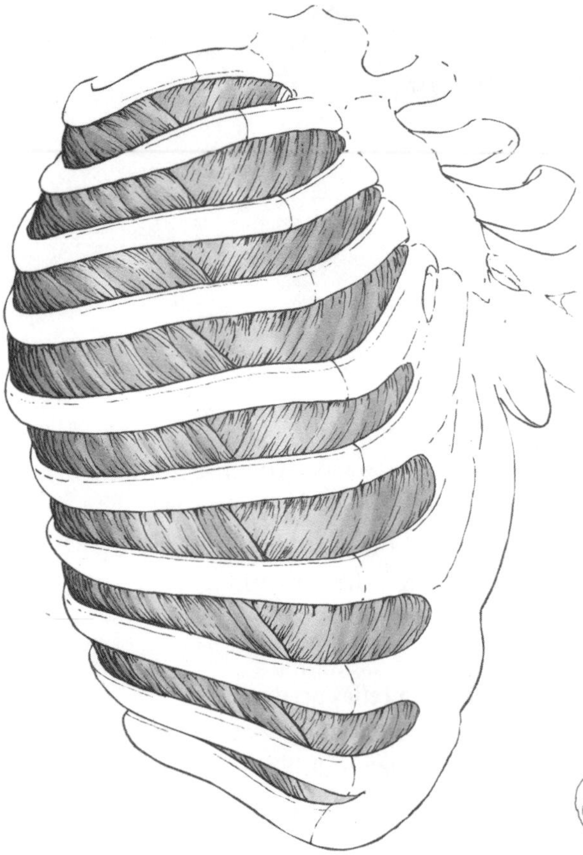

The Intercostals

These muscles are small but numerous. Situated between the ribs, they arise from each costal level, from the inferior border of the rib above to the superior border of the rib below. There are two layers of intercostal muscles:

- The **internal intercostals** comprise a deep layer of muscles that run obliquely from top to bottom and front to back.
- The **external intercostals** are more superficial and run obliquely from top to bottom and back to front.

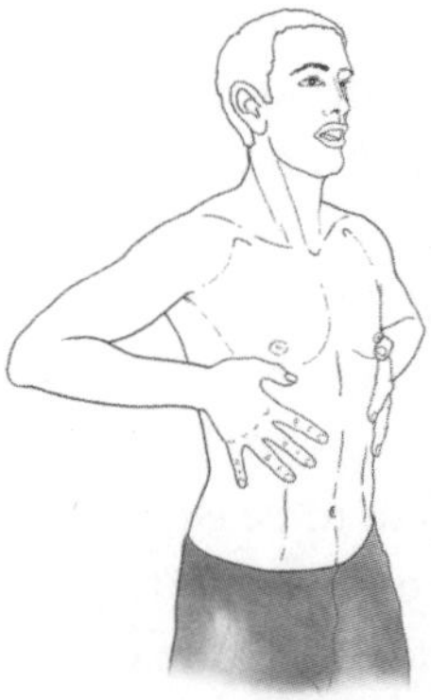

Palpation

The intercostals can be palpated at the sides of the rib cage, in the intercostal grooves.

Their Action during Respiration and Vocal Work

These muscles, located between the ribs, tend, first of all, to pull the ribs toward each other by way of the large *osteomuscular sheet* that connects them. This action can transform the whole rib cage. For example:

- If the first rib is lifted during an inhalation, all of the ribs will be lifted.
- If the twelfth rib is lowered on an exhalation, all of the ribs will be lowered.

These muscles can therefore participate in both inhalations and exhalations.

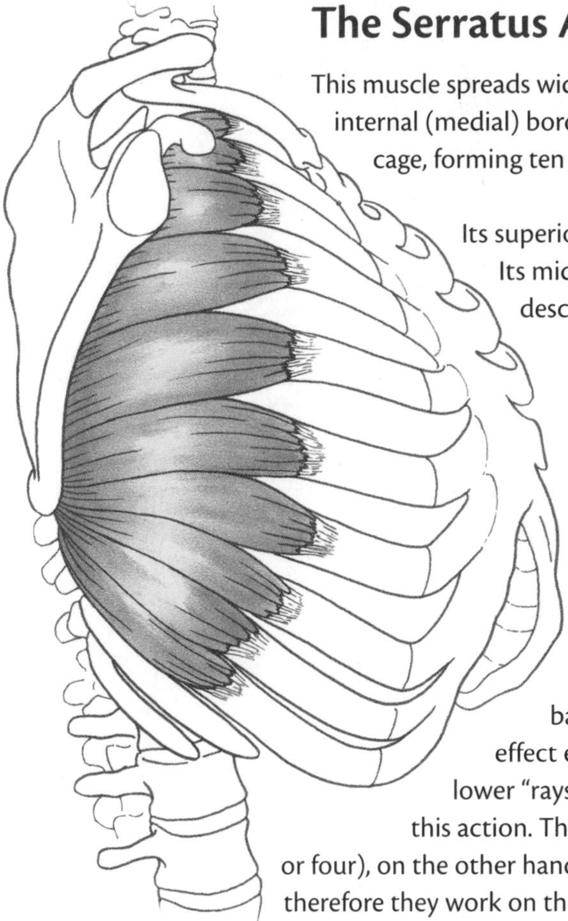

The Serratus Anterior

This muscle spreads wide on the sides of the rib cage. It arises from the internal (medial) border of the scapula and spreads around the rib cage, forming ten rays, which insert on the first ten ribs.

Its superior fibers run upward toward the upper ribs. Its middle fibers run horizontally. Its inferior fibers descend toward the lower ribs.

Its Action during Respiration

The serratus anterior pulls the ribs up and backward, and this "bucket-handle" effect encourages inhalation. It is the lower "rays" (the last five) that perform this action. The upper "rays" (the first three or four), on the other hand, do exactly the opposite, and therefore they work on the exhalation. The serratus is a powerful costal inhalation muscle, and we can easily feel its action at the sides of the ribs.

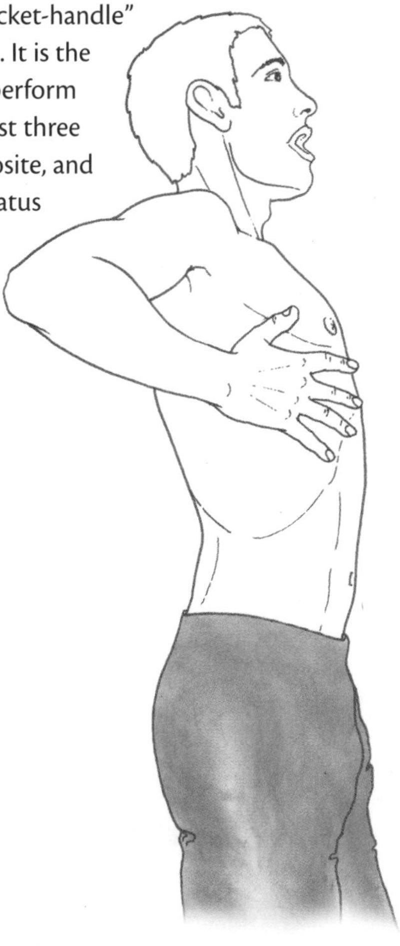

Its Action in Vocal Work

This inspiratory muscle is very efficient at "braking" the exhalation and thereby meting out the subglottal air reserve.

Watch Out for Thoracic "Rigidity"

Attention: If this muscle is overused, as it can be in certain vocal techniques, you might experience some thoracic rigidity. It's wise to alternate its use with rib mobility sequences and with the use of other inspiratory muscles.

The Pectoralis Minor

This small muscle stretches over the front of the chest, below the clavicle. It arises from the coracoid process of the scapula (see page 62) and descends toward the middle of the thorax. It terminates on the third, fourth, and fifth ribs.

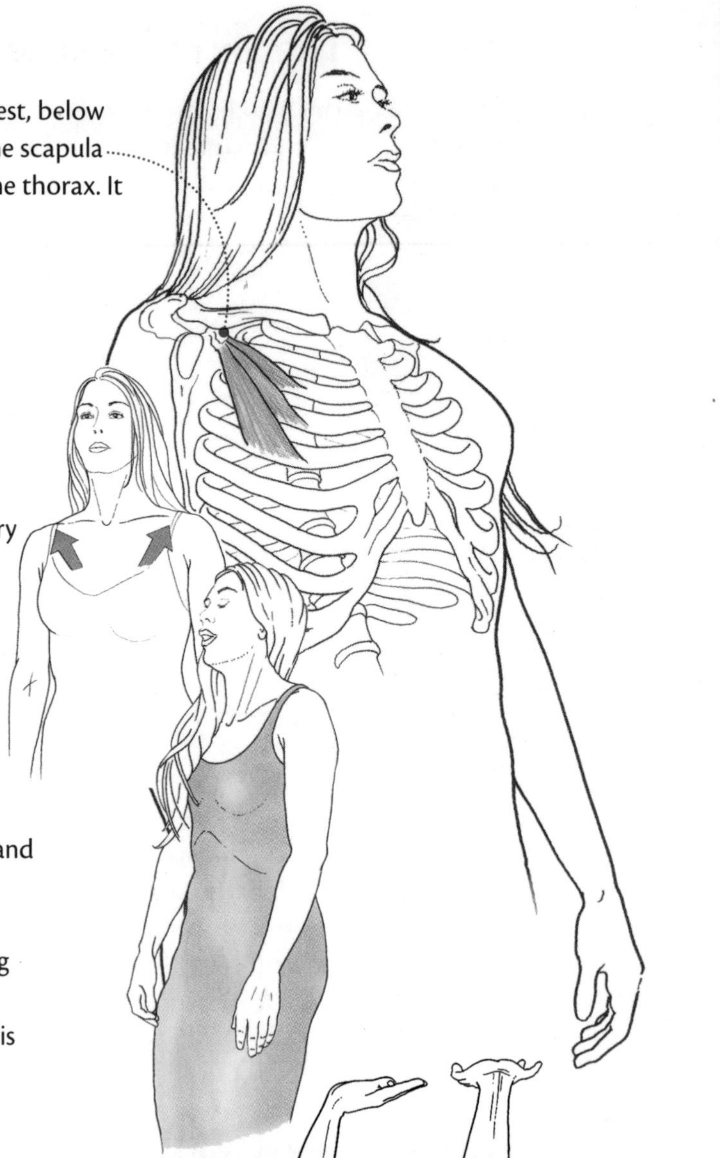

Its Action in Respiration

This muscle pulls the top ribs upward like a "pump handle," which initiates an inhalation (see page 61). The pectoralis minor is a very small costal inspiratory muscle, but its action is very interesting in terms of mobilizing the subclavian region.

Its Role in Vocal Work

This inspiratory muscle can "brake" the exhalation and can thereby mete out the subglottal air reserve.

It's common to see singers lift their shoulders during high or intense vocal passages. This results in the shoulder blades being lifted, stretching the pectoralis minor and allowing it to be recruited more easily.

Release the Shoulders

Attention: If the shoulders are rounded forward, this muscle can be shortened and can't easily be recruited during an inhalation. In this case it might be useful to stretch the muscle (in a standing or lying position) by stretching the arms away from the head to increase mobility in that area.

The Pectoralis Major

This is a large, superficial (just under the skin) muscle that spans the front of the rib cage. It originates on the sternum and the outer two-thirds of the clavicle. From there, its fibers converge toward their insertion at the top of the humerus (the anterior part).

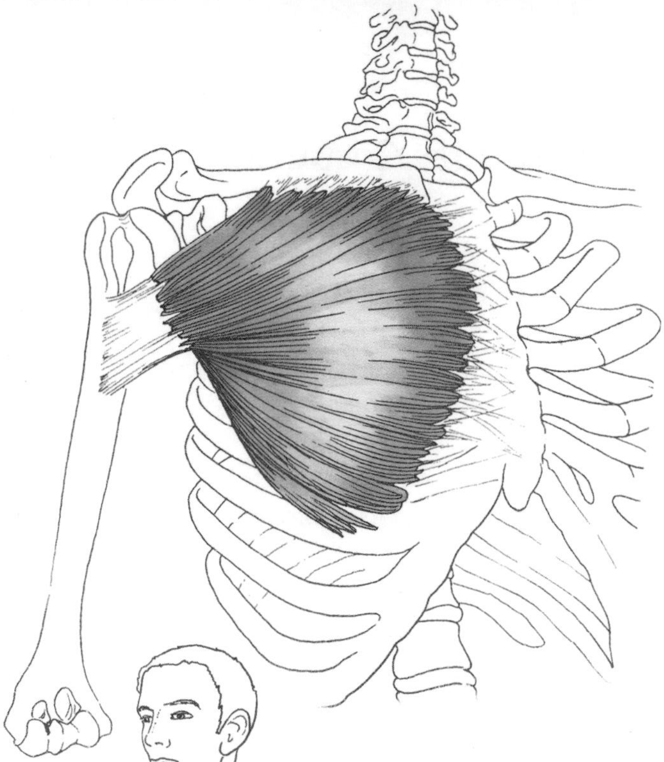

Its Action in Respiration

This muscle pulls the ribs up and to the back like a "bucket handle," which encourages an inhalation. It is the lower fibers (those that terminate on the fifth through eighth ribs) that are responsible for this action. At this level, it's a costal inspiratory muscle. We can easily feel this action at the front of the thorax, at the angle where the sternum and the ribs meet, separating and narrowing as we inhale and exhale. The upper fibers (clavicular) have the inverse effect; they serve to lower the clavicle and are therefore expiratory.

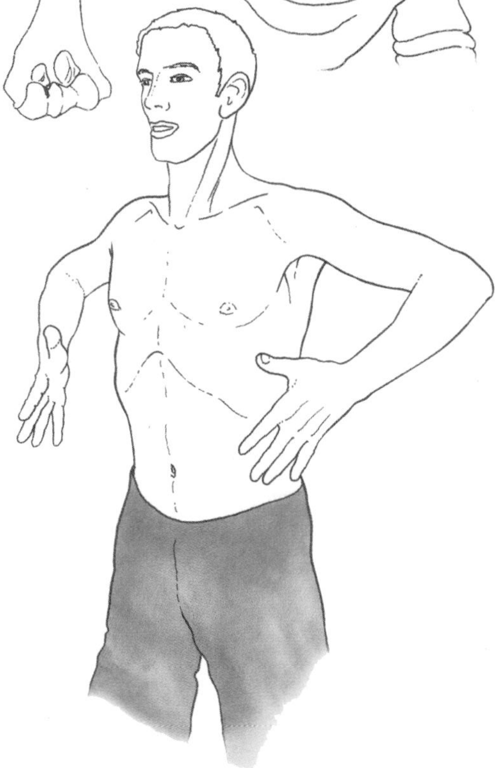

Its Role in Voice Work

This small inspiratory muscle can act to brake the exhalation and thereby mete out the subglottal air reserve.

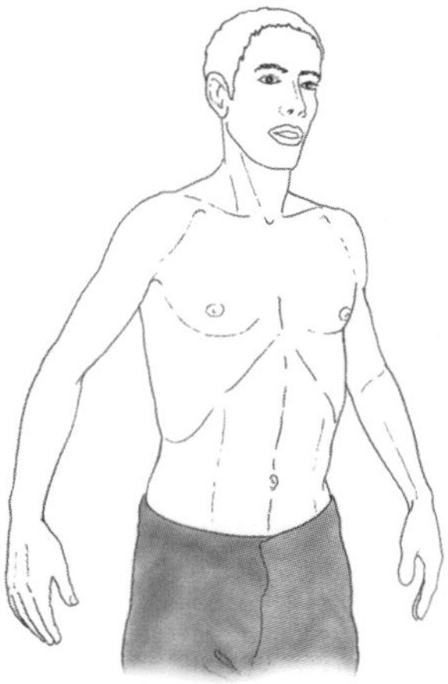

Release the Shoulders

Attention: If the shoulders are rounded forward, this muscle can be shortened and can't easily be recruited during an inhalation. In this case it might be useful to stretch the muscle by lengthening the arms in an outstretched (standing or lying) position.

The Levatores Costarum

These muscles are small and deep but numerous, and they are situated at the back of the rib cage. They arise from the transverse apophysis at each vertebral level. A short bundle of these muscle fibers descends toward the outside and terminates on the rib at the next level down, while a long bundle of fibers also descends toward the outside but terminates on the rib two levels below.

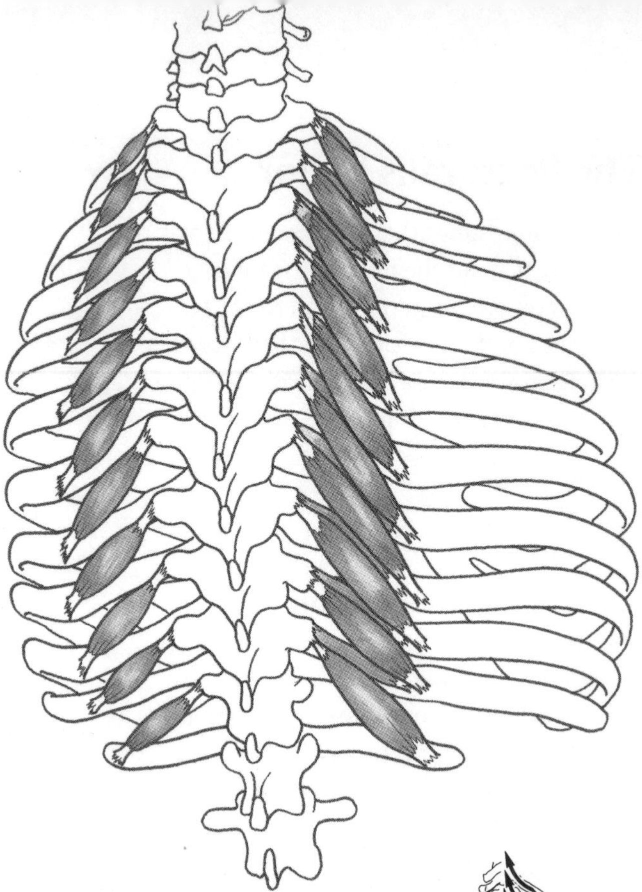

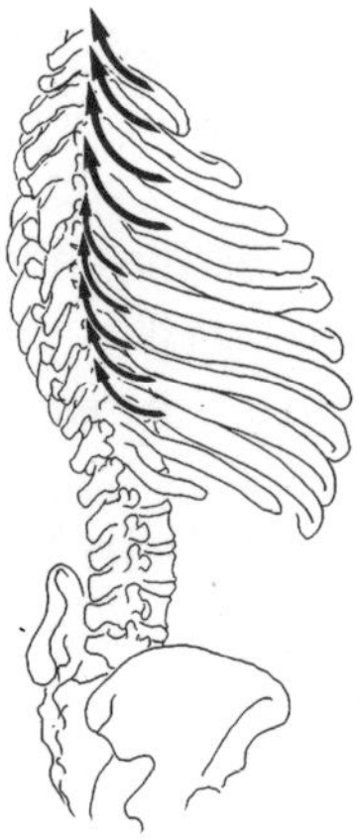

Their Action in Respiration

At each level, these muscles pull the ribs upward, and this "bucket-handle" effect encourages an inhalation. We can feel their action at the back of the rib cage close to the spine.

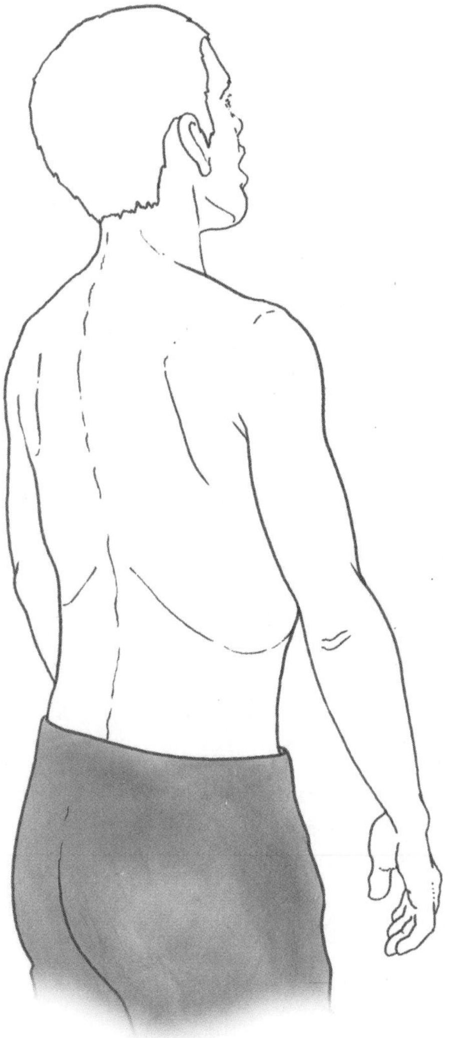

Their Role in Vocal Work

In vocal work, these inspiratory muscles can act to "brake" the exhalation and thereby mete out the subglottal air reserve.

Inhale into the Back

It's especially important to learn where these muscles are, and to develop them, if we tend to throw our chest (or even our head) forward during a vocal act. These muscles contribute to rebalancing the posture. We can find and develop them by flexing the trunk (in either a vertical or lying position), which will open the ribs in the back.

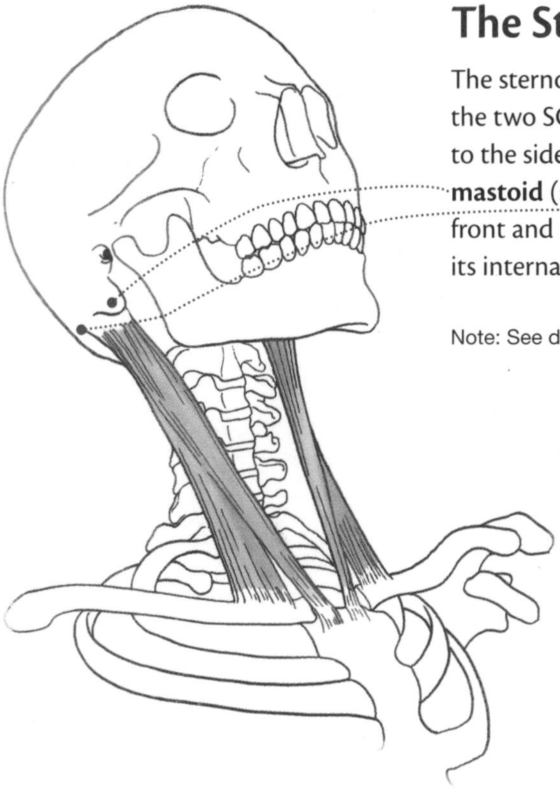

The Sternocleidomastoid (SCM) Muscles

The sternocleidomastoid is clearly visible at the front of the neck: the two SCM muscles form a "V" that runs from the sternum to the sides of the head. At the top, each SCM arises from the **mastoid** (see page 72) and the **occiput**. It descends toward the front and the midline of the neck to attach at the sternum, with its internal fibers attaching at the clavicle.

Note: See details on the palpation of the SCM on page 131.

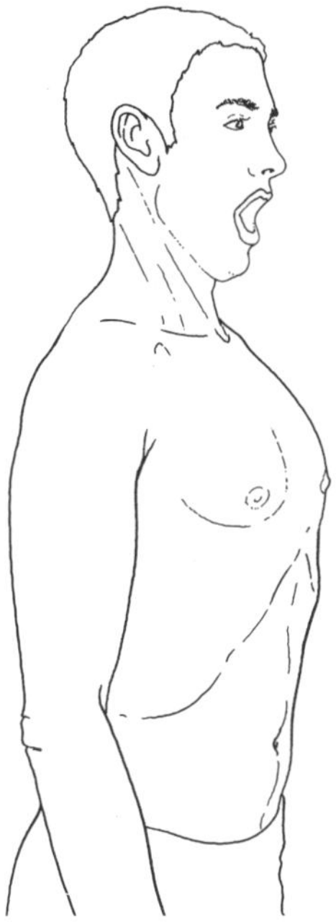

Their Action in Respiration

The SCM muscles lift the sternum, which in turn encourages an inhalation from the top part of the lungs.

Their Role in Vocal Work

In vocal work, these inspiratory muscles can act to "brake" the exhalation and thereby mete out the subglottal air reserve.

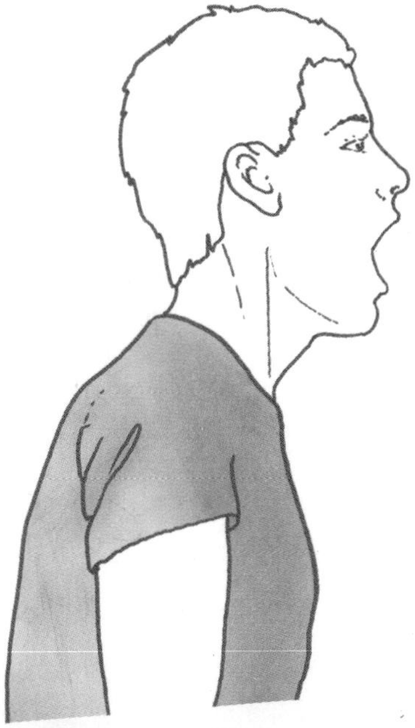

Moderate the Action of the SCM in Vocal Work

Recruitment of the SCM muscles during a vocal exhalation creates tension in the muscles of the neck, particularly in the front. And it is here that we find the extrinsic muscles of the larynx (see page 186), whose role is to balance the height of the larynx in relation to the mouth. So, when we need to put the costal inspiratory muscles into play, it is therefore preferable to recruit those that are situated lower on the rib cage.

The Scalenes

These three muscles are located just behind the SCM. They arise from the transverse processes of the cervical vertebrae from the axis to C7, and they form three bundles that run one behind the other alongside the cervical spine. The anterior and middle scalenes run obliquely and to the front and terminate on the first rib. The posterior scalene is more vertical and terminates on the second rib.

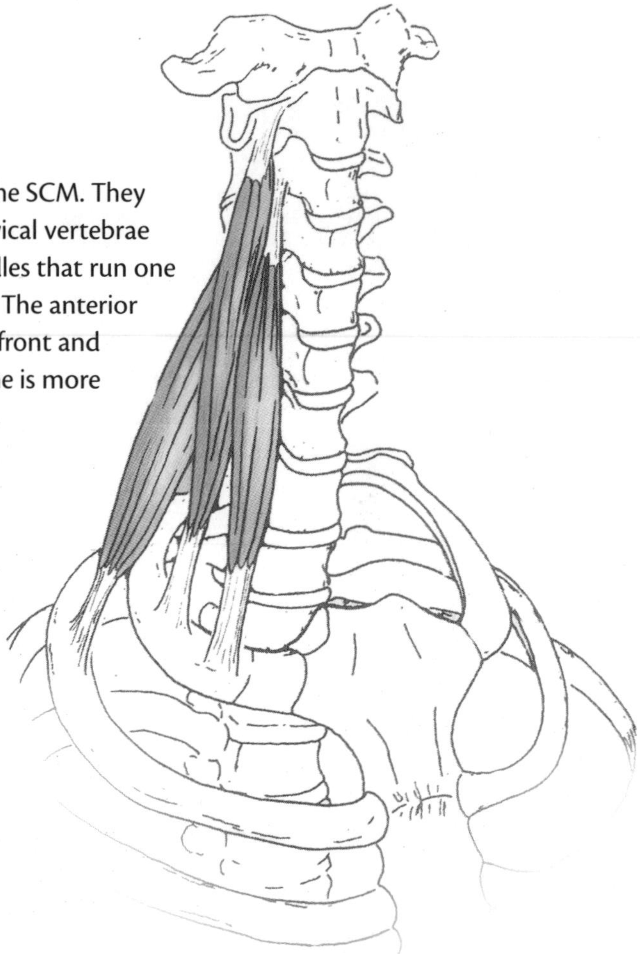

Their Action in Respiration

These muscles pull the first two ribs upward, encouraging an inhalation. There is not much pulmonary volume in this area, and so this is not a very effective movement when it comes to respiration. However, this is the starting point of a movement that is extended throughout the rib cage via the intercostal muscles.

Their Role in Vocal Work

In vocal work, as with all of the inspiratory muscles, the scalenes may act synergistically with other costal inspiratory muscles to "brake" the exhalation and thereby mete out the subglottal air reserve.

Moderate the Action of the Scalenes

Warning: In vocal work, these muscles, like the SCM, monopolize much of the cervical region and are often the cause of tension in the vicinity of the larynx.

Palpation of the SCM and Scalenes

You can feel these muscles very easily on an inhalation if you try to inhale as if you were sobbing. The SCM is very easy to identify if you tilt your head-neck-torso backward, with your fingers placed on each side of the neck. You can feel the scalenes halfway up the neck, just behind the SCM. If you palpate the neck with the contralateral hand, placing your index finger on the SCM, the middle finger will automatically be on the scalenes. Note: This is a fragile area, so palpate, but don't press!

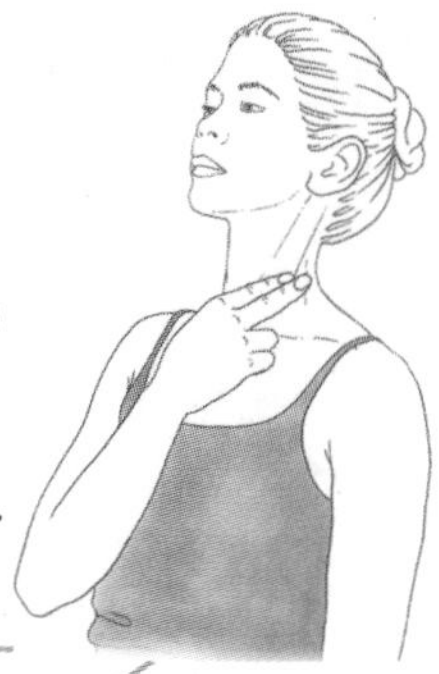

The Action of These Muscles Can Be Inverted

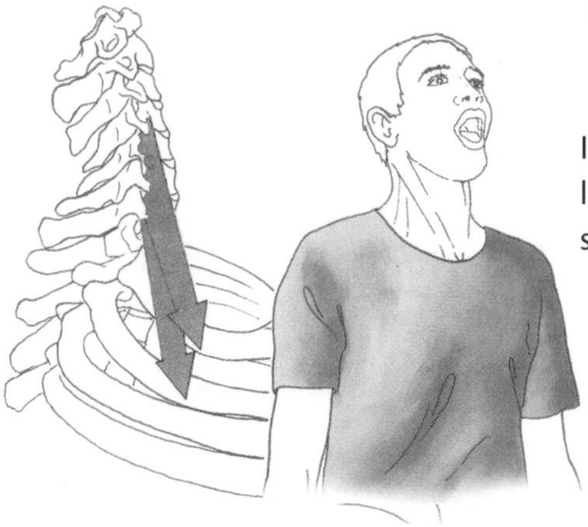

If the cervical vertebrae are already in lordosis, bilateral contraction of the scalenes will augment that lordosis.

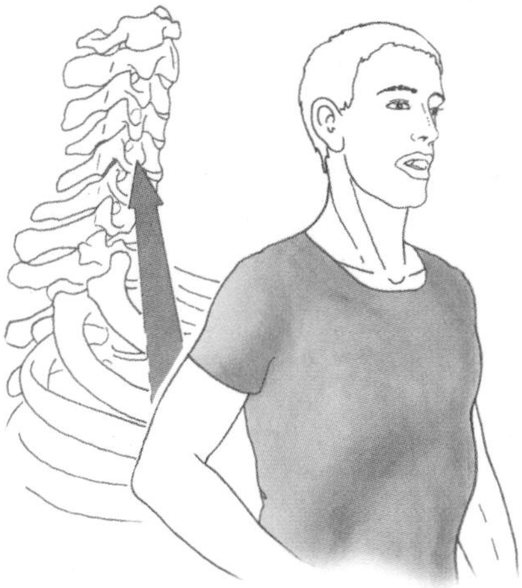

If the cervical vertebrae are aligned one on top of the other, the bilateral contraction of the scalenes will elevate the first two ribs.

Question & Answer

"I sing every day. I read that repeated inhalation high and in the front of the chest with the SCM causes tension in the neck. I know this is the case with me. I don't know how to decrease the action of the SCM and the scalenes with the inhalation before singing. Should I further develop subclavian or thoracic breathing?"

Yes, it is important to know how to take a low thoracic inhalation, or an abdominal inhalation, of course, so that the muscles of the larynx (see pages 161–70, 186–93) are very free and can properly perform their actions.

The Postural Muscles: Support for the Generator

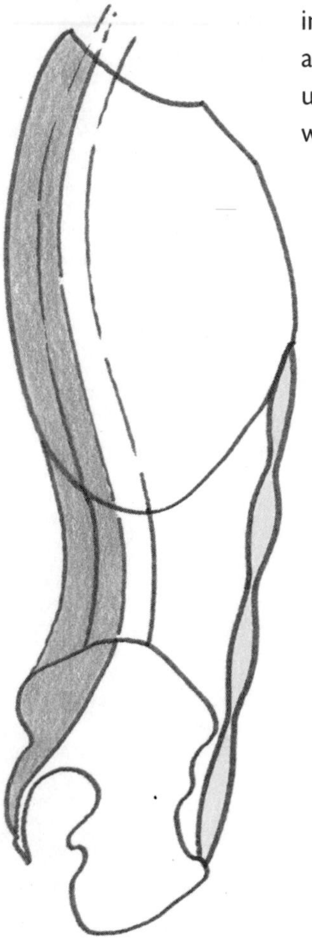

In vocalizations, the action of generating sound, as described in the previous pages, happens simultaneously with another—different—action that is required to keep the body upright. This is especially true in the trunk and the neck, where this function is performed by:

- the posterior muscles of the trunk, which we know globally as the **dorsal muscles**
- some muscles that we find in the front, but just next to the spine and very deep in the trunk*

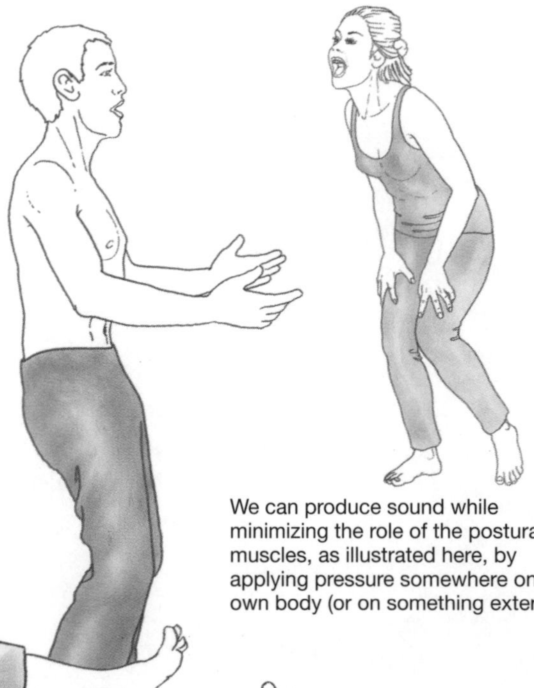

We can produce sound while minimizing the role of the postural muscles, as illustrated here, by applying pressure somewhere on our own body (or on something external).

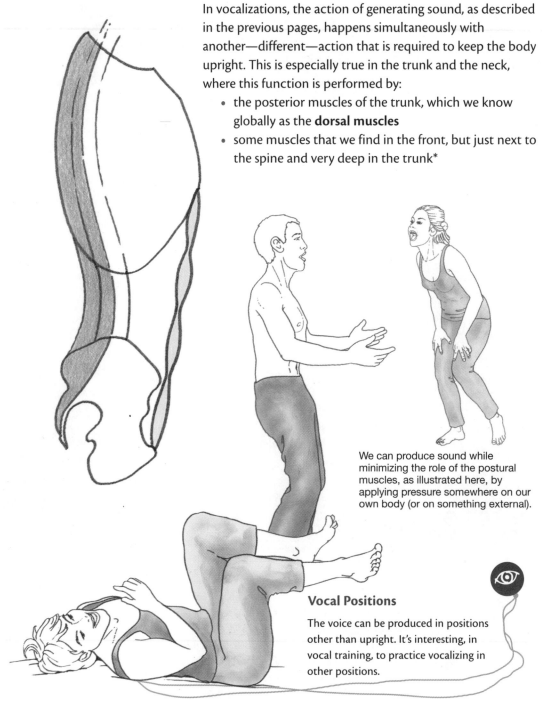

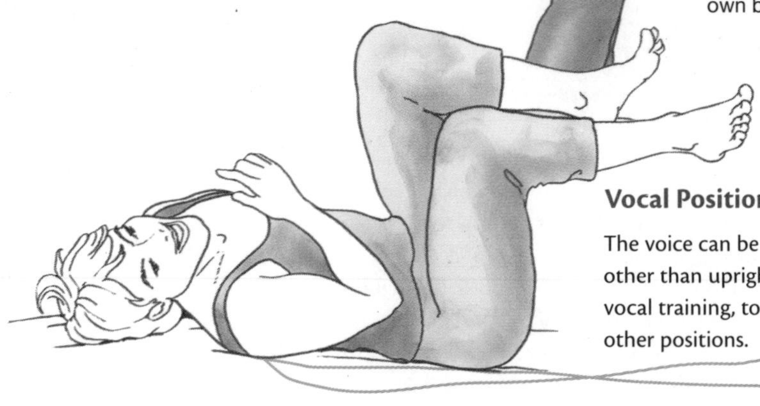

Vocal Positions

The voice can be produced in positions other than upright. It's interesting, in vocal training, to practice vocalizing in other positions.

*We will only discuss some of these muscles here. For more detailed description, see *Anatomy of Movement*, pages 74–87.

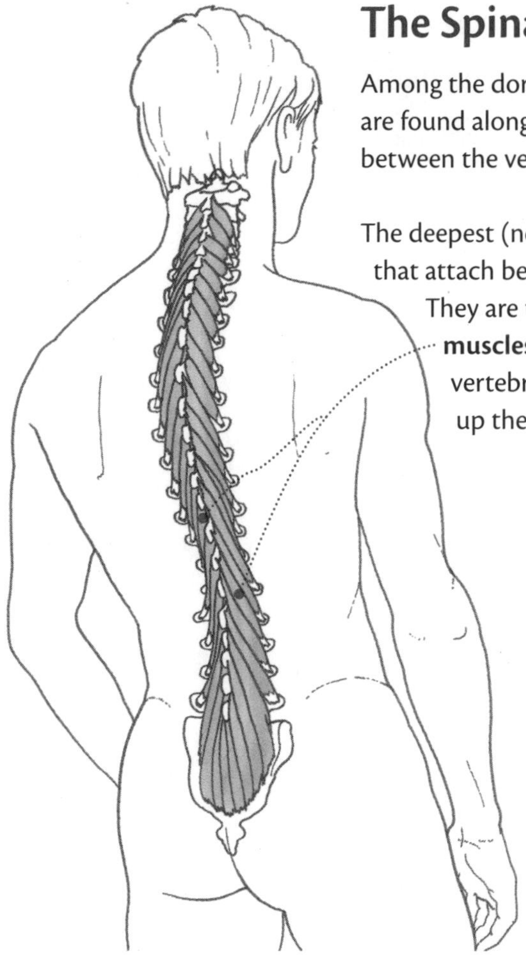

The Spinal Muscles

Among the dorsal muscles, those classified as spinal muscles are found along the length of the vertebral column and attach between the vertebrae.

The deepest (not shown) form a succession of small bundles that attach between the spinous or transverse processes. They are themselves covered by the **transversospinalis muscles,** which run from the transverse processes of vertebrae to the spinous processes of vertebrae farther up the trunk, in four overlapping layers.

The transversospinalis muscles form a chevron-like pattern along the length of the spine.

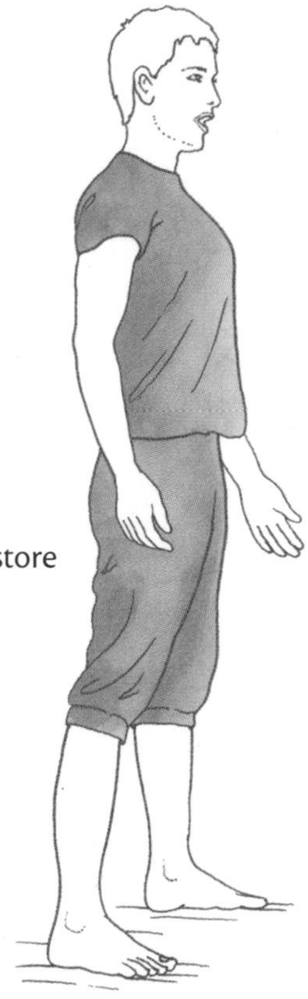

Their Action

When in a standing position, these muscles work constantly to keep and restore vertical alignment of the vertebral column (they don't serve to restore verticality of the rib cage or the pelvis; they are dedicated solely to axial retention).

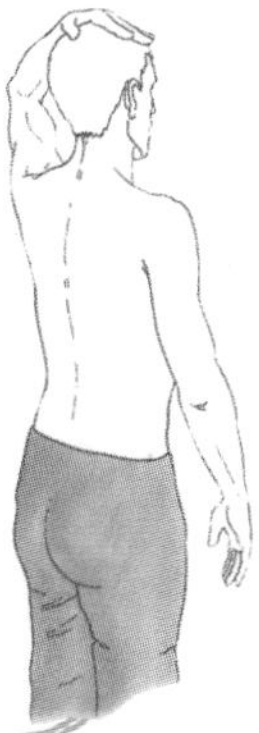

Toning the Spinal Muscles

A common exercise for toning the spinal muscles is to carry something on your head, or to place one hand (or both hands) on your head and push the top of your head against the resistance provided by your hands.

The Intermediate Muscles: The Long Muscles of the Back

Covering the spinal muscles we find the longest muscles, which arise at the bottom from a common thick muscle mass and attach on both the ribs and the vertebrae.

On each side of the spine:

- The **longissimus thoracis** runs from the transverse processes to the ribs.
- The **iliocostalis** runs more laterally toward the ribs.

These muscles are stronger than the spinal ones, and in a standing position, they are strong stabilizers of the trunk, especially when it moves out of vertical alignment.

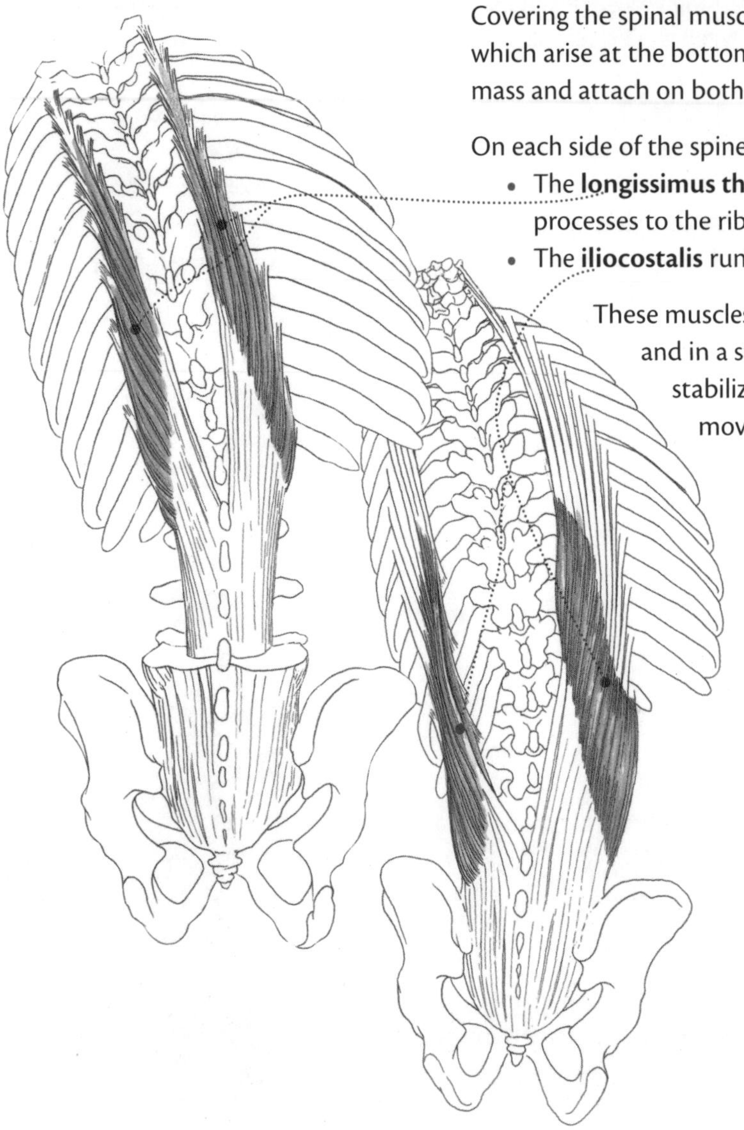

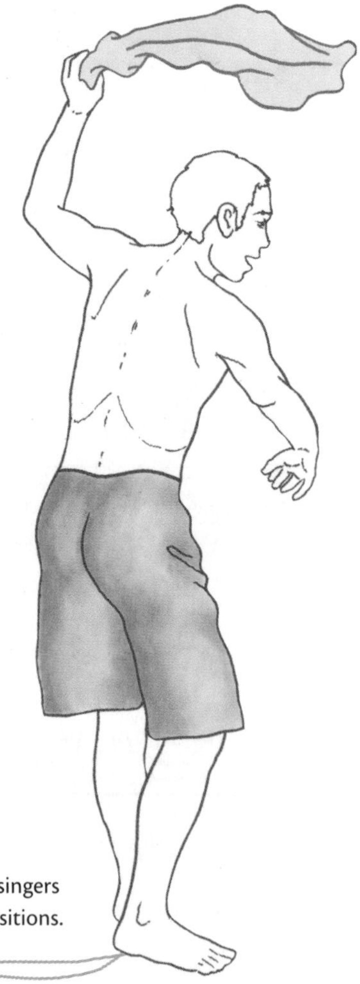

Voice and Movement

It is common today to see actors and singers working in non-vertical or inclined positions.

The Anterior Postural Muscles

At the front of the spine are muscles, closer to the spinal axis, that balance the action of the dorsal muscles.

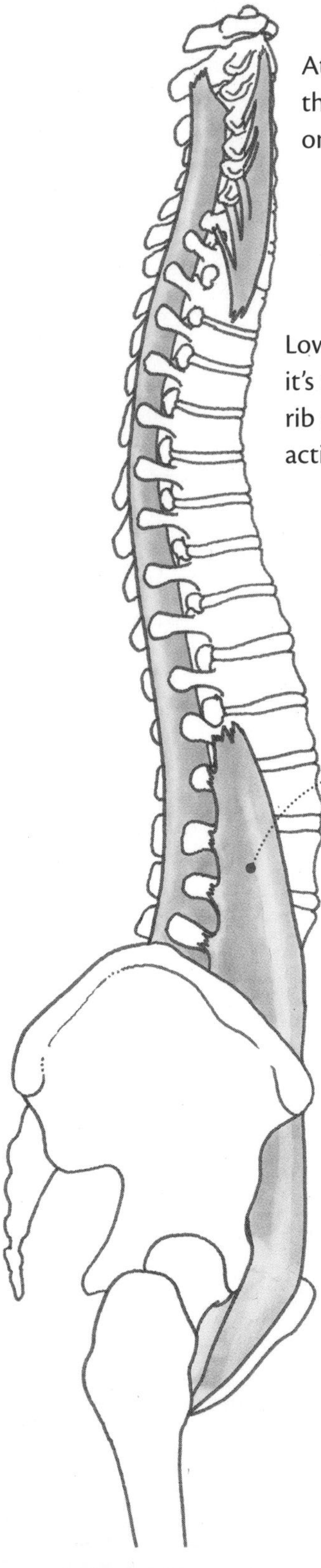

At the top, they are the **longus colli** and the **anterior cervical muscles** (detailed on pages 214–15).

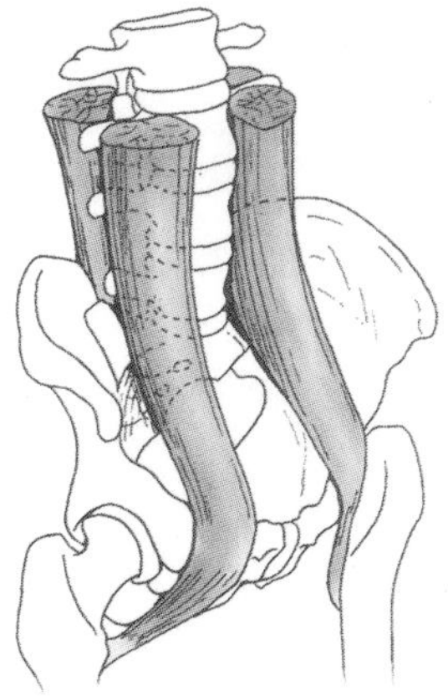

Lower, in the thoracic region, it's mainly the weight of the rib cage that balances the action of the dorsal muscles.

The dorsal muscles and the psoas essentially form four guy wires around the lumbar spine.

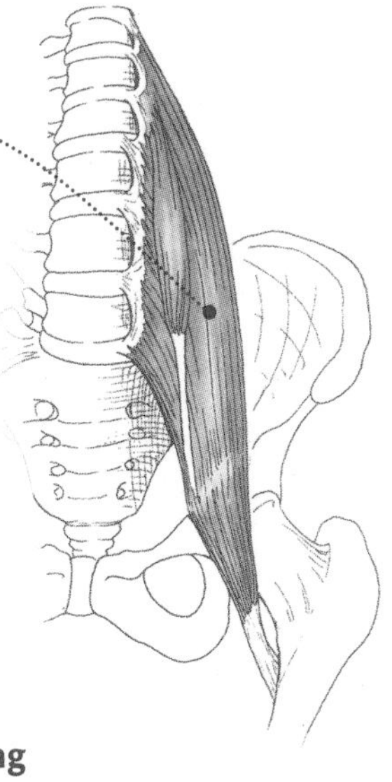

Lower, in the lumbar region, we find the **psoas** on each side. This muscle attaches to the twelfth thoracic vertebra and the five lumbar vertebrae (attaching to the sides of the vertebral bodies and on the transverse processes). It descends and crosses the pelvis, terminating on the lesser trochanter of the femur.

Toning and Relaxing

It may be worthwhile to tone the psoas, possibly even during vocal work, by lifting the knee (it's a hip flexor). But this is a muscle that's often too contracted, and in this case it's a good idea to relax or even stretch the psoas.

4
The Larynx

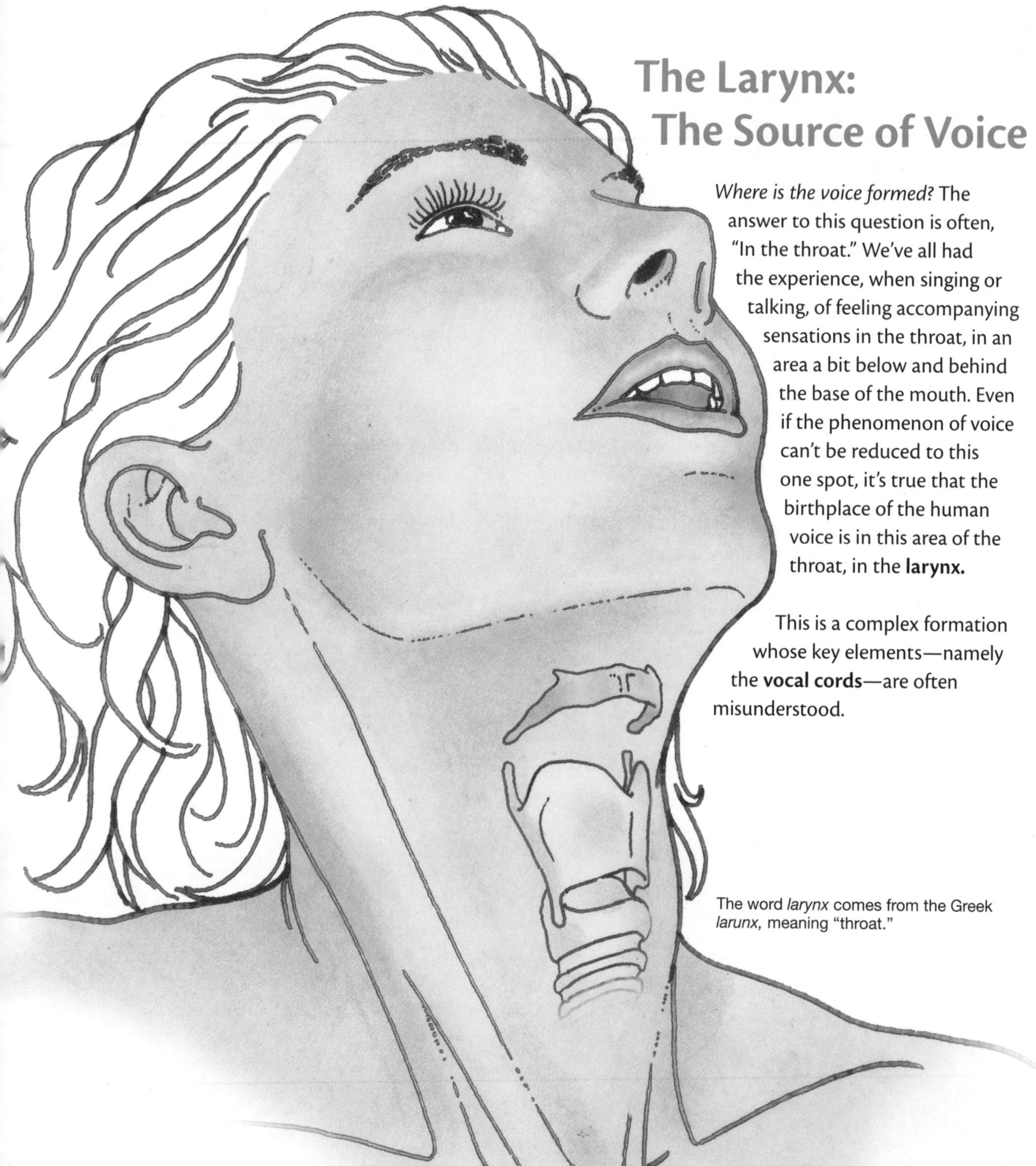

The Larynx: The Source of Voice

Where is the voice formed? The answer to this question is often, "In the throat." We've all had the experience, when singing or talking, of feeling accompanying sensations in the throat, in an area a bit below and behind the base of the mouth. Even if the phenomenon of voice can't be reduced to this one spot, it's true that the birthplace of the human voice is in this area of the throat, in the **larynx.**

This is a complex formation whose key elements—namely the **vocal cords**—are often misunderstood.

The word *larynx* comes from the Greek *larunx*, meaning "throat."

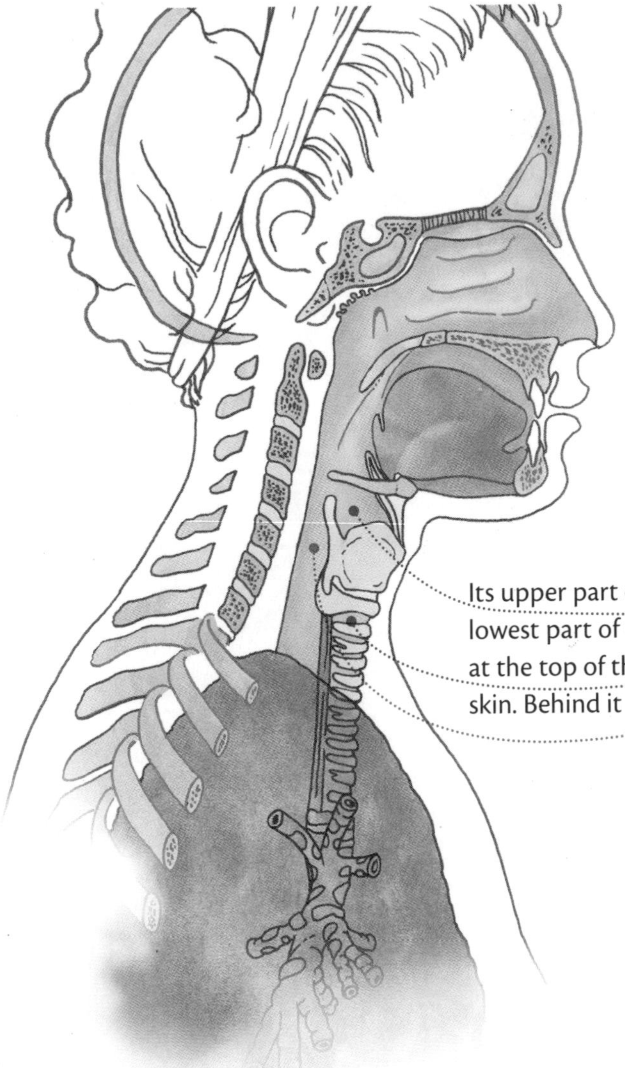

The larynx is located halfway up the neck at the level of the fifth or sixth cervical vertebra. Its exact position varies, both because the neck can change position and also because the larynx itself can move up or down in a range of about 6 centimeters in the neck.

Its upper part opens into the **pharynx** and follows the lowest part of the tongue. The lower part of the larynx is at the top of the **trachea**. At the front, it is just under the skin. Behind it is the **esophagus**.

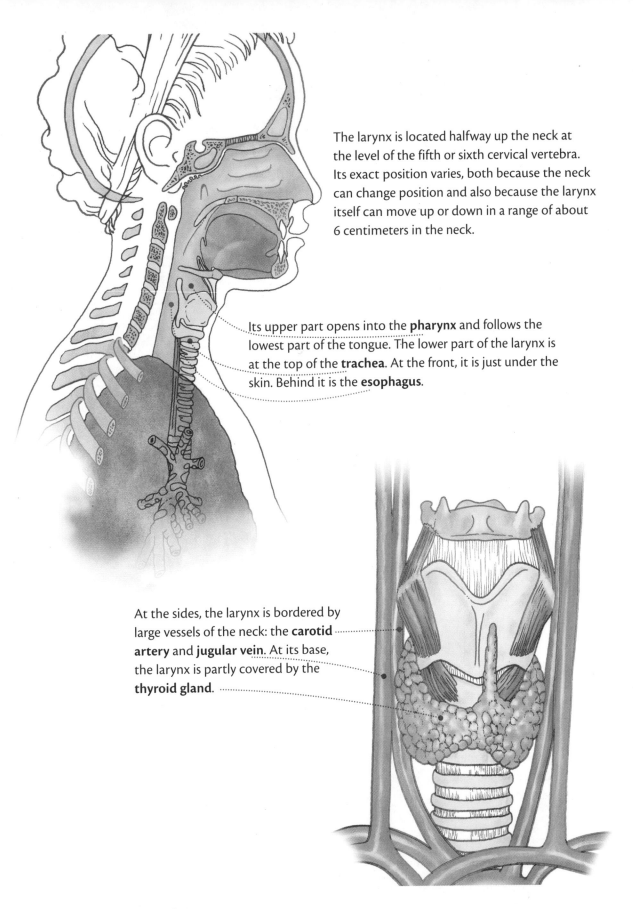

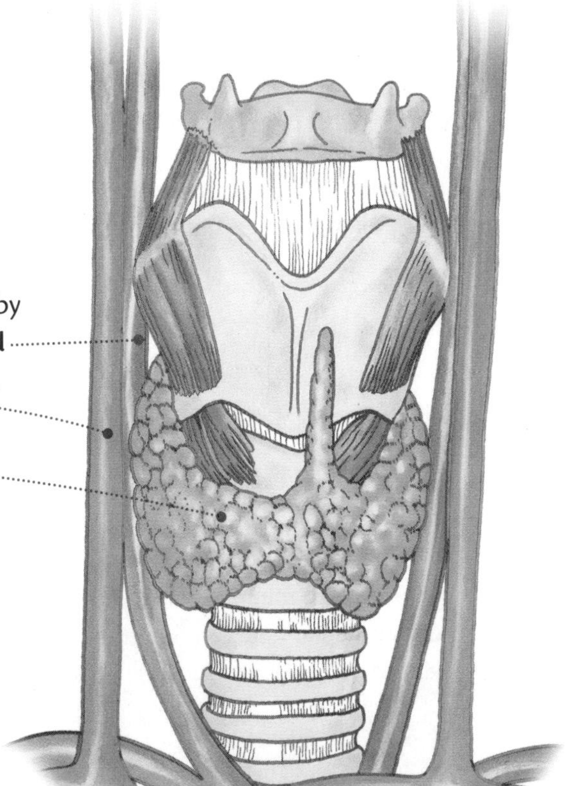

At the sides, the larynx is bordered by large vessels of the neck: the **carotid artery** and **jugular vein**. At its base, the larynx is partly covered by the **thyroid gland**.

The Laryngeal Cartilages

The larynx is shaped like a hollow organ. But unlike many viscera, it is not soft. Its shape has structure thanks to its reinforcing cartilaginous armature, which as a whole is called the **laryngeal skeleton**. It's thanks to this structure that, when the laryngeal tube is at rest, it can remain open, which is essential for breathing.

The laryngeal cartilages are more supple than bone, which gives the whole structure the quality of being rigid and at the same time flexible.* They are also articulated, and therefore the larynx is an organ whose form can be slightly modified, in particular when it comes under the influence of various pressures or from the pull of muscles that are attached to it.

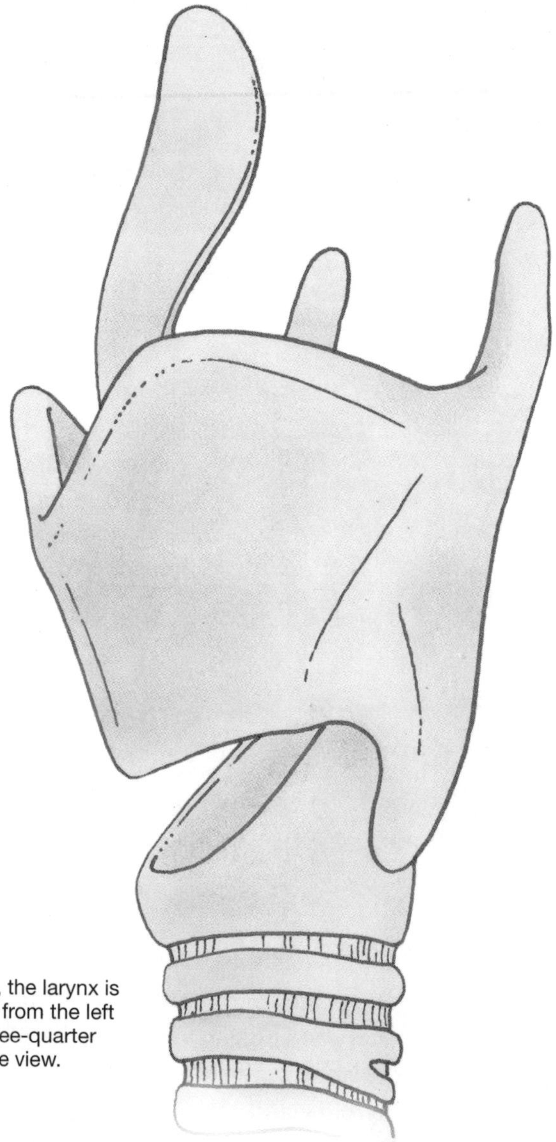

Here, the larynx is seen from the left in three-quarter profile view.

When looking at the entire larynx, the cartilages will not be visible because they are covered with membranes, muscles, and mucosa. However, we can learn to recognize, under the mucosa, certain parts of the cartilages in situ, for example when looking at a laryngoscopy (see page 185) or a sagittal section of the larynx (see following pages).

*We tend to lose this flexibility with age, because the laryngeal cartilages tend to ossify from the age of fifteen onward and become more rigid.

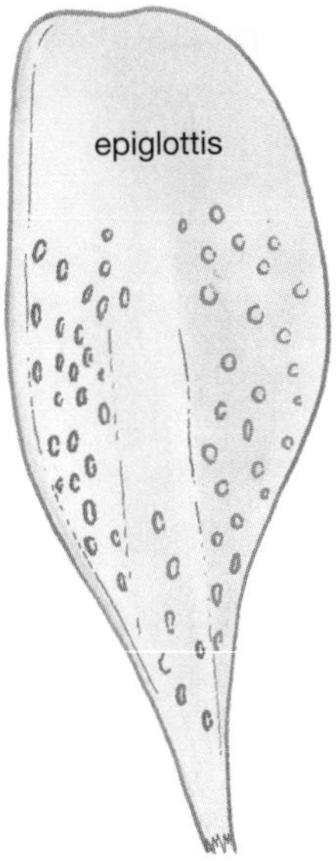

Five of the cartilages are large: the cricoid, the thyroid, both arytenoid cartilages, and the epiglottis. There are also two small cartilaginous processes and the two small cuneiform cartilages (sometimes called **Morgagni's cartilages**), which overhang the edge of the arytenoid. They are not detailed in this book.

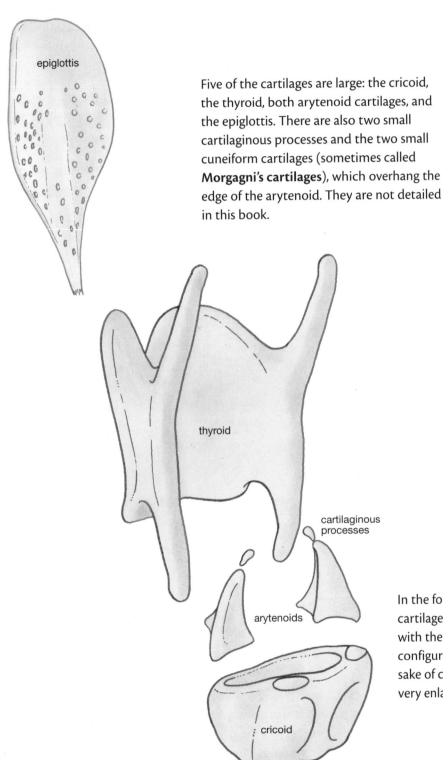

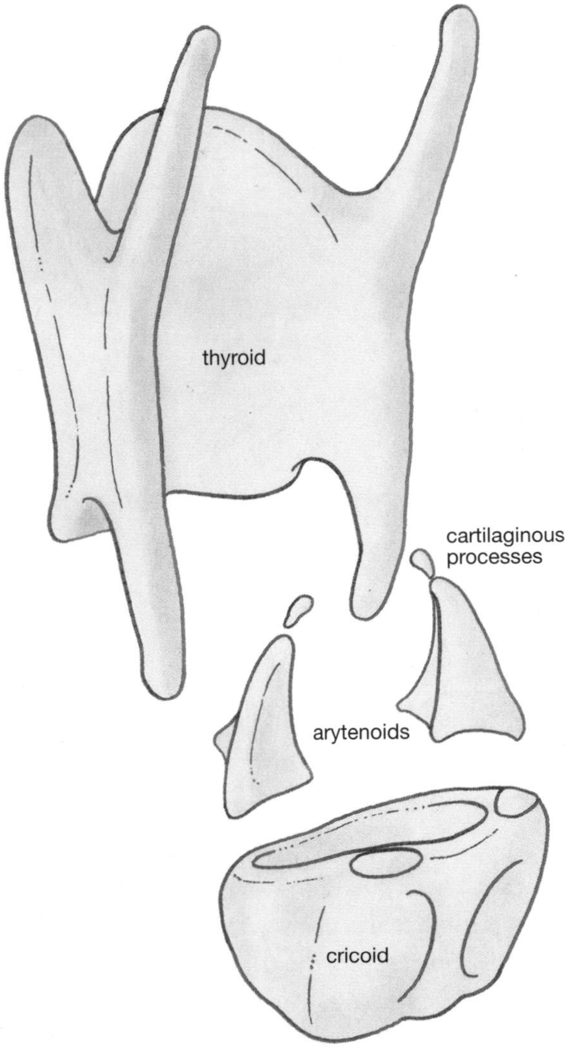

In the following pages, the cartilages are shown one by one with their varied surfaces and configurations. Note that, for the sake of clarity, the drawings are very enlarged.

The Cricoid Cartilage ("the Bottleneck")

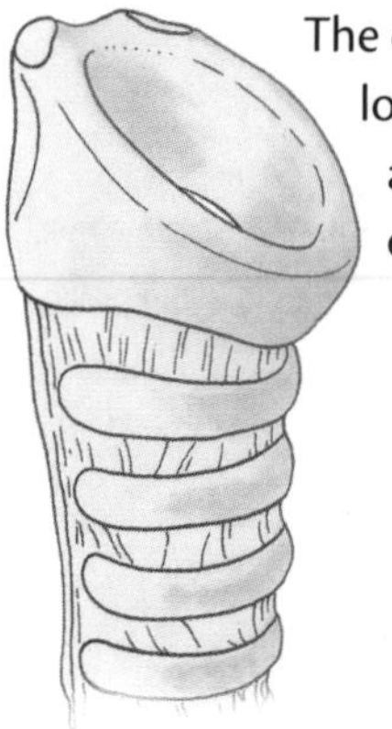

The cricoid is the lowest of the laryngeal cartilages. It's located at the top of the trachea and looks a bit like a ring around the trachea, but a little bigger, a bit like the neck of a bottle.

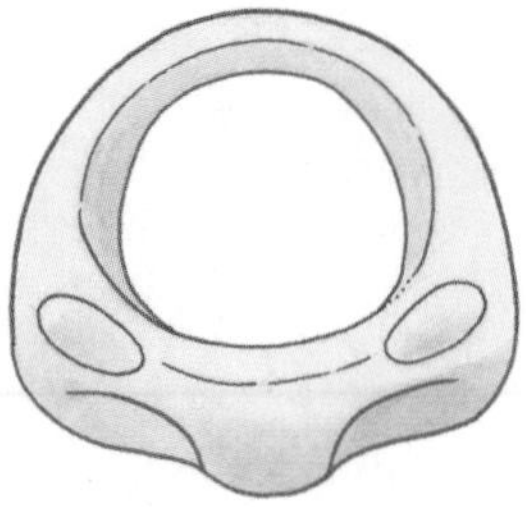

Details

Cricoid means "ring-shaped."

We sometimes compare the shape of the cricoid to that of a signet ring with the bezel toward the back. The "bezel" is higher and thicker and has two small, oval-shaped **articular surfaces** at the top.

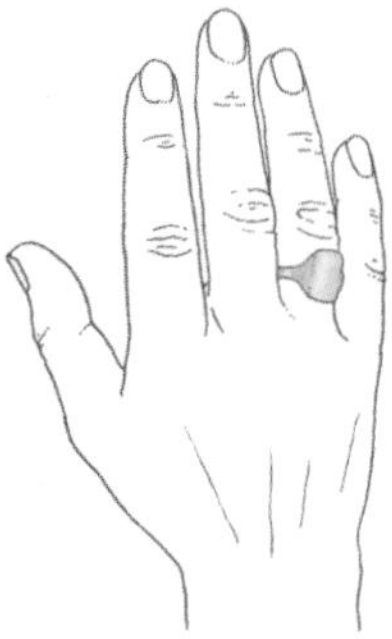

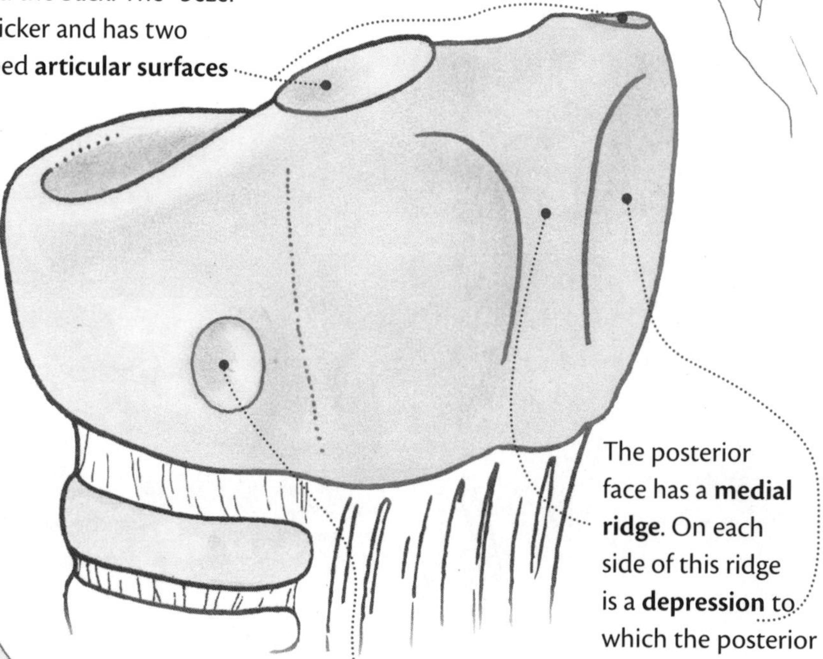

Each surface articulates with an arytenoid cartilage.

The posterior face has a **medial ridge**. On each side of this ridge is a **depression** to which the posterior cricoarytenoid muscle attaches (see page 162).

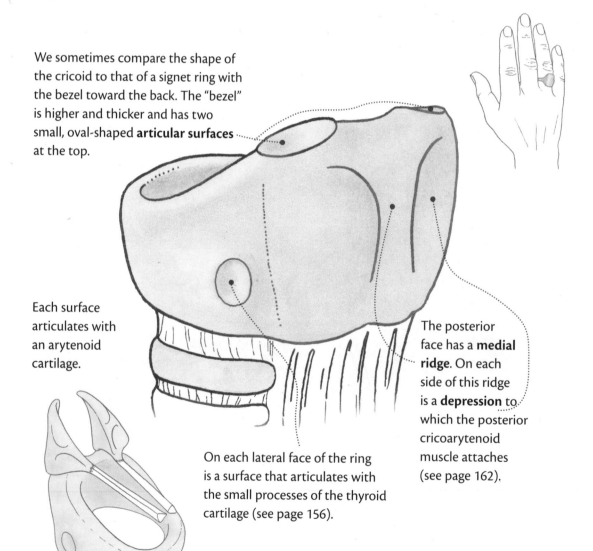

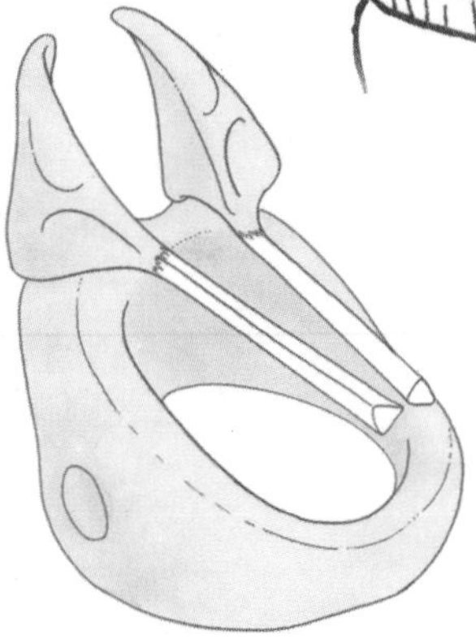

On each lateral face of the ring is a surface that articulates with the small processes of the thyroid cartilage (see page 156).

The front edge, called the **anterior arch,** has a projection on its anterior surface called the **cricoid tubercle.** The cricothyroid muscle attaches on each side (see page 166).

The lateral cricoarytenoid muscles attach on its superior borders (see page 167).

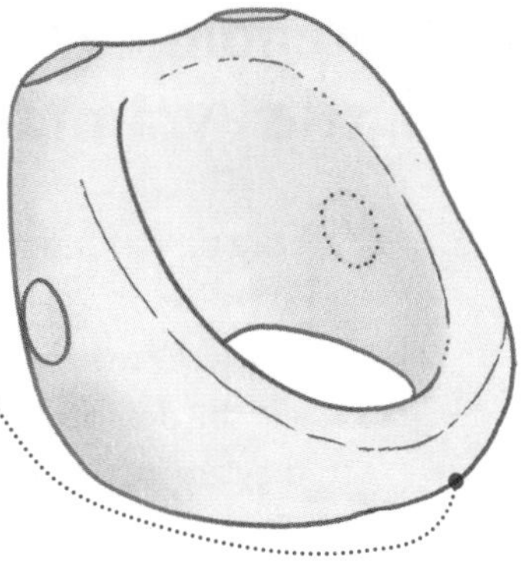

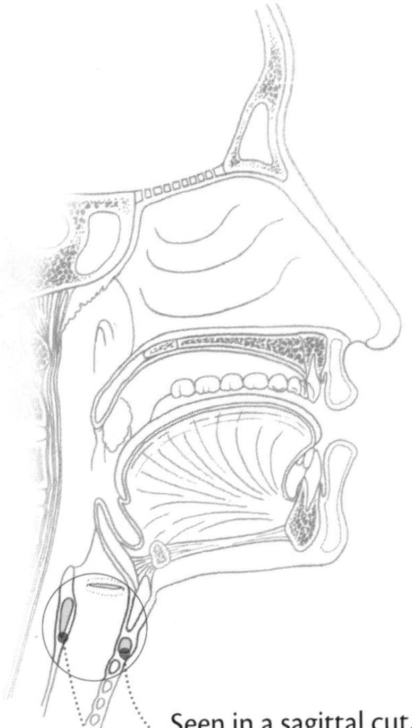

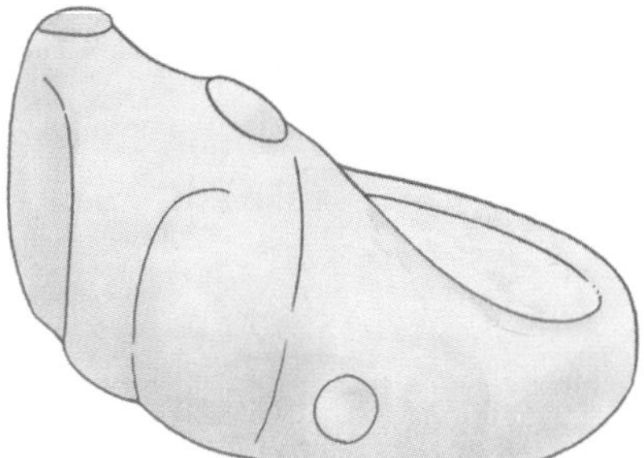

Seen in a sagittal cut, or from the top, we find the **anterior arch** (small) at the front and the **cricoid "bezel"** high and thick at the back.

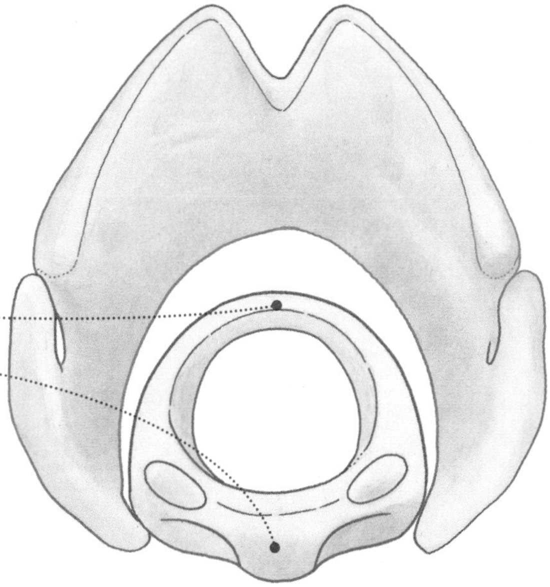

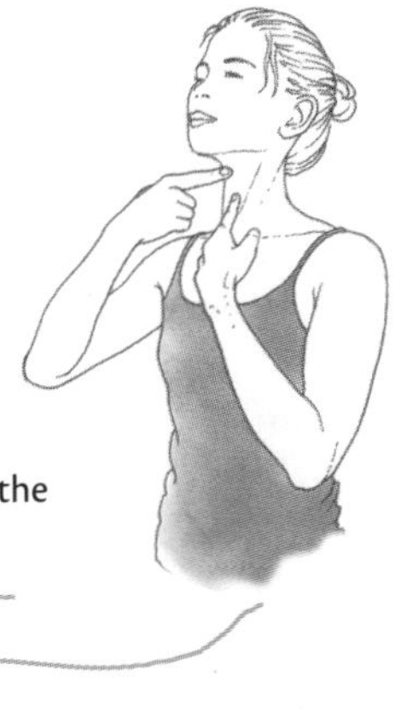

Palpate the Cricoid

The front part of the cricoid cartilage can be palpated (with caution) to the top of the trachea or below to the edge of the thyroid cartilage.

The Arytenoid Cartilages ("the Little Pyramids")

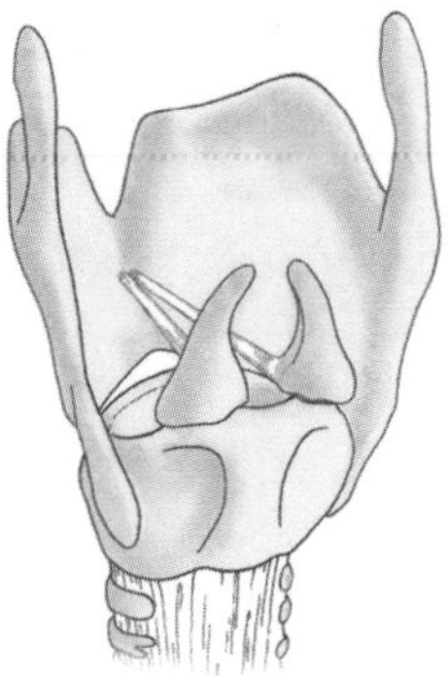

These are tiny structures arranged side by side on the upper edge of the cricoid. Note that on these two pages they are magnified to make the description easier to follow. In fact, they are only about 0.5 centimeters high. They are each shaped like a tetrahedron (four-sided pyramid), with the following:

A Funnel . . .

Arytenoid means "funnel-shaped."

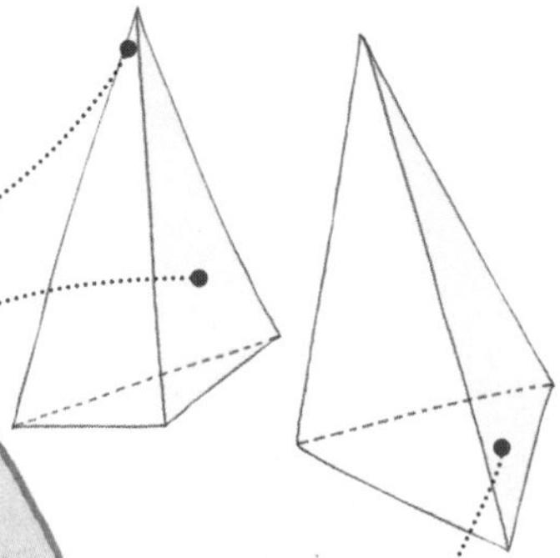

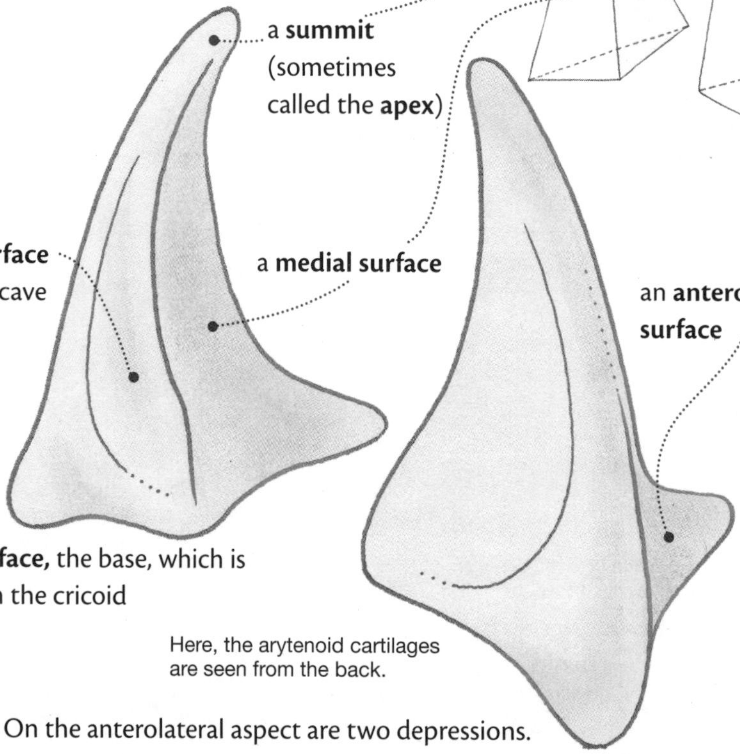

a **summit** (sometimes called the **apex**)

a **posterior surface** that's a bit concave

a **medial surface**

an **anterolateral surface**

an **inferior surface,** the base, which is in contact with the cricoid

Here, the arytenoid cartilages are seen from the back.

On the anterolateral aspect are two depressions.

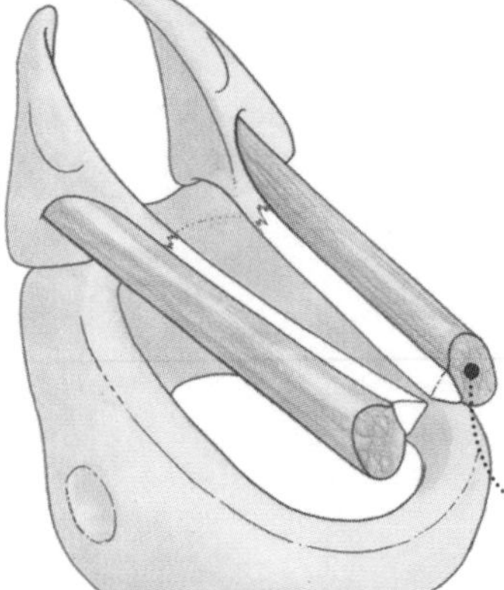

One occupies the top half, and here the vestibular ligament attaches

The other, separated by a small ridge, forms a hemispherical fossa near the base, and this is where the **vocal muscle** (the internal thyroarytenoid) attaches (see page 168).

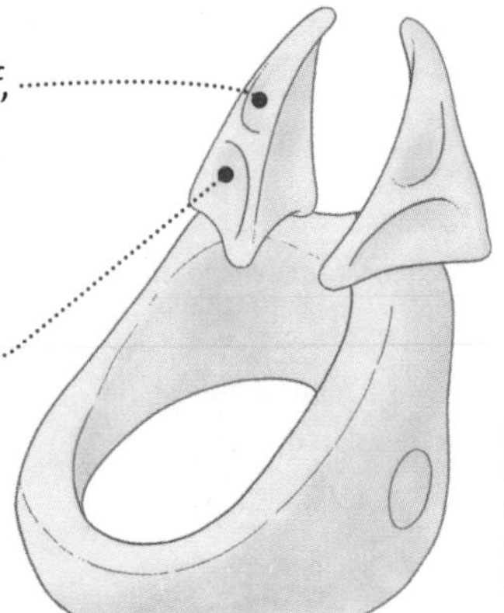

Here, the arytenoid cartilages are seen from the front.

At the base of the arytenoids are two characteristic processes (projections):

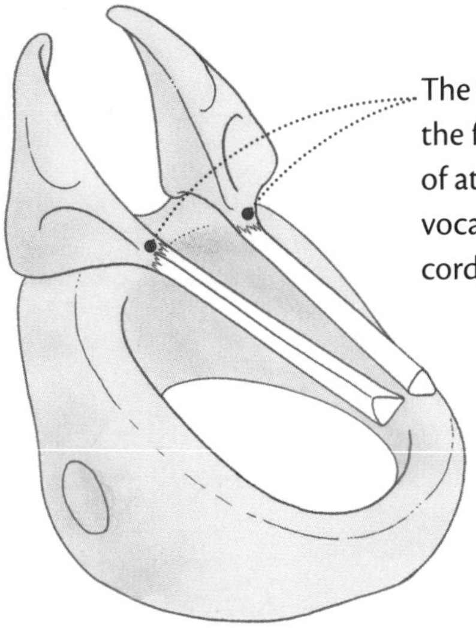

The **vocal process,** at the front, is the point of attachment for the vocal ligament (vocal cord; see page 151).

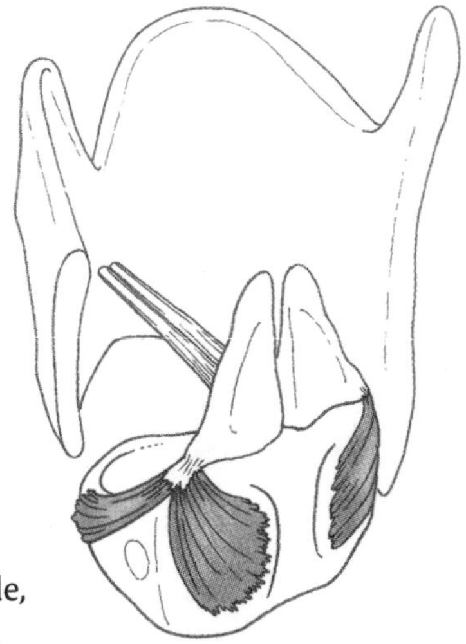

The **muscular process,** at the back and toward the outside, is the point of attachment for the cricoarytenoid muscles (see pages 162 and 167).

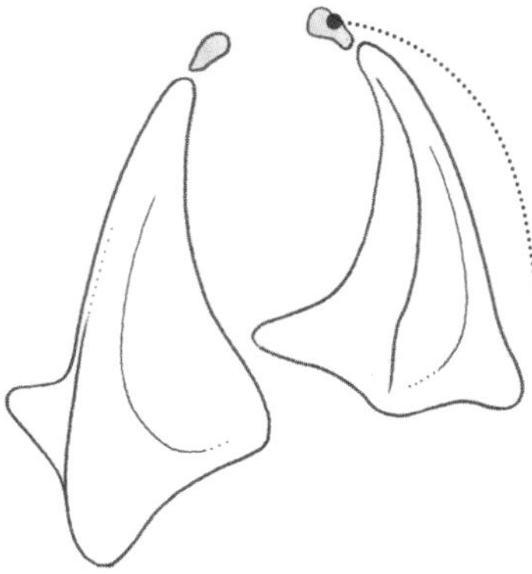

The summit articulates with a tiny cartilaginous nodule known as the **corniculate cartilage**.

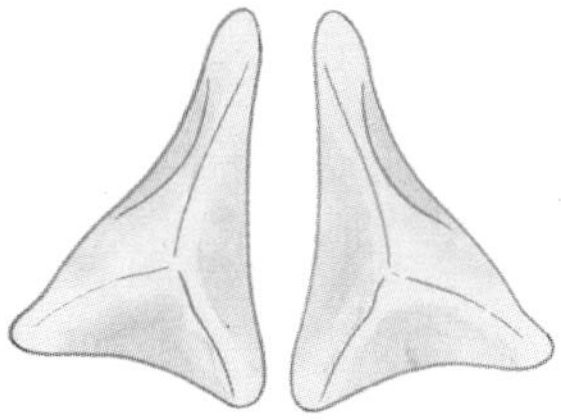

Here, the arytenoid cartilages are seen from above.

Looking at the larynx from above, we see that the laryngeal mucosa forms a small bulge where each arytenoid is located.

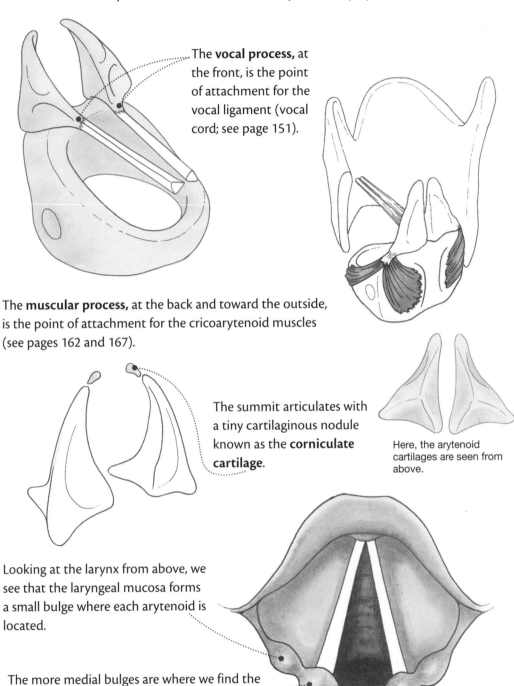

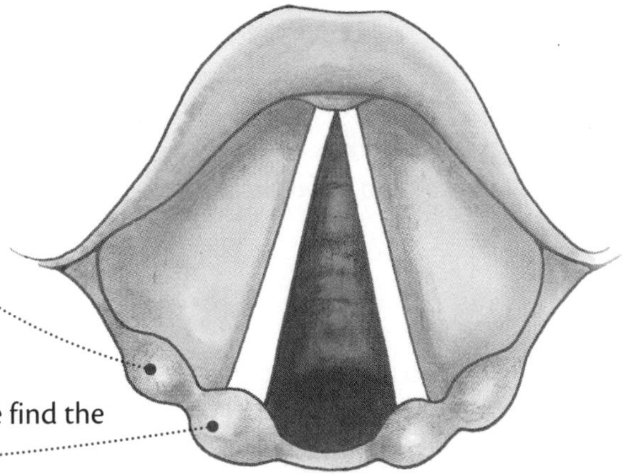

The more medial bulges are where we find the corniculate cartilages.

The arytenoid cartilages are the most mobile of the laryngeal cartilages (see page 160).

The Thyroid Cartilage ("the Shield")

This is the largest of the laryngeal cartilages, and it is shaped a bit like the cover of an open book, or a butterfly.

Attention!

Do not confuse the thyroid cartilage with the gland of the same name. The thyroid gland is lower, in front of the cricoid cartilage.

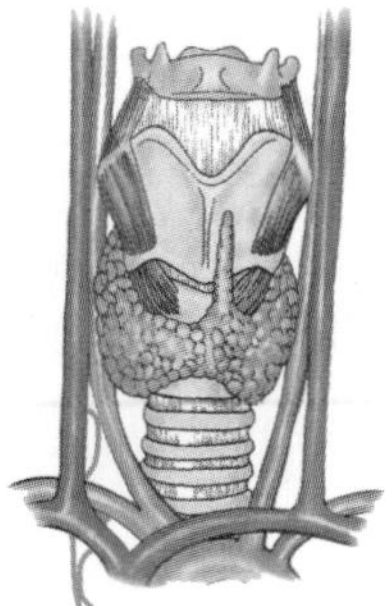

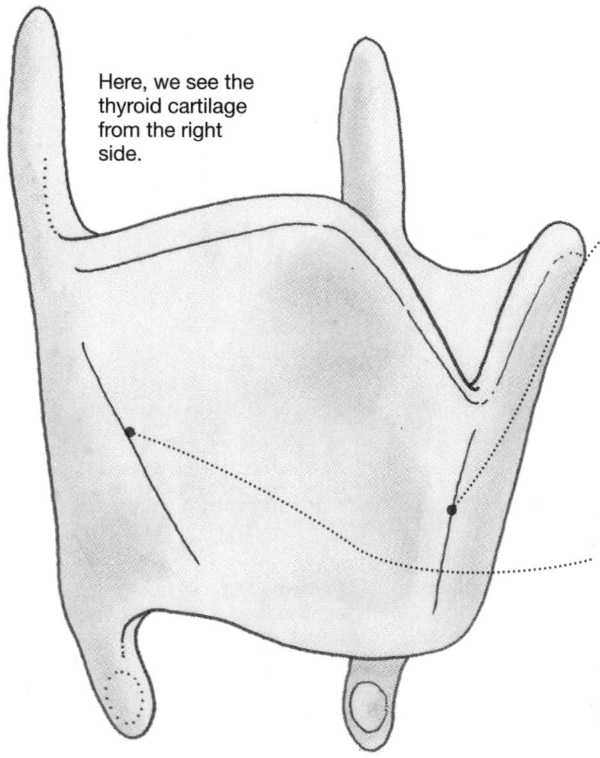
Here, we see the thyroid cartilage from the right side.

It is composed of two plates that come together at an angle in the front (which varies according to sex: 90 degrees in men, 120 degrees in women). The external face of the angle forms the "**Adam's apple,**" easily detectable in men, but it also exists in women, although it is smaller and less prominent.

The external surfaces of the plates have an oblique **ridge** that runs downward and toward the front, where two muscles are attached: the thyrohyoid muscle (which runs upward, toward the hyoid bone), and the sternohyoid (which runs downward, toward the sternum).

The size of the thyroid cartilage corresponds to the overall size of the larynx and its vocal cords.

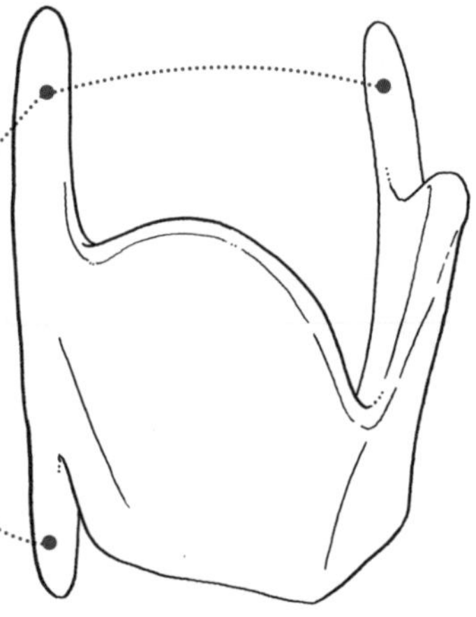

At the back, the two plates are elongated by both superior and inferior projections:

- At the top, the **superior horns** allow for articulation with the hyoid bone.
- At the bottom, the **inferior horns** have small articular surfaces on their internal surfaces that articulate with the cricoid cartilage (see page 156).

Skin Deep

We can feel the thyroid cartilage just below the skin about halfway up the neck. It is more visible in men (the angle is more prominent and the cartilage a bit bigger). You can feel the plates toward the sides of the throat and the edge in front where the plates come together.

Warning: Palpate the thyroid cartilage very carefully, without too much pressure.

The Shield

Thyroid comes from the Greek *thyreoides,* which means "in the shape of a shield."

The superior border of the thyroid cartilage has a notch, sometimes visible under the skin, called the **thyroid notch**.

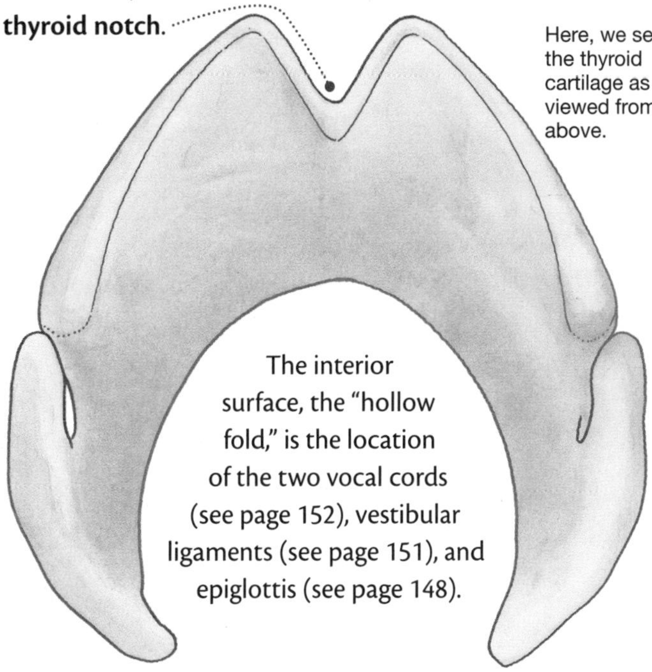

Here, we see the thyroid cartilage as viewed from above.

The interior surface, the "hollow fold," is the location of the two vocal cords (see page 152), vestibular ligaments (see page 151), and epiglottis (see page 148).

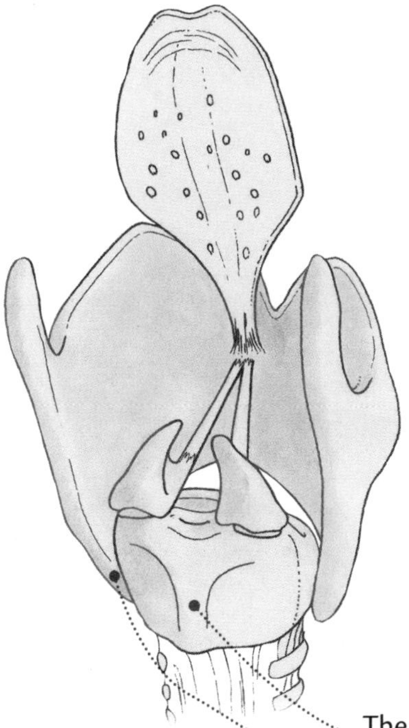

The thyroid cartilage articulates with the **cricoid cartilage** at the **inferior horns** (see page 146).

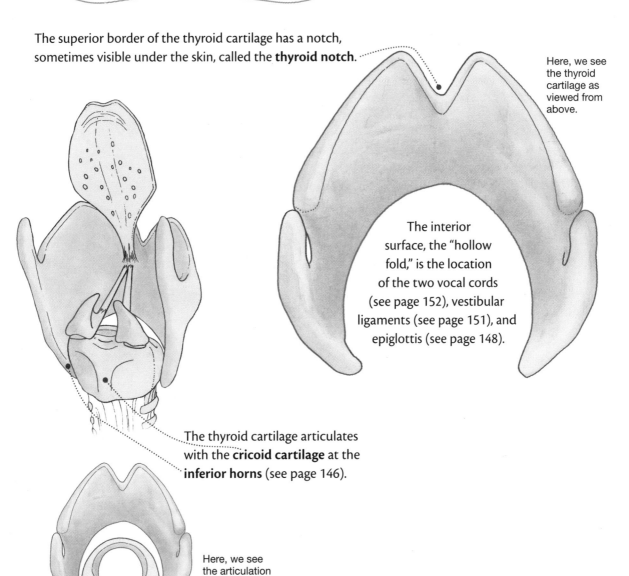

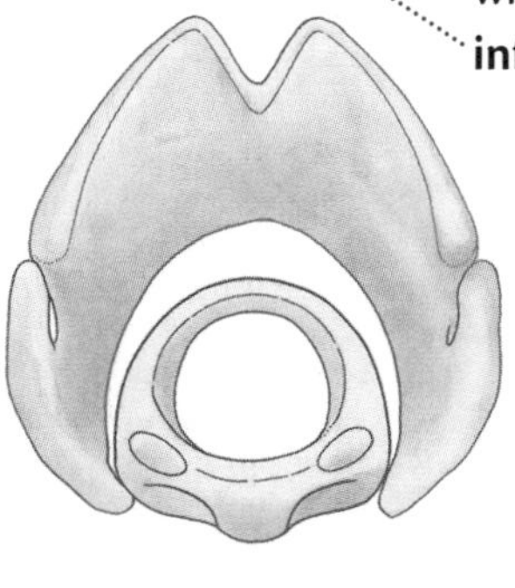

Here, we see the articulation with the cricoid cartilage from above.

The Epiglottis ("the Cover")

This is an oval cartilage, shaped like a chip or petal, that overhangs the glottis.

In profile, it is curved like a flattened "S."

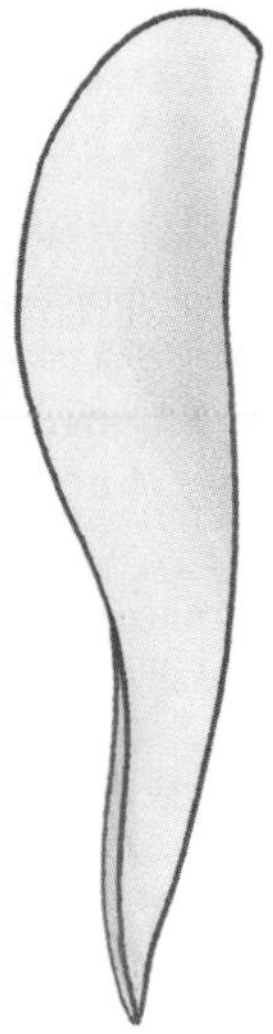

The superior part ends in two rounded points. It's mobile.

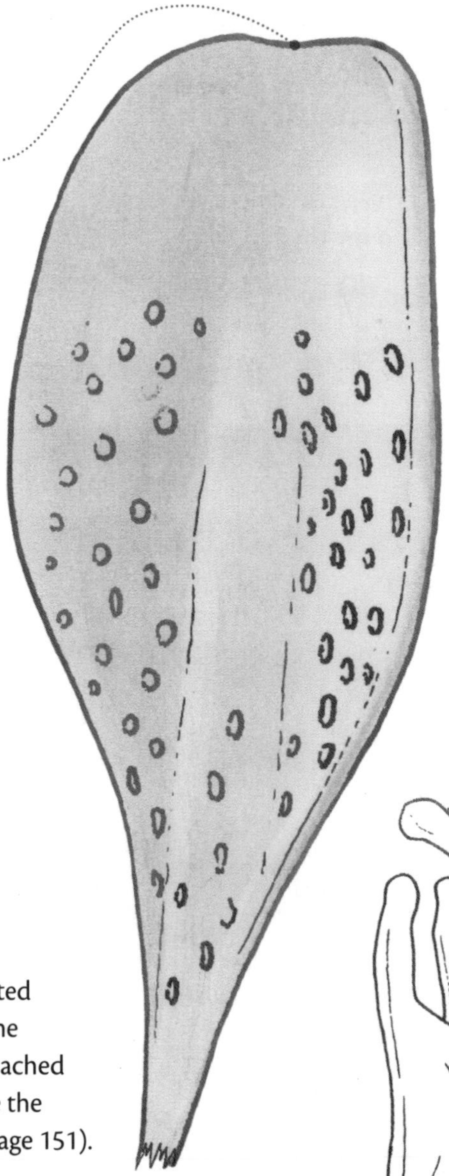

Its posterior surface (shown here) is riddled with small holes.

The lateral borders of the epiglottis provide the insertion points for the quadrangular membrane.

The inferior part is situated in the "hollow fold" of the thyroid cartilage and attached by a ligament just above the ventricular bands (see page 151).

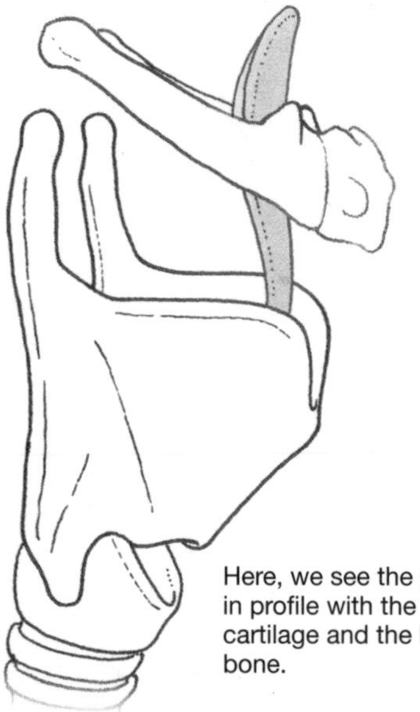

Here, we see the epiglottis in profile with the thyroid cartilage and the hyoid bone.

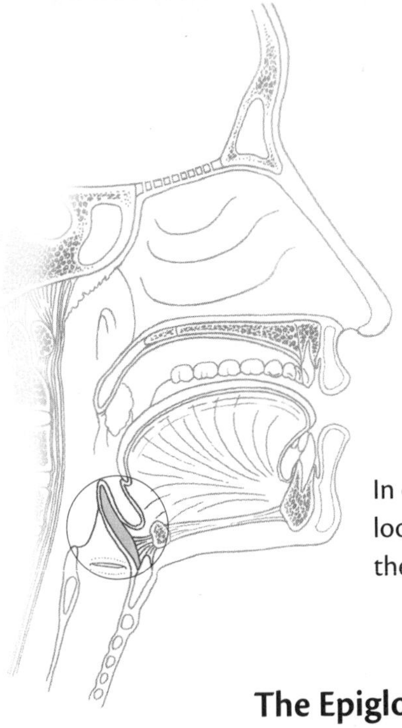

In cross-section, the epiglottis looks like a thin blade behind the hyoid bone.

The Epiglottis and Swallowing

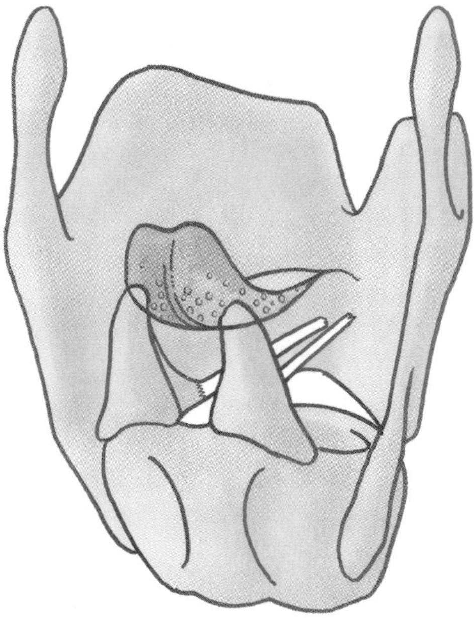

When we swallow (saliva, water, food) . . .

- We pull the tongue back (especially the back part).
- The larynx rises 2 to 3 centimeters.

These actions move the epiglottis, which folds horizontally over the laryngeal orifice, like a lid. It then prevents anything from passing into the larynx, especially food (see page 170).

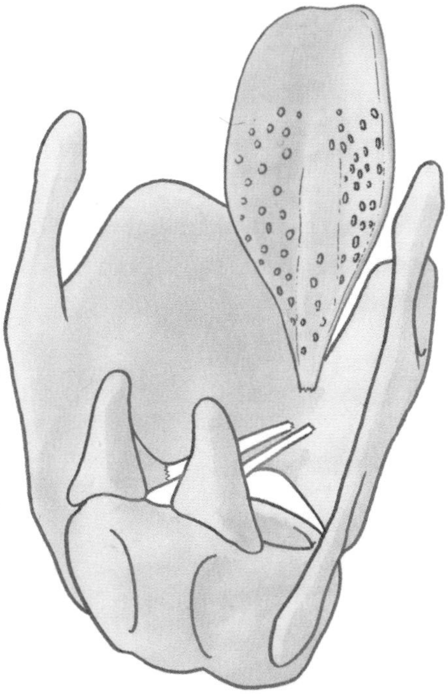

When we're not swallowing . . .

- The back of the tongue moves forward.
- The larynx descends a little, and the epiglottis moves into an almost vertical position, like an open lid.

Now air can pass.

When food or water accidentally passes through the larynx, when the epiglottis is in the "open" position, we say it went down the "wrong way."

Ligaments and Membranes

The cartilages of the larynx are held in place by ligaments and membranes.

Specific Ligaments Link the Larynx to Neighboring Structures

Some ligaments and membranes connect the larynx to neighboring structures. The **glossoepiglottic ligaments** connect the epiglottis to the mucous membranes at the rear of the tongue. The **pharyngoepiglottic ligaments** (not shown) connect the lateral edges of the epiglottis to the mucosa of the pharynx.

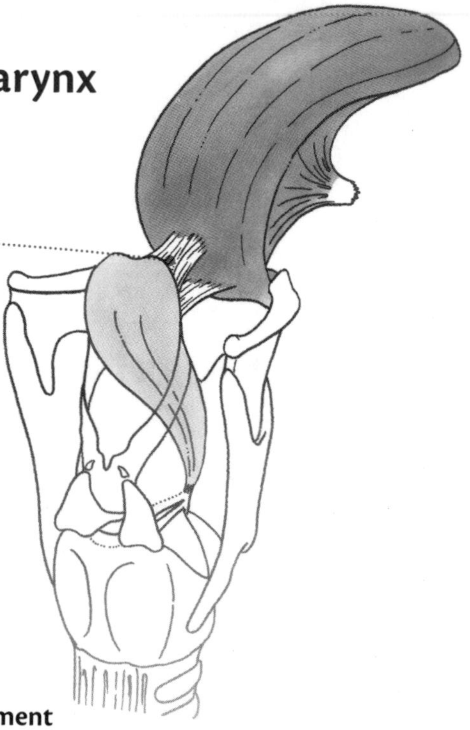

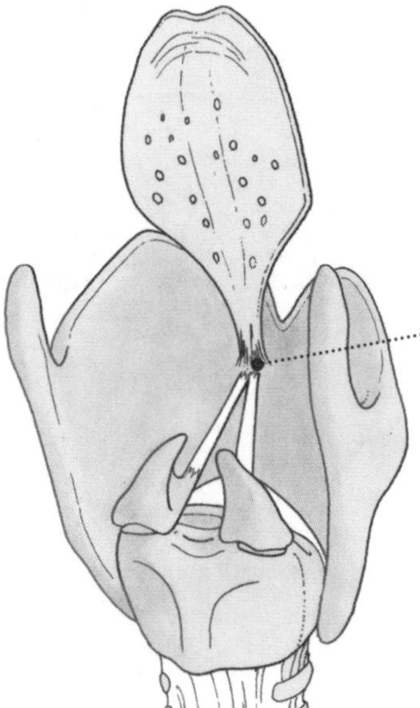

The **thyroepiglottic ligament** attaches the base of the epiglottis to the interior of the thyroid cartilage at a re-entry angle.

The fibrous **thyrohyoid membrane** connects the thyroid cartilage to the hyoid bone.

A fibrous band joins the inferior border of the cricoid cartilage to the highest ring of the trachea.

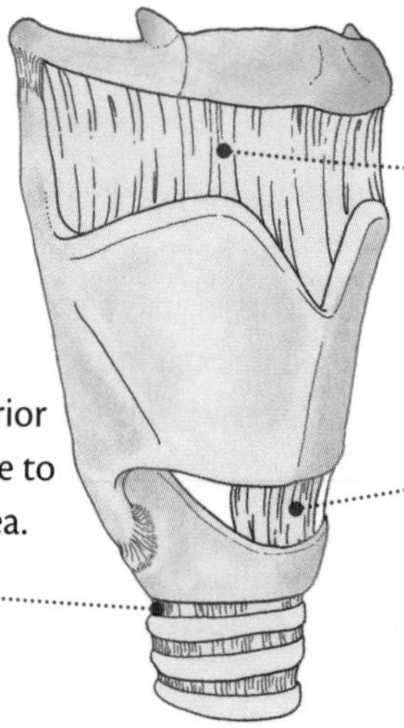

Specific Ligaments Link Structures of the Larynx to Each Other

A fibrous, elastic membrane lines the inside of the laryngeal cartilages. It is thicker and reinforced at three points:

- between the lateral borders of the epiglottis and the anterior border of the arytenoids; these are the **aryepiglottic ligaments**
- from the hemispherical fossa of the arytenoid cartilages to the "hollow fold" of the thyroid cartilage; these are the **vestibular ligaments,** thickened strands that travel in the **ventricular bands** (see page 182)
- from the vocal processes of the arytenoid cartilages to the "hollow fold" of the thyroid cartilage, below the vestibular ligaments; these are the **vocal ligaments** (see pages 152–53)

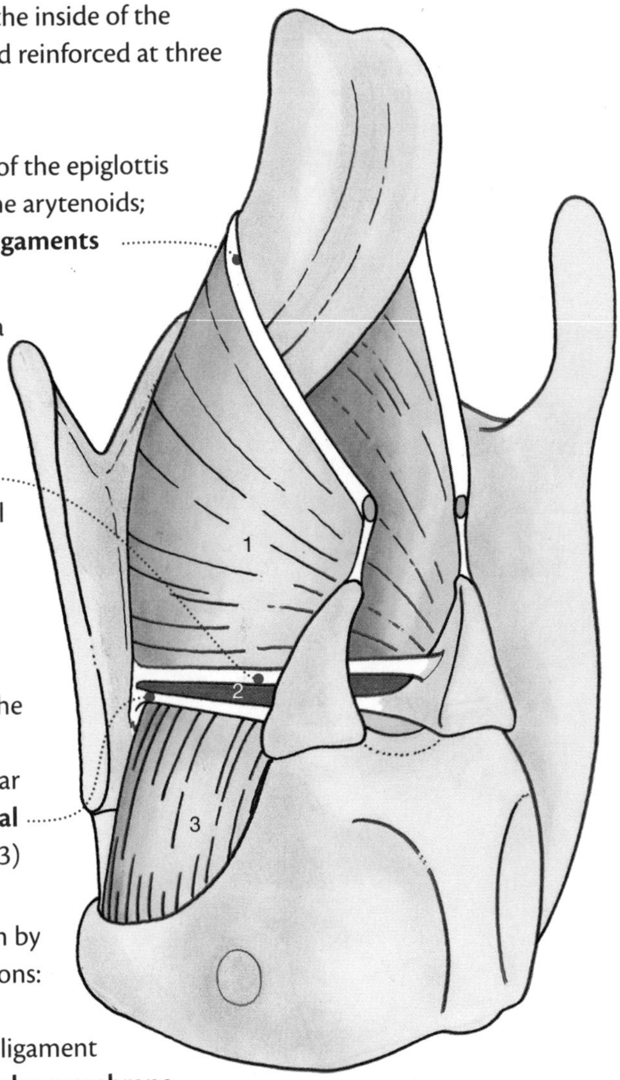

The laryngeal membrane is known by different names in different locations:

- Between the aryepiglottic ligament and the vestibular ligament (1), it's called the **quadrangular membrane**.
- Between the vestibular ligament and the vocal ligament (2), it's the laryngeal vestibular membrane.
- The lower part of the membrane extends from the superior border of the cricoid to the vocal ligament (3); it's called the **cricothyroid membrane**. It lies in an oblique plane, more lateral at the bottom, more medial at the top; its lateral portion is sometimes called the **conus elasticus** (elastic cone).

This membrane is completed by a membrane that runs from the superior border of the cricoid cartilage to the inferior border of the thyroid cartilage. It is thickened at the front; this is the **median cricothyroid ligament** (shown on page 150).

The Vocal Cords Are Almost Ligaments

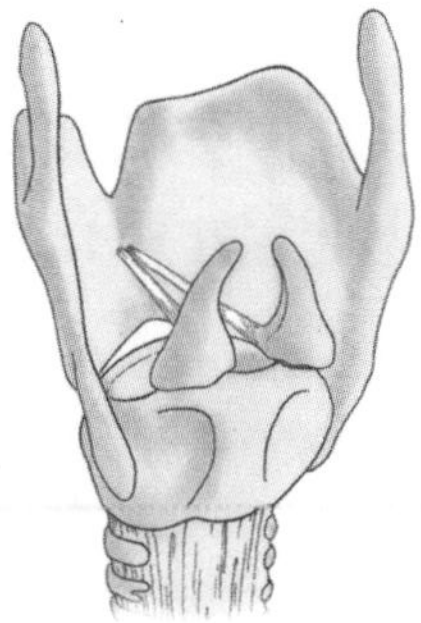

Poetry . . .

We often envision them as very different than they are in reality. We think of them as vertical, or like the strings of a guitar, or as numerous . . .

Details

Each vocal ligament attaches to the vocal processes on the arytenoid cartilages. On the thyroid cartilage, it attaches to the inward-pointing angle, about midway up. Here the right and left ligaments make contact with each other.

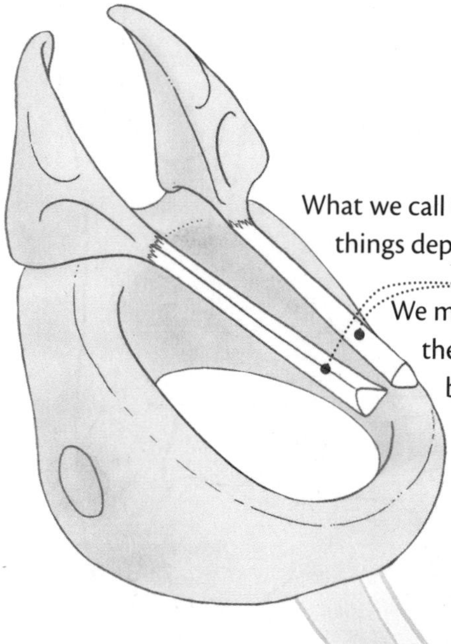

What we call the vocal cords can, in fact, be two different things depending on the context.

We may mean the **vocal ligaments,** which resemble cords; these fibrous, whitish cords are held under tension between the arytenoid cartilages and the thyroid cartilage.

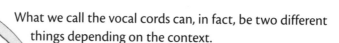

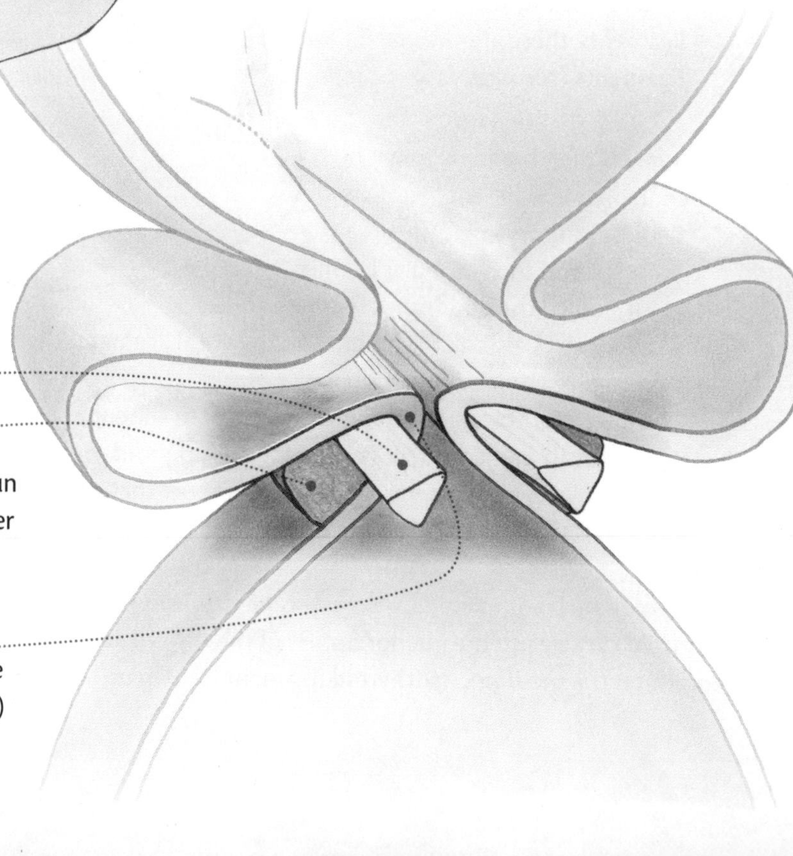

Or we mean **the entire mechanism,** which includes:

- the *vocal ligaments*
- the *vocal muscles* (a.k.a. the internal thyroarytenoids) that run along the external border of the ligaments (see page 168)
- the *mucosa* that cover the medial border of the ligaments (see page 171)

Each vocal ligament is itself composed of three parts, a bit like three layers:

- The **innermost layer** (located against the mucosa) consists of loose tissue containing some elastic fibers.
- The **middle layer** is made up mainly of elastic fibers.
- The **outermost layer** (located near the vocal muscle) is composed of collagen fibers that are virtually inextensible.

Therefore, when looking at the vocal cord in its entirety, we find five different layers. From the inside outward, they are: the mucosa, the three layers of the vocal ligament (inner, middle, and outer), and the vocal muscle.

This enlarged view of the illustration on page 152 (right side) shows details of the vocal cord.

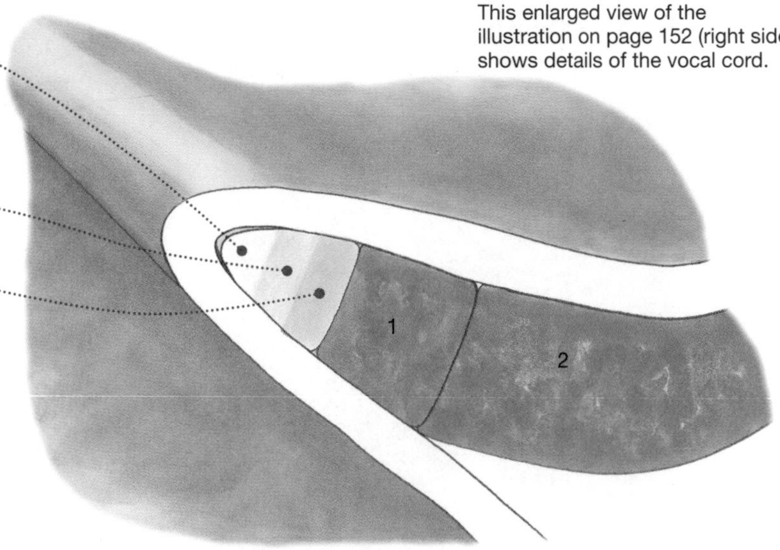

On the outside of the vocal muscle, or internal thyroarytenoid (1), is the external thyroarytenoid muscle (2), which itself does not belong to the vocal cord mechanism but rises outward to form the wall of the laryngeal vestibule.

From Low to High and Vice Versa

To maintain and exercise the fibers of the vocal ligaments, it is interesting to practice voice exercises that gradually go from low notes (unstretched fibers) to high notes (fibers under tension), and vice versa, many times.

Anatomical Details

The border of the vocal ligaments that faces the glottis, and the corresponding mucosa, make up the "free border" of the vocal cords. With a laryngoscopy, we see two white stripes, which correspond to where the mucosa meets the vocal ligaments.

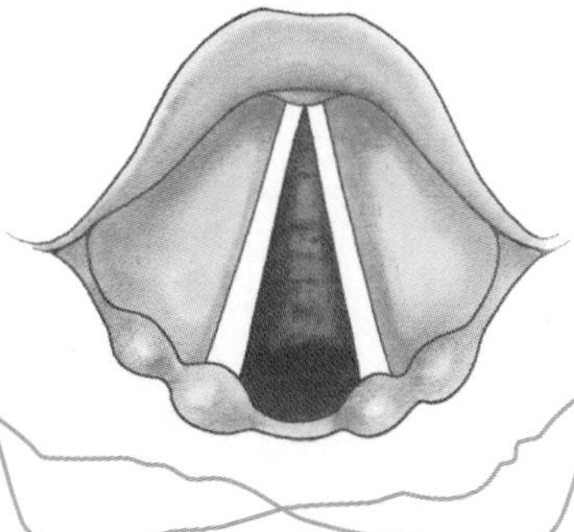

Vocal Cord Thickness Can Change

The thickness of the vocal cords changes depending on the tension they are under. Most commonly, the vocal cords are thin in "head voice" (upper register) and thick in "chest voice" (low register); see page 159 for details. However, when you're really belting it out, this rule may not always hold true.

The Cycle of Opening/Closing the Vocal Cords (the Myoelastic Theory)

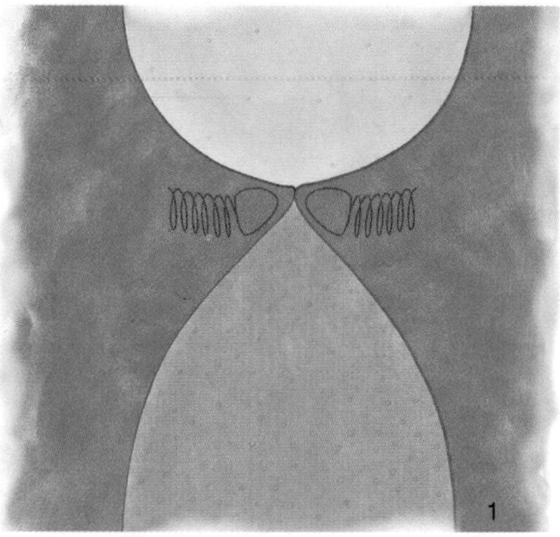

When the Glottis Is Closed

We increase the subglottal pressure until it is sufficient to open the vocal cords.

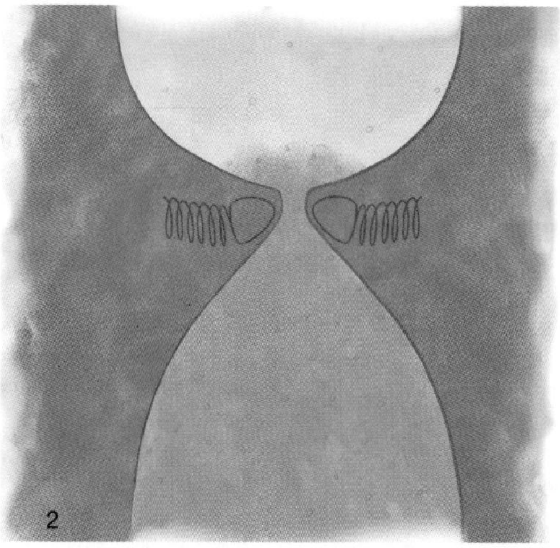

The vocal cords open under the pressure.

Attention

This model of myoelastic theory does not explain the production of airy sounds or high-pitched sounds of low intensity. This is where we need to look at the myoelastic aerodynamic theory (see pages 172–73) for an explanation. It is also important to note that the mechanisms described here take place very quickly (for example, 440 cycles per second with a tuning fork).

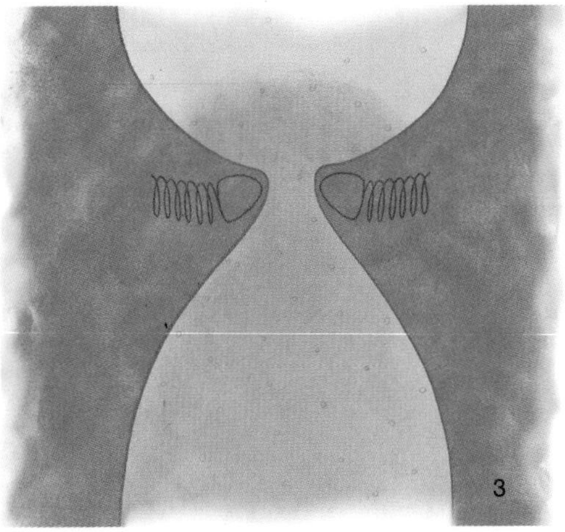

When the Glottis Is Open
Only part of the air below the glottis passes between the vocal cords.

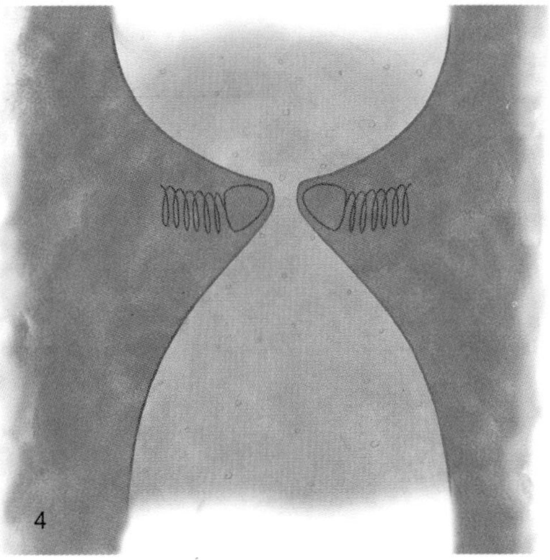

Gradually, as the air escapes, the subglottal pressure drops. The vocal cords stay in contact until the glottis closes.

The Role of the Vocal Cords in the Production of Sound

Pitch Control
The pitch of the sound is defined by the speed at which the vocal cords open and close: the faster they vibrate, the higher the sound, and vice versa. When the vocal cords are stretched, they vibrate faster as they seek to return to their original position. Thus, the tighter the vocal cords, the higher the pitch.

Intensity Control
Sound intensity depends mainly on the subglottal pressure: the higher the pressure, the stronger the sound.

The Laryngeal Joints

Mobility between the laryngeal cartilages is possible because of four joints.

The Cricothyroid Joints

There are two cricothyroid joints (one on the right, one on the left). Each one has the following:

a surface located on the **tip of the inferior horn**

a surface located on the **lateral face of the cricoid ring** (see page 142)

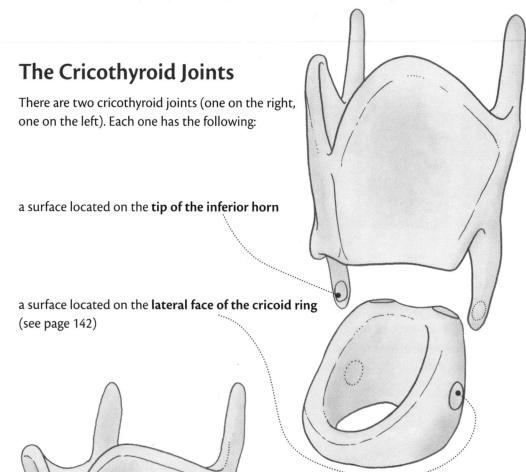

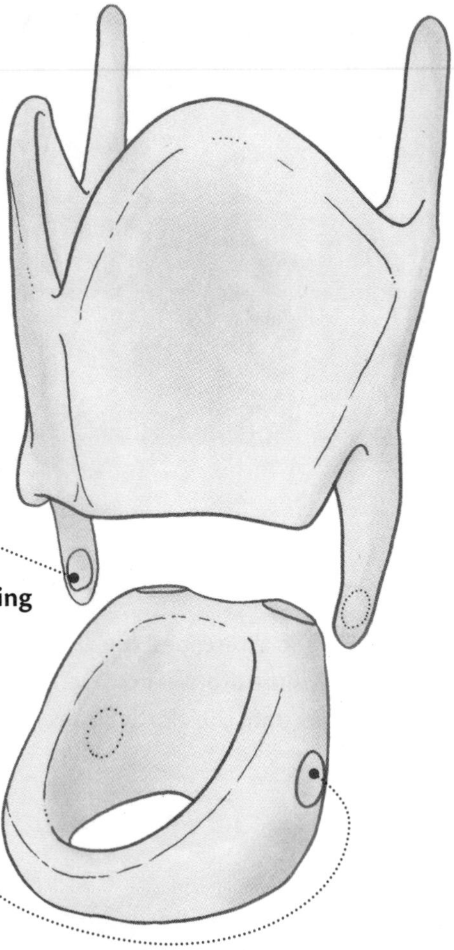

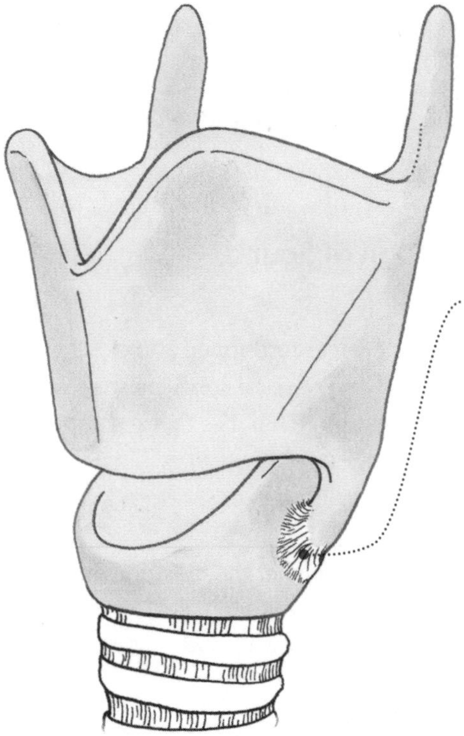

A fibrous sleeve (joint capsule) unites these surfaces, while allowing small movements. It is reinforced by two ligaments: one anterior, one posterior (not shown).

These articulations allow the thyroid cartilage to rock on the cricoid cartilage, which allows us to stretch the vocal cords (see page 158).

The Cricoarytenoid Joints

There are also two cricoarytenoid joints (one on the right, one on the left). Each has the following:

a surface located on the **base of the arytenoid cartilage** (see page 144)

a surface located on the **superior border of the cricoid "bezel"** (see page 142).

Here again, a fibrous sleeve (capsule) holds these surfaces together, while allowing small movements of the arytenoid.

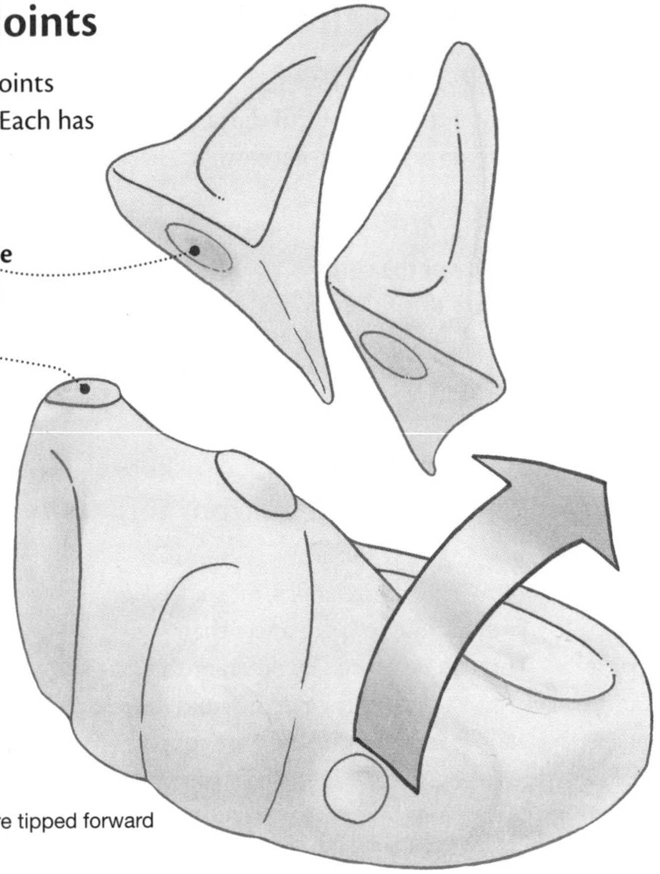

Here, the arytenoid cartilages are tipped forward to show the articular surfaces.

These joints allow the arytenoid cartilage to pivot around on the cricoid cartilage, taking the different positions that we see on page 160.

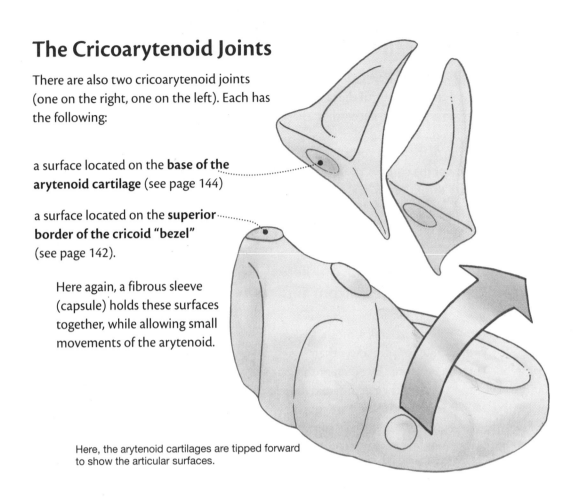

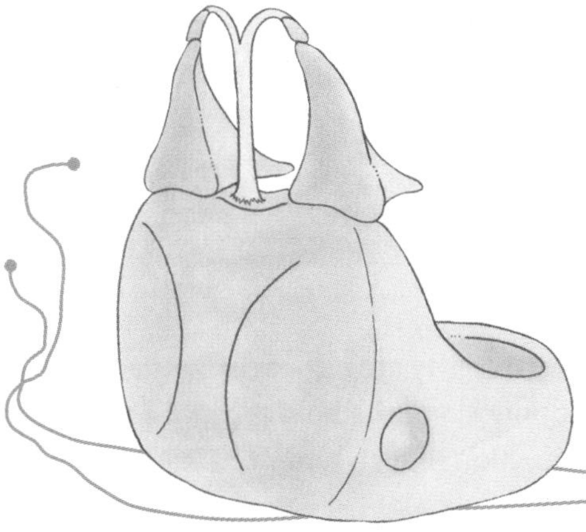

The Jugal Ligament

It should be noted here that a Y-shaped ligament, the **jugal ligament,** joins the tops of the two arytenoid cartilages and two cartilaginous horns. This ligament helps stabilize the top of the arytenoids.

Movements of the Laryngeal Cartilages

Because of their joints, the laryngeal cartilages can all move between each other. It is a fundamental characteristic of the larynx: it's not just a solid passageway, and its structures can move between each other. These movements can transform, at once, and in many different ways:

- the position of the vocal cords
- the tension of the cords
- the form of the intralaryngeal spaces

Thyroid Cartilage Can Rock on the Cricoid Cartilage at the Bottom and Front

Instead of remaining horizontal, it moves obliquely, with its anterior part lower than the posterior. This can be caused by different things, in particular the action of the cricothyroid muscle (see page 166). It acts to bring the anterior parts of two large laryngeal cartilages closer together.

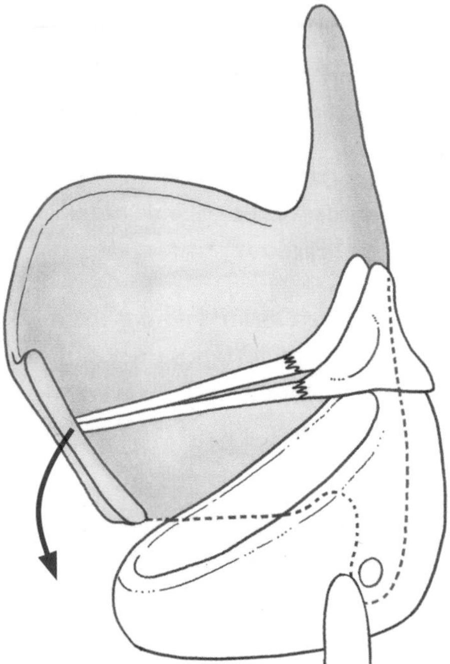

Cricoid Can Rock at the Top and in Back under the Thyroid

Instead of remaining horizontal, it moves obliquely, its anterior part having lifted a bit. There is also a coming together of the front parts of two large laryngeal cartilages, this time "from below."

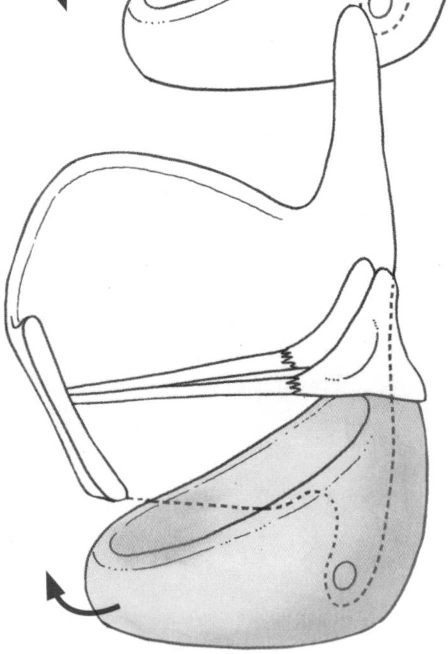

The effect on the vocal cords is the same in both cases: they are put under tension, and as a result their vibrating mass tends to become thinner, and the border between the cords becomes thinner. This can occur in the production of high notes or what would be called the *head voice.*

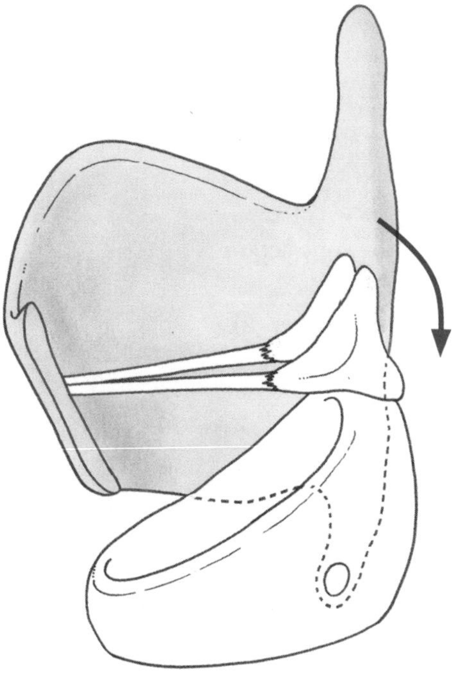

The Thyroid Can Rock at the Bottom and in the Back on the Cricoid

Instead of moving more obliquely, as we've seen on the previous page, it ends up more vertical, its anterior part moving up relative to its posterior part. This acts to bring the posterior parts of the two big laryngeal cartilages closer together.

The Cricoid Can Rock on the Thyroid at the Bottom and Front

Instead of remaining horizontal, it moves obliquely, with its anterior part lower than its posterior part. This also acts to bring the posterior parts of the two large laryngeal cartilages closer together, but "from below."

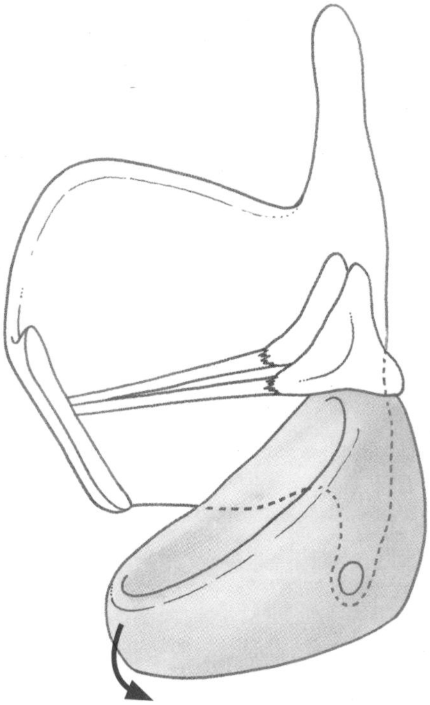

The effect on the vocal cords is the same in both cases: they are no longer under tension, and as a result their vibrating mass is thicker than in the previous example, and the border between the cords is thicker and rounded. This can occur in the production of low notes or what could be called the *chest voice*.

Arytenoid Cartilages Can Move Closer to Each Other or Farther Apart

When they come together, this movement is called *adduction* (1). This movement is not necessarily symmetrical between the two arytenoids. The posterior parts of both vocal cords can also move closer together; this is called *vocal cord adduction*. In this movement, the vocal cords can be more or less tense, depending on the intensity of the contraction of the interarytenoid muscles that cause this action (see page 164). Conversely, the two cartilages may separate slightly from each other (2). In this movement, the posterior parts of the vocal cords are also separated.

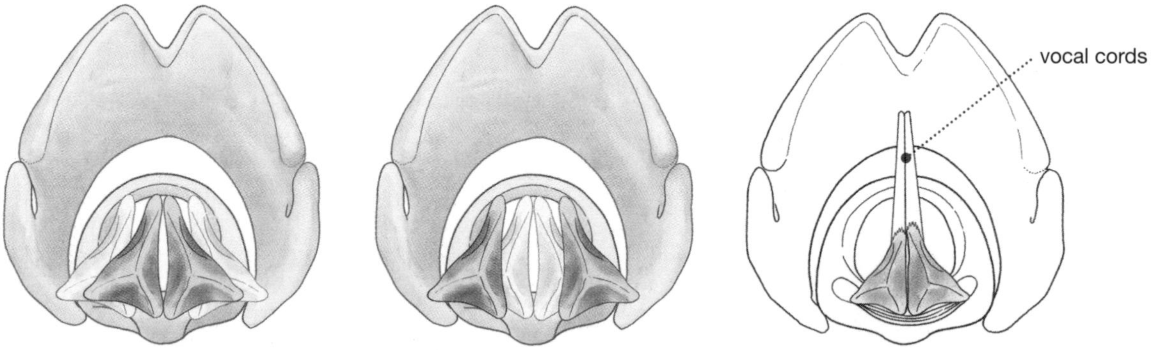

1 2

Arytenoid Cartilages Can Pivot on Each Other

This movement occurs around a longitudinal axis. The vocal processes (see page 145) can rotate inward (1); this movement brings the middle part of the vocal cords closer. The vocal processes can also pivot outward (2); this separates the posterior parts of the vocal cords.

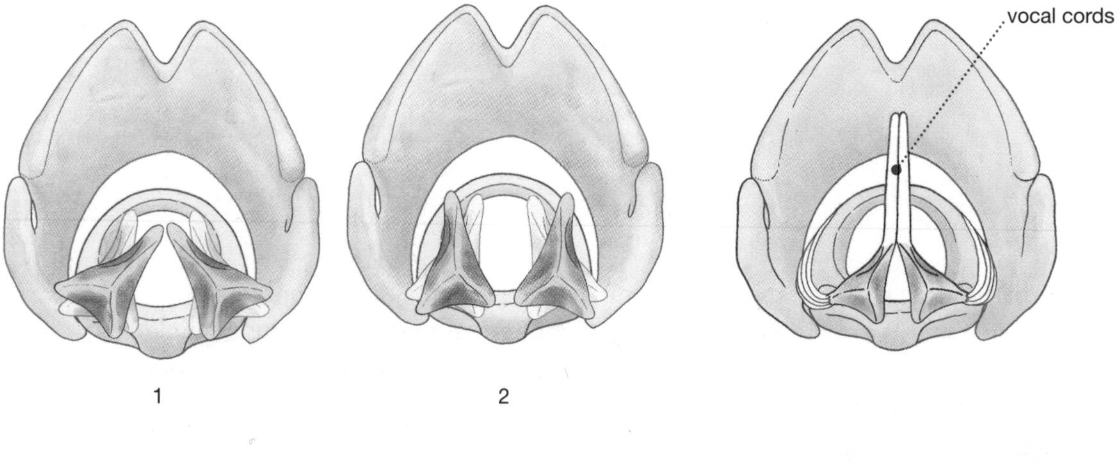

1 2

The Intrinsic Muscles of the Larynx

The laryngeal cartilages are mobilized by several kinds of forces. In particular, they are mobilized by *muscular contraction.* These muscles are of two kinds:

- Some are between the cartilages. They only attach to the larynx. They mobilize the laryngeal cartilages among themselves. They are called the **intrinsic muscles** of the larynx and they are very small.
- Some connect laryngeal cartilages to neighboring structures: the base of the skull, mandible, sternum, clavicle, scapula. They mobilize laryngeal cartilages by pulling on them from these structures that lie above or adjacent to the larynx. They are called the **extrinsic muscles** of the larynx, and they are longer than the intrinsic muscles. (See pages 186–94 for a discussion of the extrinsic muscles.)

All these muscles are symmetrical and attach from both the right and the left. In the following pages, the illustrations, for the most part, show these muscles on both sides of the body.

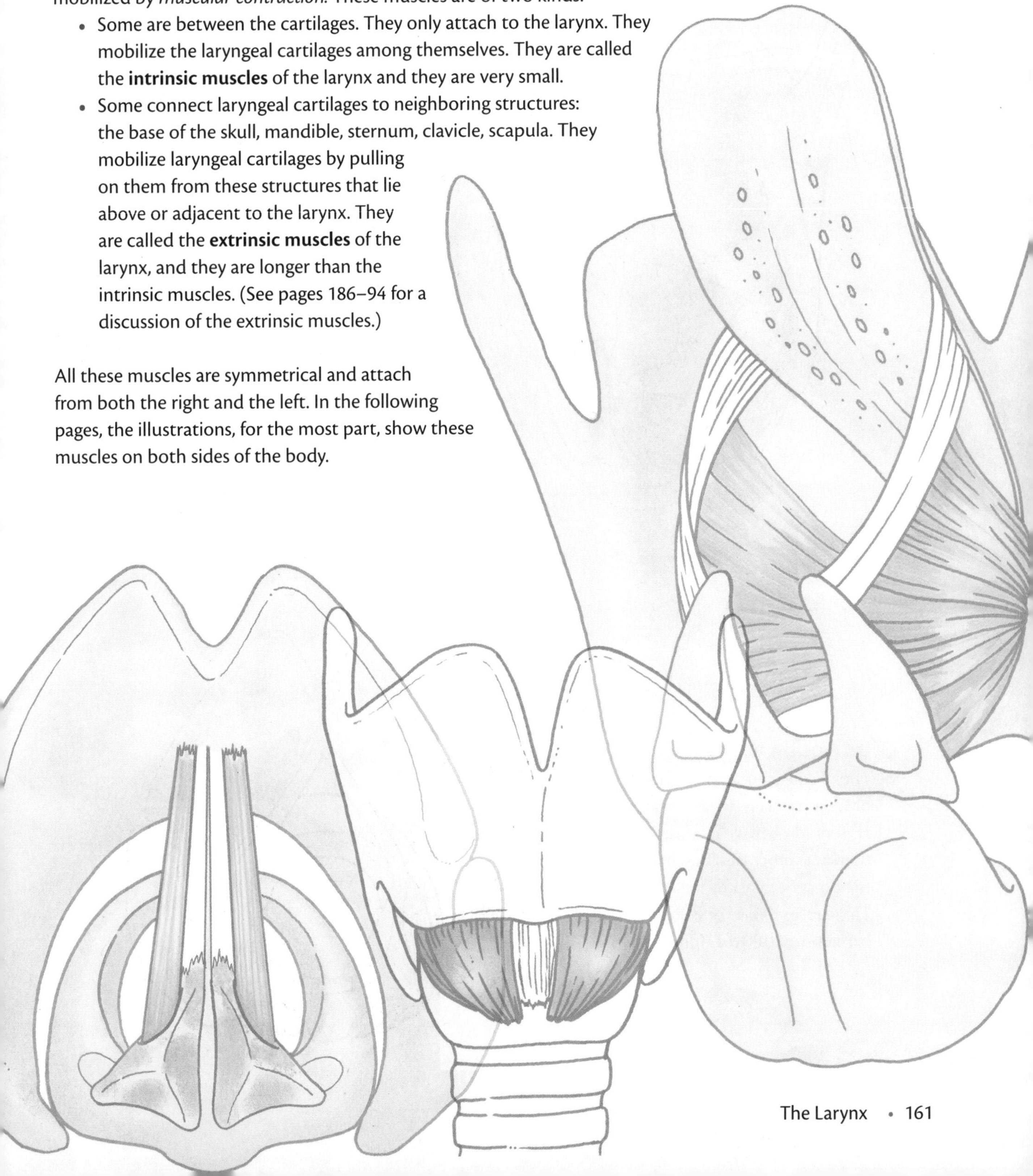

The Muscle That Opens the Glottis: The Cricoarytenoid

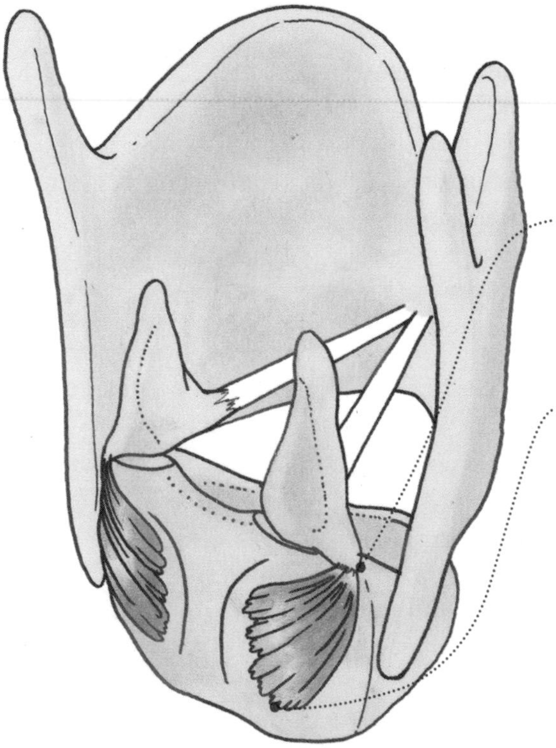

This little muscle attaches as follows:

At the top
it attaches at the back of the arytenoid cartilage on the muscular process (see page 145).

At the bottom
it attaches on the posterior surface of the cricoid cartilage, on the cricoid "bezel" (except on the crest in the middle).

It's shaped like a small fan that flares out as it moves downward and toward the middle.

Its Action

The cricoarytenoid pulls the muscular process medially, which rotates the arytenoid cartilage in a way that separates the vocal processes; this, in turn, spreads the posterior part of the vocal cords. This causes a wide opening of the glottis (sometimes called a "forced opening").

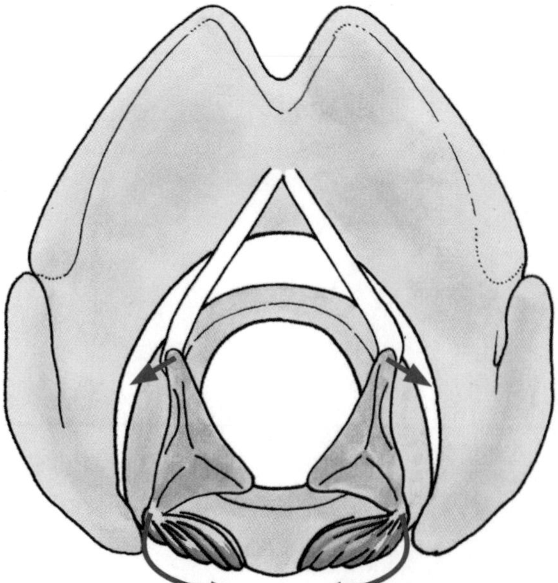

The action of this muscle is most noticeable on the inhalation or exhalation.

When the glottis is in a relaxed position, it is not really closed but slightly open. Respiratory air *rubs* against the walls and because of this friction there is a slight noise.

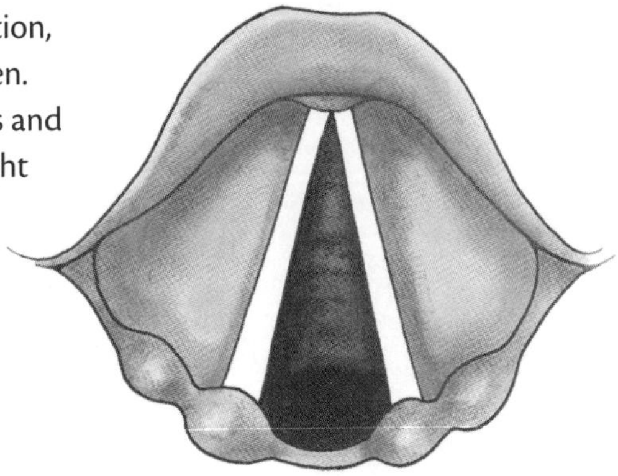

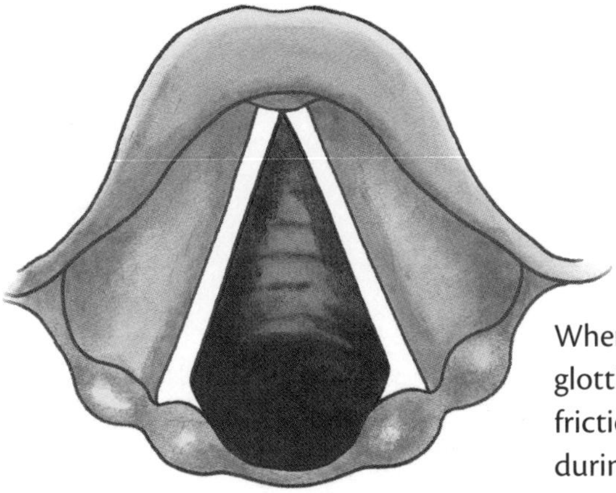

When the cricoarytenoid muscle is contracted, the glottis is open and air flows freely, without creating friction on the walls; therefore we do not hear noise during the flow of the breath.

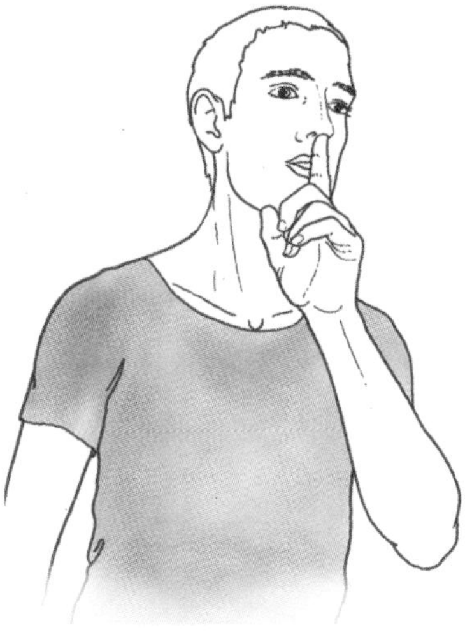

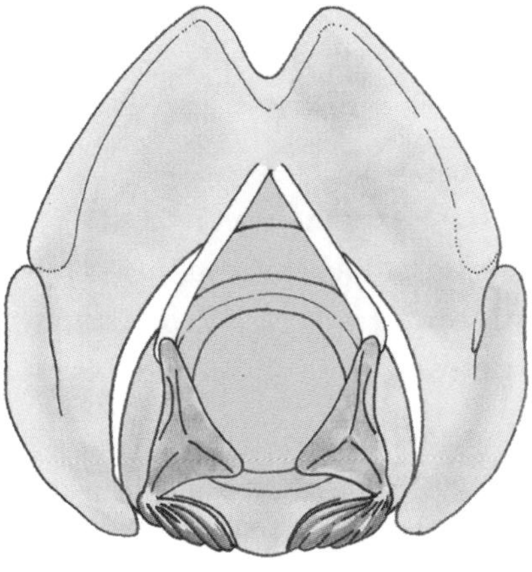

The glottis takes the shape of a pentagon.

Listen to the Silence

When it's necessary to use the voice for long periods of time, it can be useful to have the inhalations between the vocal exhalations be as silent as possible. This not only produces better sound, but it's also healthier for the vocal cords and the mucous membranes, which will be better lubricated. Practice listening to speakers and singers to see if their inhalations are silent or not.

The Muscle That Brings the Vocal Cords Together: The Interarytenoid

This tiny muscle is composed of three bundles, one behind the other, behind the arytenoid cartilages.

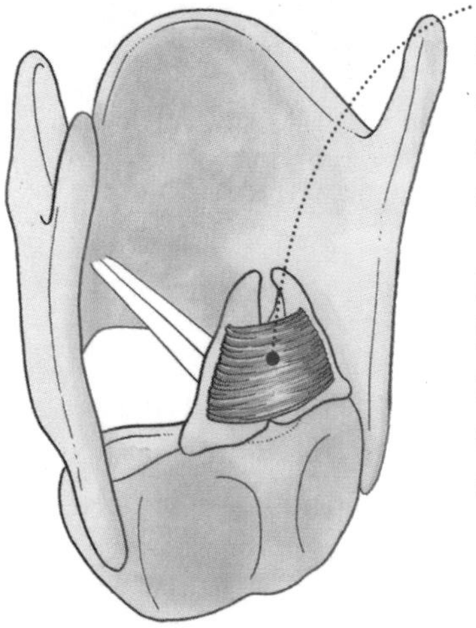

The **transverse bundle** takes a quadrilateral form and is stretched between the rear surfaces of the two arytenoid cartilages.

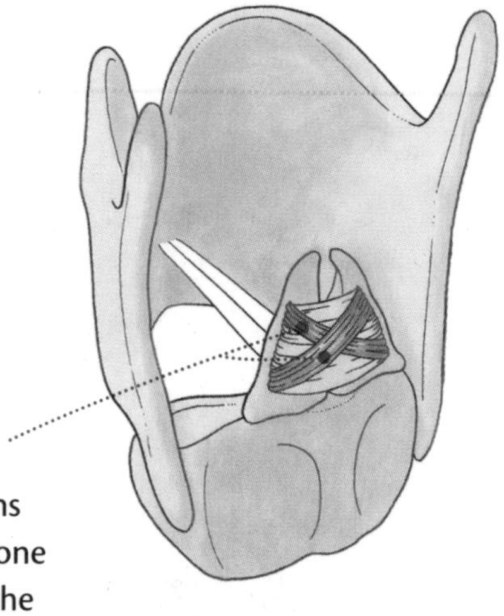

Farther back, the two **oblique bundles** form a cross. Each runs from the muscular process of one of the arytenoid cartilages to the top of its symmetrical partner.

Its Action

This muscle brings the arytenoid cartilages closer together and therefore the vocal cords as well. It is a *constrictor* of the glottis.

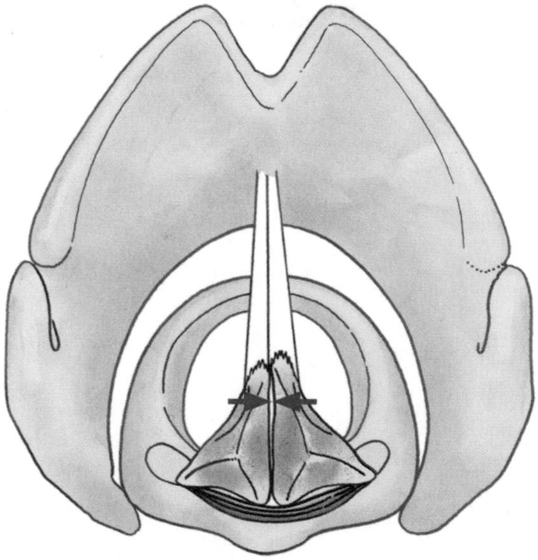

A Noisy Inhalation

Habitually keeping this muscle too tense means that the glottis will be too closed. Some people keep this muscle contracted when singing or speaking, which makes for a noisy inhalation between the vocal exhalations. This is caused by friction during the passage of air (which is greater than when the glottis is in a resting position, as described on page 180). This can be a factor in the drying out of the vocal folds, especially if you breathe through your mouth (unlike during a nasal inhalation, which moistens the air), and if the ambient air is dry.

The Drawing Together of the Vocal Cords and Phonation

It's the action of the interarytenoid muscle drawing the vocal cords toward each other that allows for phonation. We also call this *vocal cord adduction*.

If this action is accompanied by vocal cord vibration, it allows for the transmission of a **voiced sound** (we say a sound is voiced when it comes with laryngeal vibration).

The action of this muscle is required for the production of vowels (although it is also possible to whisper vowels without full vocal cord adduction).

The action of this muscle also allows for the emission of voiced consonants.

We can feel the vocal cord adduction and vibration of the larynx if we make an unvoiced consonant that moves into a voiced consonant—for example, passing:

from "fffff" to "vvvvv"

from "sssss" to "zzzzzz"

from "chchchchch" to "jjjjjjj"

The Ujjayi Breath

This muscle is used in the ujjayi breathing technique in certain yoga bandhas where the breath produces glottal friction.

The Cricothyroid Muscle

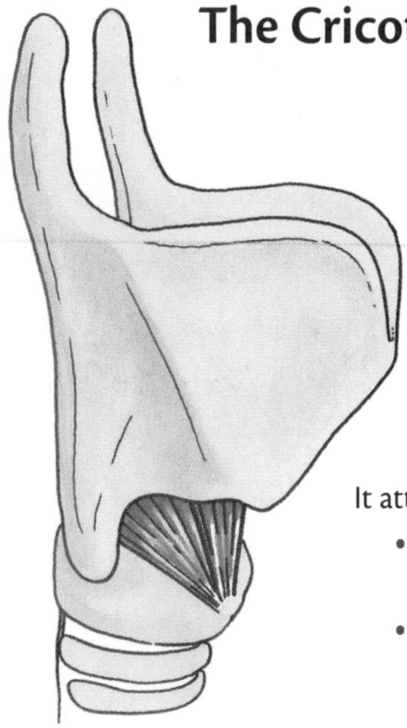

This muscle is shaped like a fan that opens upward.

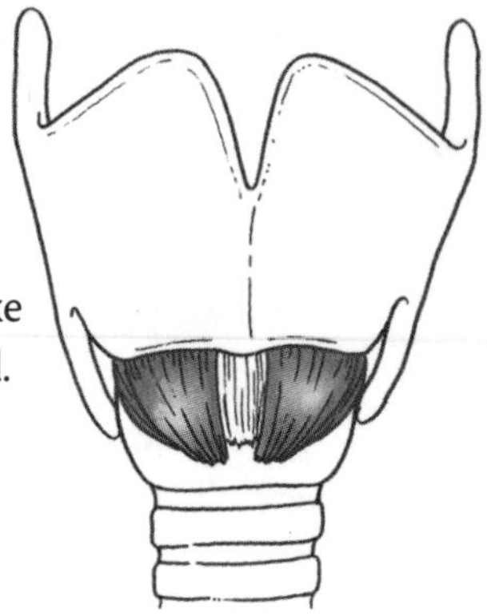

It attaches as follows:

- at the top, on the inferior border of the thyroid cartilage, up to the inferior horn
- at the bottom, on the front of the cricoid cartilage

Its Action

It tips the thyroid cartilage forward and backward. As well, it causes the anterior part of the cricoid cartilage to move back and upward.

Action on Voice

Both of these actions can happen simultaneously. They have the same effect. The vocal cords are placed under tension: they stretch, their vibrating mass becomes thinner, the "free border" becomes thinner. This happens during the production of high notes, singing in what we call *head voice* (see page 158).

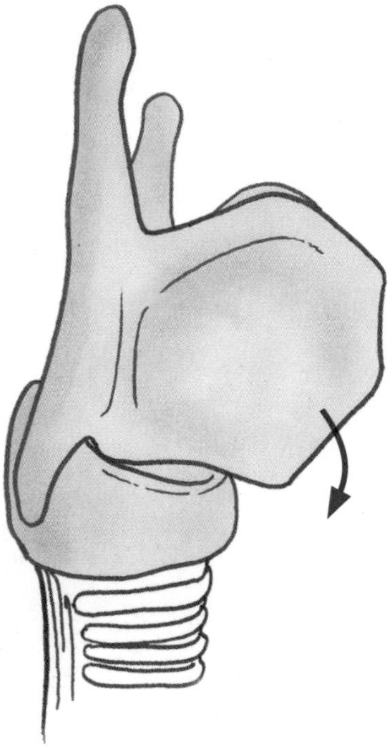

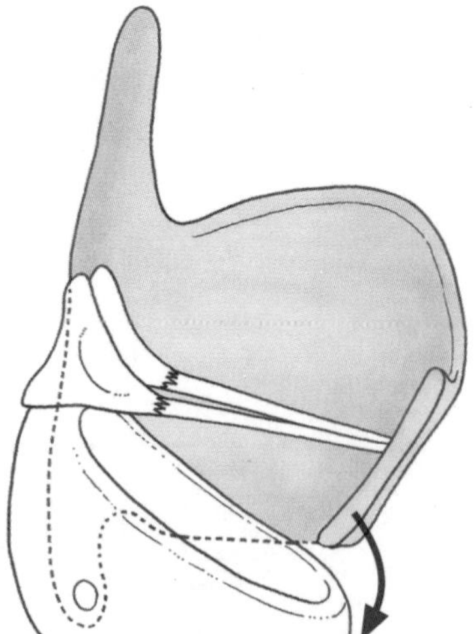

Details

This action is accompanied by a small displacement of:

- either the thyroid backward on the cricoid
- or the cricoid forward under the thyroid

The Muscle That Closes the Glottis: The Lateral Cricoarytenoid

This muscle arises from the muscular process of the arytenoid cartilage. It descends forward and toward the exterior, to terminate on the lateral part of the cricoid arch.

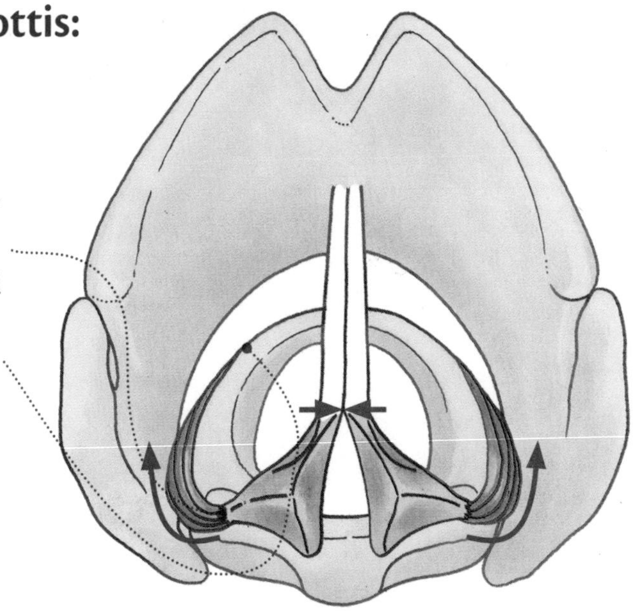

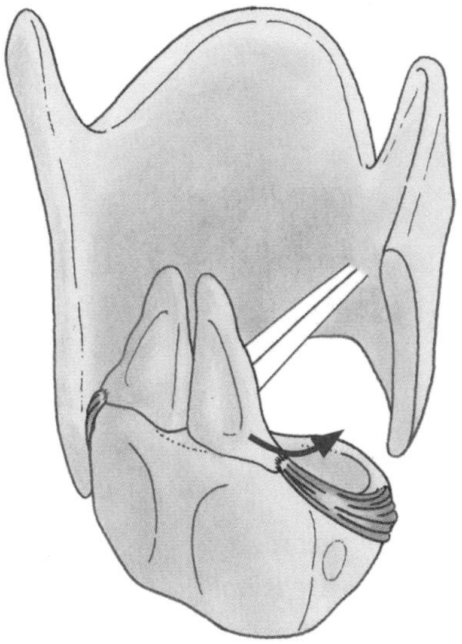

Its Action

This muscle brings the muscular process forward and outward. This causes the arytenoid cartilages to turn in a way that strongly pulls the vocal processes toward the midline. It is a constrictor of the glottis. Specificially, it closes the posterior third of the glottis.

When the lateral cricoarytenoid's action is added to that of the interarytenoid, the glottis is firmly closed. This is what happens in the *glottal stop*, specifically during intense exercise—for example, when we block the breath (apnea) when doing a pushup.

This muscle shortens the vibrating portion of the cord, which helps elevate the vocal tone (this is called *damping*).

Back-to-Back with the Vocal Cords: The Vocal Muscle

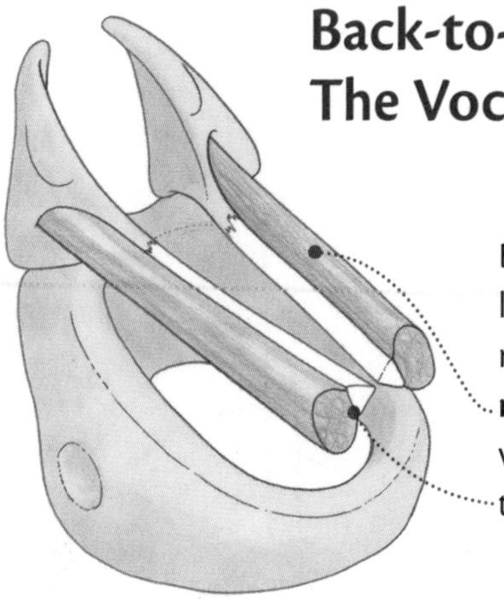

Each vocal ligament (vocal cord) is "doubled," over the entire length of its outer border, by a very small muscle whose fibers run parallel to the ligament: the **internal thyroarytenoid muscle,** also known as the vocal muscle or the muscle of the vocal cord. It is an integral part of the vocal cord and adheres to the outer edge.

Its Action

The internal thyroarytenoid only slightly moves the cartilage to which it's attached (the cartilage is often "fixed" by other small muscles). It therefore most often contracts *statically,* that is to say, *isometrically,* which means that its length doesn't change. This action serves mainly to tighten the muscle itself.

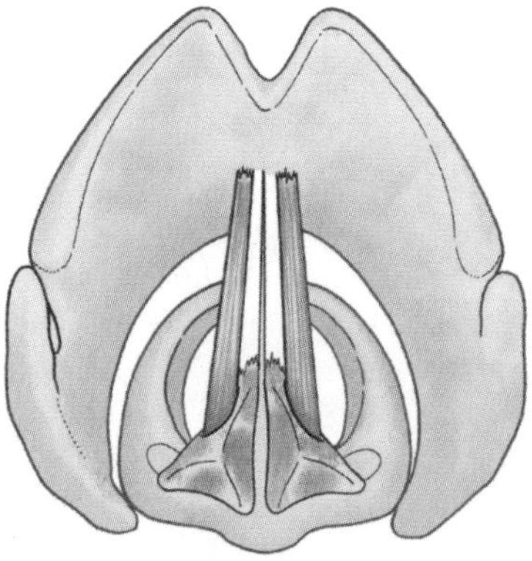

The Role of the Vocal Muscle in the Production of Sound

When the vocal muscle contracts, it contributes to the rigidity of the vocal cord, and this helps elevate the tone (it raises the pitch). Unlike the cricothyroid muscle, it does not alter the thickness of the cords, and therefore it allows the mucosa to remain relaxed and undulate with the pressure (the Venturi effect; see page 172). For these reasons, it is a primary regulator muscle in what is referred to as the "heavy mechanism," or the "chest mechanism."

The Vocal Cord Adapts to the Air Pressure . . .

Here, we have what we might call a "composite beam," where two materials with different properties are combined in a single structure. The resistance of the vocal cord in traction (passive resistance) is combined with muscle contractility, which more actively resists by tightening isometrically. The combination of these two features allows multifaceted adaptation of the vocal cord to subglottal air pressure.

The Muscles of the Ventricular Band

The vocal muscle (internal thyroarytenoid) is extended laterally by the **external thyroarytenoid** muscle, which pulls it a bit to the side before forming a fan-shaped vertical sheet that rises upward to the lateral edge of the epiglottis. In the back, this sheet is limited by the thin **aryepiglottic muscle.** All of these muscles border laterally the area that runs from the vocal cords to the epiglottis. They give this part of the larynx the form of a mouthpiece. A portion of this muscle makes up a part of the ventricular band.

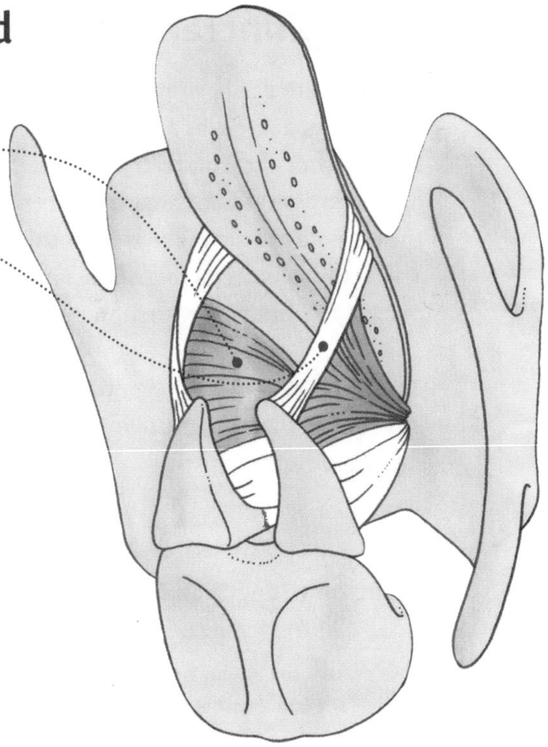

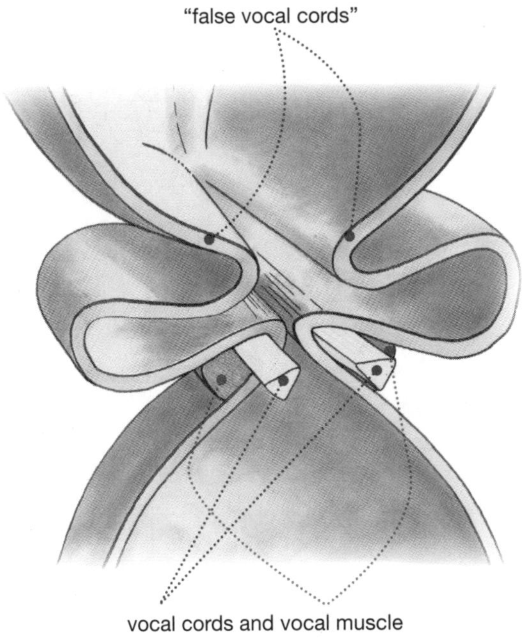

The **ventricular bands,** or folds, are located above the vocal cords (see page 152) and glottis (see page 182). They are also called **vestibular folds** or **false vocal cords**. There is a band on each side (the right and the left). These mucosal folds contain thickened strands of fibrous membrane (the ventricular ligaments; see page 151) that give them their structure. They are also supported by a thickened part of the external thyroarytenoid muscle. The contraction of this muscle can actively change the shape of the ventricular bands, bringing them closer together, or making them thicker or thinner. They also play a sphincteral role in the larynx (see the next page).

For more details on the ventricular bands, see page 182.

The action of the muscles of the ventricular bands can change the shape of the laryngeal ventricles, hollowing or flattening them as necessary. These ventricles are the first resonance chambers above the vocal cords, and changes here may well change the timbre of the voice.

The Sphincteral Role of the Larynx

The larynx does not primarily serve the voice; rather, its primary role, more ancient and vital, is to serve as a passageway for breathing. For this function, it possesses a framework of cartilage, thanks to which the laryngeal tube at rest has a certain openness essential for breathing. Its second role, equally vital, is to intermittently close the entry to the airway so that nothing but air (neither liquid nor solid food) enters. This is the *sphincteral role* of the larynx,* and it is made possible by three successive sites of closure. These may function in isolation, but they often close synchronously.

The epiglottis covers the glottis like a hinged lid. When we swallow, this action is initiated by the tongue pulling backward. This movement can also be initiated by the aryepiglottic muscle.

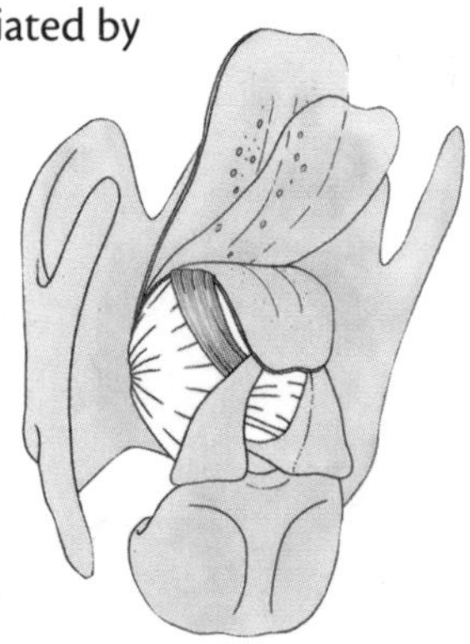

Stick Out Your Tongue and Try to Swallow

When we prevent the tongue from pulling backward—for example, when we stick out our tongue and try to swallow—we can feel how hard it is to lower the epiglottis; this makes the aryepiglottic muscle, a depressor of the epiglottis, work. It may be interesting to develop this area.

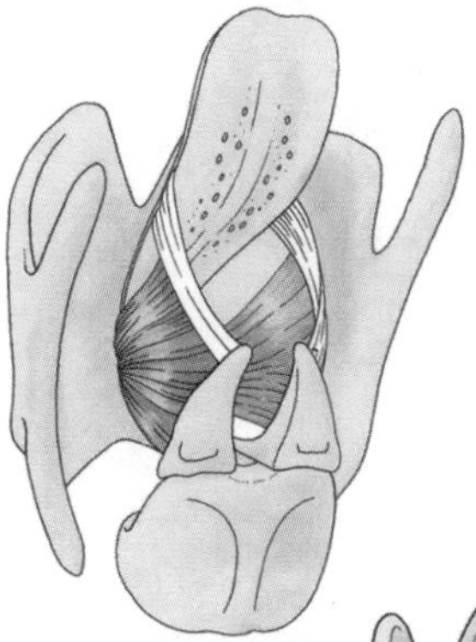

The ventricular bands move closer together, while thickening. Note that this movement is often very asymmetrical.

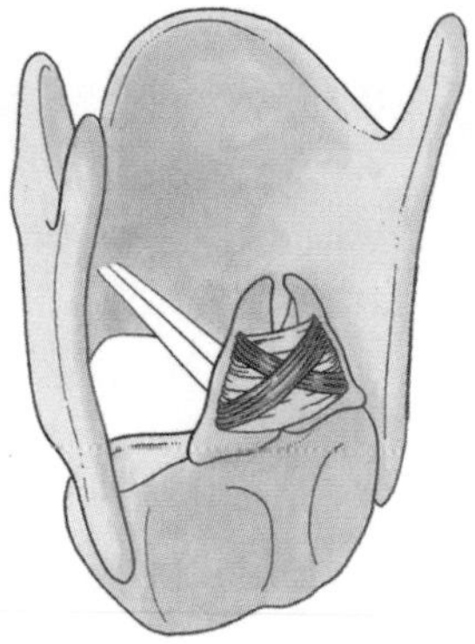

The vocal cords move closer together and touch. Of the three levels of closure, this level is often not the most powerful.

Closure of the larynx always occurs during swallowing, when the bolus of food passes from the back of the mouth to the hypopharynx.

*The word *sphincter* is not used literally here. A sphincter is a circular muscle that can contract to tighten an orifice or duct to prevent the passage of elements. The larynx is therefore not a sphincter in the strict sense of the word, but it can function as a sphincter through the three mechanisms described on this page.

The Laryngeal Mucosa

The cricoid cartilage and the thyroid cartilage above it, together with their ligaments and muscles, form a cylinder. The interior of this cylinder is lined with a mucous membrane that completely covers all of the cartilage. (Any open spaces we might see when looking at the simple skeletal structure would be covered, and the skeletal structure would be only partially discernable.) Just as we saw on page 152, the mucosa molds to the contours of the cartilage, transforming the interior of the larynx into a sequence of folds and valleys.

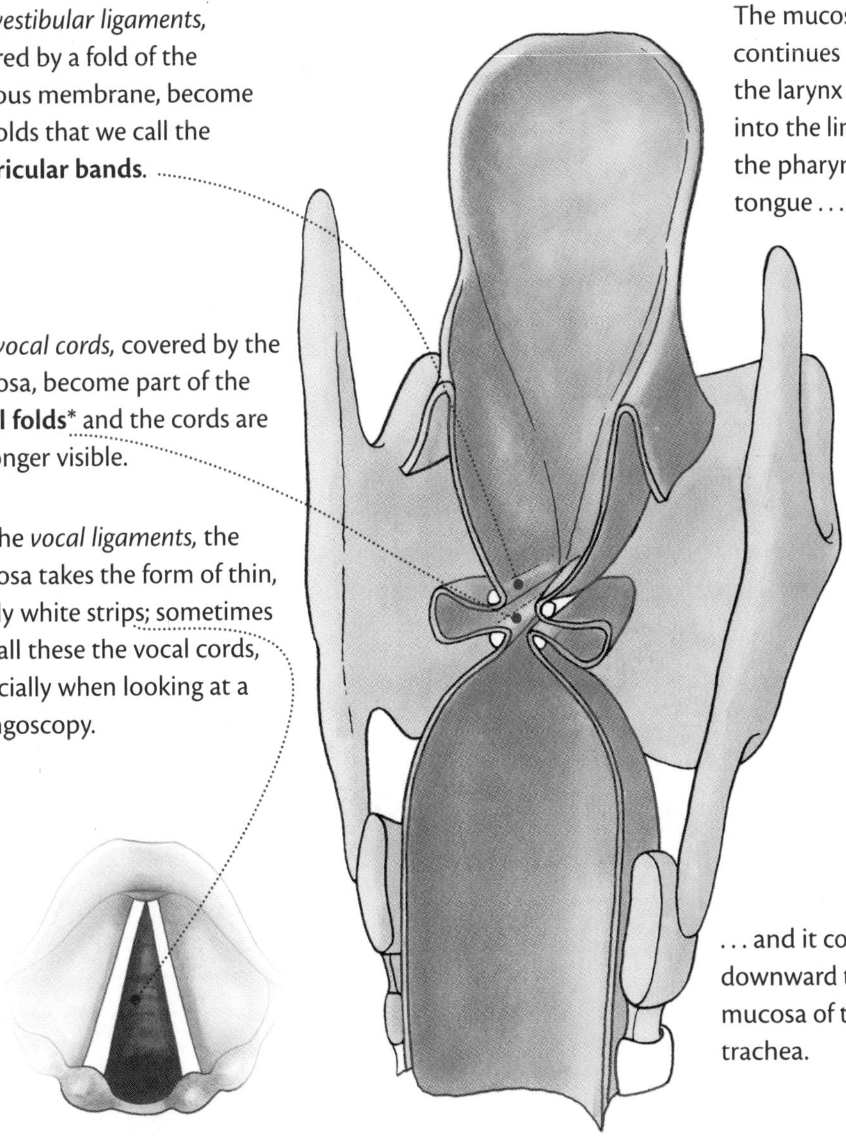
The *vestibular ligaments,* covered by a fold of the mucous membrane, become the folds that we call the **ventricular bands**.
The mucosa continues from the larynx upward, into the lining of the pharynx and tongue . . .
The *vocal cords,* covered by the mucosa, become part of the **vocal folds*** and the cords are no longer visible.
On the *vocal ligaments,* the mucosa takes the form of thin, pearly white strips; sometimes we call these the vocal cords, especially when looking at a laryngoscopy.
. . . and it continues downward to the mucosa of the trachea.

*Logically we should call these "vocal folds." However, the practice remains to use the word "cords."

The Role of the Mucosa in the Production of Sound

The Venturi Effect (or Bernoulli Effect)

This is a phenomenon that can be observed when a fluid flows through a conduit with areas of varying diameters. It can be explained as follows:

The narrower, the faster . . .

We can consider that the rate of flow does not vary during the flow (the principle of conservation of the flow). So, the speed (S) is determined by the diameter (D) of the section: S×D. When the diameter decreases, the speed increases and vice versa.

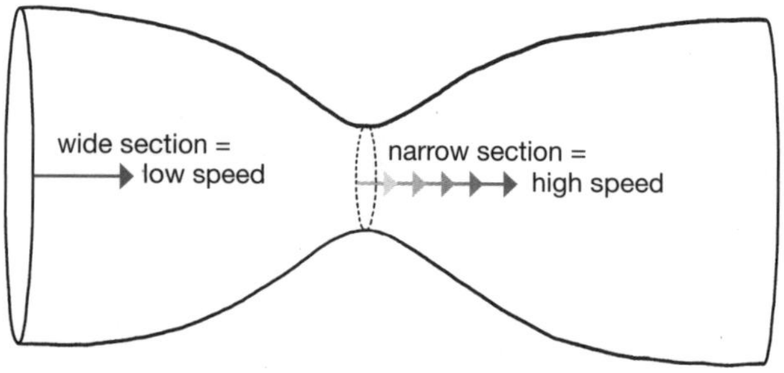

The greater the speed, the lower the pressure . . .

The energy required to set the fluid in motion is the same at all points of the flow. So, this is determined by two factors: the pressure and the speed of the flow. It can therefore be inferred that, as the speed increases, the pressure of the flow decreases, and vice versa.

low speed =
high pressure

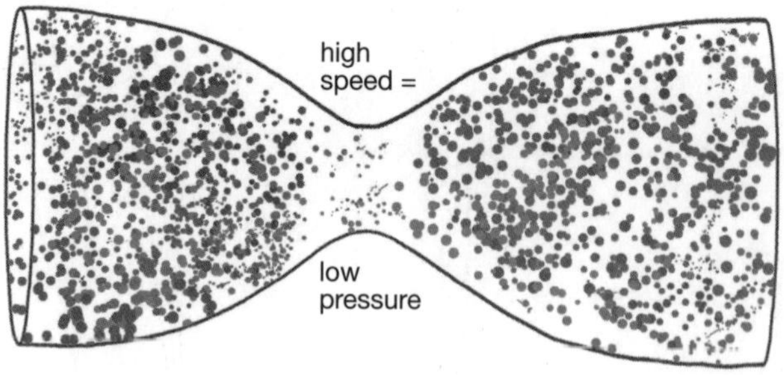

So, the narrower the conduit, the greater the speed and the lower the pressure

When a fluid flows through a conduit that is increasingly narrow, its speed increases while its pressure decreases. This is the Venturi effect.

The Role of the Mucosa in Phonation (the Myoelastic Aerodynamic Theory)

1. The vocal cords are close together but not touching. We increase the subglottal pressure so that a thin stream of air passes between the cords.

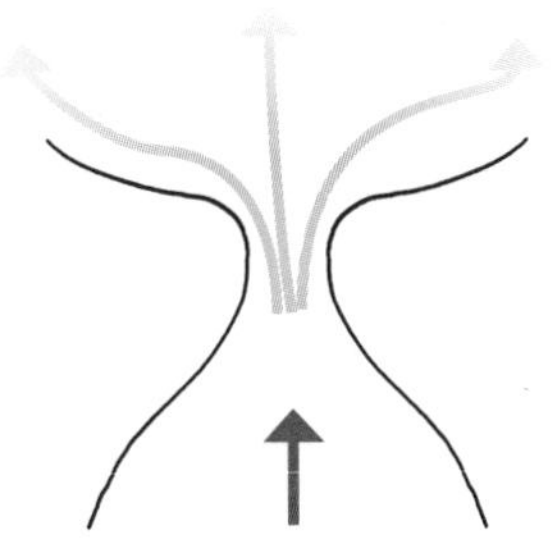

2. The narrow space between the vocal cords produces a depression of the airflow at this location (Venturi effect). This depression causes a suction effect on the mucosa, which then brings the vocal cords together.

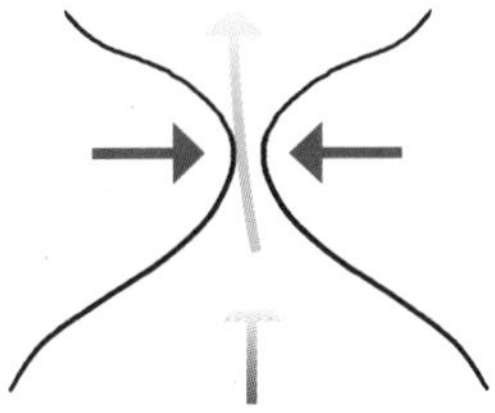

3. The complete closure of the vocal cords causes the airflow to stop. Suction forces related to the Venturi effect stop while the subglottal pressure increases.

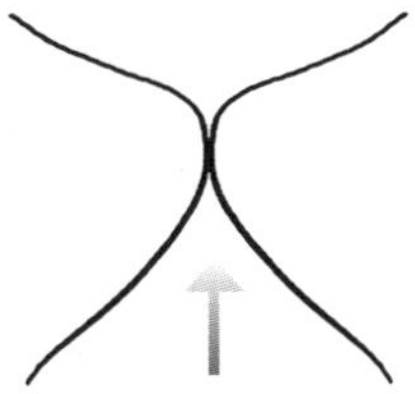

4. The vocal cords gradually open because of the subglottal pressure. We return to position 1.

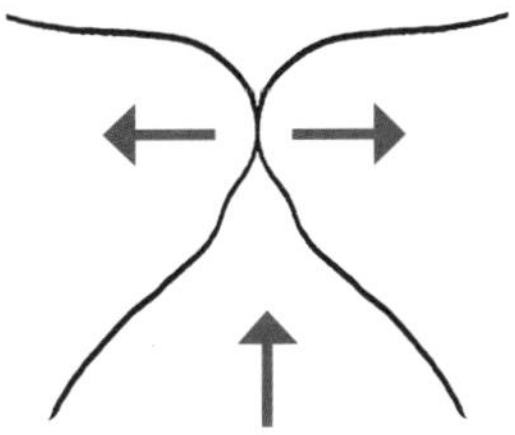

This phenomenon of the undulation of the mucosa explains why the human voice can make high sounds of low intensity. Indeed, it is possible to produce sounds with the vocal cords tensed (high sounds) and the pressure low (weak intensity) by keeping the vocal cords just barely open.

The Three Levels of the Larynx

If we look at a frontal section of the larynx, we see three sections aligned vertically, one on top of the other. Each of these areas has specific features that are important for the production of sound. We'll now look at the three levels in the order in which the expiratory air passes through them—that is to say, we will describe in succession, from the subglottal level to the glottis level to, finally, the supraglottal level.

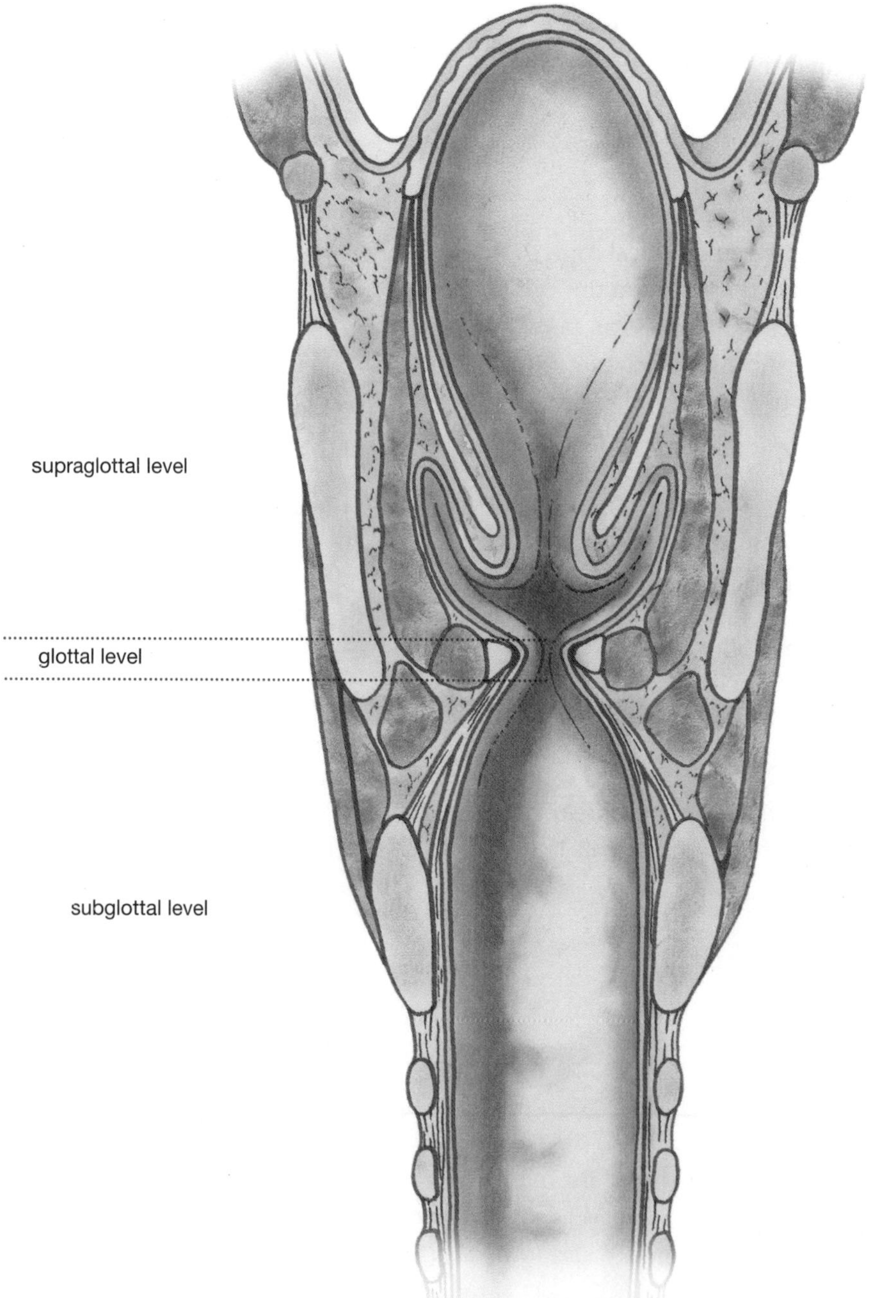

The Subglottal Level

This is the uppermost area of the trachea, just below the vocal cords. At this level, the mucosa runs from the walls of the trachea to the inner border of the vocal cords. Therefore, seen from the inside, the area has a shape that tapers upward.

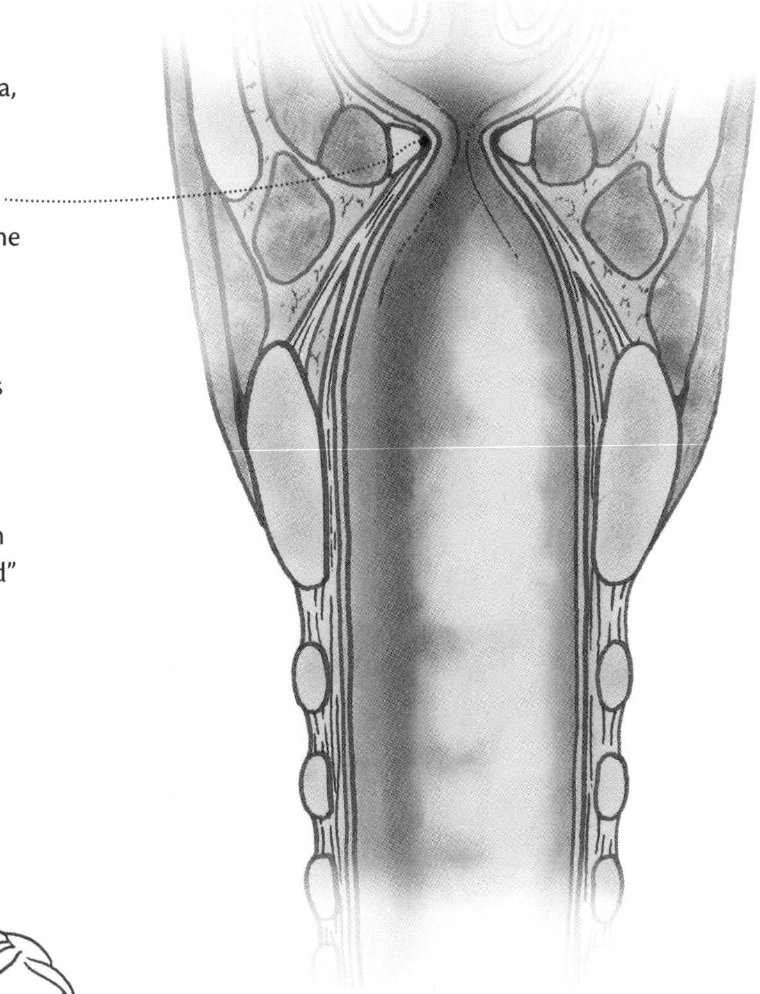

Here, deep in the mucosa, there are a large number of sensory nerve receptors that are sensitive to pressure (these are called **baroreceptors**). These receptors are stimulated every time there is a change in pressure. Thus, each change in the pressure of exhaled air is "recognized" by the sensory nervous system.

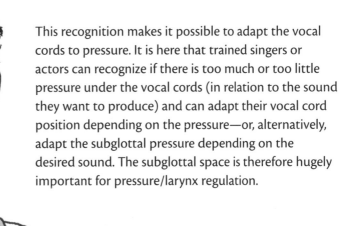

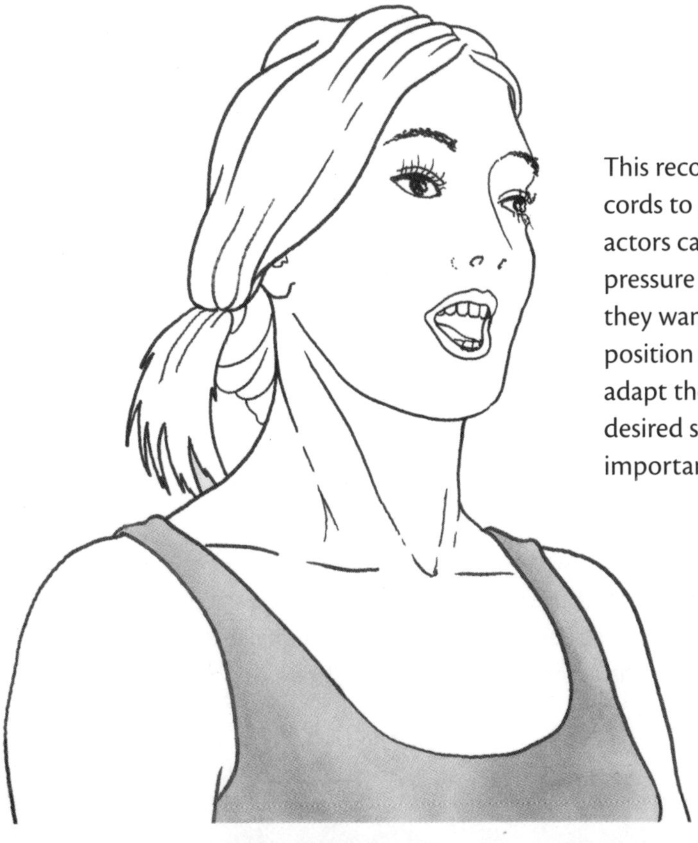

This recognition makes it possible to adapt the vocal cords to pressure. It is here that trained singers or actors can recognize if there is too much or too little pressure under the vocal cords (in relation to the sound they want to produce) and can adapt their vocal cord position depending on the pressure—or, alternatively, adapt the subglottal pressure depending on the desired sound. The subglottal space is therefore hugely important for pressure/larynx regulation.

The Glottal Level

This is the level of the **vocal cords** (see page 152) and the mucosal area where the **vocal folds** take shape. It has two interesting characteristics:

- The mucosal structure here differs from that of the subglottal mucosa, which is a ciliated, respiratory type (ciliated pseudostratified epithelium) like that found in the bronchi. Instead, the mucosa of the vocal fold is the same type as that of the mouth: stratified squamous epithelium.

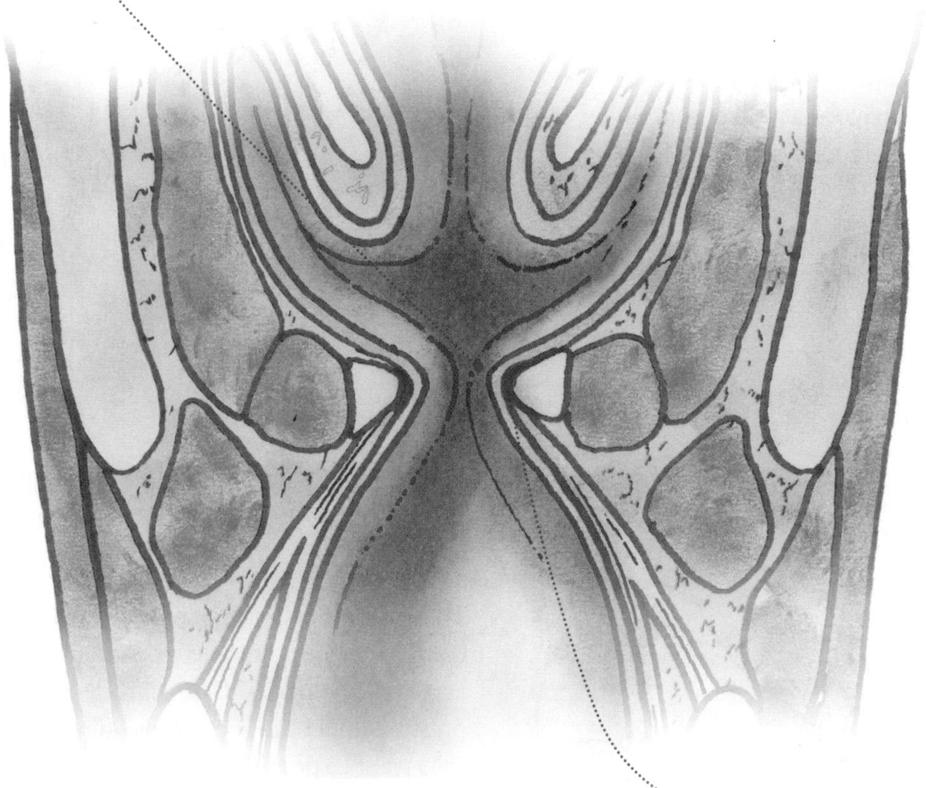

- In the area of the vocal folds, the mucosa doesn't adhere to the underlying tissue. There's a virtual space, called **Reinke's space,** between the vocal ligament and the mucosa. Here the mucosa is separable, meaning that it can slightly detach, allowing it to glide on the ligament. This is what makes undulations possible in certain vocal techniques (see page 173). This space has little or no vascularization. It can be the site of pathologies, specifically edema (swelling), which makes undulation difficult and jeopardizes the ability of the larynx to vibrate.

The Glottis Is a Space

We often think of the glottis as an organ. It's not. In fact, it's a space, defined by the limits of the vocal folds. The shape of the glottis is variable, and this has nothing to do with the production of sound. However, the fact that the form of the glottis is variable directly influences the vocal act.

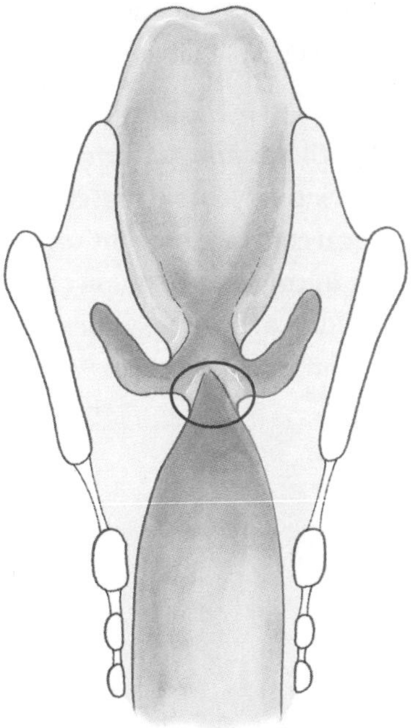

The Glottis Can Be *Closed*

When the glottis is "closed," it means that the vocal folds have been brought together to the point of touching. We call this *adduction of the vocal cords* (or vocal folds), or *glottal adduction*.*

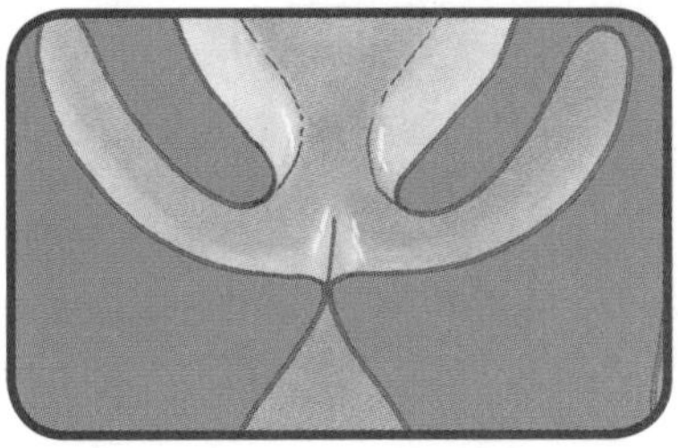

A priori, when the glottis is closed, air, liquid, and solids cannot pass. Closure of the glottis is not a resting position but is the result of muscular action, which can be of greater or lesser intensity (see the next page).

"A Lump . . ."

Sometimes the glottis will close in an undesirable way, even unconsciously or involuntarily: we might say, "I have a lump in my throat."

The Glottis and Phonation

Concerning the voice, the coming together of the vocal folds is a prerequisite for phonation. However, there is not necessarily a complete closure. In a frontal sectional view, the vocal folds are often even somewhat concave toward the glottal opening. That's when the Bernoulli effect (see page 172) takes place and attracts the vocal folds toward each other as the air passes.

*The term *adduction* refers to a movement in the frontal plane (see page 16) in which a structure is mobilized toward the middle of the body. Here it refers to the movement in which the back of each vocal cord moves toward the midline of the glottis.

There are several ways of closing the glottis, each of which will correspond with the production of different sounds.

The Glottis Just Adducted ("Just Closed")

The interarytenoid muscles are contracted, so that the vocal cords are brought together and the vocal folds are in contact. Air doesn't pass, and there is a buildup of pressure under the cords, creating a sound wave (see pages 94, 154, and 290).

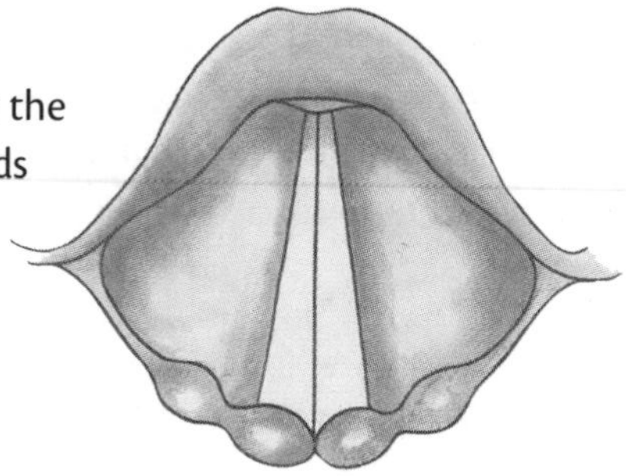

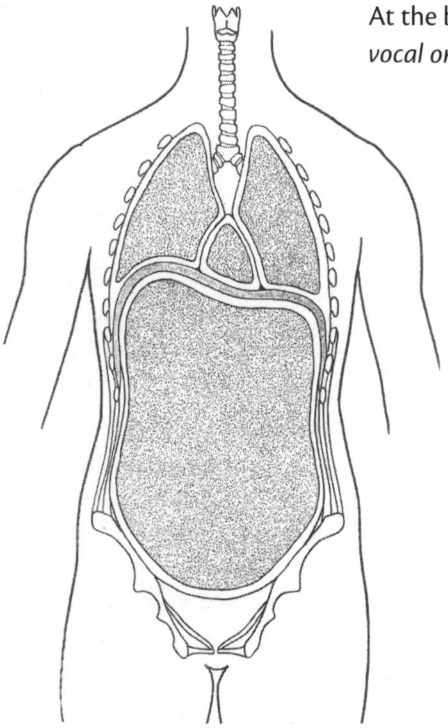

At the beginning of the vocal sound (called the *attack* or *vocal onset*), this position of closure can work in three ways.

It may precede the vibration of the vocal cords. In this case the glottis behaves first like a sphincter, which interrupts the expiratory airflow, thereby creating subglottal air pressure before the vibration. In this case, when the cords start vibrating the closure is very abrupt, like a small explosion. This is sometimes called a *hard attack*. An extreme example of this is when we cough.

The closure may take place gradually during the exhalation. Unlike the first example, this closing will both interrupt the flow of expiratory air and turn it into sound waves, but the two coexist. In this case we hear the breath at the initiation of the sound. This is sometimes called the *soft attack* or *aspirate onset*. A lot of air is lost with this.

Or it can be exactly synchronous with the vibration. In this case, the flow of air is immediately converted into a sound wave. This is sometimes called *balanced onset*.

The closure of the vocal cords and/or the ventricular bands is not necessarily symmetrical in any of the glottal positions, and therefore neither is the closure of the glottal slot.

Closed Glottis with Medial Compression

With the glottal opening closed, and the interarytenoid muscles contracted as before, we add the contraction of the posterior cricoarytenoid muscles. They rotate the arytenoid so that the vocal processes are in very close contact. This effect is called *medial compression*. As a result, the arytenoids are tightened along their full length. The vibrating portion of the vocal cords is then shortened, and this allows for the emission of very high notes.

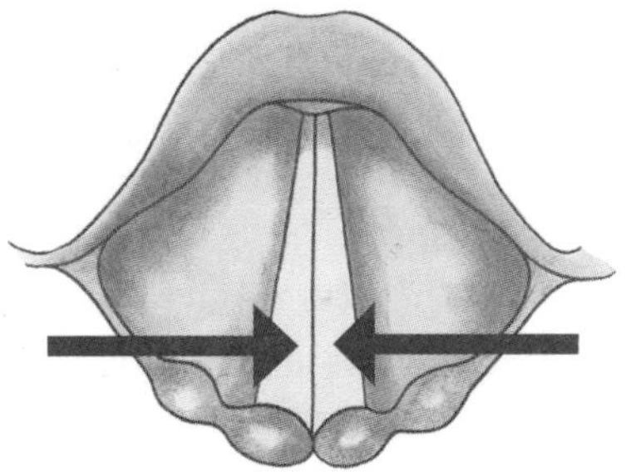

Strong Closure of the Glottis

The cricoarytenoid and posterior interarytenoid muscles are strongly contracted. This strong closure is necessary when the subglottal pressure is very high, such as:

- when we produce a very powerful sound
- when we rise to very high notes, because there the vocal cords are elongated and thin, and more difficult to close; if the glottis isn't closed tightly enough, there will be a bit of air leakage during phonation
- when we cough

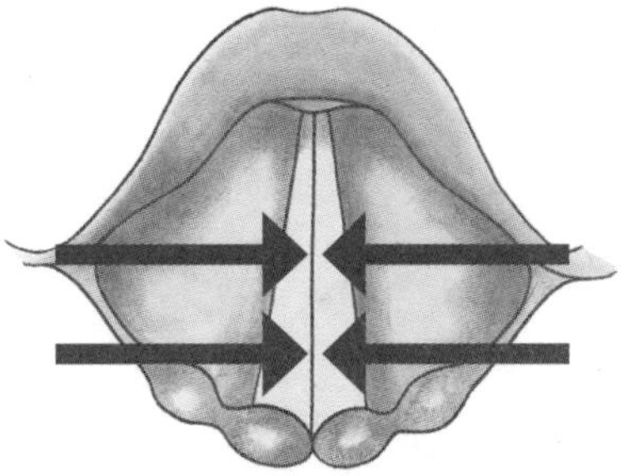

Strong Closure of the Glottis with Ventricular Band Closure

The cricoarytenoid and posterior interarytenoid muscles are strongly contracted. The glottal opening is closed, with the vocal folds in very close contact. Air does not pass. Even if there is high air pressure in the glottis, air, in most cases, does not pass (neither in the anterior two-thirds nor in the posterior third). In a view of the larynx from above, we do not see the vocal cords, as they are hidden by the two folds of the ventricular bands, which are in contact.

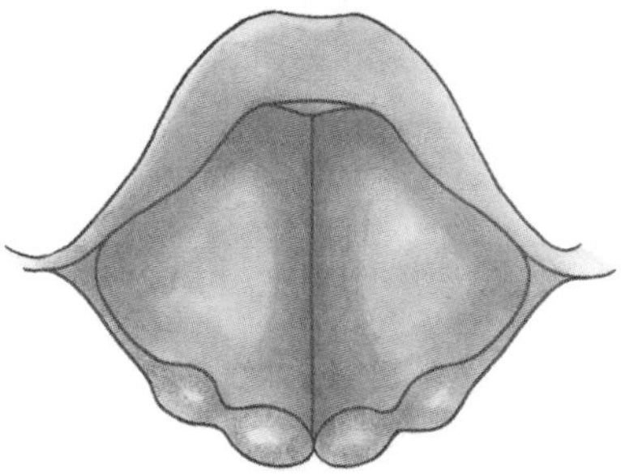

Positions of "Effort"

We take certain positions when we cough strongly or scream or when we have the hiccups. Sometimes we take them during strenuous exercise, when we want to block the exhalation.

The Glottis Can Be *Open*

We can observe two key positions, with all of the intermediate positions possible between them.

The Glottis Slightly Open

All of the intrinsic laryngeal muscles are relaxed. The vocal cords are not in contact but slightly apart. The air passes with a slight noise caused by the friction.

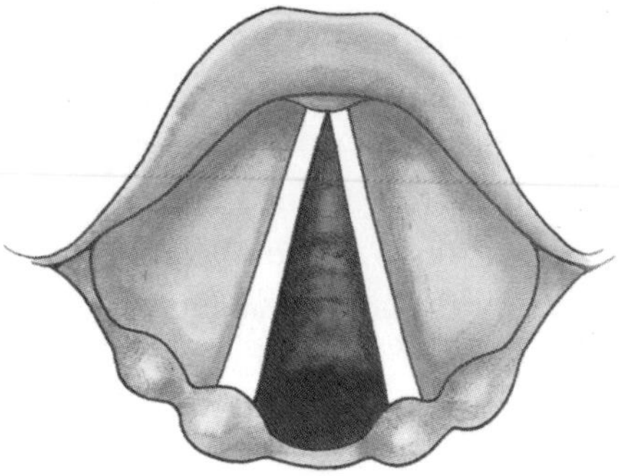

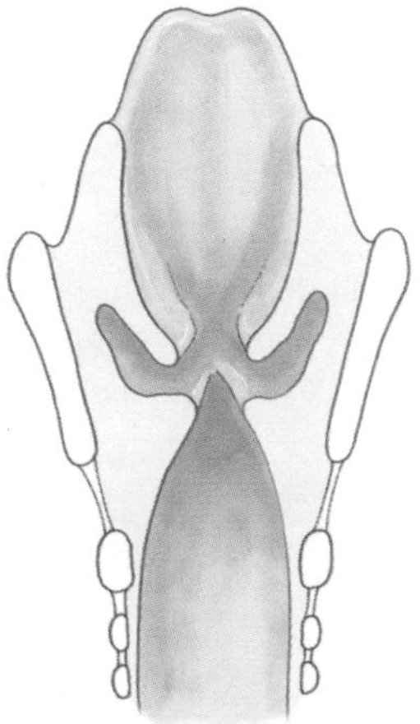

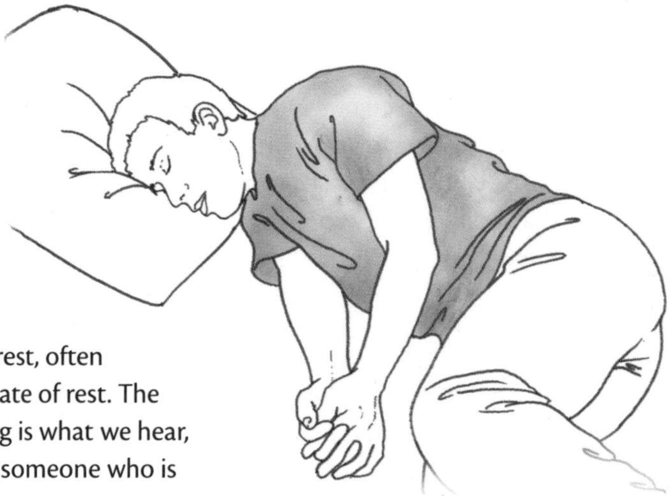

This is a position of glottal rest, often associated with a general state of rest. The frictional noise of air passing is what we hear, typically, when we listen to someone who is sleeping.

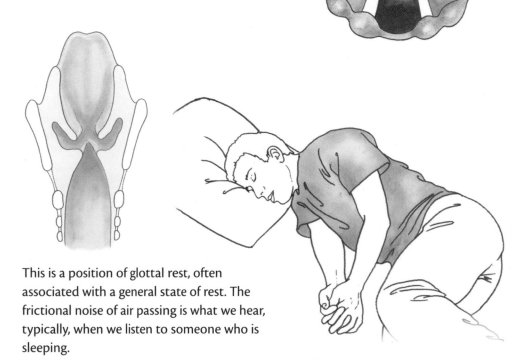

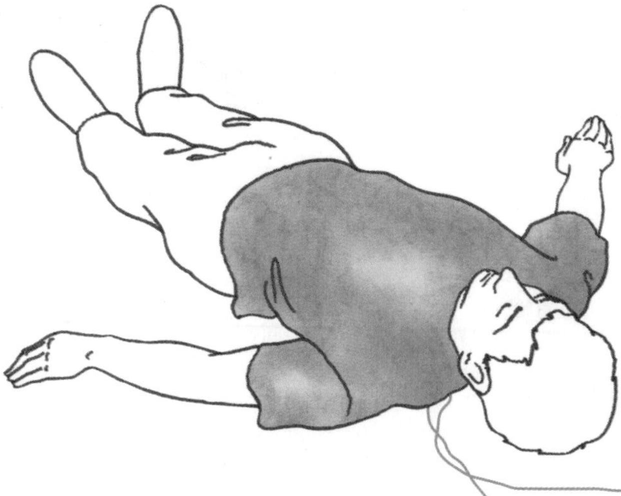

An Indication of Relaxation of the Entire Body

This slight rubbing sound can be used as an indicator to check the state of release in a relaxation exercise (easiest in a prone position or yogic "rest" position). It can also be used in any position to check the level of glottal release in voice work.

The Glottis Wide Open (Also Called a Forced Opening)

The interarytenoid muscles (see page 164) are relaxed (with no adduction, or bringing together, of the cords). The lateral cricoarytenoid muscles (see page 162) are contracted and cause the arytenoids to rotate. The vocal cords are separated. The glottal opening takes the shape of a pentagon, and the air flows without the noise of glottal friction.* You can't hear any sound, neither the inhalation nor the exhalation, even if the singer or speaker is using a microphone.

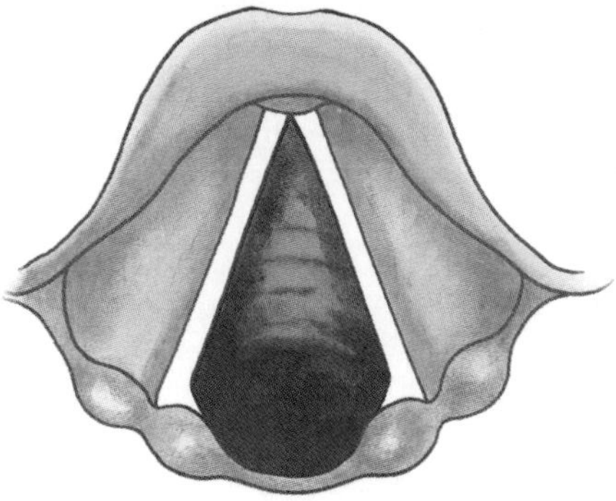

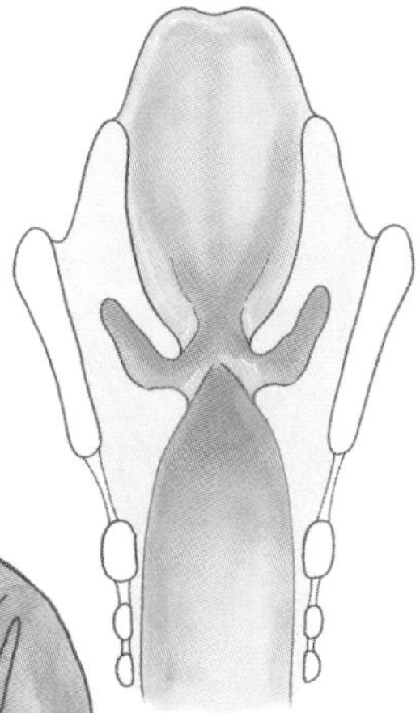

Detail Work

You can practice recognizing the difference between these two open positions using a microphone. Start from a very relaxed position, with only the sound of frictional noise. From there, try to not make any noise at all and observe what happens when breathing silently. You will most likely feel yourself having to make a bit of effort, specifically in the area of the larynx and thyroid—the place you palpated on page 147.

*Note: Frictional noises may also arise from other parts of the vocal tract (see part 6).

The Open Glottis and Phonation

The open glottis creates the conditions for whispered vowels. It allows the production of voiceless consonants, also called unvoiced consonants. You can feel the opening at the glottis when a voiced consonant (voiced) passes to a voiceless consonant (unvoiced)—for example, passing:

from "vvvvv" to "ffffff"
from "zzzzzz" to "sssss"
from "jjjjj" to "chchchchch"

The Supraglottal Level

Here we find the **ventricular bands:** the mucosal folds located above the vocal cords in the supraglottal area. They are also called the false vocal cords. There is one on the right and one on the left.

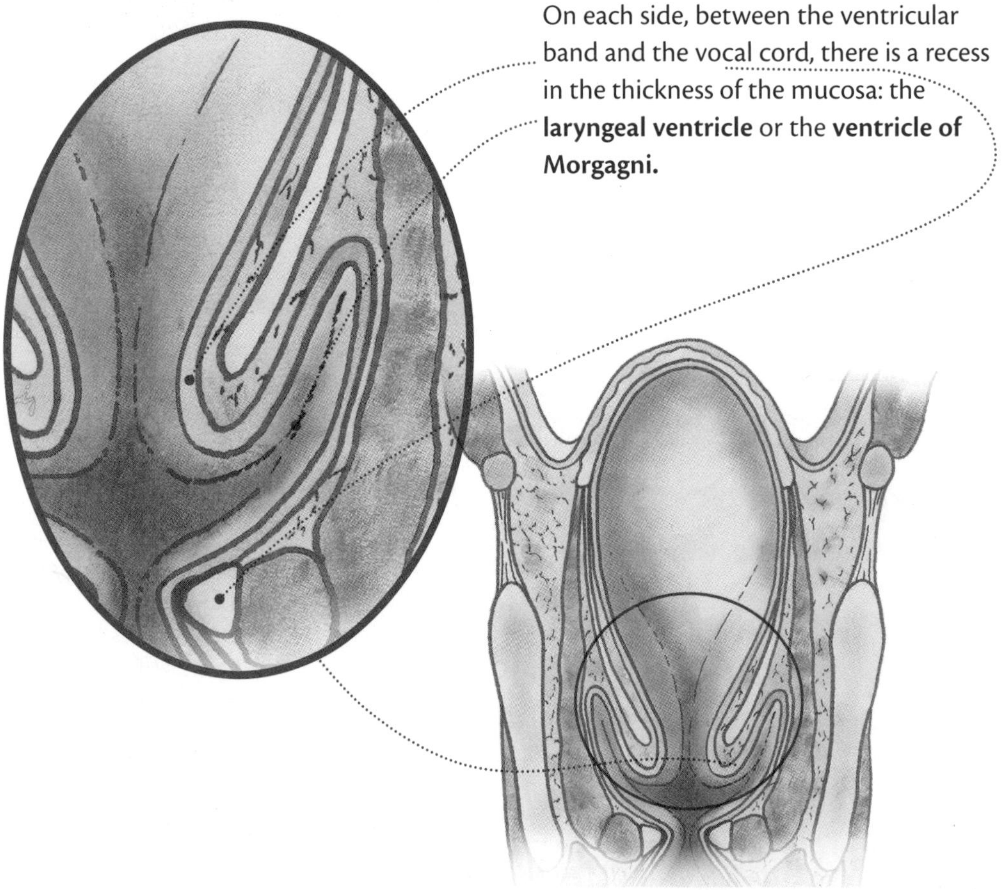

On each side, between the ventricular band and the vocal cord, there is a recess in the thickness of the mucosa: the **laryngeal ventricle** or the **ventricle of Morgagni.**

The ventricular bands not only are mucosal folds (as they are often described) but are underpinned by a thickening of the lateral or external thyroarytenoid muscle. They can actively change shape with the contraction of the muscle. Therefore, these ventricular bands are not inert but can become thicker or thinner, and closer together or farther apart. They contribute to the sphincteral role of the larynx (see page 170). As previously mentioned, because of the muscular action, the laryngeal ventricles can change shape, hollowing or flattening when it's advantageous. They are the first resonance chambers above the vocal cords, and these changes can change the timbre of the voice.

The Ventricular Bands and Laryngeal Timbre in Upper Registers

There is a "kinship" between the internal and external thyroarytenoid muscles. When the glottis closes forcefully, the internal contracts and the external tends, by its proximity, to join this contraction. This is what happens spontaneously when the voice enters the high registers: subglottal pressure increases, the vocal cords are increasingly under tension, the internal thyroarytenoid muscles contract to assist them, and the external thyroarytenoid muscles tend to accompany them. The result is that the ventricular bands come closer together as the notes climb. This reduces the size of the first supralaryngeal cavity, which changes the timbre. Classical singing often calls for evening out the tone and smoothing out these differences; when climbing to higher registers, there's an attempt to leave the musculature of the ventricular bands decontracted while the muscles of the vocal cord contract.

Habitual Closing

It has to be noted that regardless of singing styles that attempt modifications of timbre, closing the ventricular bands is not harmful for the larynx and we do it without problem many, many times a day.

The Vibration of the Ventricular Bands

The ventricular folds can vibrate, but much less rapidly than the true vocal cords, and the resulting sound is much deeper. This is the mucous membrane vibrating, a bit like the lining of the soft palate. Sometimes this vibration joins the vibration of the vocal cords.

It is not true that the vibration of the ventricular bands produces a falsetto (head voice in a male), as is sometimes said. The sound of the ventricular folds is not a high pitch but, rather, a low one.

The Aryepiglottic Space and the "Twang"

Above the ventricular bands we find a space that is still under the epiglottis and not really the pharynx. This space is bounded by:

- the epiglottis at the top
- the ventricular bands at the bottom
- both aryepiglottic muscles at the sides and top
- both external thyroarytenoid muscles at the sides and bottom

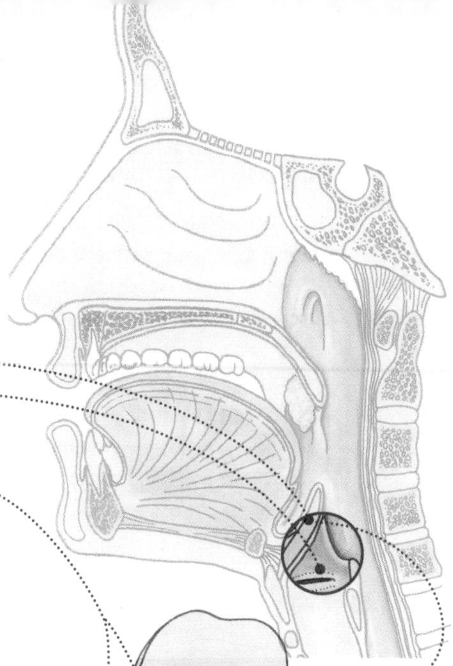

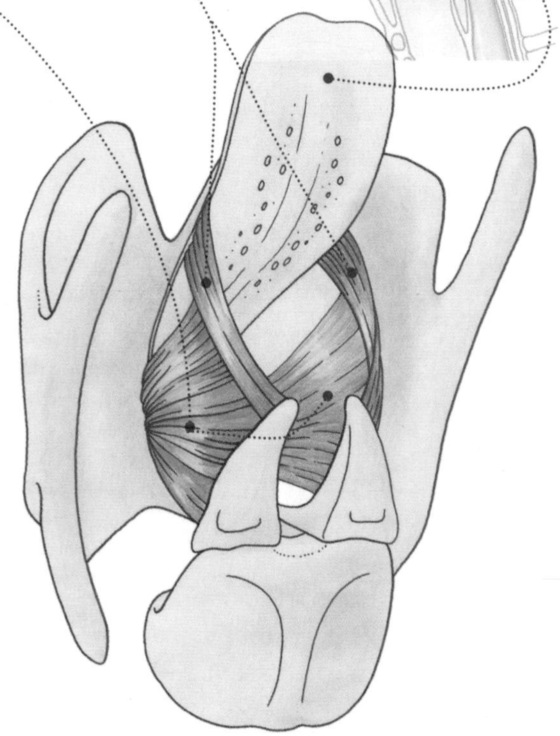

This area can change shape in two ways. The contraction of the lateral muscles can make it narrower, and the contraction of the aryepiglottic muscles has the effect of lowering the epiglottis, reducing the space. (This lowering of the epiglottis is not related to the pulling back of the tongue, as takes place in swallowing.)

When this space is reduced, it favors certain high harmonics. This creation of high harmonics is called the *twang*.

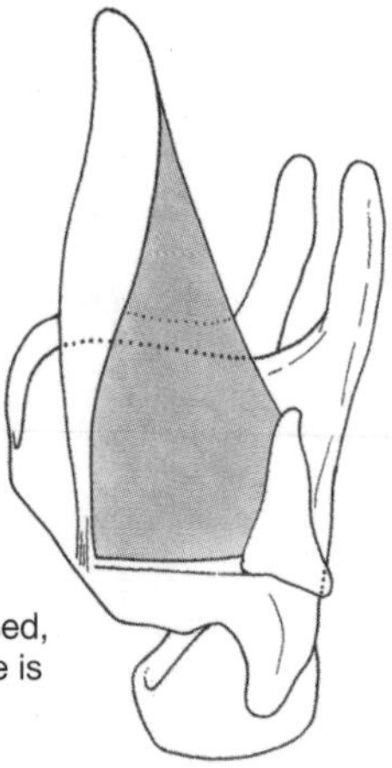

With the epiglottis raised, the aryepiglottic space is enlarged.

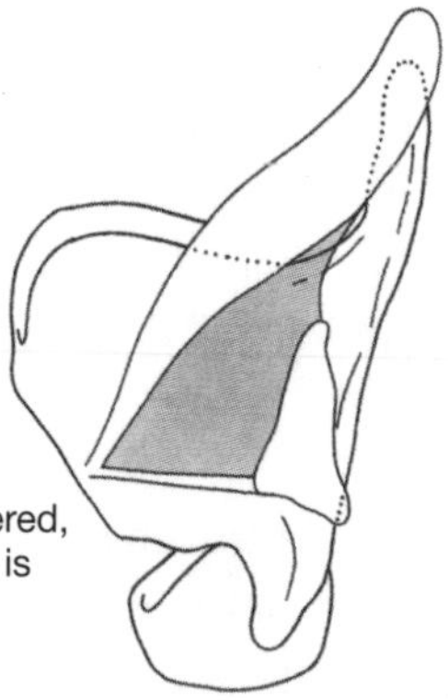

With the epiglottis lowered, the aryepiglottic space is reduced.

Views of the Larynx by Laryngoscopy and Nasofibroscopy

Several kinds of exams are used to see the larynx from above. We cite them here just to give you an understanding of the orientation of the image depending on the method of examination (the images can easily be seen on the Internet).

An indirect laryngoscopy. This is the oldest method, but it's still sometimes used. A mirror is placed at the end of a rod that is held at the back of the throat. The observer then has a reflected view in the mirror. Note that this is a reversed view.

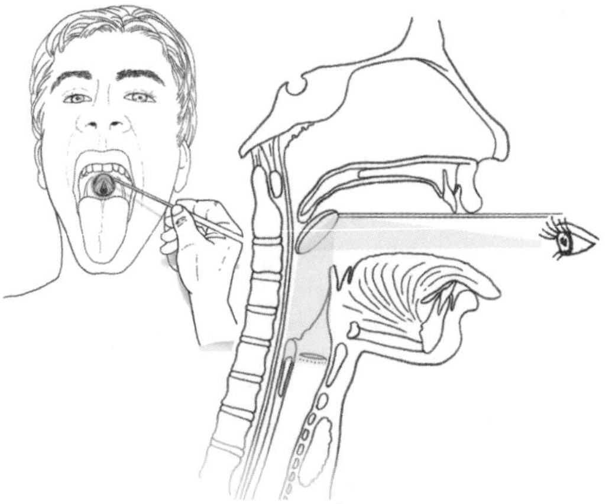

We see the epiglottis at the top of the image. The vocal cords are seen as two white lines. If the glottis is open, they form as a "caret." Apart from the vocal cords, we see the ventricular bands. We can guess at where the tracheal rings are. We see the space of the glottis if it is open. The arytenoids are at the bottom of the image, where they form two rounded bulges. Two more median bulges are formed by the cartilaginous horns.

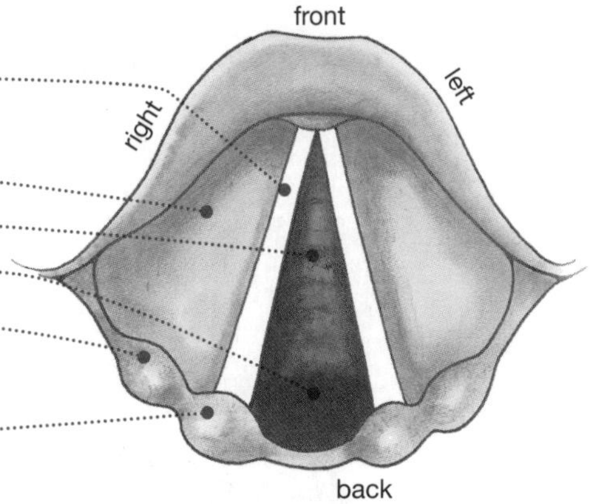

A laryngoscopy. This method uses a rigid endoscope inserted into the mouth. We see the same landmarks, but in the reverse order. The arytenoid cartilages are at the top of the image, and the epiglottis at the bottom.

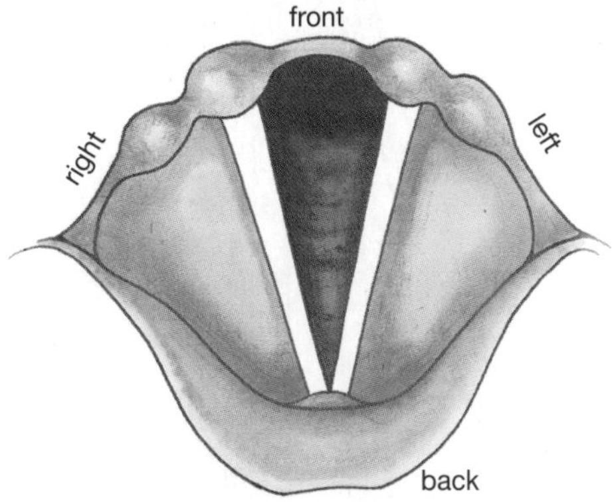

A nasofibroscopy. A flexible tube connected to an optical device and a camera is introduced through a nostril to just above the larynx. The image is not inverted; the arytenoids are at the top of the image, and the epiglottis is at the bottom.

The Extrinsic Muscles of the Larynx

The muscles of the larynx that we studied earlier (pages 161–70) stretch between the laryngeal cartilages. They are called the **intrinsic laryngeal muscles**.

The larynx is also mobilized or stabilized in its environment by muscles that connect to neighboring structures:

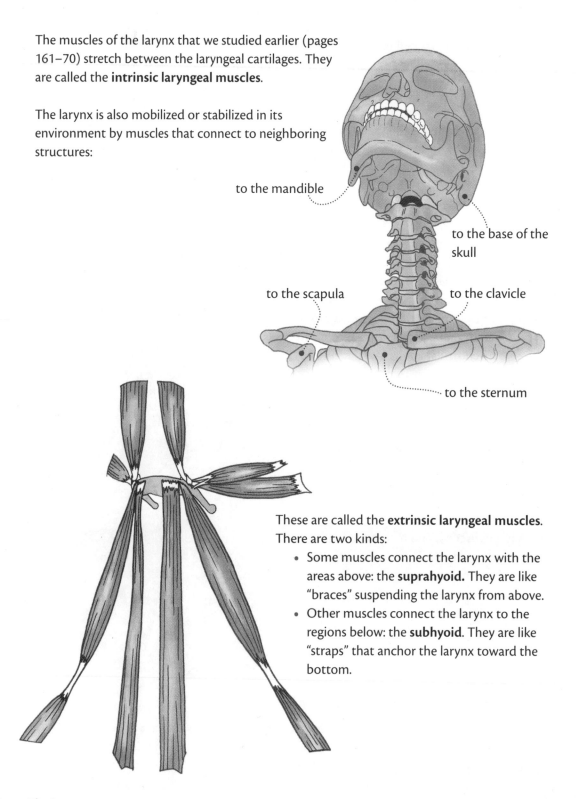

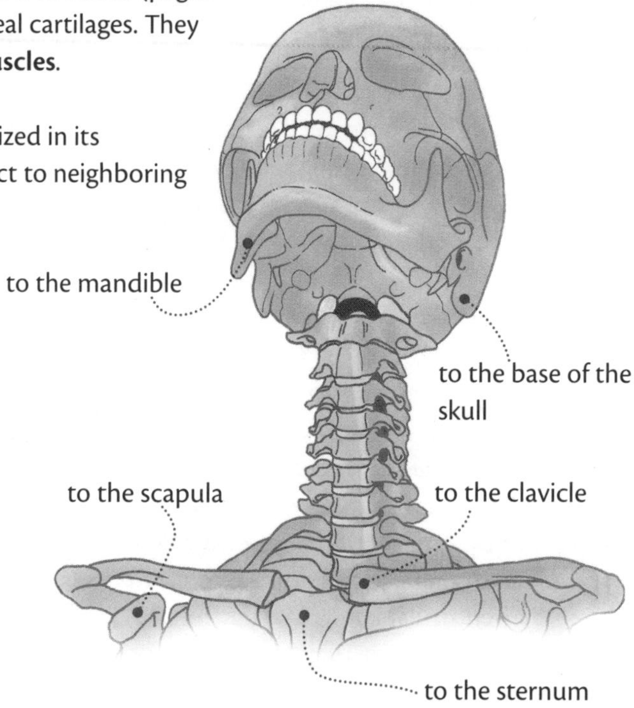

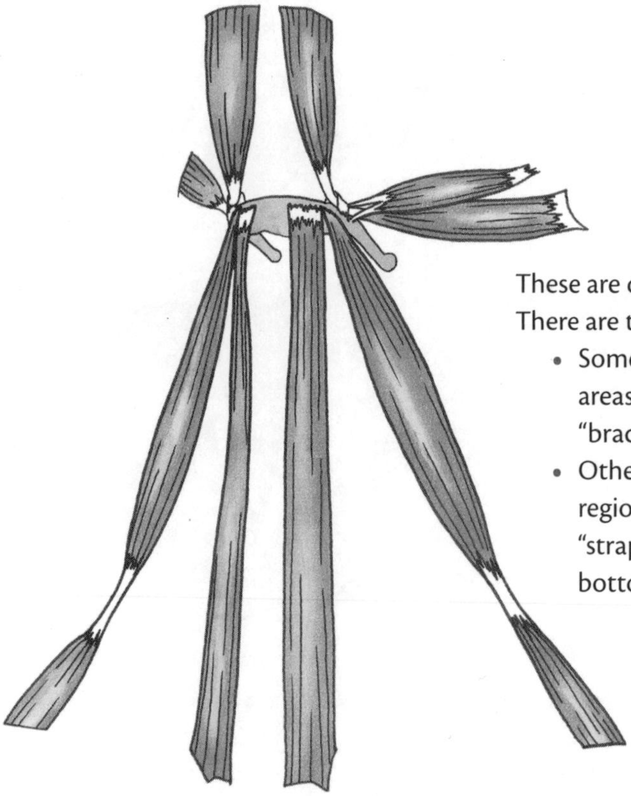

These are called the **extrinsic laryngeal muscles**. There are two kinds:

- Some muscles connect the larynx with the areas above: the **suprahyoid.** They are like "braces" suspending the larynx from above.
- Other muscles connect the larynx to the regions below: the **subhyoid**. They are like "straps" that anchor the larynx toward the bottom.

The Suprahyoid Muscles

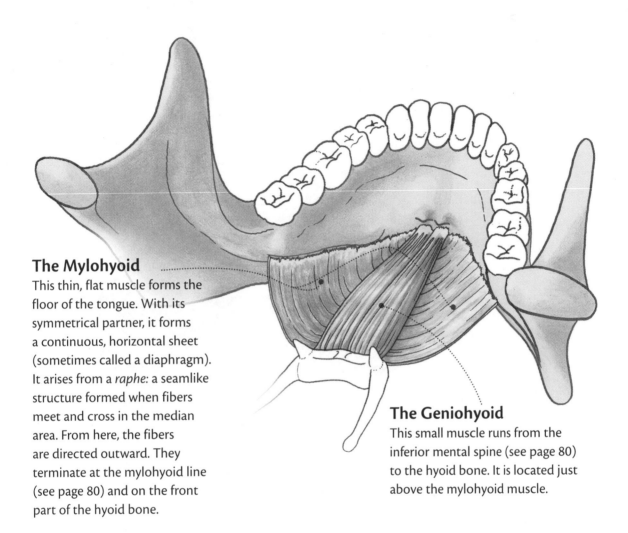

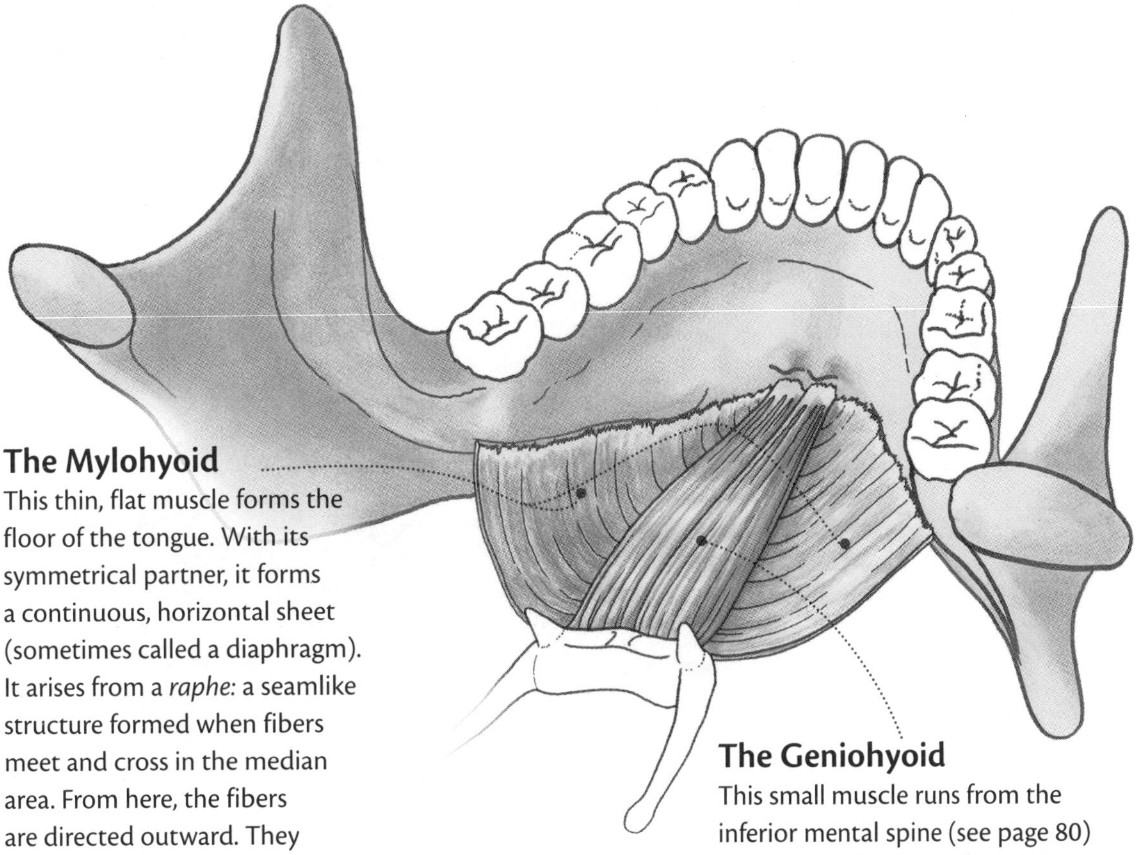

The Mylohyoid

This thin, flat muscle forms the floor of the tongue. With its symmetrical partner, it forms a continuous, horizontal sheet (sometimes called a diaphragm). It arises from a *raphe:* a seamlike structure formed when fibers meet and cross in the median area. From here, the fibers are directed outward. They terminate at the mylohyoid line (see page 80) and on the front part of the hyoid bone.

The Geniohyoid

This small muscle runs from the inferior mental spine (see page 80) to the hyoid bone. It is located just above the mylohyoid muscle.

Action of These Muscles

When they contract, they become flatter. They support the weight of the tongue. They participate in the elevation of the hyoid bone and its forward traction. They contribute to the lowering of the mandible.

Before Clicking Your Tongue

We can tone these muscles by lifting the tongue to the hard palate—for example, just before we "click" the tongue.

The Digastric Muscle

This muscle gets its name because it is in the form of two bundles, or "bellies."

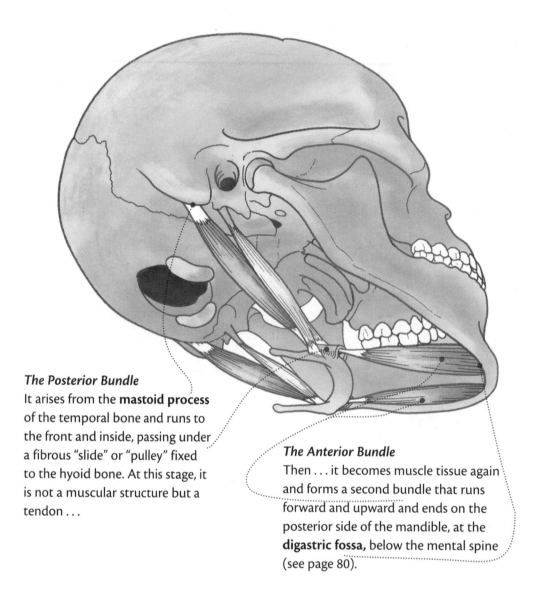

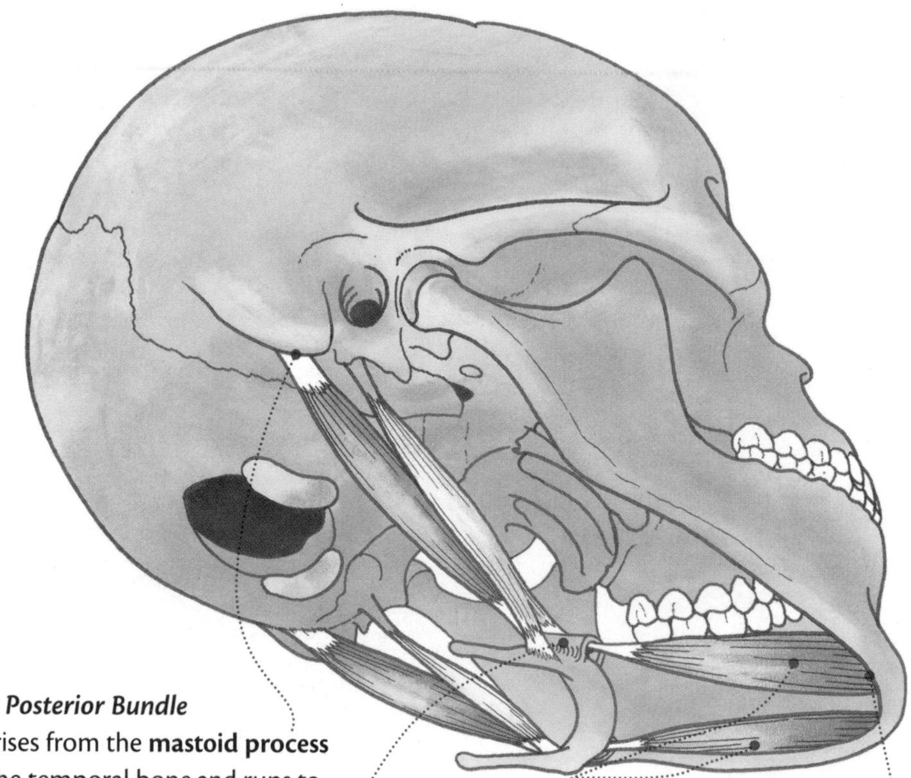

The Posterior Bundle

It arises from the **mastoid process** of the temporal bone and runs to the front and inside, passing under a fibrous "slide" or "pulley" fixed to the hyoid bone. At this stage, it is not a muscular structure but a tendon . . .

The Anterior Bundle

Then . . . it becomes muscle tissue again and forms a second bundle that runs forward and upward and ends on the posterior side of the mandible, at the **digastric fossa,** below the mental spine (see page 80).

The Stylohyoid

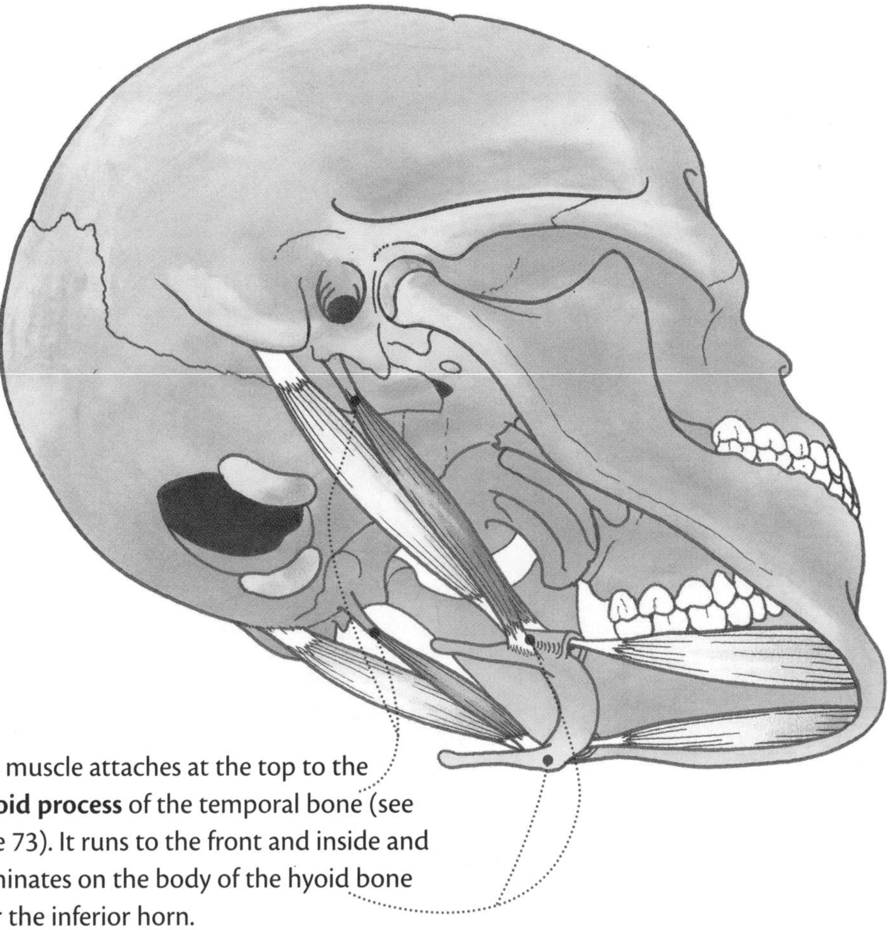

This muscle attaches at the top to the **styloid process** of the temporal bone (see page 73). It runs to the front and inside and terminates on the body of the hyoid bone near the inferior horn.

The Action of the Digastric and Stylohyoid

These two muscles elevate the hyoid bone. They also raise the base of the tongue, which is important in swallowing. If they act more on one side than the other, they draw the hyoid bone to that side. The horizontal portion of the digastric muscle is the muscular floor of the tongue, along with the mylohyoid and geniohyoid discussed on page 187.

Head, Neck, Larynx

The movements of the head backward or to the side put under tension one or another of the bundles of the digastric or stylohyoid, which, in turn, elevate the hyoid bone directly upward or to one side.

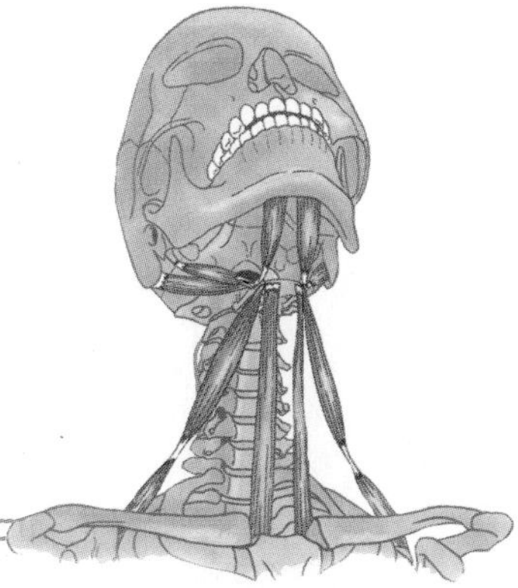

The Subhyoid Muscles

Two subhyoid muscles are deep and follow each other:

- the sternothyroid
- the thyrohyoid

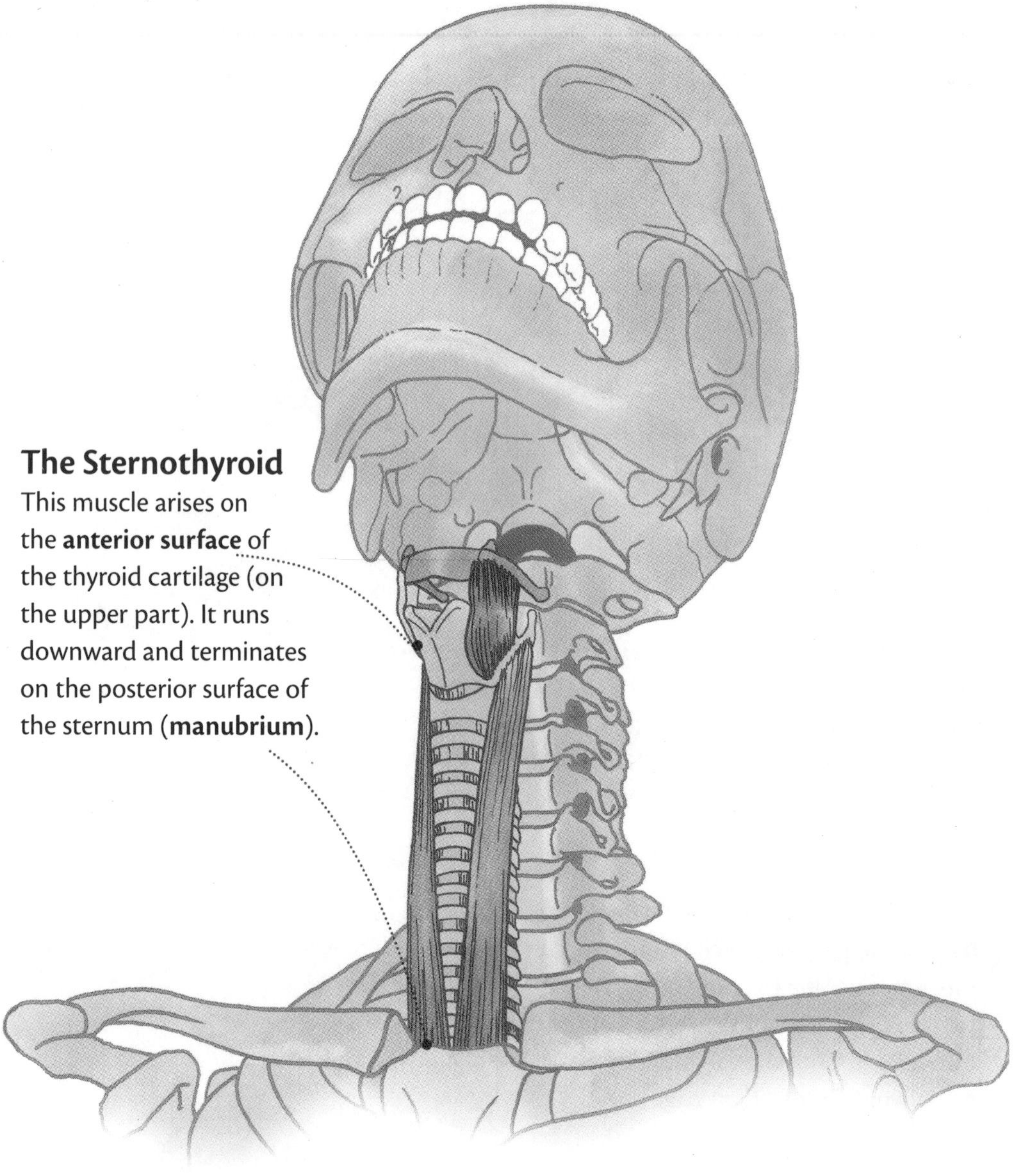

The Sternothyroid

This muscle arises on the **anterior surface** of the thyroid cartilage (on the upper part). It runs downward and terminates on the posterior surface of the sternum (**manubrium**).

Its Action

It lowers the thyroid cartilage. If it acts more strongly on one side, it lowers that side more.

Vocal Application

The sternothyroid makes up part of the group of muscles that lower the larynx.

The Thyrohyoid

This muscle arises from the superior horn of the hyoid (protruding slightly forward on the body). It runs downward and terminates on the anterior surface of the **thyroid cartilage** (in the superior part).

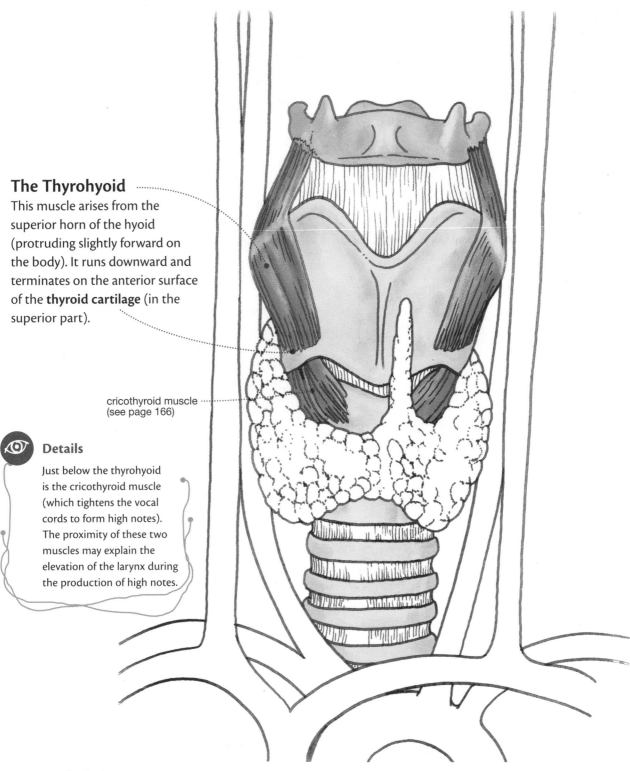

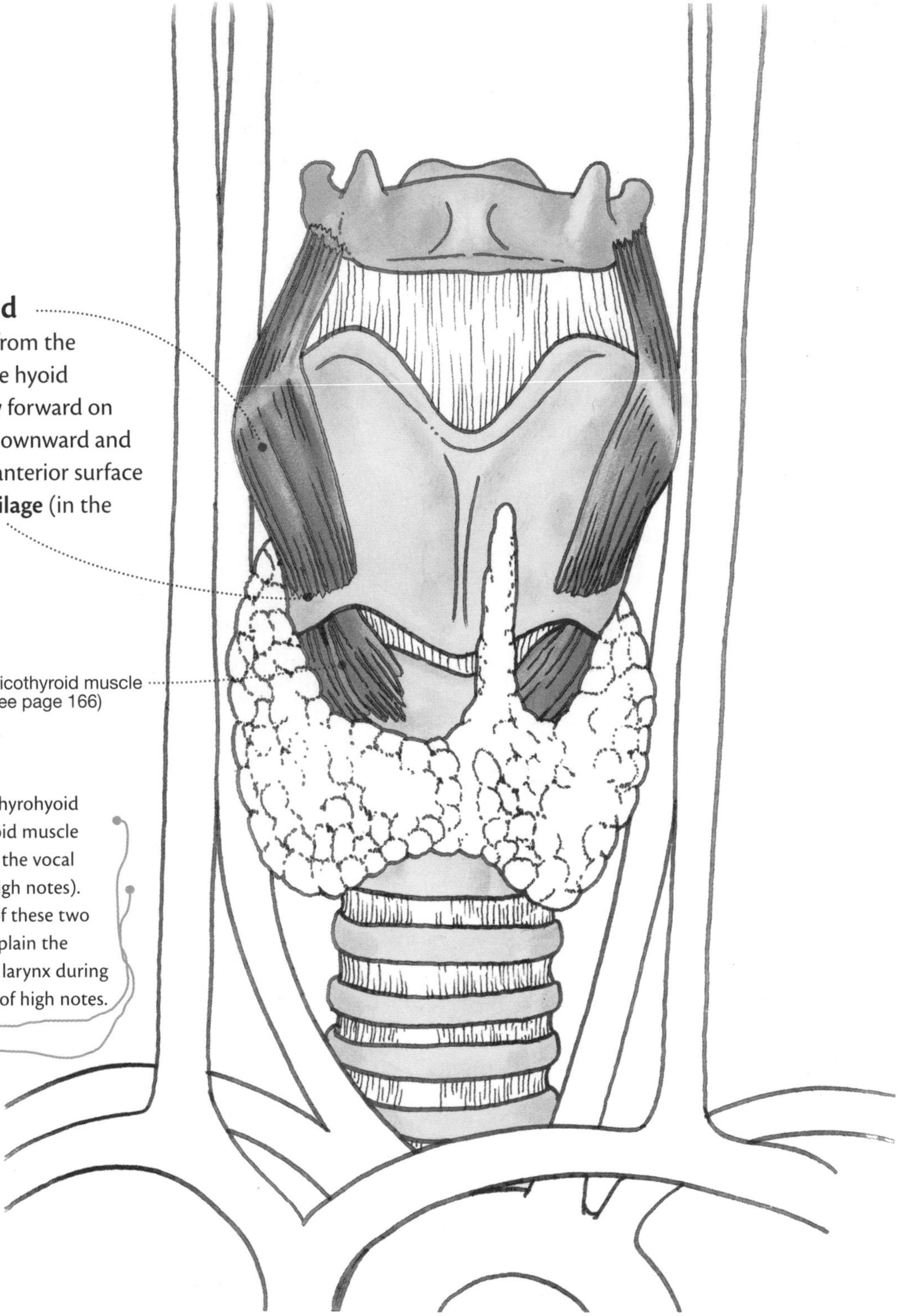

Details

Just below the thyrohyoid is the cricothyroid muscle (which tightens the vocal cords to form high notes). The proximity of these two muscles may explain the elevation of the larynx during the production of high notes.

Its Action

It can lower the hyoid bone. But if the hyoid bone is fixed, it can raise the thyroid cartilage. It actively contributes to the suspension of the larynx from the hyoid bone.

There are two more superficial subhyoid muscles:

- the sternohyoid
- the omohyoid

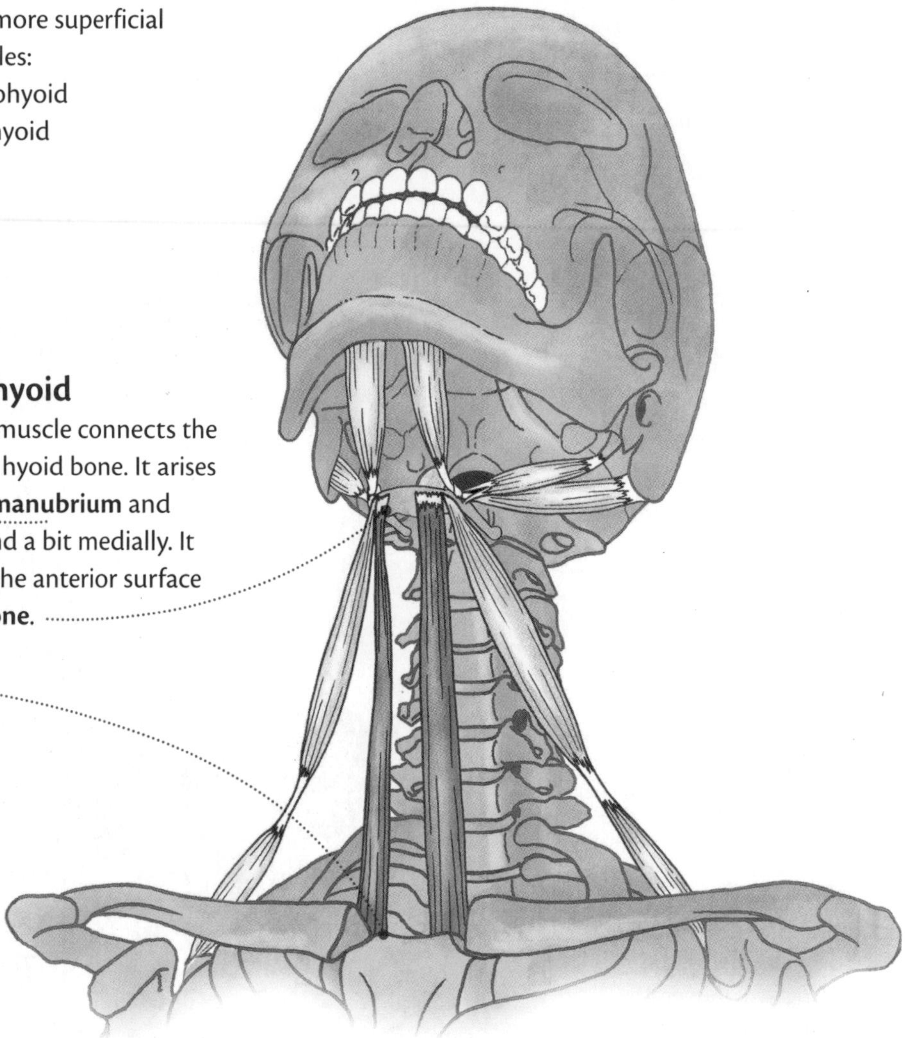

The Sternohyoid

This long, thin muscle connects the sternum to the hyoid bone. It arises on the sternal **manubrium** and runs upward and a bit medially. It terminates on the anterior surface of the **hyoid bone**.

Its Action

It lowers the hyoid bone. If it acts more strongly on one side, it lowers that side more.

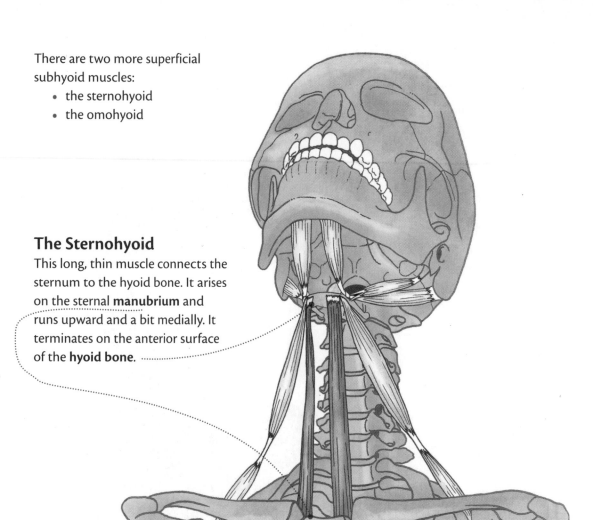

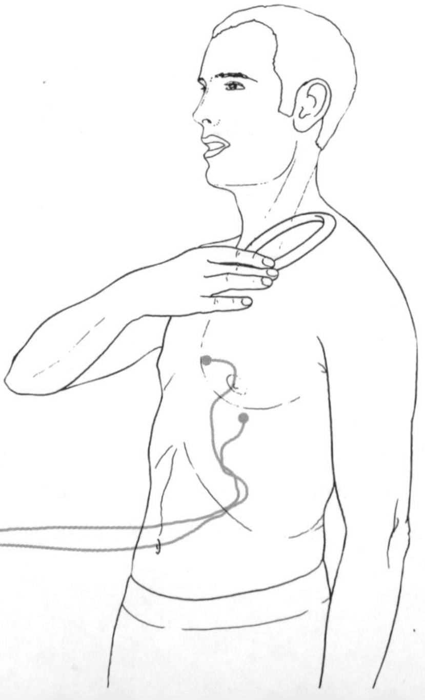

The Rib Cage, Sternum and the Hyoid

Movements that lower the sternum and rib cage put this muscle under tension and passively pull the hyoid bone downward.

The Omohyoid

This long, thin muscle connects the scapula to the hyoid bone. It arises on the anterior surface of the **hyoid bone** and runs downward and to the exterior.

It has two muscular bodies that are separated by a tendonous area. It terminates at the upper edge of the **scapula.**

Its Action

It lowers the hyoid bone. If it contracts more stronly on one side, it lowers that side more.

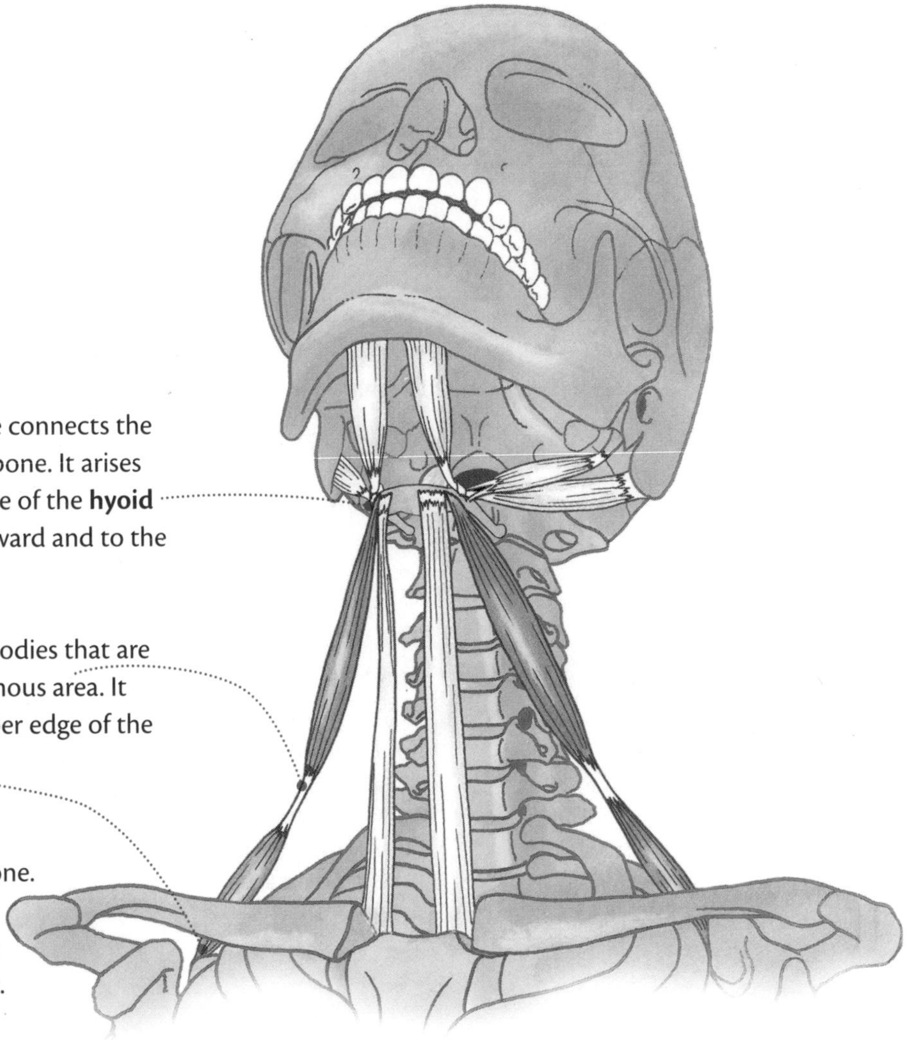

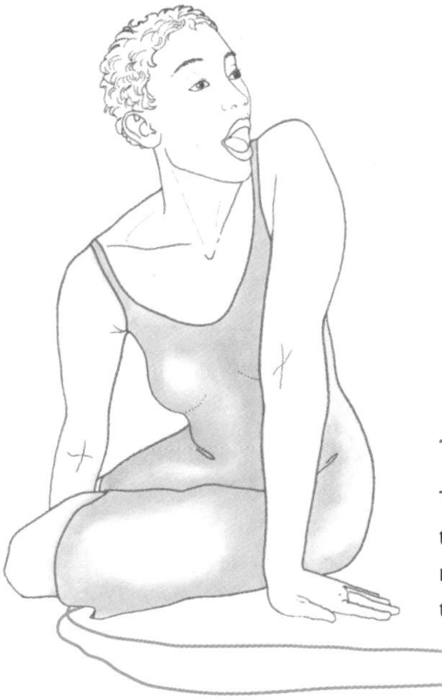

The Scapula and the Hyoid Bone

The action of lowering and pulling back the scapula (here, on the right) puts this muscle under tension and passively pulls the hyoid bone down.

The Larynx Can Be Pulled *Upward*

It can be pulled upward by:

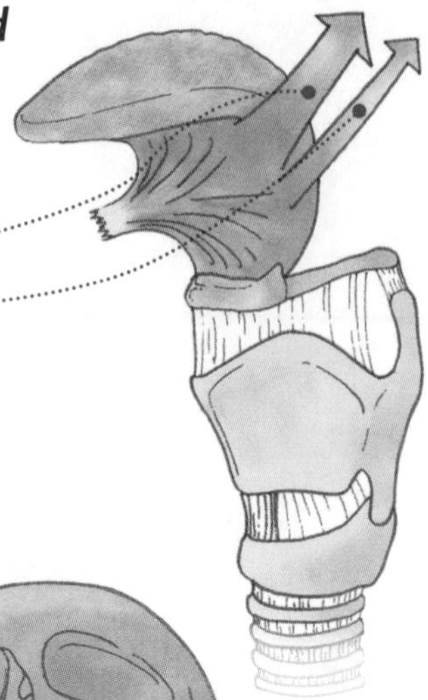

- the muscles that lift the tongue: the **palotoglossuss and styloglossus**
- the muscles that lift the pharynx: the superior and middle **pharyngeal constrictor muscles**
- the **suprahyoid muscles**

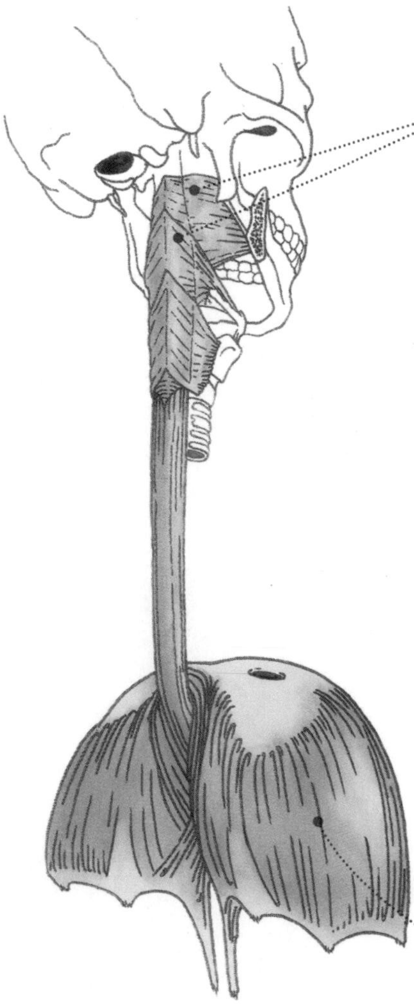

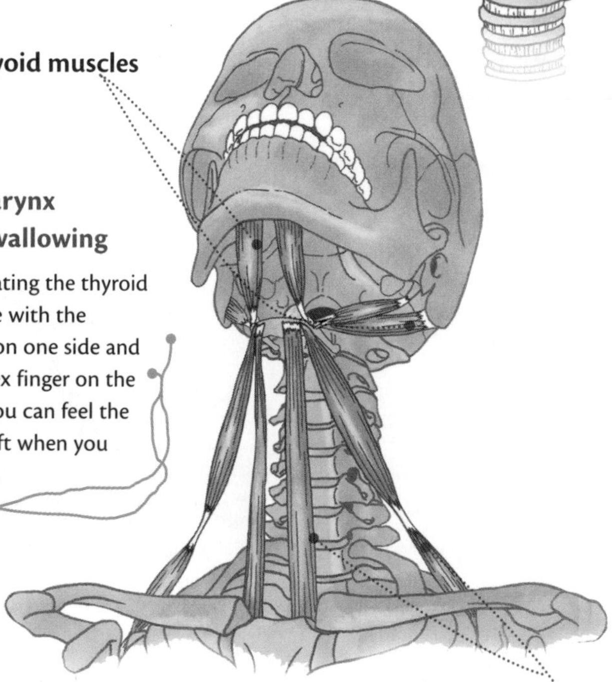

The Larynx and Swallowing

By palpating the thyroid cartilage with the thumb on one side and the index finger on the other, you can feel the larynx lift when you swallow.

The Larynx Can Be Pulled *Downward*

It can be pulled downward by:

- the **subhyoid muscles**
- the **diaphragm** (see page 121).

Tongue and Larynx

Place a finger at the back of the tongue and inhale; it is possible you'll be able to feel the tongue lower, because it is attached to the larynx, which is itself being pulled down by the diaphragm.

Stabilizing the Larynx

The "manufacturing" of treble notes takes place at the vocal cords: their vibrational frequency determines the pitch of the note (see page 293). There is no anatomical need for the larynx to lift when we produce a high note. Yet many people lift the larynx spontaneously and can be bothered by it. Why is that?

There can be a spontaneous synkinesia* between the internal and external thyroarytenoid muscles (see page 183), or between the cricothyroid and thyrohyoid muscles (see page 191).

There is also a neuromotor factor: When singing a high note, we begin by thinking about the height of the note. Many people then have a tendency to establish a synkinesia between what the brain plans to do (create a "rising" sound) and bringing the larynx "up." This lifting of the larynx can be performed by any muscle that elevates the hyoid bone (the suprahyoid) or the tongue, pharynx, or palate. Lifting the larynx can become a habit and, through repetition, an automatic neuromotor response. It may also be related to the fear of singing in high registers, involving brain areas that control the emotional memory of fear.

In light of this, it is interesting to know that you can exercise to keep the larynx stable, especially when singing high notes. Be careful, however, not to overcompensate and drop the larynx too much during high notes, which would lengthen the pharynx and change the resonance (unless that's what you want, of course).

Also, be careful not to let this cause a tensing of the muscles around the larynx, which could lead to tension of the intrinsic muscles as well. The stability of laryngeal height is achieved by a balance between the forces that lower and the forces that lift the larynx. This is done through progressive training that should leave the laryngeal region more relaxed than tensed.

The same phenomenon also exists in the other direction: we may systematically lower the larynx when singing a low note. Again, it is interesting to work to keep some stability in the pharynx, by recruiting, this time, the muscles that elevate it. Again, this action must be balanced so that the larynx is stabilized without tension.

**Synkinesia:* involuntary movement of one body part when another is moving.

5
The Vocal Tract

Anatomy of the Vocal Tract

In anatomy, a *tract* is the interior of an organ that is shaped like a hollow tube. The **vocal tract** includes all of the structures through which air passes as it travels to and over the vocal cords.

It includes the areas from above the glottis to the lips or nostrils. It is sometimes called the **pharyngobucconasal space.**

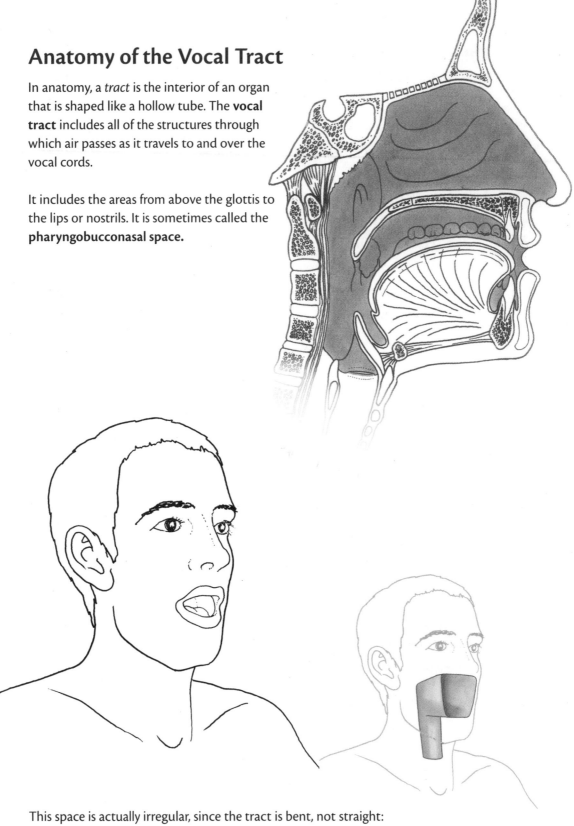

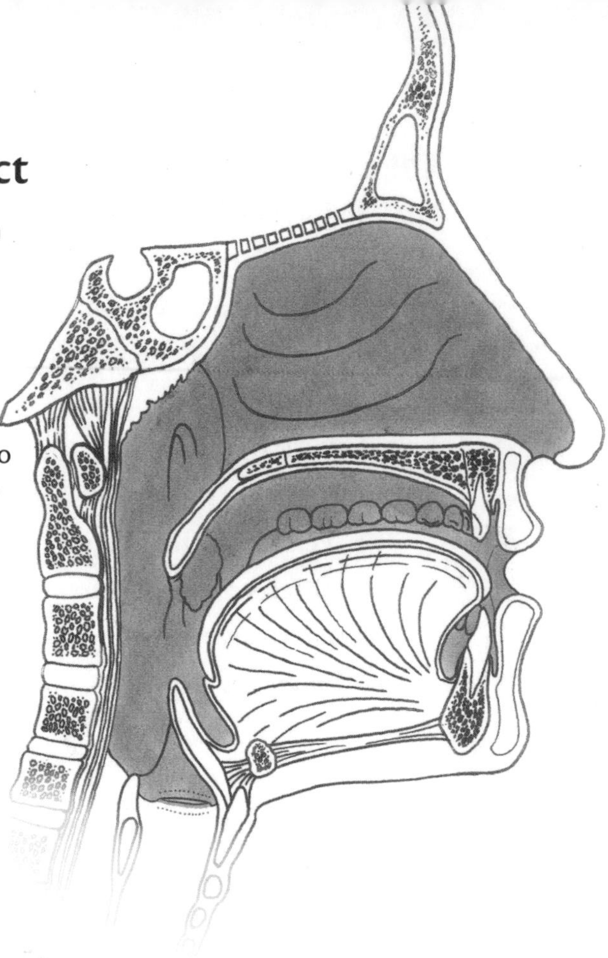

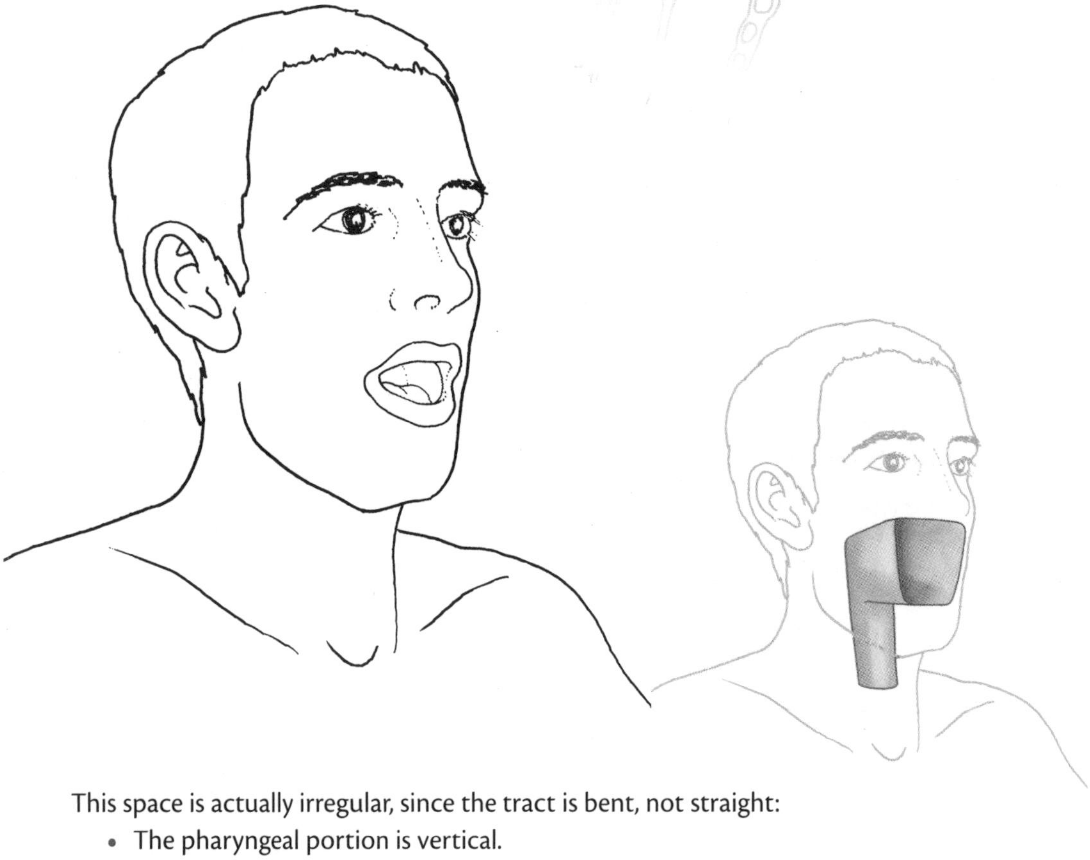

This space is actually irregular, since the tract is bent, not straight:

- The pharyngeal portion is vertical.
- The oral or nasal portion is relatively horizontal when the person is standing.

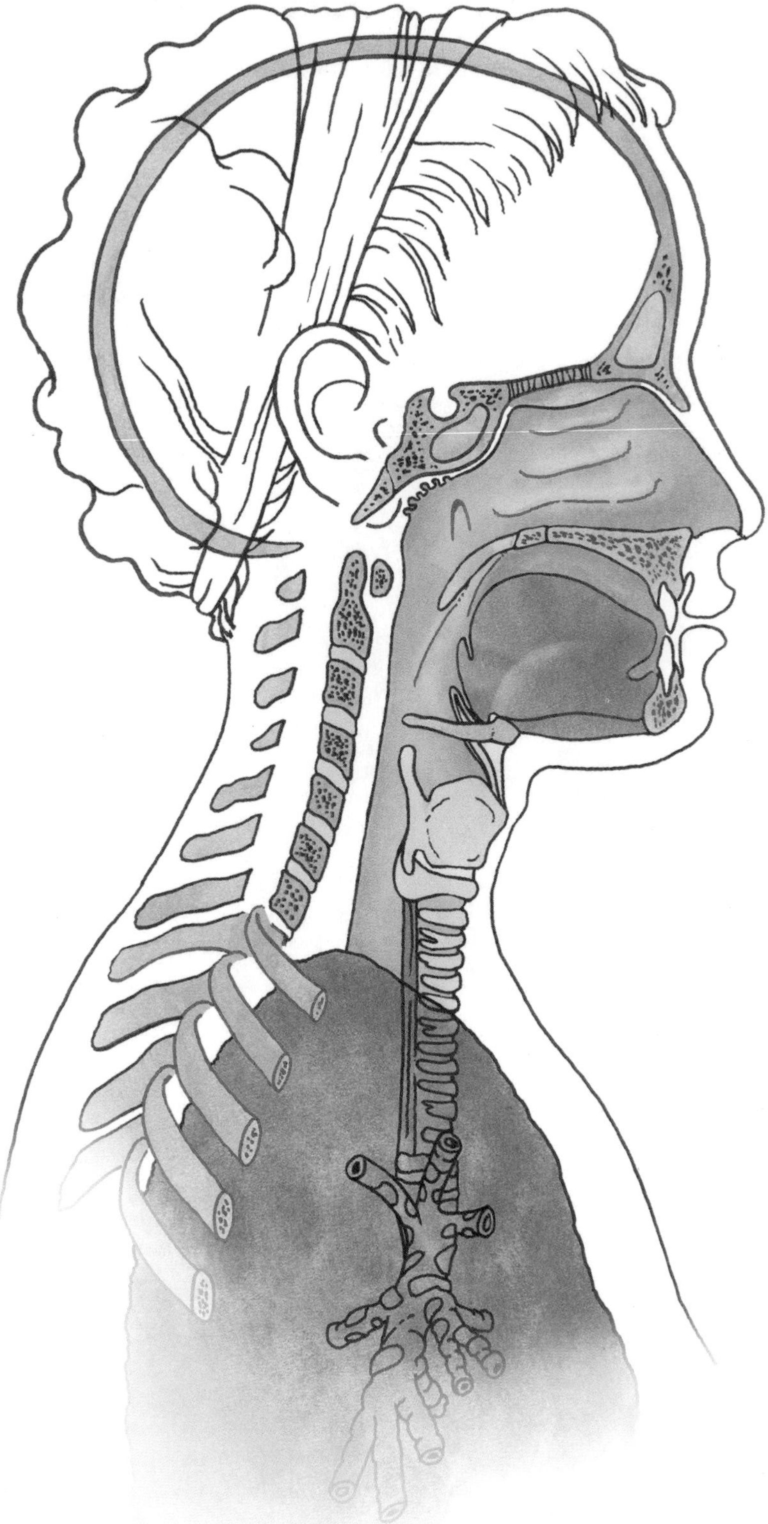

The vocal tract can be divided into regions with names that are more or less known. Above the glottis, we find:

the **nose,** with the nasal cavity and the nostrils

the **mouth,** with the tongue, the dental arches, and the lips

the **soft palate**

the **pharynx**

the **aryepiglottic space** (seen with the larynx on page 184)

In this part we will look at all of these areas, in the order in which the expiratory air travels through them. We will give as much detail about the bones and the muscles attached to them, especially in the cranium, as is necessary to understand these regions.

The Skeletal Framework of the Vocal Tract

The bones that make up this framework are just mentioned here. They are detailed in the pages specific to the various areas.

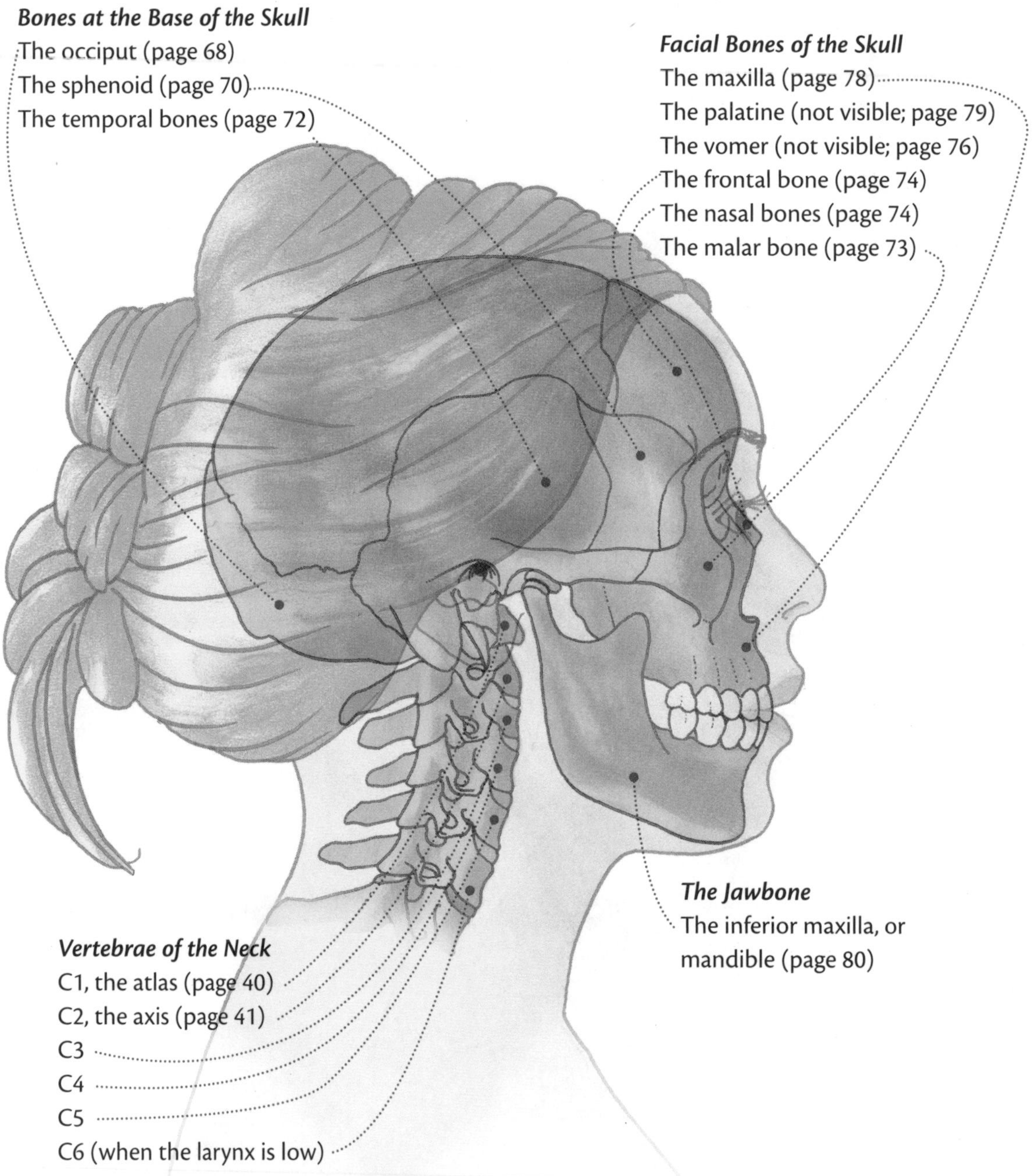

The skeletal framework reflects the shape of the vocal tract:

- The cervical vertebrae form a more or less vertical structure.*
- The mouth and nose form two horizontal structures.*

*If the person is standing

The bones can articulate, which can modify the shape of the vocal tract in several key regions:

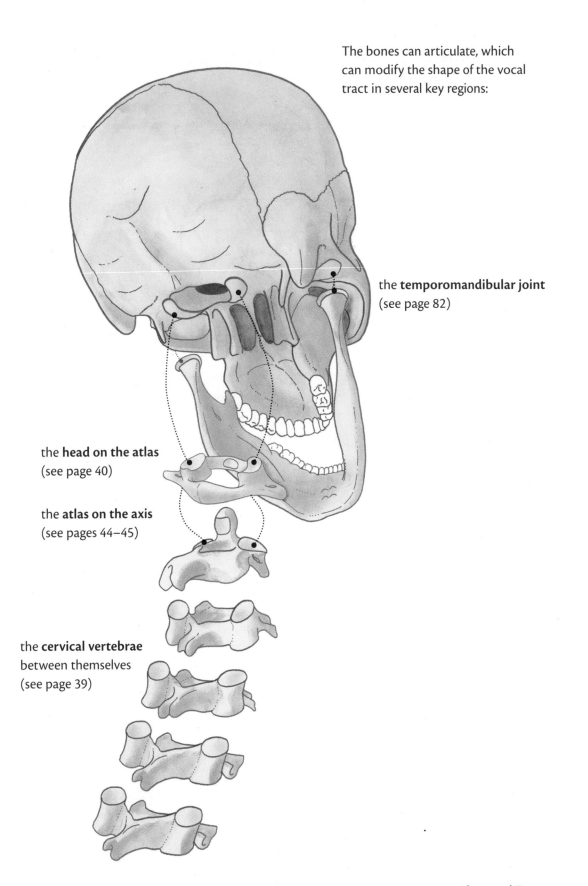

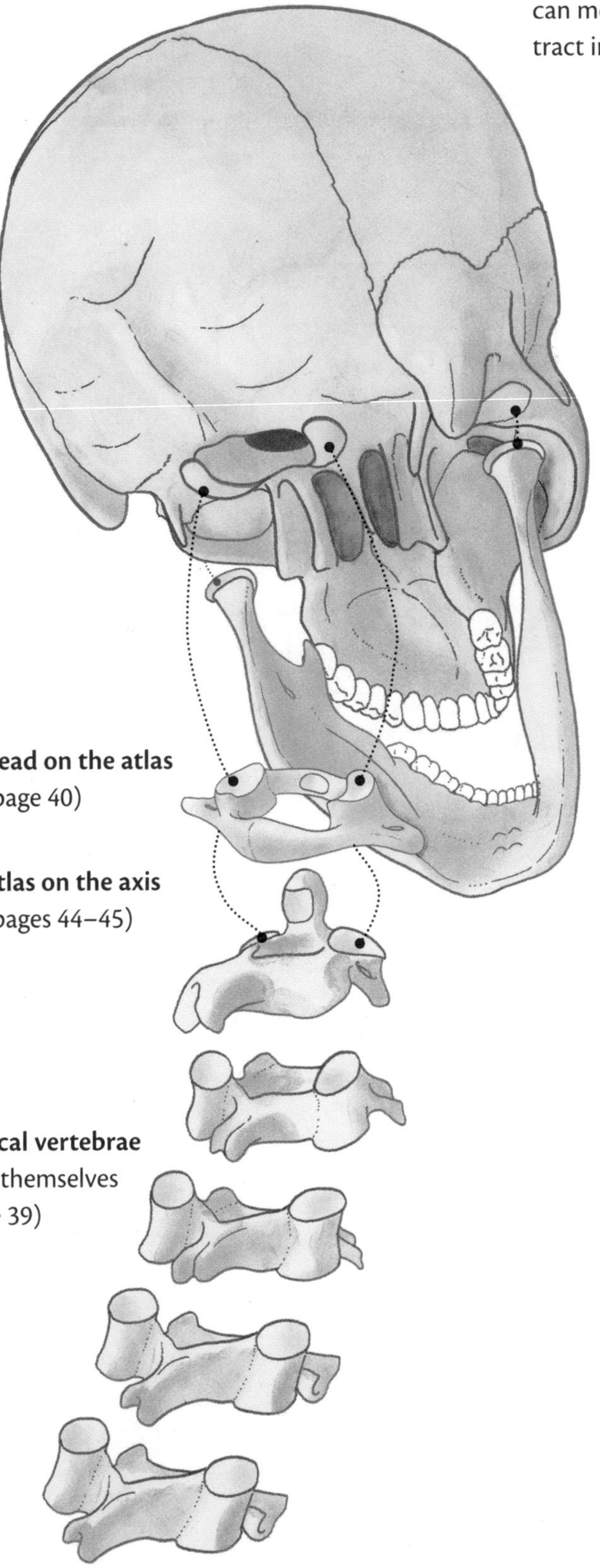

The Role of the Mouth and Pharynx in Resonance

What Is a Resonator?

All cavities are resonators. Resonators have specific excitation frequencies called *eigenmodes* (or "resonant frequencies"). When a sound of the same frequency enters a resonator, it will be amplified.

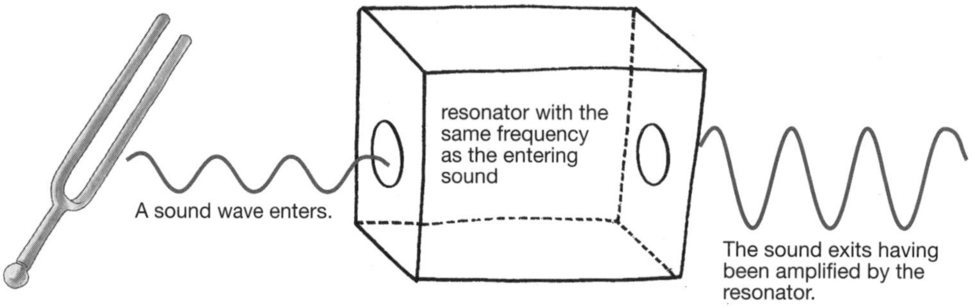

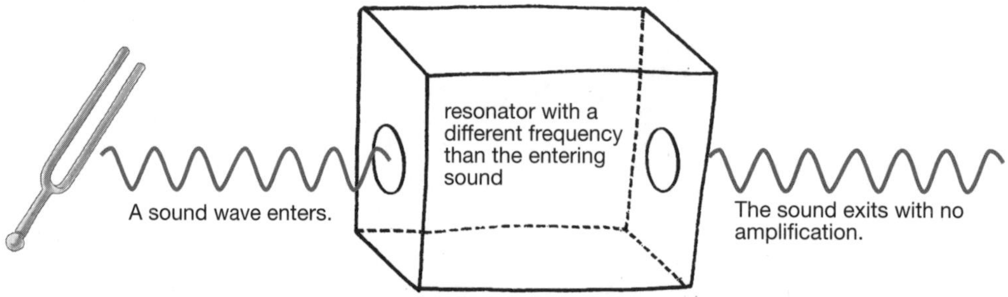

The frequencies that are resonated depend on the dimensions of the cavity (the bigger the cavity, the lower the frequency, and vice versa).

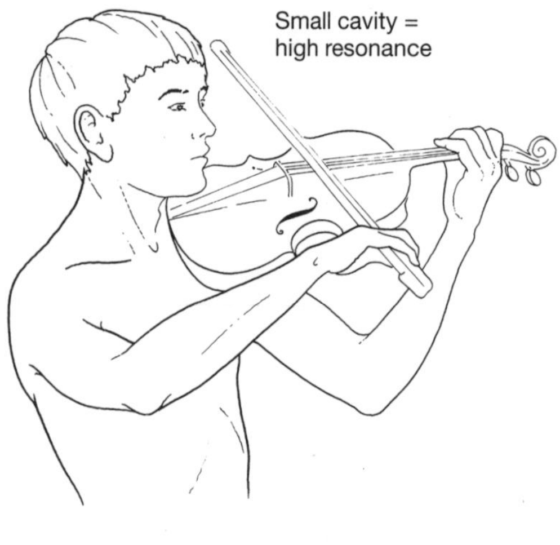

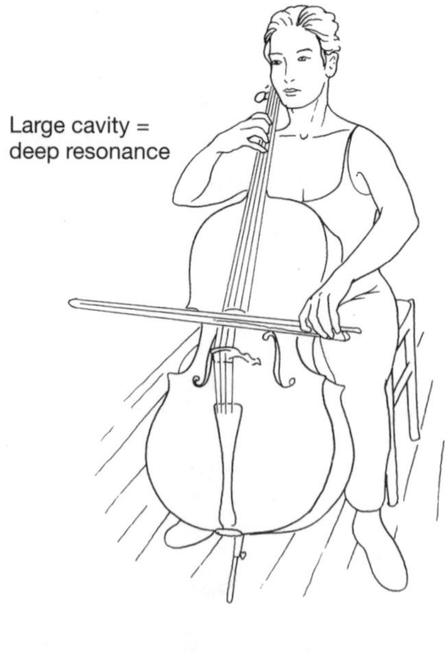

The pharynx and the mouth can be seen as cavities with flexible and movable walls. They therefore have a range of resonant frequencies, which amplify some of the sounds from the larynx. The pharynx amplifies the bass (250–500 Hz), and the mouth the treble (700–2500 Hz).

The Vowels

Each vowel is formed in two zones of frequency reinforcement, or *formants*, abbreviated as F1 and F2:

- F1 is associated with the pharynx.
- F2 is associated with the mouth.

The frequencies of F1 and F2 are related to the size of their respective resonators.

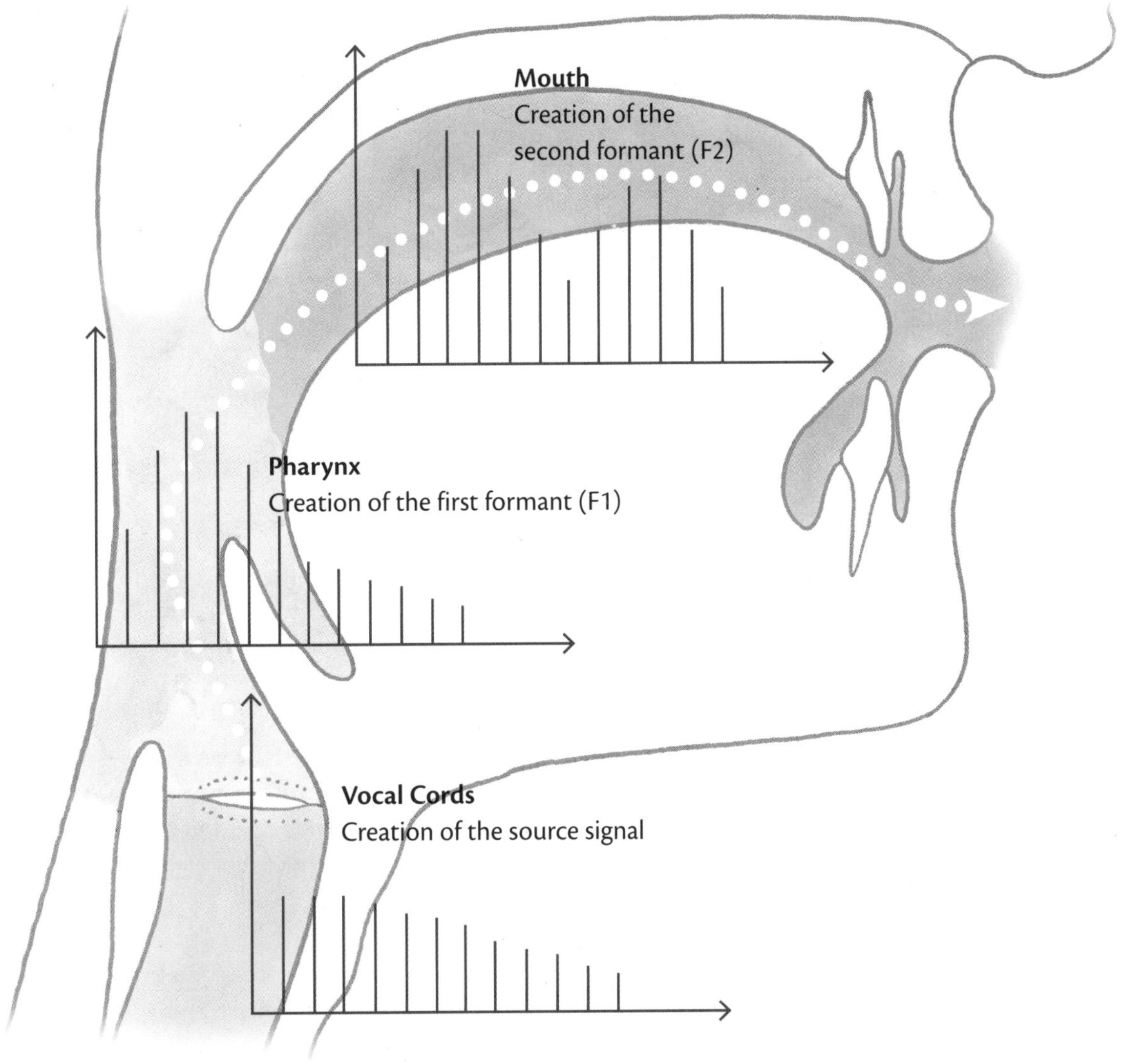

The Vocal Tract in the Neck

The vertical portion of the vocal tract is located in the neck, or, more precisely, in the upper two-thirds of the neck.

The upper part of this area merges with the head (when looking at a person from the front, the base of the head covers this part of the neck). It is composed of the following:

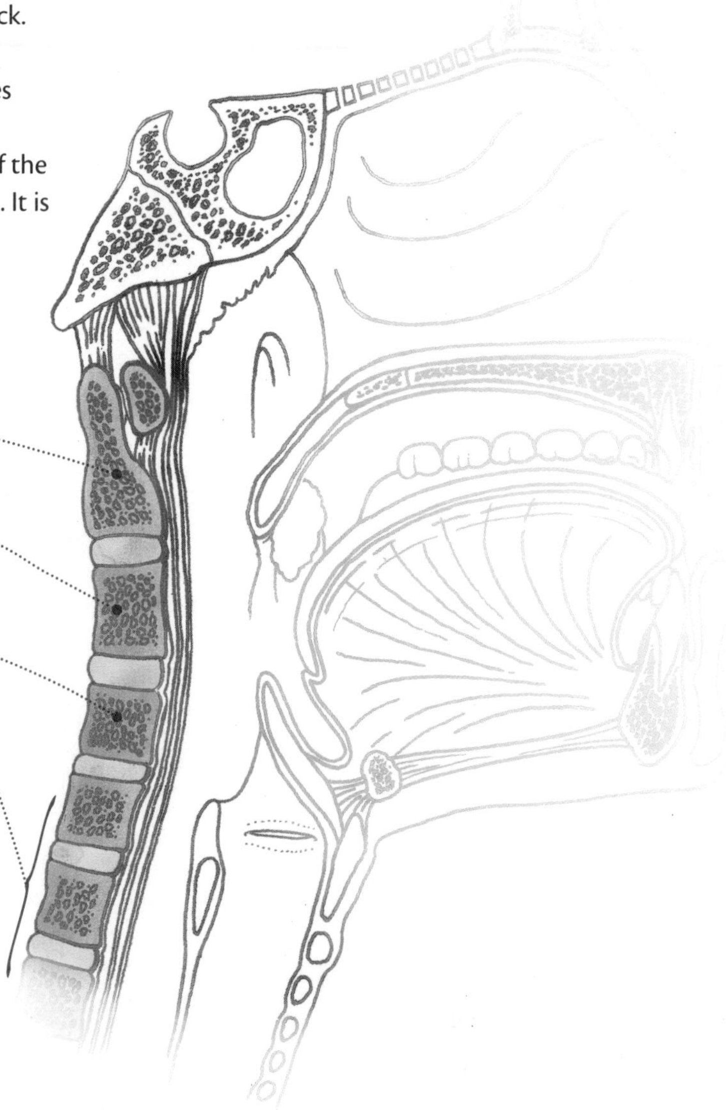

The **atlas** is behind the nose.
The **axis** is behind the mouth.

C3 is located at the mass of the tongue.

C4, C5, and sometimes **C6** are behind the larynx (this landmark moves as the larynx moves up and down).

The voice can be produced when the body is in motion, or even when the neck is moving—for example, when we vocalize in a supine position while rolling the head to the side. But more often we sing or speak in a vertical position. In this case, the cervical region is needed to provide support for the vocal tract. This region, which is highly mobile, is prone to stiffness because of muscle imbalances.

In vocal work, mobility is sometimes useful, and stability is needed at other times. We try to balance these two qualities by working on neck joint mobilization and strengthening of the postural muscles.

When the Body Is Vertical

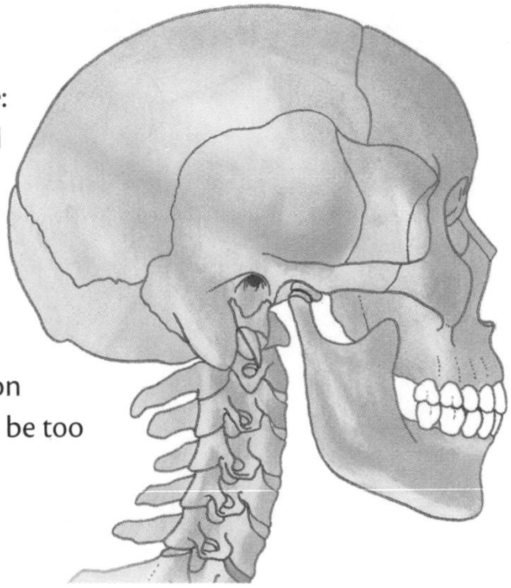

At the time of phonation, the neck is very busy. All at the same time:

- It must support the head, keeping it balanced on the vertebral column. In the following pages, we will discuss the problem of keeping the head balanced.
- It must support the larynx and pharynx, which, like the tongue, are suspended and are structures for which the neck acts like a mast to stabilize or anchor their movements.
- It must support itself, that is to say, balance its vertebrae one on top of the other, even in extreme positions where its base may be too far forward or it may be stacked up too straight.

These functions require that the neck have the ability to alternately mobilize or stabilize itself depending on the circumstances. Active mobilization and stabilization of the neck is achieved by the interplay of several sets of muscles:

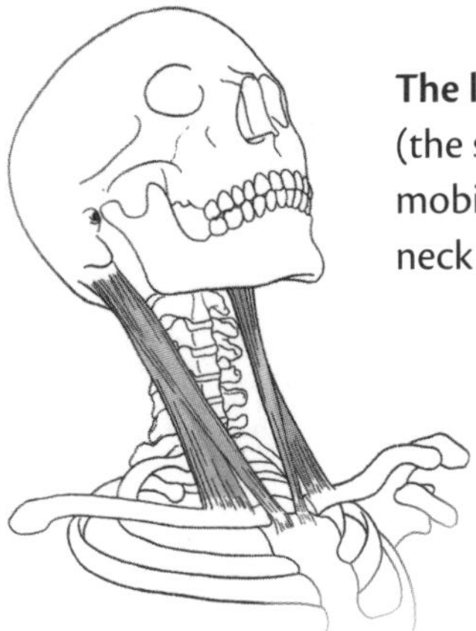

The lateral muscles (the scalenes and SCM) mobilize or stabilize the neck laterally.

The posterior muscles, strong and numerous, both realign the head on the neck and realign the neck on the trunk.

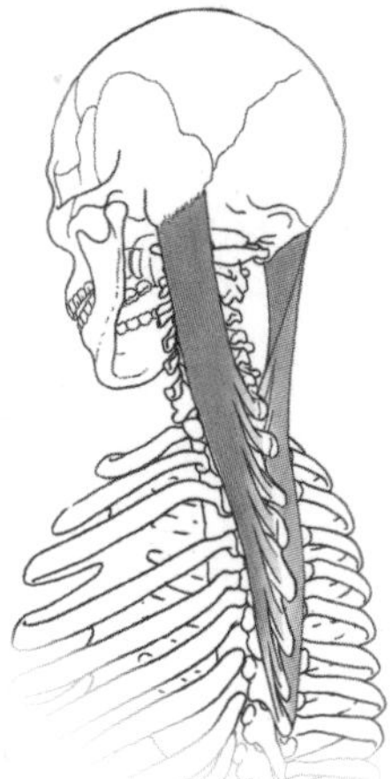

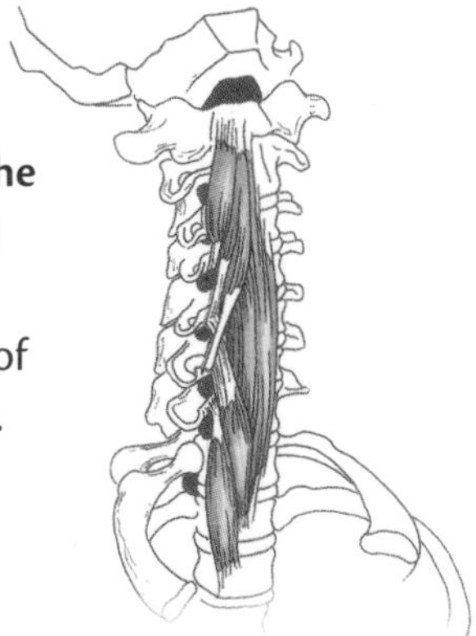

The muscles just at the front of the vertebral column ensure the rectilinear alignment of the cervical vertebrae.

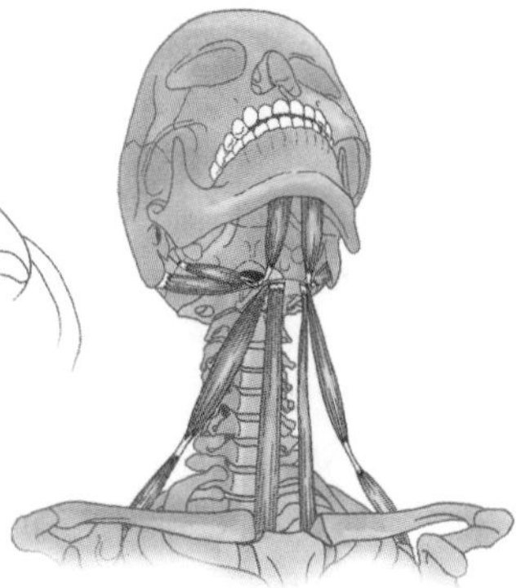

At the front of the neck are the hyoid muscles that balance the larynx and also help mobilize the neck. (These muscles are covered in our discussion of the larynx; see, in particular, the discussion of the extrinsic muscles of the larynx, beginning on page 186.)

"Belting It Out" and the Neck

In some forms of vocal work, like when we belt out the sound, we need a strong and stable neck.

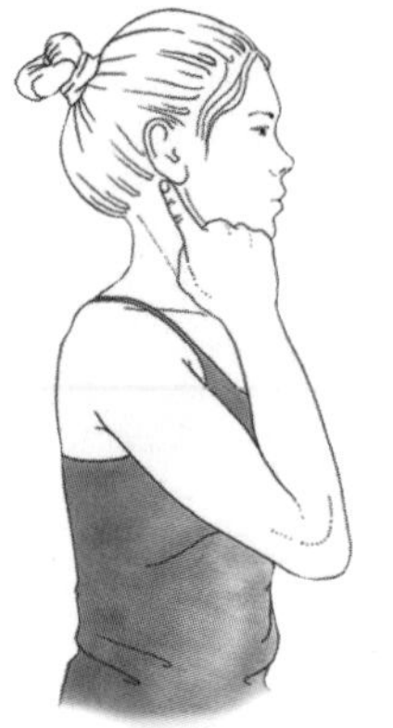

Balancing the Head

As we saw on page 42, the head meets the neck at the **occipital condyles,** which articulate with the surfaces on the lateral mass of the **atlas**. We find this joint just in front of the mastoid process, which can be palpated below and behind the ear.

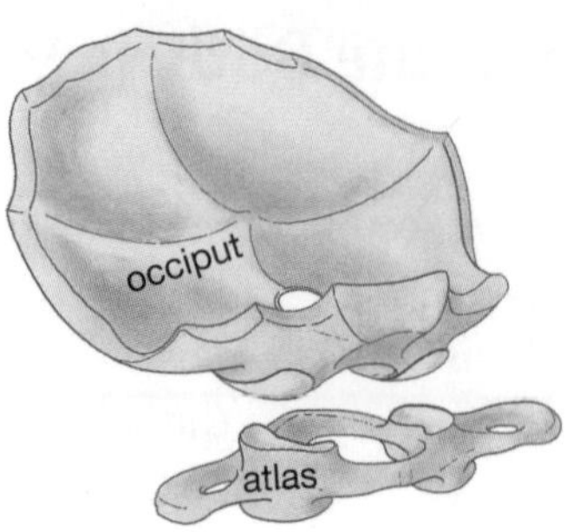

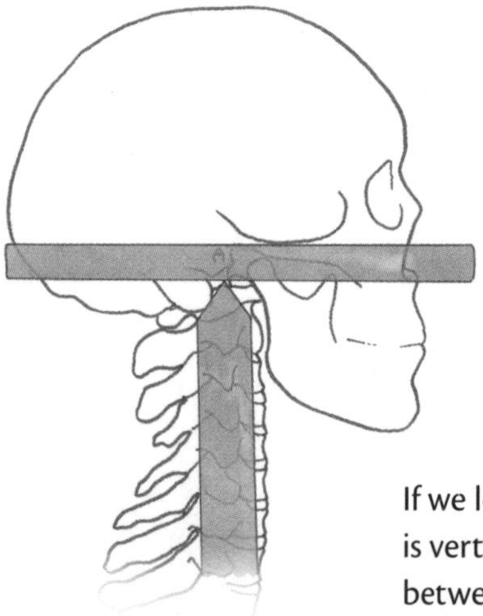

If we look at the head "block" in profile when the neck is vertical, we can compare it to a pendulum balanced between the weight of the structures located forward and behind the atlas. When the head is balanced, the maxilla is approximately horizontal.

A point of detail that is often overlooked: the center of gravity of the "block" of the head* is not above the atlas joint. It is forward of the condyles of the occiput, and therefore in front of the atlas.

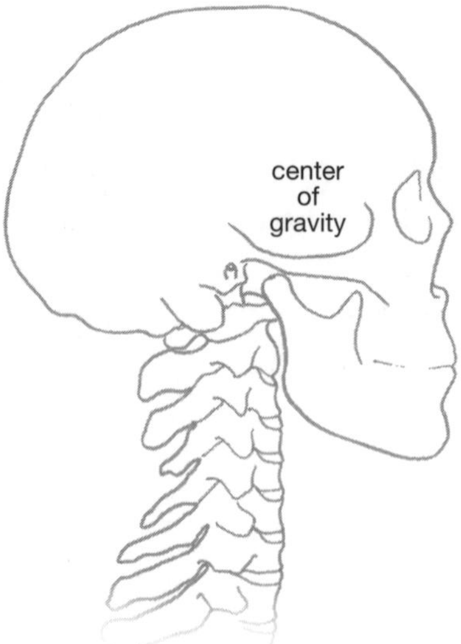

The Center of Gravity

This is the point where the forces exerted on the object come together.

*The "block" of the head includes the skull, the mandible, and all the soft structures located around these bony parts.

When the neck is vertical, because of the location of the center of gravity, the head tends to fall forward. (When the head falls forward, its tends to pull the upper part of the neck, then the lower part, into flexion.)

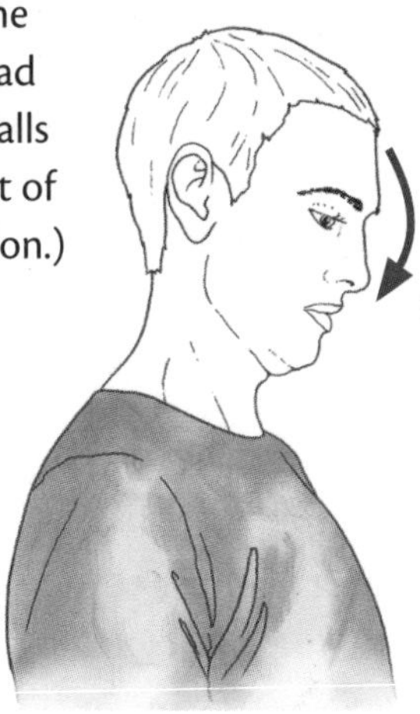

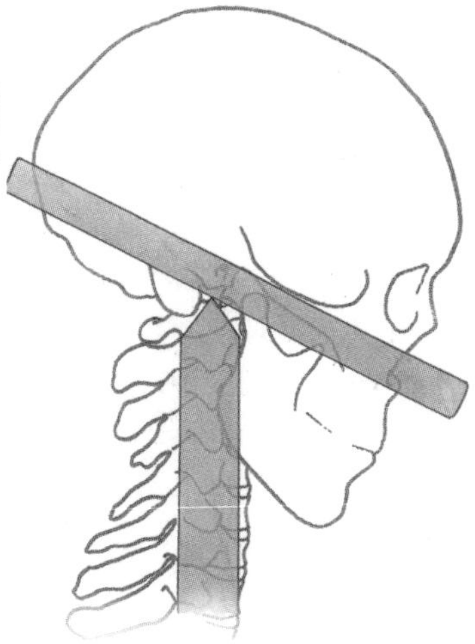

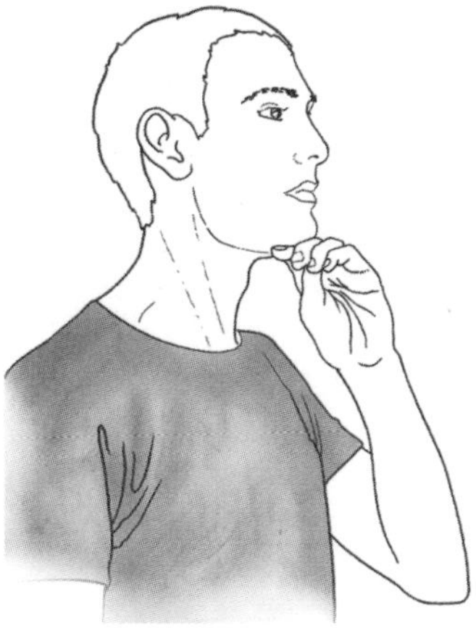

We can easily feel this tendency of the head to fall forward if we support our head under our chin: we feel the weight on our hand.

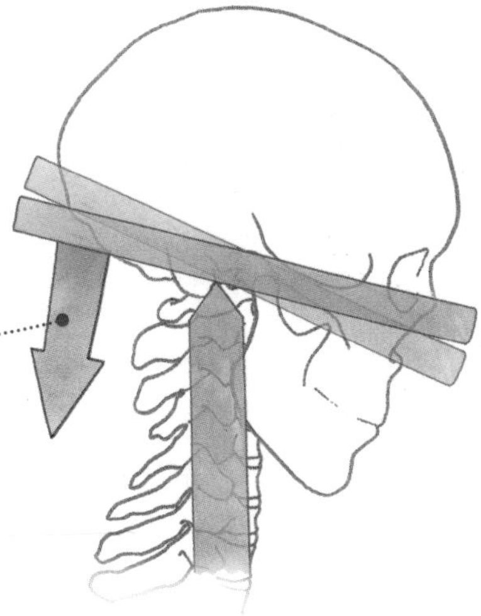

For the neck to stay vertical, there needs to be a force that brings the head back into alignment. That force is the contraction of the muscles at the back of the neck, of which there are several types: the small muscles, which adjust the level of the head on the neck, and the longer and more powerful muscles, which extend at the back of the cervical vertebrae, sometimes to the thorax, and prevent the neck from following the head when it falls forward.

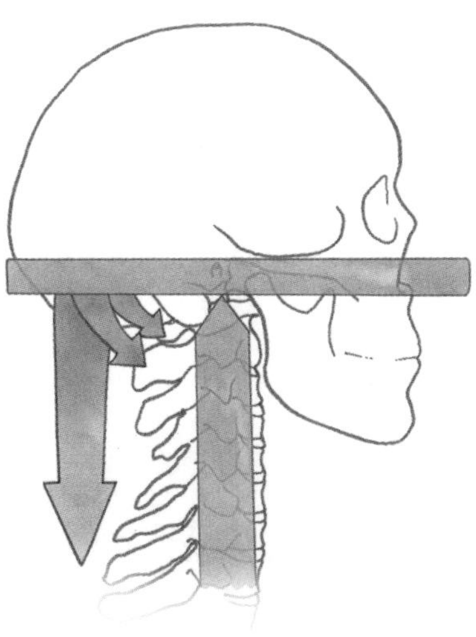

These muscles are constantly working when we are in a vertical position; in other words, from the time we get up until we go to bed, they are among the muscles of the locomotor system that work the hardest. They are frequently overworked and unbalanced.

Just knowing this often helps us understand what we are feeling when there is too much or too little force in this area. The most important thing to remember is that the balance of the head/neck is achieved from *behind*.

The Suboccipital Muscles

These small muscles are located deep in the neck, very close to the skull. They are very important for the orientation and mobility of the base of the "vocal" skull. They give the head an ability to move that is comparable to that of a bird. In vocal work, one should always look at the state of tonicity of these muscles. If they are too contracted, they can make the upper regions of the neck and head rigid, hindering vocal flexibility. If, instead, they are too lax or poorly coordinated, the longer, superficial, and less accurate stabilizers will be recruited, which will sometimes jeopardize the play of the larynx. It is therefore very important to know how this interplay works both during vocal work and independent of it.

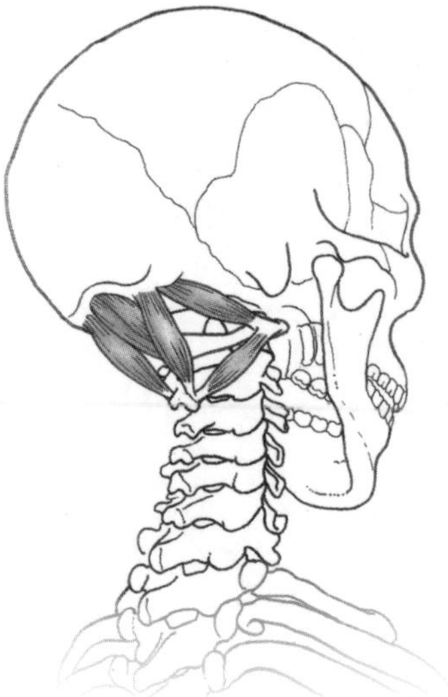

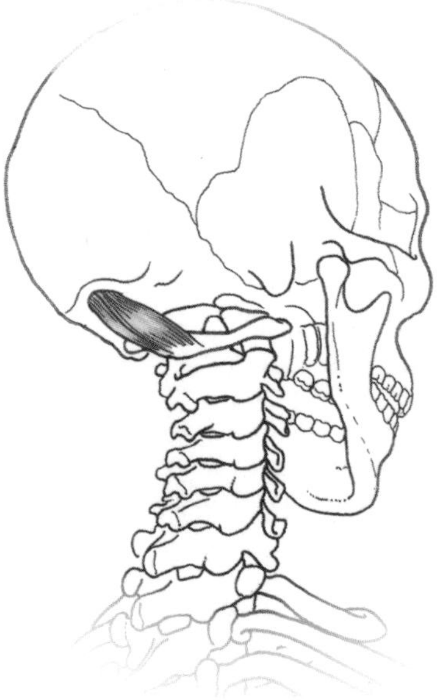

The Rectus Capitis Posterior Minor

At the top, this small muscle attaches at the occiput (at the inferior nuchal line; see page 68), near the midline. It descends vertically and ends on the posterior arch of the atlas.

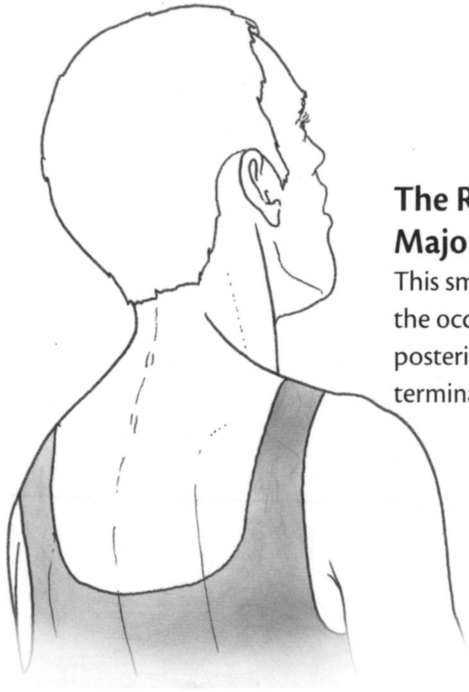

The Rectus Capitis Posterior Major

This small muscle attaches at the top to the occiput, just outside the rectus capitis posterior minor. It descends obliquely and terminates on the spinous process of the axis.

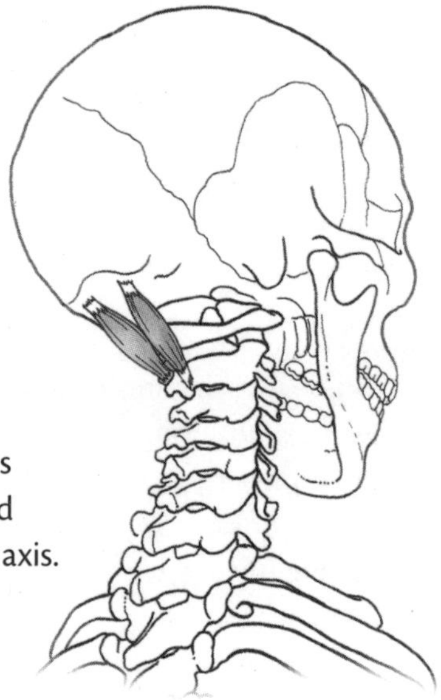

Their Action

They move the head into extension on the axis.

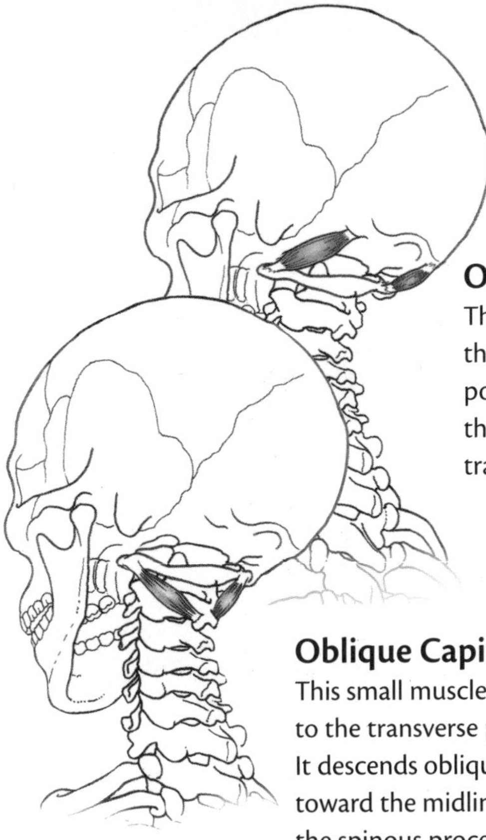

Oblique Capitis Superior

This small muscle attaches at the top to the occiput, just outside the rectus capitis posterior major. It descends obliquely toward the front and forward and terminates on the transverse process of the atlas.

Oblique Capitis Inferior

This small muscle attaches at the top to the transverse process of the atlas. It descends obliquely backward and toward the midline and terminates on the spinous process of the axis.

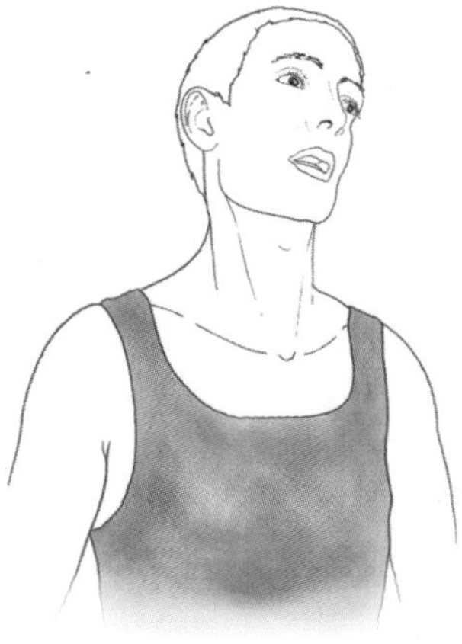

Their Action

If they contract on one side, they cause a lateral tilt of the head on the atlas, and a rotation to the opposite side.

It should be noted that these muscles are close to those of the jaw, though they differ in function. This proximity means that there can be a muscular synkinesia between these two regions. We can make the same observation about the proximity of the suboccipital and suprahyoid muscles; for example, it is common in high registers for the suboccipital muscles to contract synchronously with the suprahyoid, but these contractions are not useful and can inhibit proper action.

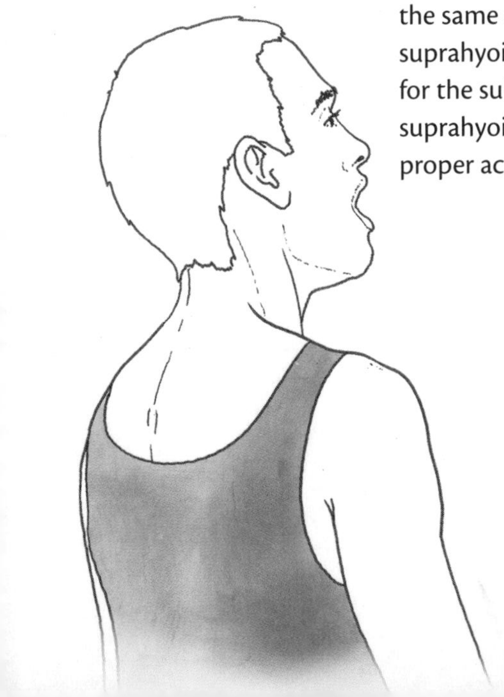

Observation

The suboccipital muscles are strongly influenced by eye movements (see page 215).

The Long Neck Extensors

These muscles are much larger than the suboccipital muscles, and they provide a link between the head and all of the cervical vertebrae.

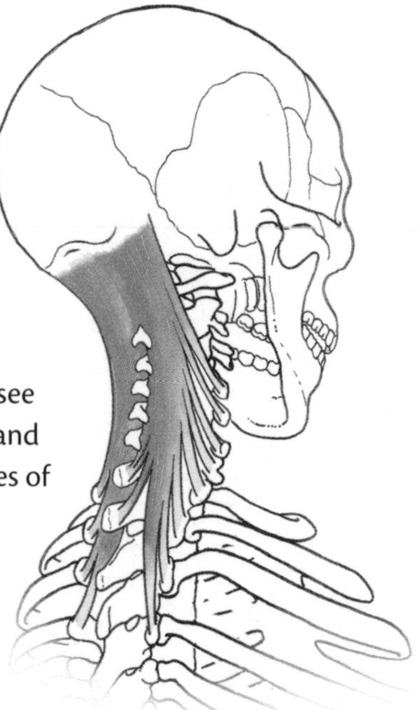

The Semispinalis Capitis

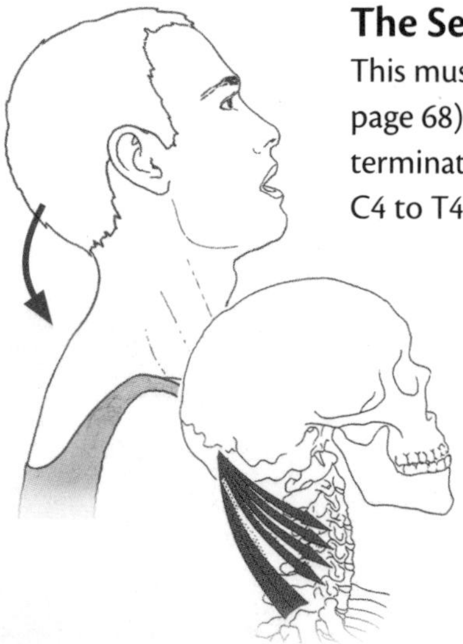

This muscle attaches at the occiput (see page 68). It descends along the neck and terminates on the transverse processes of C4 to T4.

Its Action

It extends the head on the neck and the neck on itself.

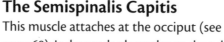

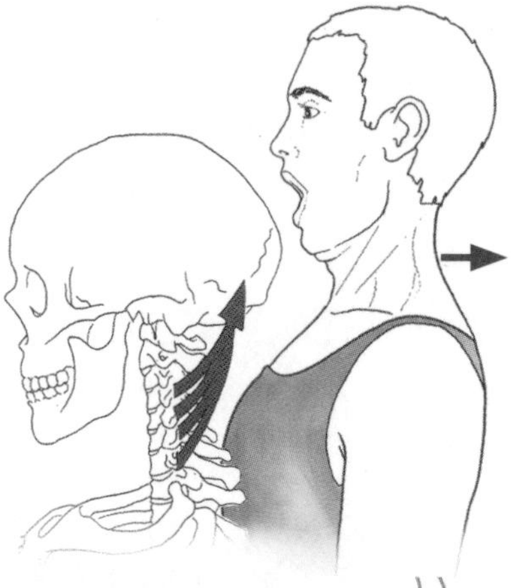

Attention

If the head is fixed, the muscle acts on the cervical vertebrae, pushing them posteriorly and straightening the cervical lordotic curve. This action can even be made by simply tensing the semispinalis capitis, which then passively pulls the vertebrae posteriorly (for example, when we pull in our chin). But this way of straightening the neck while retracting the chin greatly reduces the suprahyoid space and the space around the larynx; it is not desirable for good vocal agility. It is best to try to straighten the neck using the longus colli muscle (see page 214).

The Trachelomastoid

This muscle (not shown) is attached at the top of the mastoid process (see page 72). It descends along the neck and ends on the transverse processes of C3 to C4.

Its Action

If both sides contract at once, it extends the head on the neck and the upper half of the neck. If just one side contracts, it contributes to the lateral inclination of the head and upper neck and rotation to the same side.

The Levator Scapula

This muscle attaches at the top to the transverse processes of the first four cervical vertebrae. It descends to terminate at the superior and medial angle of the scapula.

Its Action on the Neck

It lifts the neck up from the shoulder blades.

Observations

It is common to see this muscle being overused in a person who has kyphosis in the upper spine and a forward projection from the base of the neck (forward head posture). This person will then try to stack up the vertebrae of the neck by lifting the shoulder blades and bringing them closer together. This is effective to bring the center of gravity backward, but not really to straighten the neck.

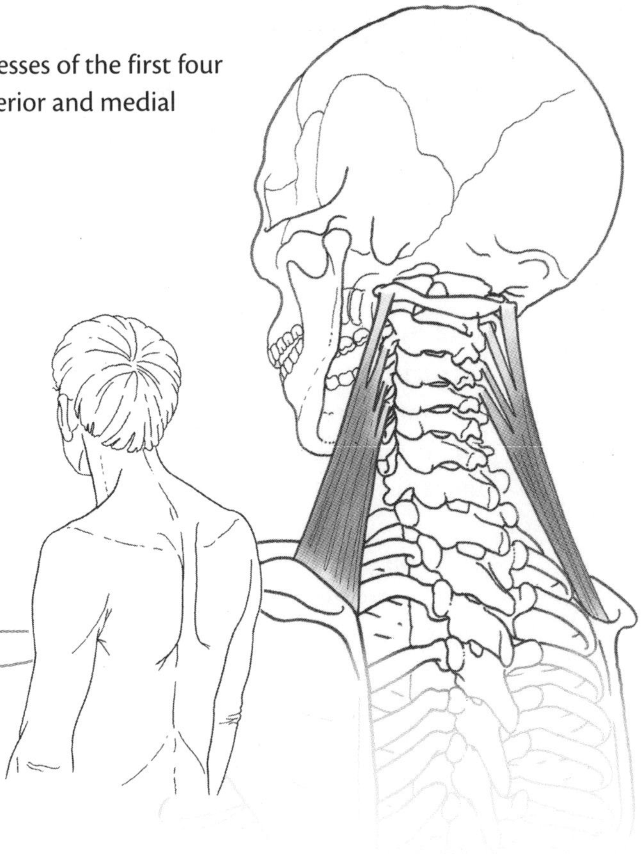

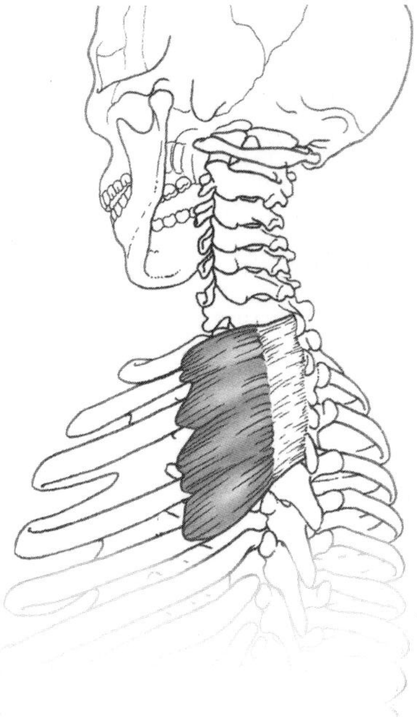

The Serratus Posterior Superior

This muscle is attached at the top to vertebrae C6 (or C7) to T2. It terminates on the first four ribs.

Its Action on the Neck

It straightens the base of the neck from the first rib. This is an important action to counteract the tendency of forward projection at the base of the neck.

To find the action of the serratus, get the spine in a horizontal position. From here, let the base of the neck "slip" down. Then, from the area of the serratus, try bringing the trunk back into alignment.

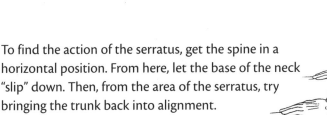

The Splenius

The splenius muscles, like the **long neck extensors,** are larger than the suboccipital muscles and are situated obliquely in the neck. They join the region of the mastoid to the cervicothoracic region.

The Splenius of the Head: The Splenius Capitis

This muscle attaches at the top to the mastoid and the adjacent part of the occiput. It descends along the neck and terminates on the spinous processes of C6 to T7.

Its Action

If both sides contract at the same time, they bring the head and the upper half of the neck into extension. If one side contracts, it rotates and laterally tilts the head and neck to the same side.

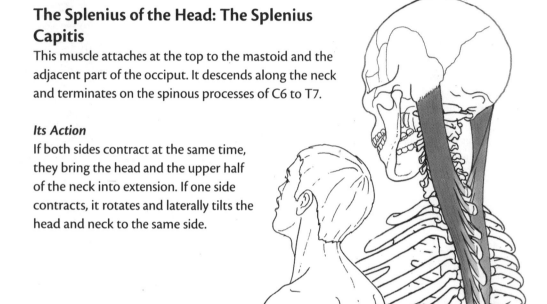

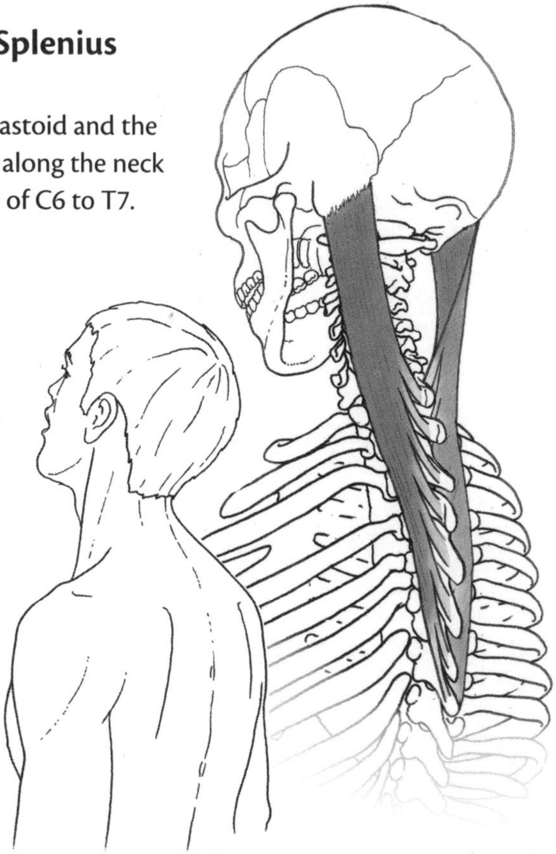

The Splenius of the Neck: The Splenius Cervicis

This muscle (not shown) attaches at the top to the spinous processes of C1 to C3. It descends along the cervical vertebrae and terminates on the transverse processes of T1 to T3.

Its Action

If both sides contract at the same time, they bring the neck and the upper part of the thoracic region into extension. If one side contracts, it contributes to the lateral inclination of the lower cervical spine.

The Trapezius

This is one of the largest muscles in the body. It attaches at the top to the occiput (see page 68) and then to the cervical and thoracic spinous processes. From there, the fibers fan outward to terminate on the outer third of the clavicle and the spine of the scapula.

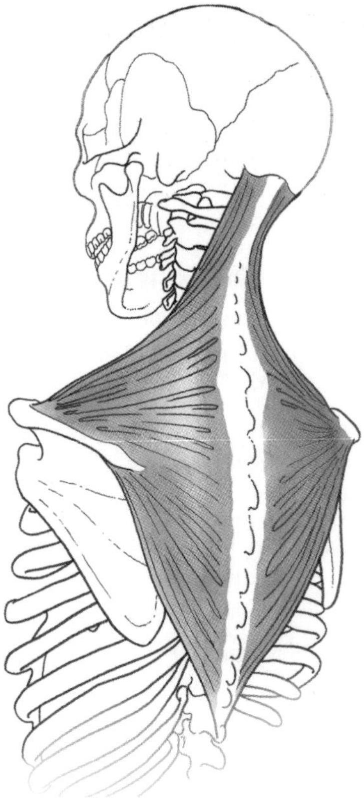

Its Action

It brings the shoulder blades together. Its upper fibers elevate the scapula (and the trapezius brings the shoulders up and closer together, which is something we often see in singers when they are hitting high notes; see page 195). The upper part of the trapezius has a significant effect on the neck: if both sides contract at the same time, they extend the head and the neck. If one side contracts, it contributes to lateral inclination of the cervical spine and rotation of the neck and head to the side opposite the contraction.

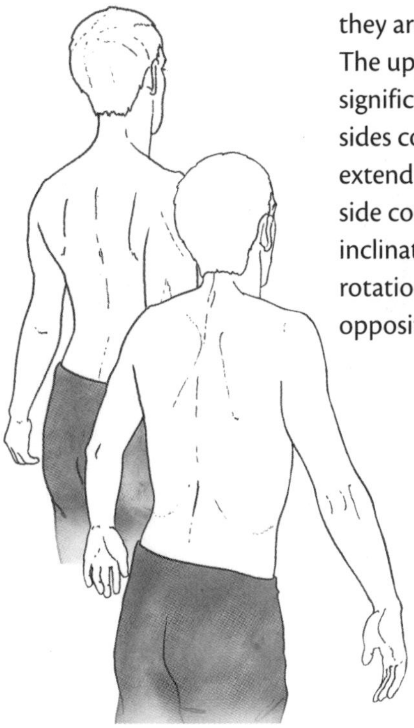

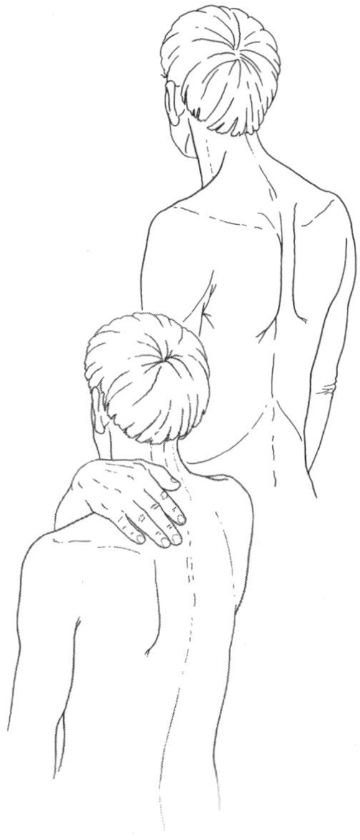

The Upper Part of the Trapezius Is Often Contracted

It is even more pronounced if the neck muscles are not well coordinated in keeping the neck erect; in this case, the upper trapezius acts to stabilize the head/neck/trunk. But because it has no effect on individual vertebrae it tends to telescope the neck, as the SCM does (see page 216) and sometimes lifts the shoulders more than is necessary. To relax the trapezius, of course, we can massage it, but ultimately we need to reorganize the vertical position of the neck and head by using the deeper muscles that are more appropriate for the task.

More details on the neck muscles can be found in *Anatomy of Movement*, pages 76–87.

The Anterior Cervical Muscles

These muscles are called "anterior cervical" because they are in front of the cervical spine. However, because the cervical spine is in lordosis (it forms a convex curve to the front and a concave one to the back), the principal role of these muscles is to stack up the cervical vertebrae, straightening them with their contraction.

The Longus Colli

This muscle is made up of three bundles of fibers. A superior oblique bundle runs from the anterior arch of the atlas to the transverse processes of C3 to C6.

A longitudinal bundle, quite median, runs from the vertebral bodies of C2 to T2 to the transverse processes of C4 to C7.

An inferior oblique bundle runs from the bodies of T1 to T3 to the transverse processes of C5 to C7.

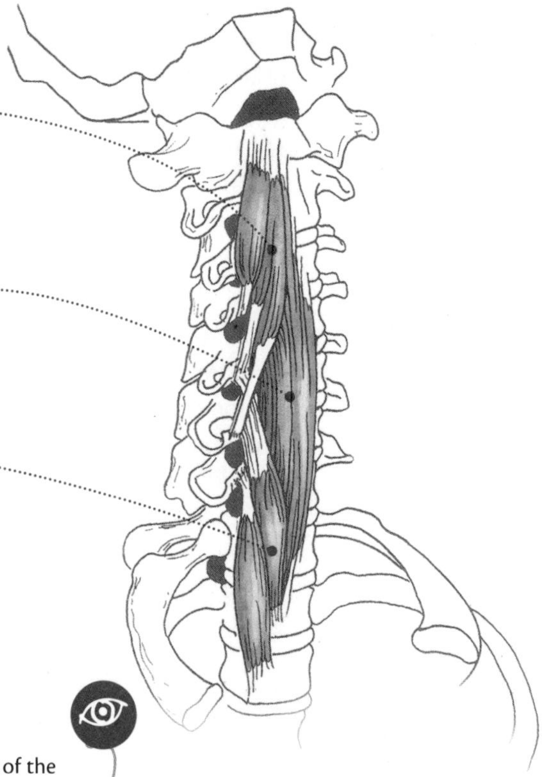

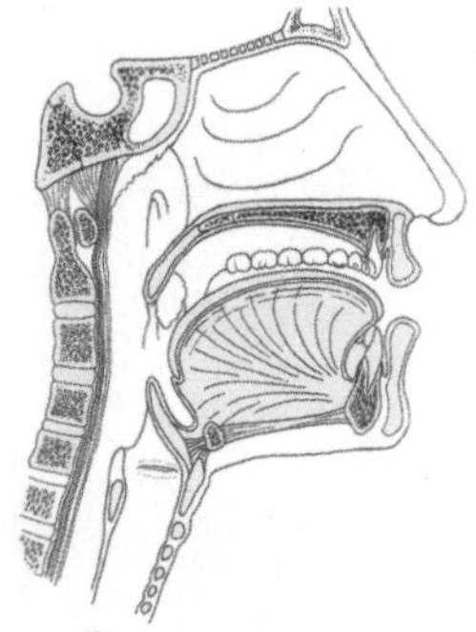

Feel the Longus Colli

The longus colli lies just in front of the cervical vertebral bodies, just behind the pharyngeal constrictor muscles. You can feel the constrictors when you swallow saliva, and you can imagine the longus colli, which is just behind where you feel this sensation.

Its Action

This muscle straightens the cervical lordosis, stacking up the cervical vertebrae. Important detail: it performs this action without acting on the head. It is interesting to master the act of bringing the cervical spine into a vertical position without pulling in, as you do with the semispinalis, which we looked at on page 210. To develop this muscle you can push the head up against resistance, such as the weight of the hand.

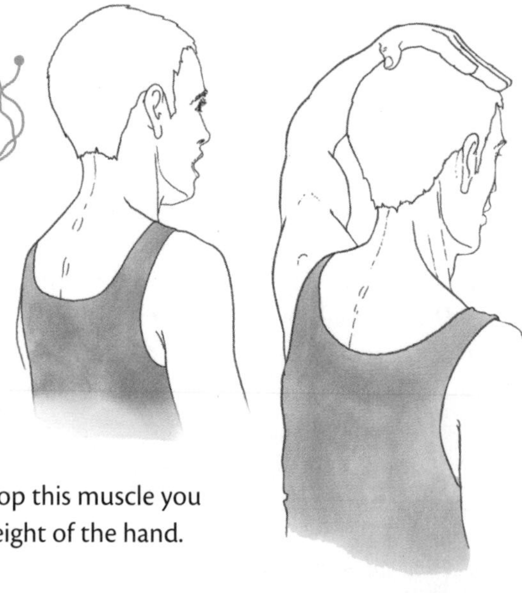

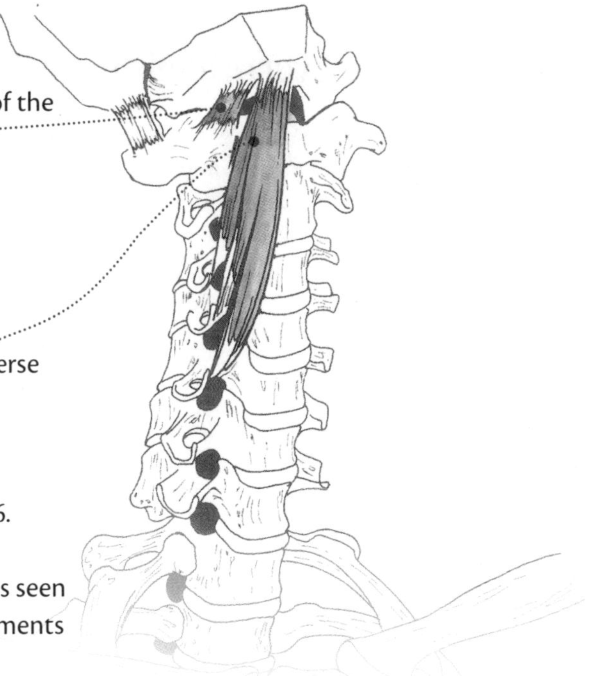

The Rectus Capitis

This muscle runs from the basion of the occiput to the front of the atlas.

Its Action

It flexes the occiput on the atlas.

The Longus Capitis

This muscle runs from the basion of the occiput to the transverse processes of C3 to C6.

Its Action

It flexes the occiput on the atlas, and this action extends to C6.

These two muscles are antagonists to the suboccipital muscles seen on pages 208–9; that is to say, they make inverse (mini) movements and therefore can balance them.

Conclusions

All of these muscles, the anterior cervical and suboccipital, contribute to the precise balance of the head on the neck. They are important in vocal techniques where we search for stability or fine-tuning of the head/neck region and for the subtle actions of the pharynx, the soft palate, and the back of tongue, especially in high note passages of sustained velocity common in classical singing.

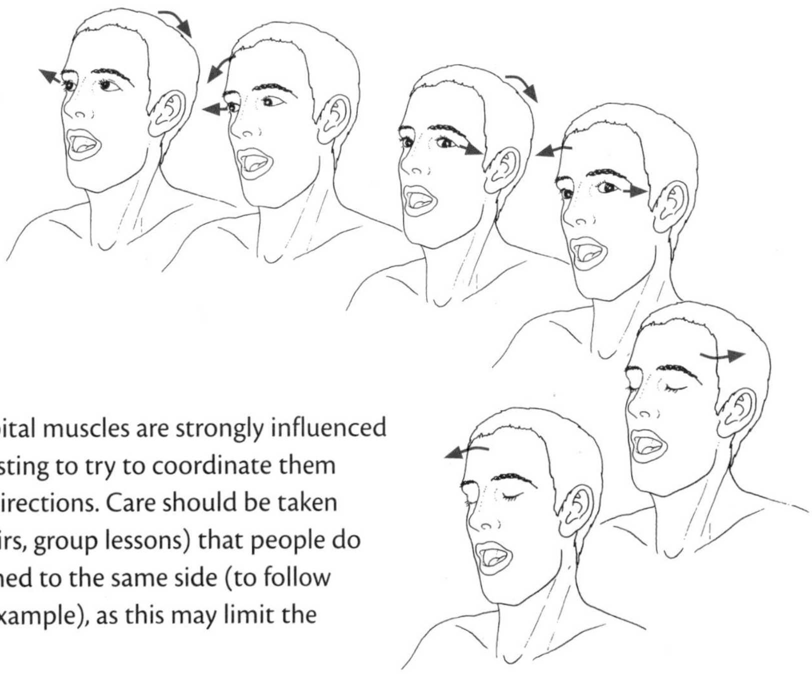

The anterior cervical and suboccipital muscles are strongly influenced by eye movements, and it is interesting to try to coordinate them while moving the eyes in various directions. Care should be taken when singing in vocal groups (choirs, group lessons) that people do not have their eyes constantly turned to the same side (to follow the conductor or the pianist, for example), as this may limit the coordination of these muscles.

The Sternocleidomastoid or SCM

We covered the sternocleidomastoid and its role in the inhalation earlier (page 129). Here we will look at its role in the verticality of the neck.

Its upper insertions are on the head: on the mastoid process (see page 72) and the occiput. From there, it travels down to the front and middle of the neck and ends on the sternum and the inner part of the clavicle. It has no attachment to the skeleton between the upper and lower inserts.

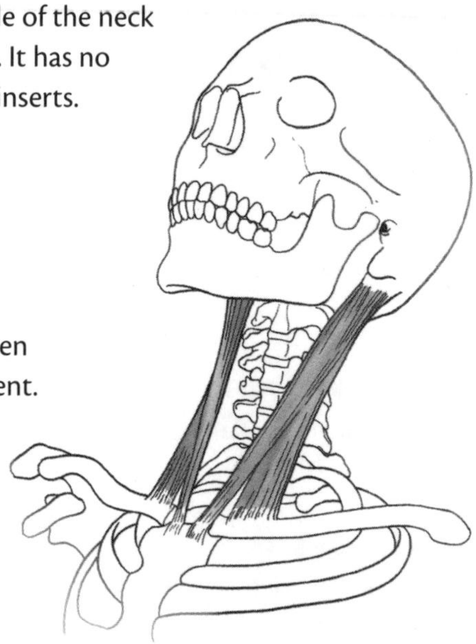

Its Action

The action of the SCM on the neck is powerful and can often disorganize our posture by "telescoping" the spinal alignment. That is, the SCM tends to pull the occiput:

- forward—the "block" of the head is no longer balanced above the chest but ahead of it
- in extension to the back, which can provoke an action by the posterior suboccipital muscles

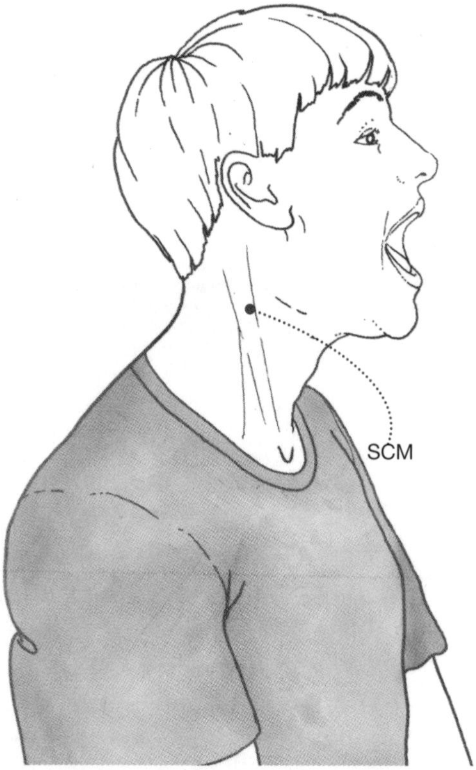

When the SCM Becomes Vertical . . .

It is common to see dominant action of the SCM during vocal work. If this action is not balanced by that of the deep muscles of the neck, it brings the atlas and head strongly forward, which doesn't make for good laryngeal suspension (see page 186). In this case the SCM in profile will be vertical and not oblique.

The Scalenes

We have already seen these three muscles in terms of their role in the inhalation (see page 130). Here we will look at their role in keeping the neck vertical.

The scalenes are located just behind the SCM. Their upper insertions are lower than those of the SCM: they arise from the transverse processes of the axis to C7. They do not attach to the head. They then descend obliquely and to the front and terminate on the first and second ribs.

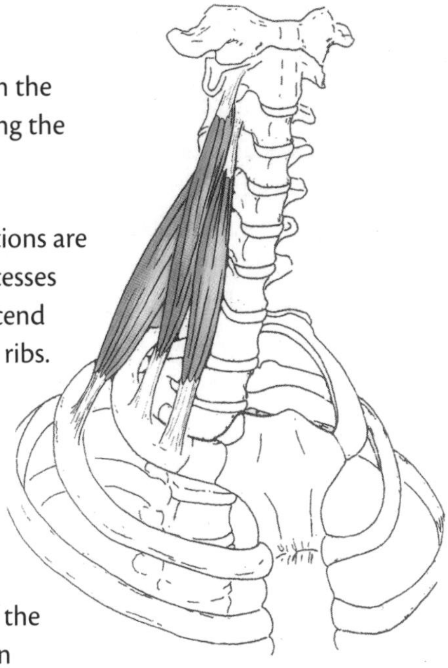

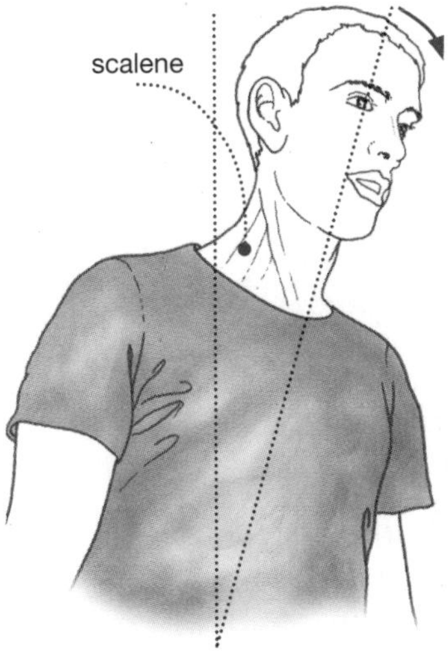

Their Action

The scalenes, located on the sides of the neck, stabilize it laterally, and you can see and feel them when you tip your head to the side.

As we saw on page 131, their action can bring the cervical vertebrae forward.

A Little Test for Alignment of the Cervical Spine and Head

If the SCM and scalenes are too contracted and short, it is difficult to align the neck and head vertically. (If this is the case, it is impossible, or at least it is an effort, to lean on a headrest when you sit on a seat that has one.)

This alignment can be tested by lying on your back, without a pillow under your head. Your ribs should be on the floor. Here, some people will struggle to get their head on the floor, or can only do it by bringing their upper neck into a strong lordosis, so that their eyes may be aimed backward.

If this test is positive, before seeking to realign the neck and head "forcefully," you should try sessions of passive realignment. Lie in the same position as for the "test," with a wedge placed under your head. Over time, reduce the thickness of the wedge as the muscles relax and your alignment improves.

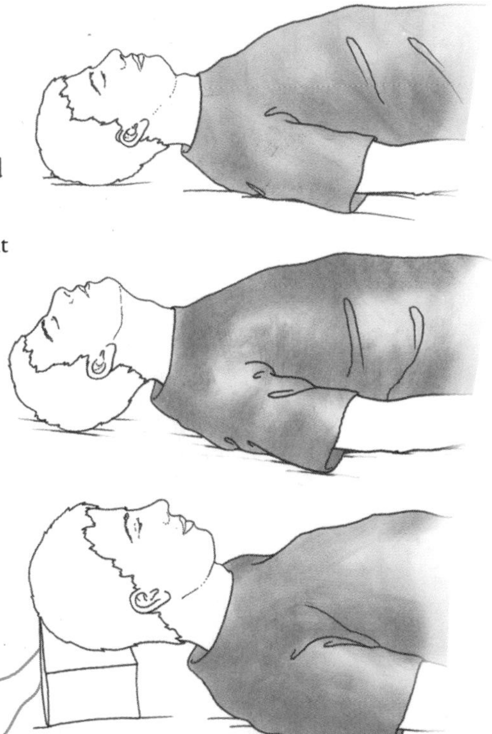

The Pharynx

The pharynx is a flexible tube, a dozen centimeters in length and about 2 centimeters from front to back. It is located in front of the cervical spine. It's extended downward by the esophagus. Its anterior surface has openings through which it communicates, from top to bottom, with the back of the nasal cavity, with the back of the mouth, and with the top of the larynx and esophagus. The pharynx allows communication between these zones.

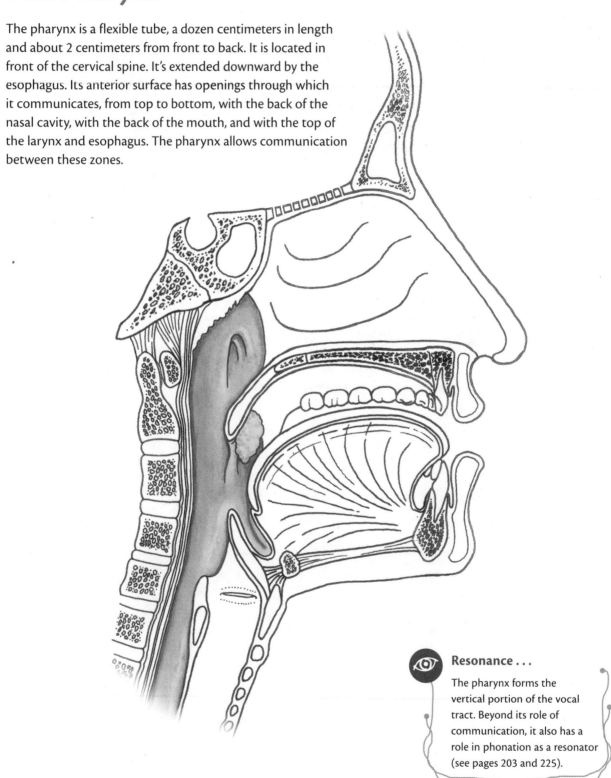

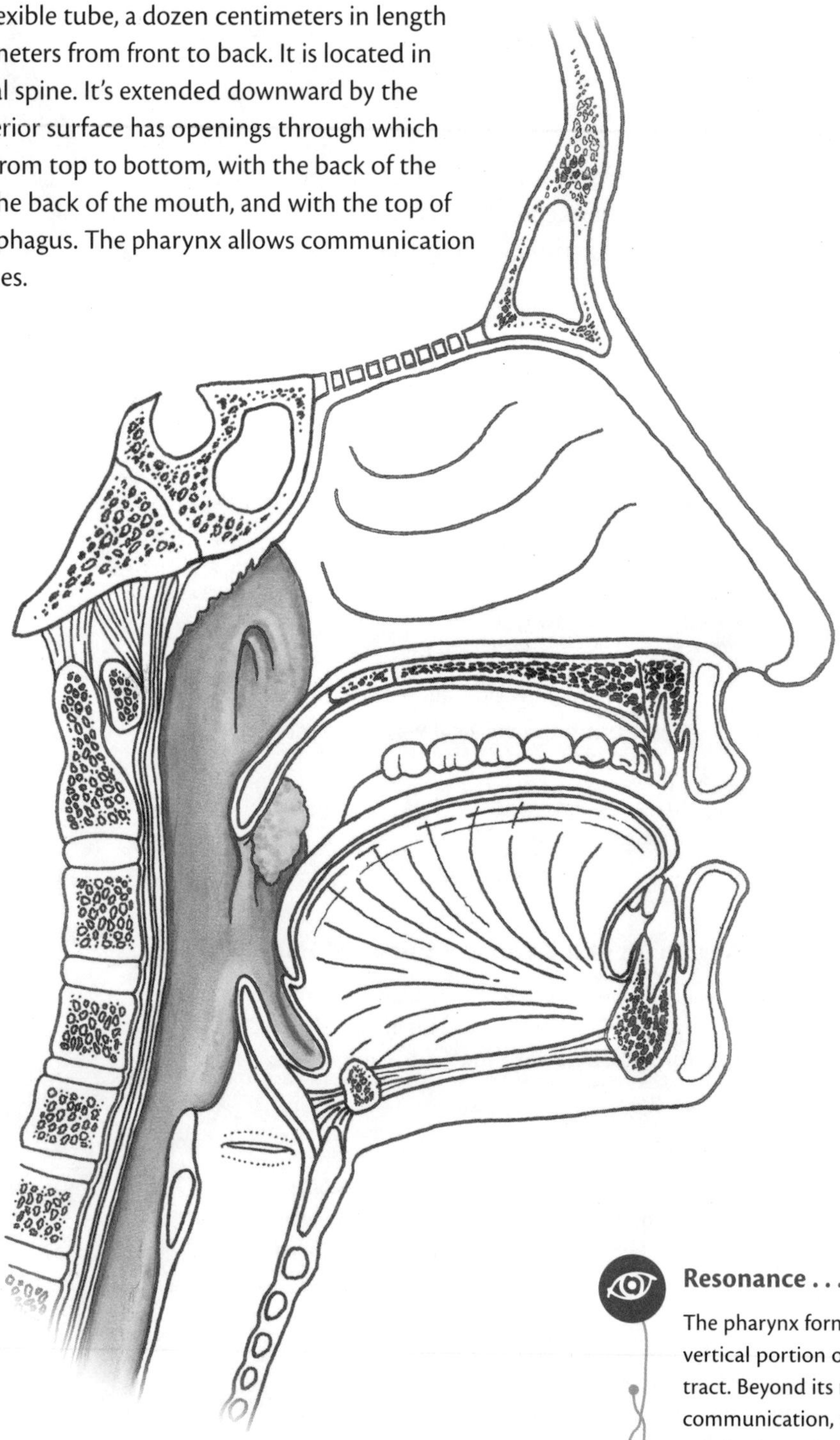

Resonance . . .

The pharynx forms the vertical portion of the vocal tract. Beyond its role of communication, it also has a role in phonation as a resonator (see pages 203 and 225).

The Regions of the Pharynx

From top to bottom (following the pathway of inspiratory air), we find:

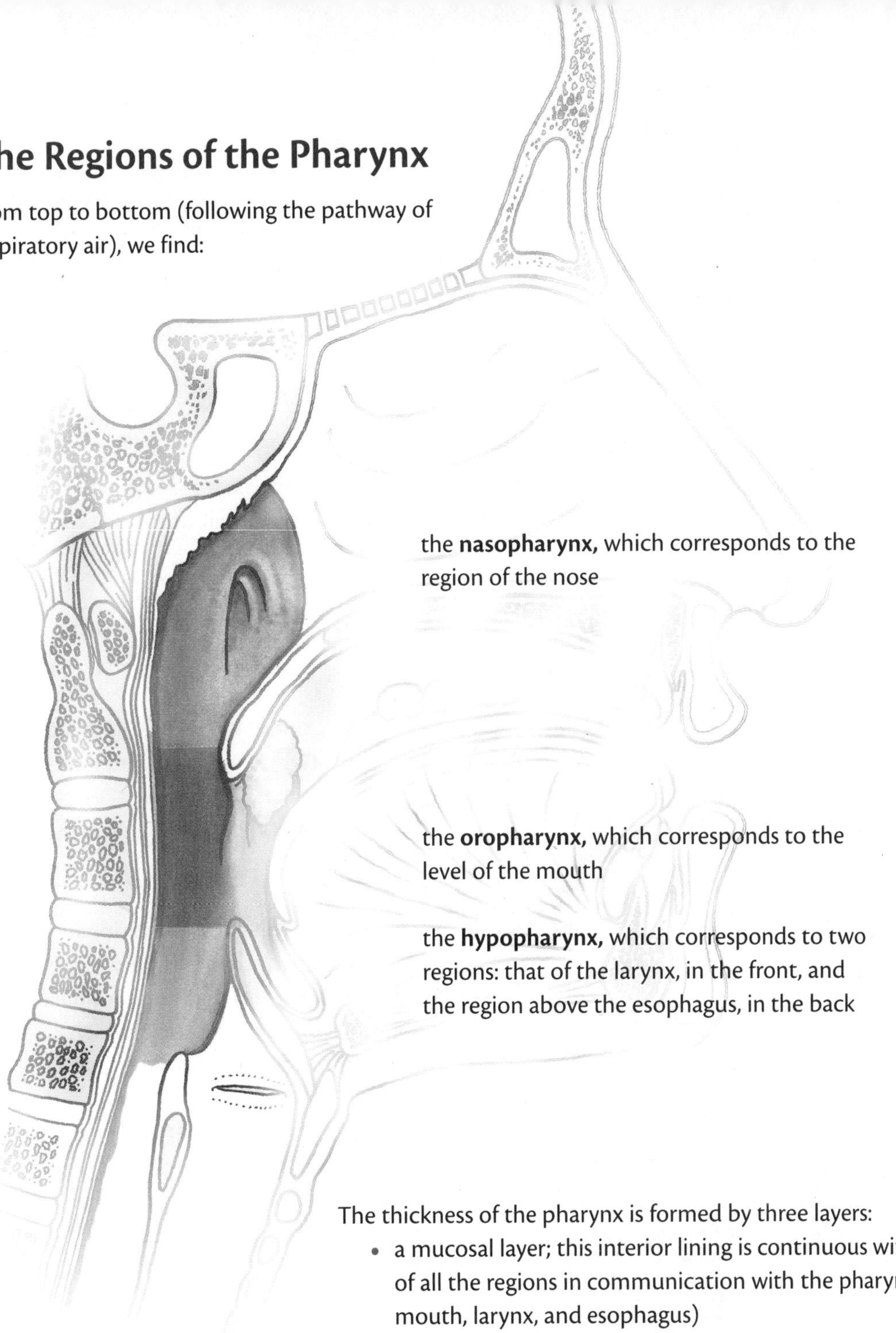

the **nasopharynx,** which corresponds to the region of the nose

the **oropharynx,** which corresponds to the level of the mouth

the **hypopharynx,** which corresponds to two regions: that of the larynx, in the front, and the region above the esophagus, in the back

The thickness of the pharynx is formed by three layers:

- a mucosal layer; this interior lining is continuous with the mucosa of all the regions in communication with the pharynx (the nose, mouth, larynx, and esophagus)
- a muscular layer, which is itself composed of three successive muscles called the **pharyngeal constrictors** (see pages 222–23)
- a fibrous layer

At the top, the muscular layer is absent, and reinforcement is provided by a fibrous layer: the pharyngobasilar fascia, from which the pharynx is suspended from the basion of the occiput, by a projection called the **pharyngeal tubercle** (see page 68).

The Nasopharynx

Also called the rhinopharynx, this is the part of the pharynx that is at the back of the nasal cavity. It is approximately 5 centimeters wide. It has a rather cuboid shape, with six faces, all lined with the mucosa of the pharynx.

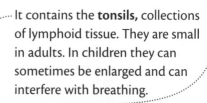

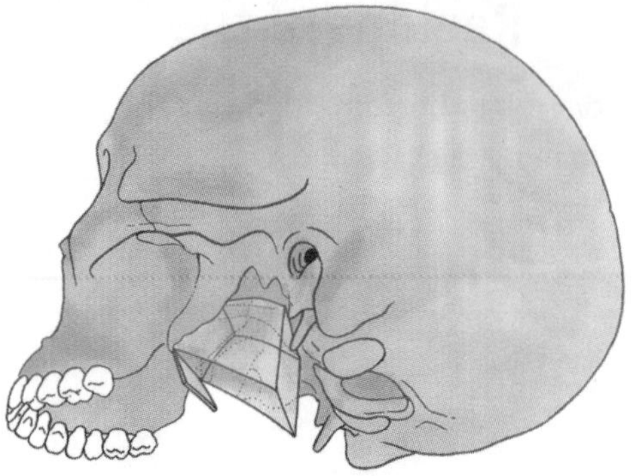

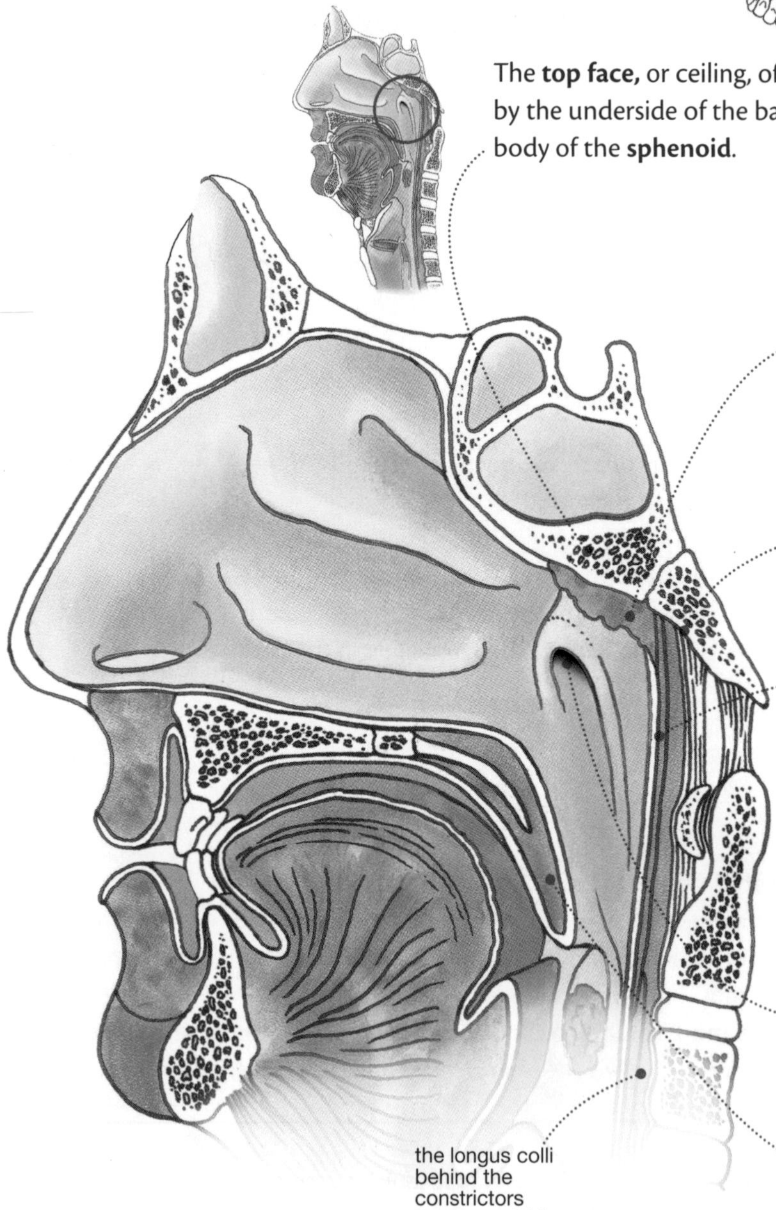

The **top face,** or ceiling, of the nasopharynx is formed by the underside of the basion of the **occiput** and the body of the **sphenoid**.

It contains the **tonsils,** collections of lymphoid tissue. They are small in adults. In children they can sometimes be enlarged and can interfere with breathing.

The top of the **posterior surface** is made up of the pharyngobasilar fascia and the **superior pharyngeal constrictor muscle,** which continues on the lateral surfaces.

The **lateral faces** (here we see the right side), on the inside of the constrictor muscle, consist of muscles* that tense and lift the soft palate; between them runs the inferior extremity of the **Eustachian tube** (see page 283).

The **inferior face** has a shape that varies and is made up of the **soft palate** (see page 236).

*These muscles are not visible here, as they are covered by the mucosa.

The Oropharynx

This is what we call the throat, the part of the pharynx behind the oral cavity. It is about 4 centimeters wide. It is separated from the oral cavity by a closure at the back of the mouth: the isthmus of the fauces (or the oropharyngeal isthmus), formed by the **palatoglossal arches** (see page 237), between which lies the **palatine tonsil**.

This is the part of the pharynx that can be observed, whether by looking in another person's throat or at one's own throat in a mirror.

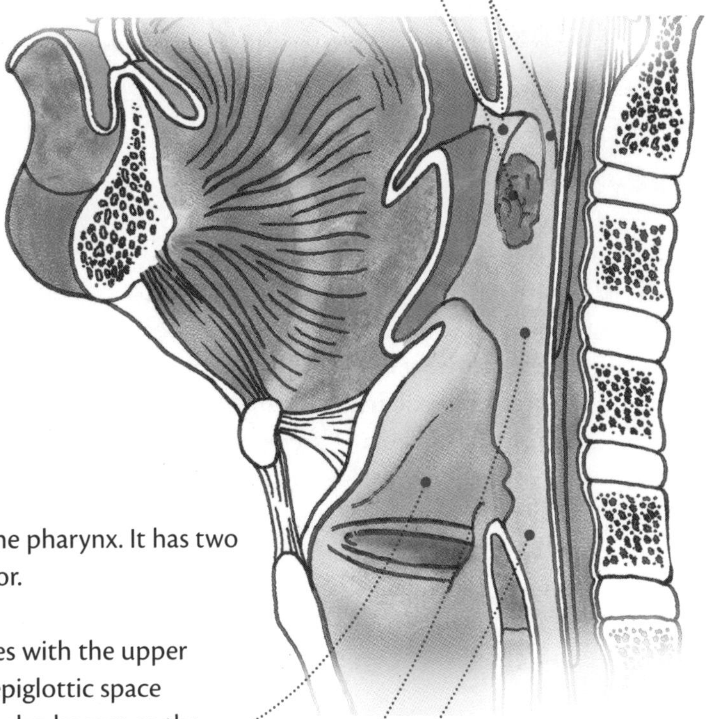

The Hypopharynx

This is the lowest part of the pharynx. It has two parts: anterior and posterior.

Anteriorly, it communicates with the upper part of the larynx, the aryepiglottic space (see page 184). This area is also known as the **laryngopharynx**.

Posteriorly, the upper part (2 centimeters wide) is continuous with the oropharynx. Its lower part is narrower and communicates with the top of the **esophagus**.

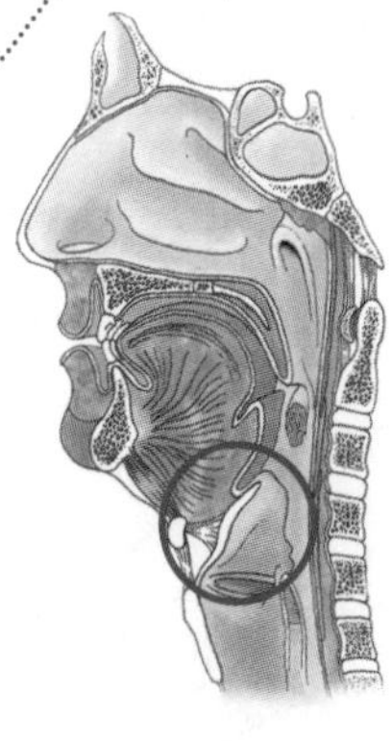

The Muscles of the Pharynx

These are the pharyngeal constrictors, three muscles that follow one another from top to bottom. They constitute the side faces and the rear face of the tract. Their fibers are oblique and converge in the back at the midline.

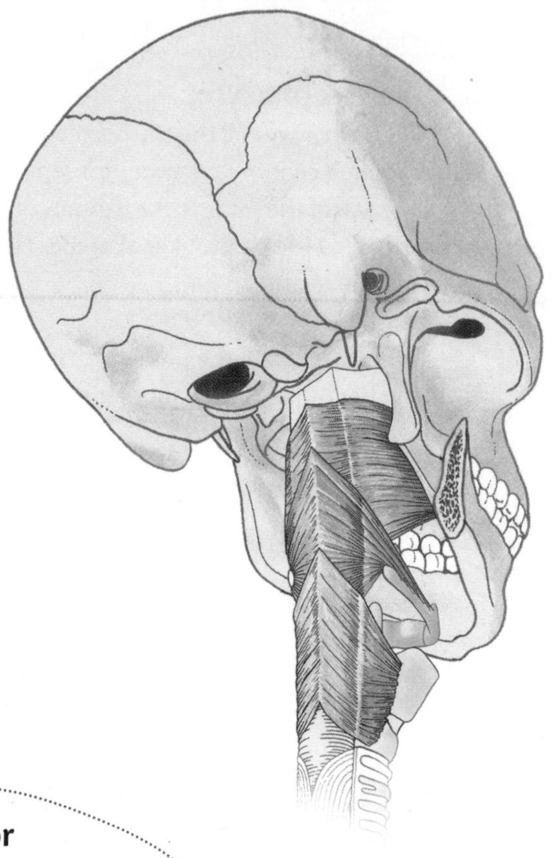

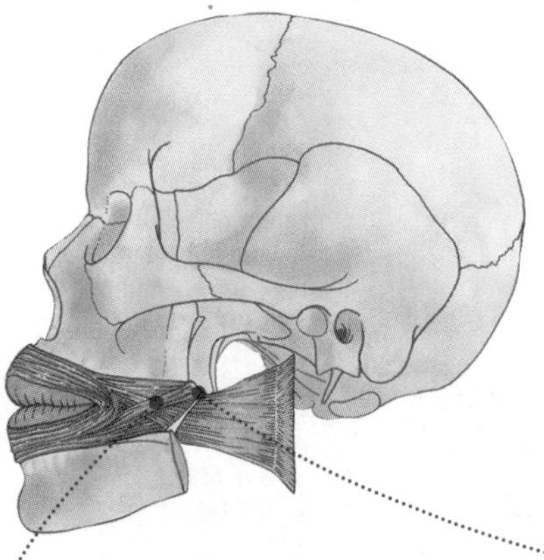

The Superior Pharyngeal Constrictor

This muscle is attached at the front to a succession of regions:

- the external wing of the **pterygoid process**
- the **pterygomaxillary ligament** (the ligament that runs from the hook of the external wing of the pterygoid process to the posterior extremity of the mylohyoid line; see page 80); at this level, it is connected to the buccinator muscle (see page 269)
- the lateral and posterior part of the genioglossus muscle, forming the pharyngoglossus muscle (see page 256)

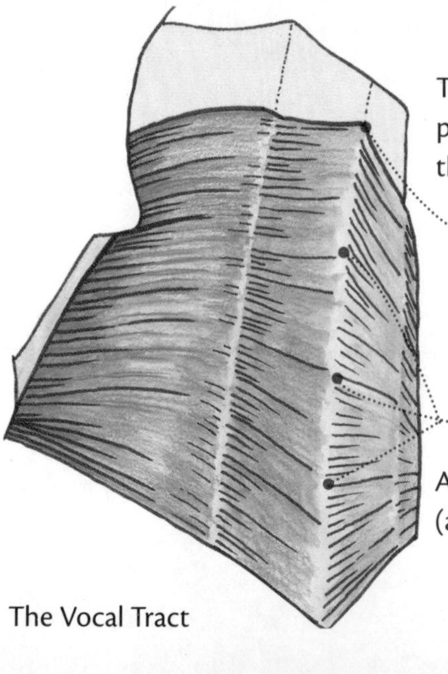

The fibers run to the back, forming the lateral wall of the pharynx, and then to the midline; from there, they form the posterior wall. At this point:

- The upper fibers run toward the top to terminate under the pharyngeal tubercle (the projection located on the inferior surface of the basion of the occiput; see page 68).
- The middle fibers run horizontally and increasingly downward.

All of the posterior fibers terminate in the form of a **raphe** (a crossing of the left and right fibers).

The Middle Pharyngeal Constrictor

This muscle is located at the back of the mouth, below the superior constrictor. It attaches in the front on the greater horn of the hyoid bone. From there, its fibers fan out and run backward. They continue to widen as they run to the midline.

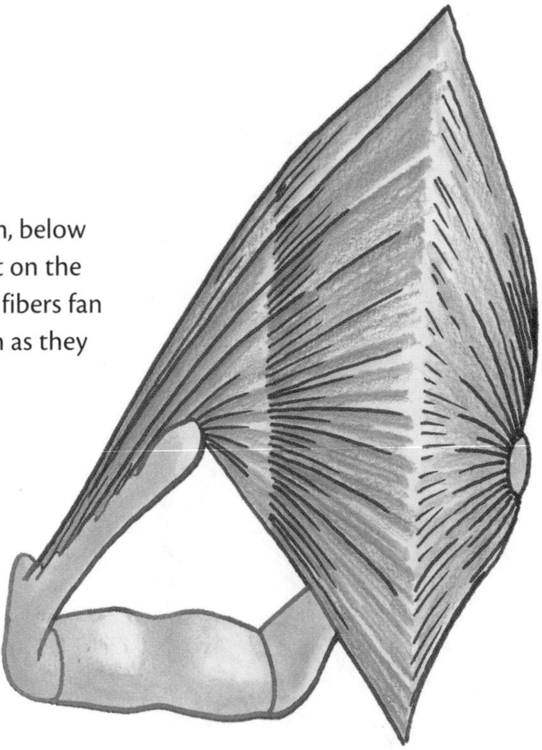

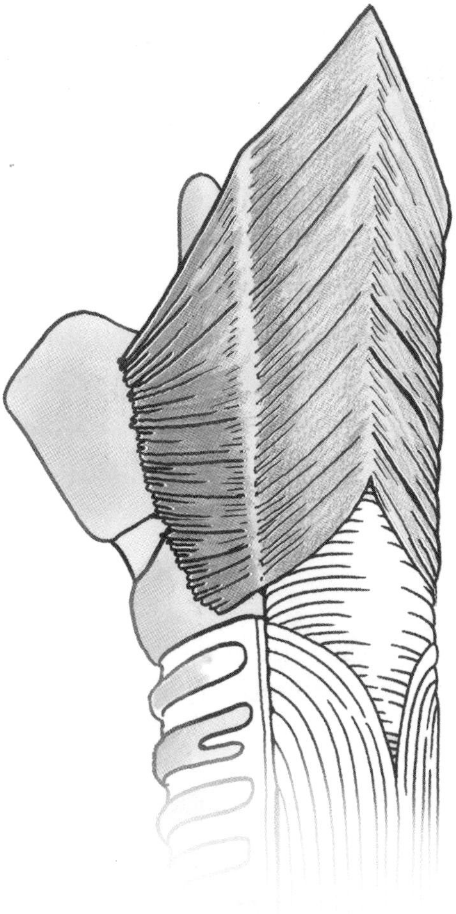

The Inferior Pharyngeal Constrictor

This muscle is located at the rear of the larynx. It attaches in the front to the side of the cricoid cartilage, to the back of the lateral surface of the thyroid cartilage, and to the ligament that links them. From there, the fibers run first backward, fanning out, and then moving again upward to the posterior midline, where they intersect. The lowest fibers form the lower limit of the esophagus, with which they continue.

The Constrictors Overlap

In the back, the middle constrictor covers the lower part of the superior constrictor. The inferior constrictor covers, in the back, the lower part of the middle constrictor.

The contraction of each constrictor causes, at its level, a tightening of the pharynx.

The upper and middle constrictors lift the larynx upward. When we swallow, we can feel the larynx rise.

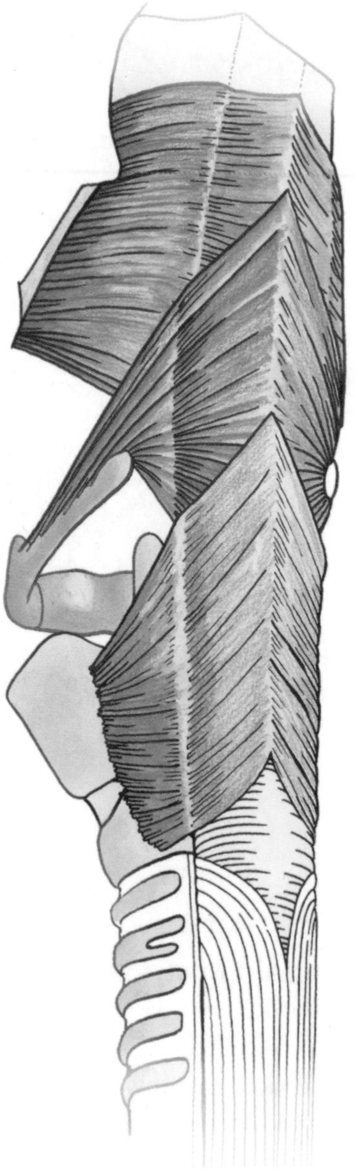

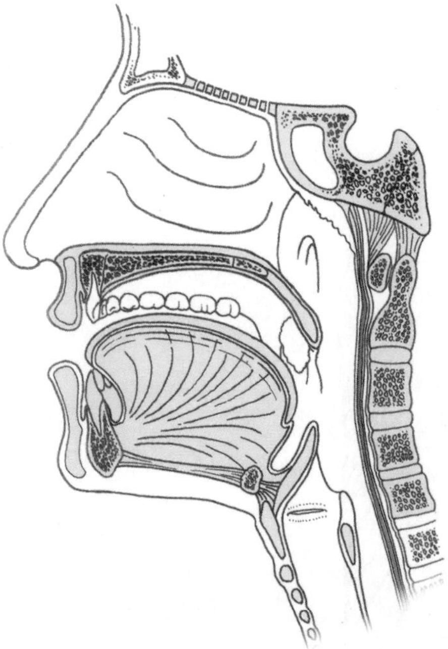

Looking at a sagittal cut of the vocal tract, we see that the constrictor muscles form a thickening at the front of the cervical spine, from which they are separated by the longus colli (see page 214).

The constrictors contract successively from top to bottom at the pharyngeal phase of swallowing. This allows the bolus of food or the liquid to move from the rear of the mouth to the esophagus. This sequence is a reflex action (that is to say, it's involuntary), while the preceding sequence, called the *buccal period,* is performed by the tongue and is a voluntary action (though in most cases we don't pay attention, we can control it if we do pay attention).

This action occurs when we eat and drink and, more often, when we just swallow saliva (which happens between 1,500 and 2,000 times per day).

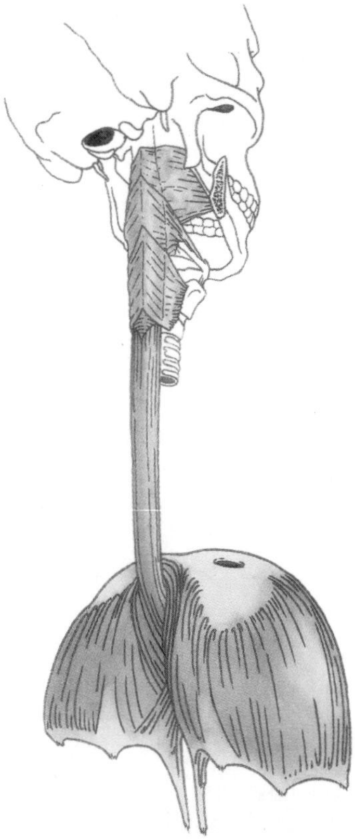

The Pharynx: Articulation and Resonance

Articulation

We can make sounds by moving the pharynx and the back of the tongue closer together (like when we make an "r" sound way back in the mouth). Similarly, it is possible to create sound-producing vibrations in this area or between the soft palate and the pharynx when we snore.

Resonance

The lowering of the larynx, for example, by the diaphragm (see page 121), extends and modifies the pharyngeal conduit, increasing the deep resonance harmonics.

A tonic state of the pharynx muscles, which changes the diameter of the pharyngeal passageway, will modify the resonance.

The muscles of the pharynx may tend to contract at the same time as those in the region of the larynx, just by virtue of their proximity, in particular:

- when the vocal emission is very intense
- when the vocal emission is very high

This tightens the pharyngeal conduit, which is not always desirable for vocal production. It is important to know how to relax the pharyngeal muscles, and to recognize when they are relaxed. (If we contract them afterward, we want the appropriate contraction.)

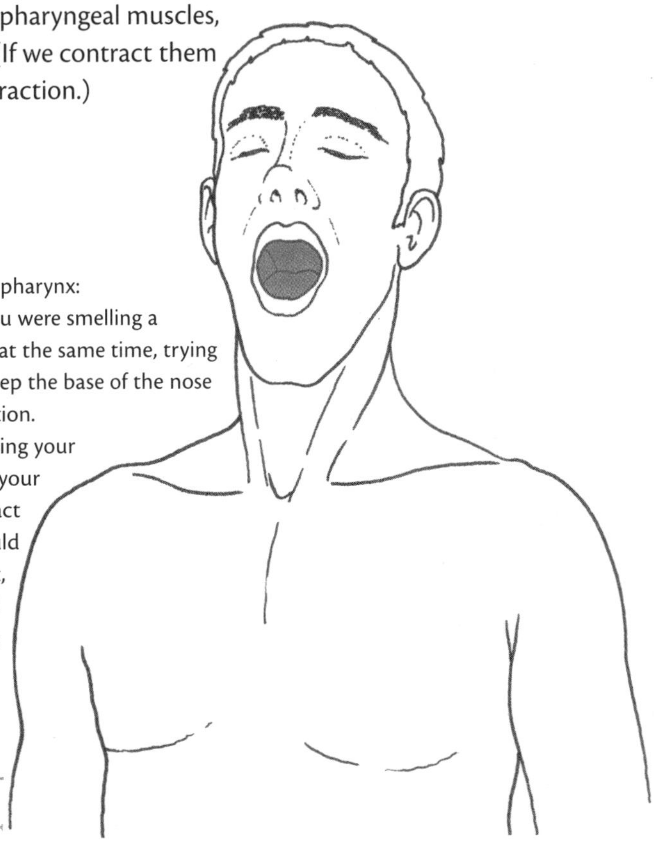

Relax the Muscles of the Pharynx

Several exercises are used to relax the muscles of the pharynx:

- At the level of the nasopharynx, inhale as if you were smelling a fragrance, an inhalation that is light and deep at the same time, trying to expand the base of the nose. Then try to keep the base of the nose open during an exhalation, and during phonation.
- For the oropharynx, chew a dozen times, keeping your lips in contact and trying to open the back of your mouth as wide as possible (so as not to contract the masseter/temporalis too much). This should relax the constrictors at the back of the throat, just by virtue of their proximity. This is a good vocal preparation, and as an added bonus you will salivate, which will hydrate the pharynx/larynx.
- For the entire pharynx, just yawning will encourage relaxation.

The Mouth

The mouth is the main cavity that air passes through after making its way through the pharynx. The mouth performs many different functions: it is one of the breathing airways, the entrance to the digestive tract, and the organ of tasting.

Regarding the voice, it is the fundamental location for the articulation of sounds.

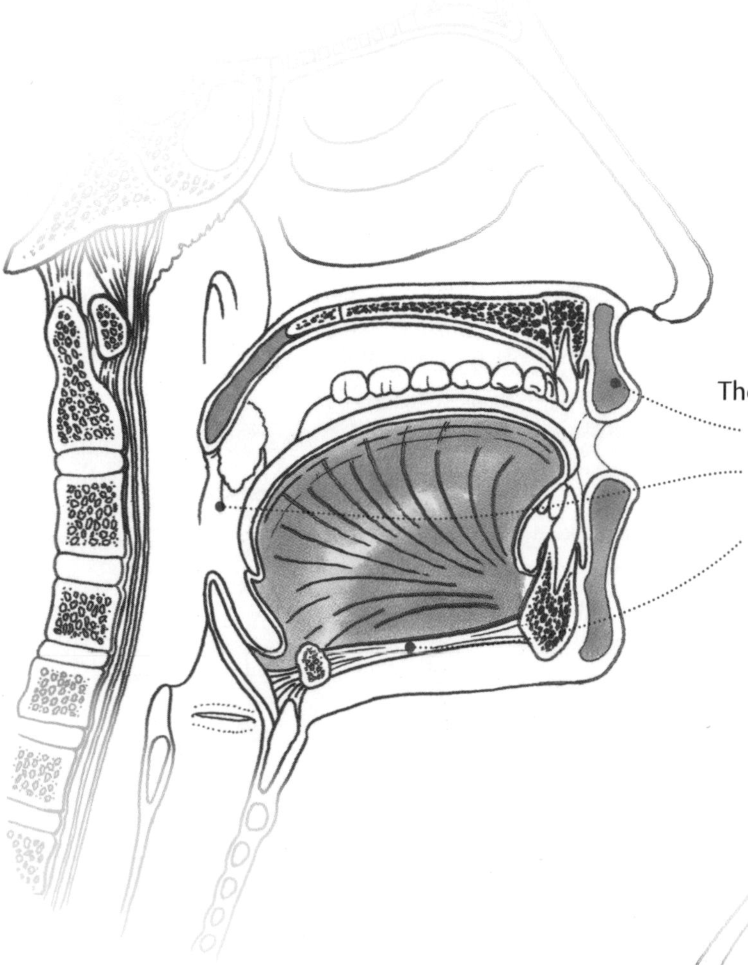

The limits of the mouth are:

- at the front, the **lips**
- at the back, the **isthmus of the fauces**
- at the sides, the **cheeks** (not visible here)
- at the bottom, the **geniohyoid and mylohyoid muscles**

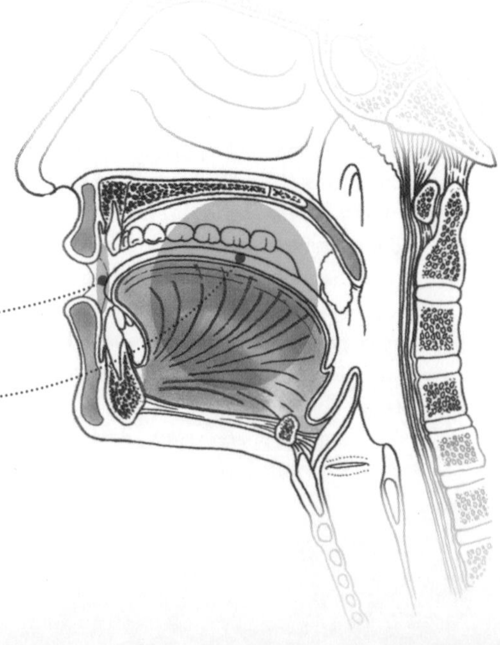

Within the oral cavity, the prominent dental arches define two areas:

- in front of the dental arches and behind the lips, the **oral vestibule**
- at the back and inside of the dental arches, the **buccal or oral cavity**

In this section we'll describe several parts of the mouth with easily recognizable roles:

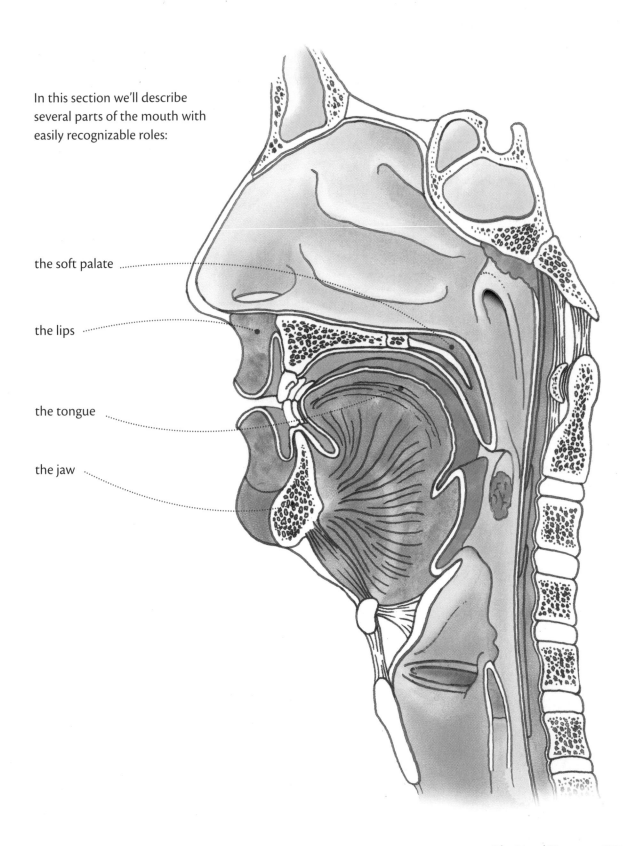

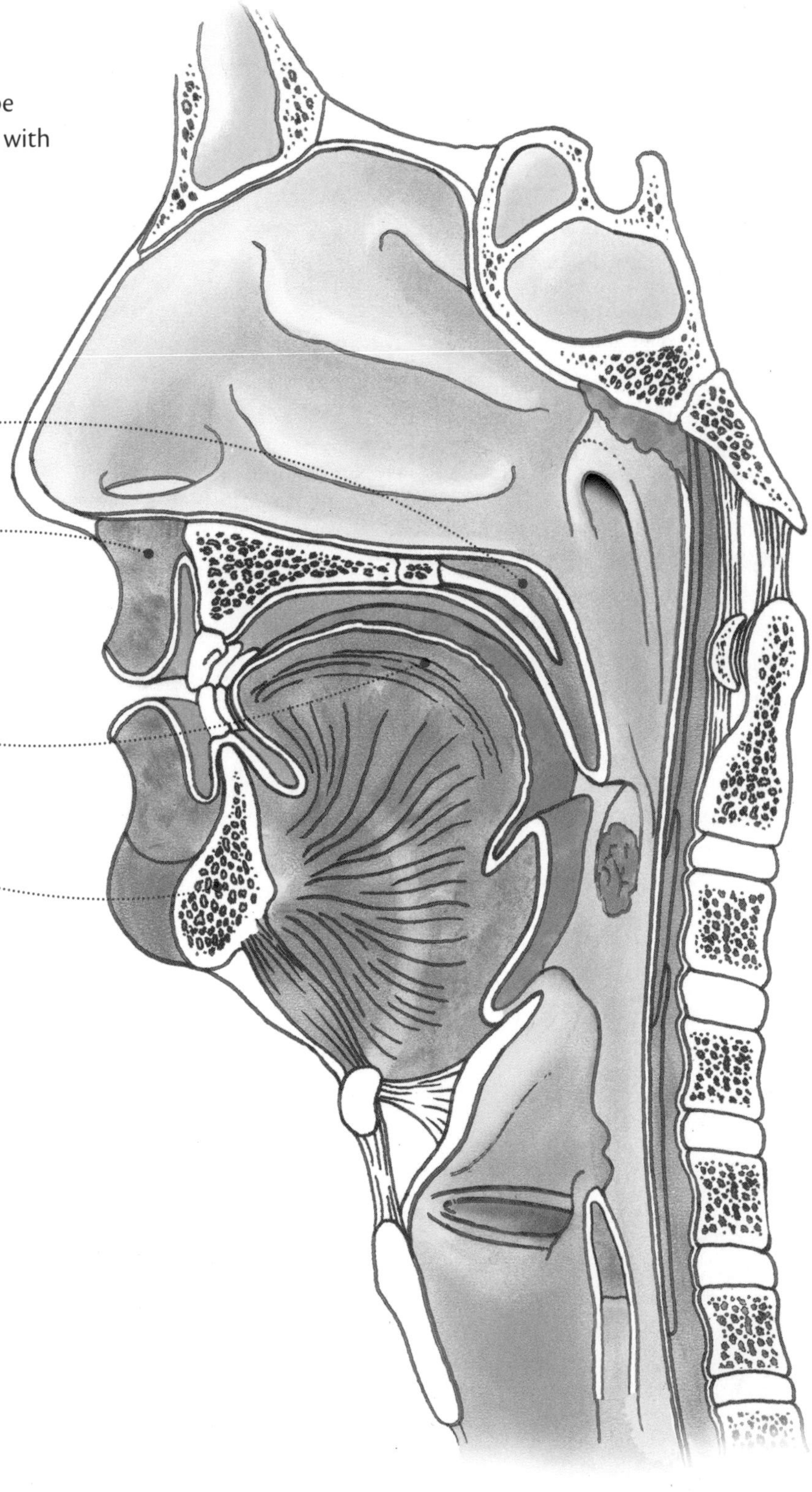

Opening/Closing the Mouth: The Muscles of the Mandible

The Masseter

This muscle of the jaw is located in front of the ear. It is very thick and made up of two layers: one deep, one superficial.

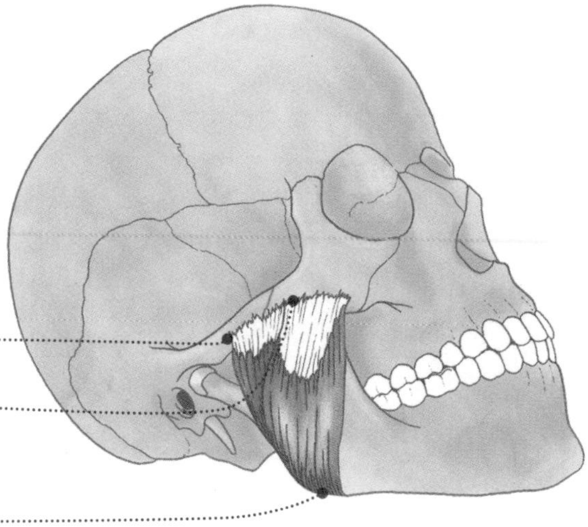

At the top, the deep layer attaches to the **zygomatic arch,** and the superficial layer attaches to the zygomatic process of the **malar bone**. The fibers descend toward the back and terminate at the angle of the mandible (the **gonion**) and on the ramus and back part of the body, up to the oblique crest situated on the external surface.

Its Action

It elevates the mandible, helps close the mouth, and contributes a little to bringing the jaw forward.

Zooming In on the Zygoma

The masseter attaches to the zygoma or zygomatic arch. You can easily feel this bony arch, which runs from the cheekbone to the ear canal, by placing three fingers at the top of the cheek. The zygoma is divided into three parts, defined by three different bones:

- the superior maxilla in the front (its anterior "root"); see page 78
- the malar bone, or cheekbone; see page 73
- the zygomatic process of the temporal bone at the back (its posterior "root"); see page 73

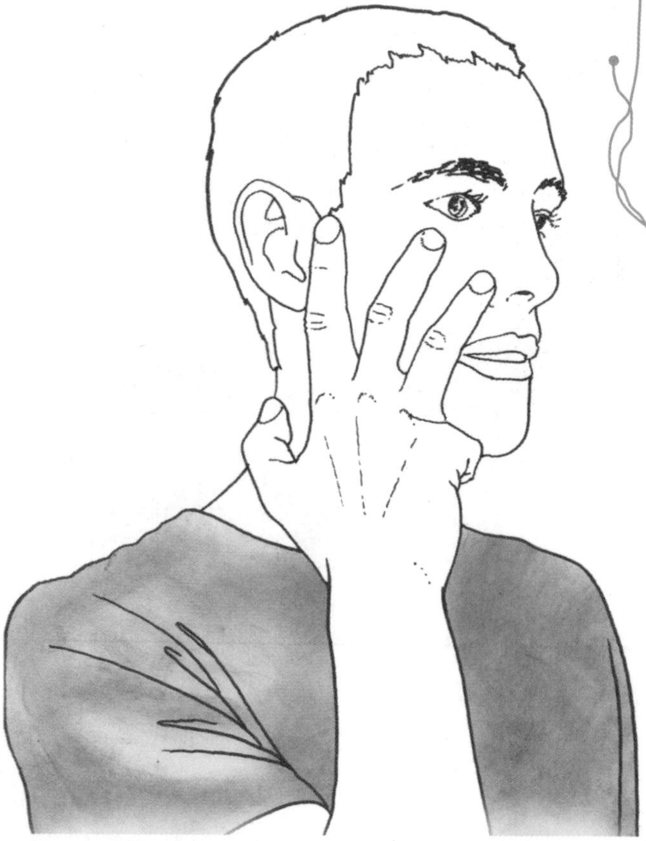

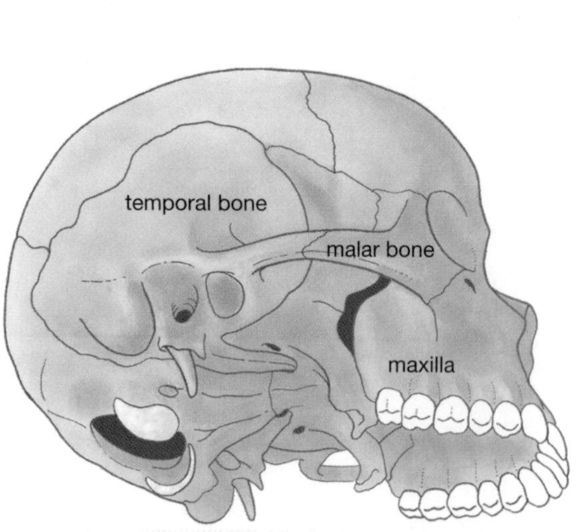

The Temporal Muscle

This muscle attaches at the top in the area of the temple and spans four bones: the parietal, frontal, sphenoid, and temporal.

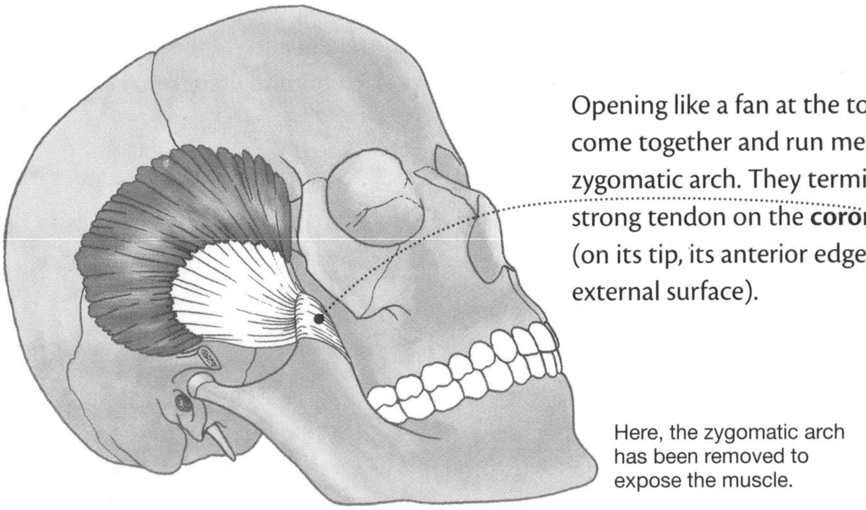

Opening like a fan at the top, its fibers come together and run medial to the zygomatic arch. They terminate in a strong tendon on the **coronoid process** (on its tip, its anterior edge, and its external surface).

Here, the zygomatic arch has been removed to expose the muscle.

Its Action

It elevates the mandible by contributing to the closure of the mouth. It pulls it a bit toward the back.

Zooming In on the Temple

The temple is a hollow area behind the ridge of the frontal bone. It consists of four bones:

- the greater wing of the sphenoid
- the lower and lateral part of the frontal bone
- the lower and lateral part of the parietal bone
- the front portion of the temporal petromastoid; when the masseter is contracted, instead of a recess, we see a bulge here.

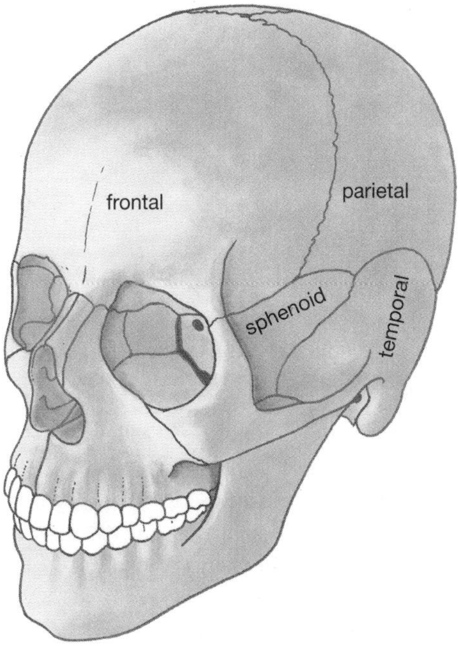

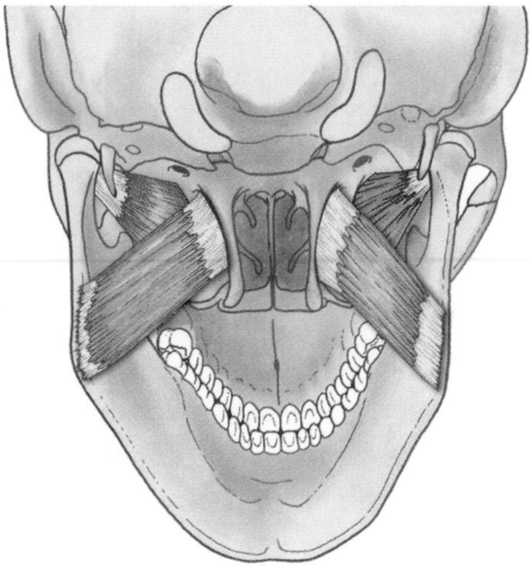

The Pterygoid Muscles

These muscles are so called because they attach in part on the pterygoid. They are located at the back of the mouth, just medial to the ramus of the mandible, in an area called the **pterygopalatine fossa,** which is formed in part by the sphenoid (see page 70) and the superior maxilla (see page 78).

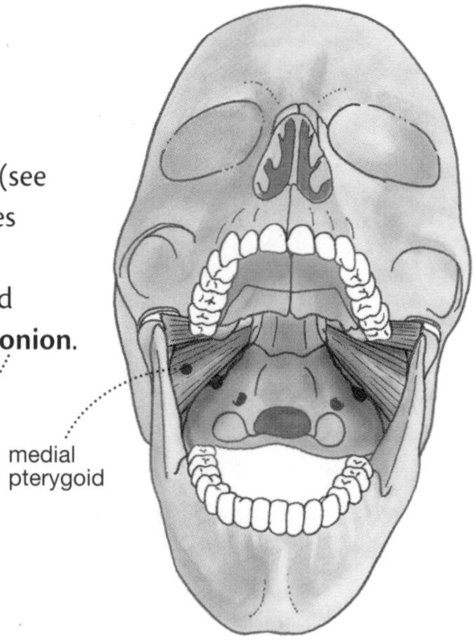

The Internal or Medial Pterygoid

This muscle is the equivalent of the masseter muscle (see page 228) in the deep part of the mandible. It attaches to the angle formed by the inner and outer wings of the **pterygoid process**. Its fibers descend outward and backward and terminate on the deep surface of the **gonion**.

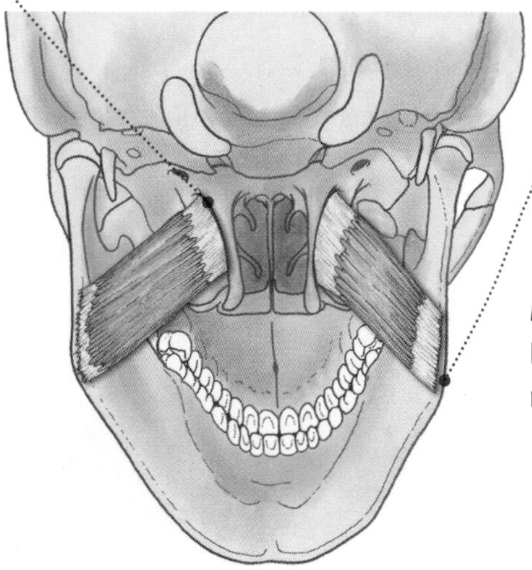

Its Action

It elevates the mandible. It pulls the mandible inward and also forward.

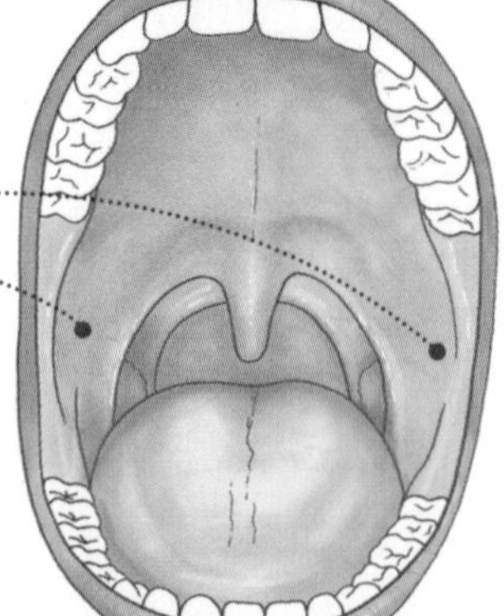

A Muscle That Often Hurts

The internal pterygoid forms a vertical mass between the last molars at the back of the mouth, just outside the anterior arch of the soft palate. Palpation at this location often shows a contracted and painful muscle (see page 240).

The External or Lateral Pterygoid

This muscle is made up of two bundles: inferior and superior. It attaches to the external wing of the pterygoid process (on the external face).

The superior fibers arise from higher up and run horizontally toward the back. The inferior fibers rise toward the back. All the fibers meet in a tendon that attaches to the deep surface of the neck of the **mandibular condyle**.

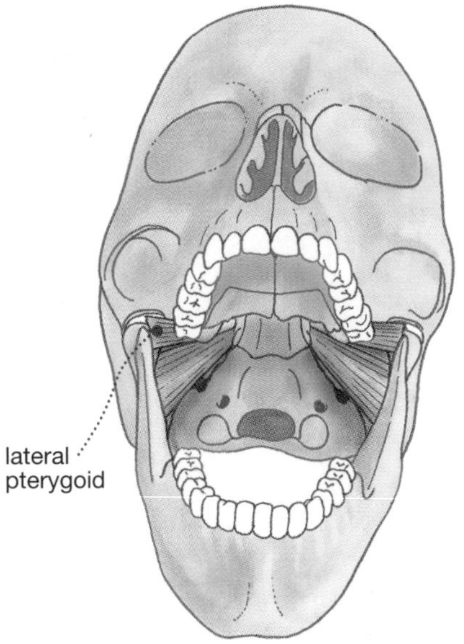

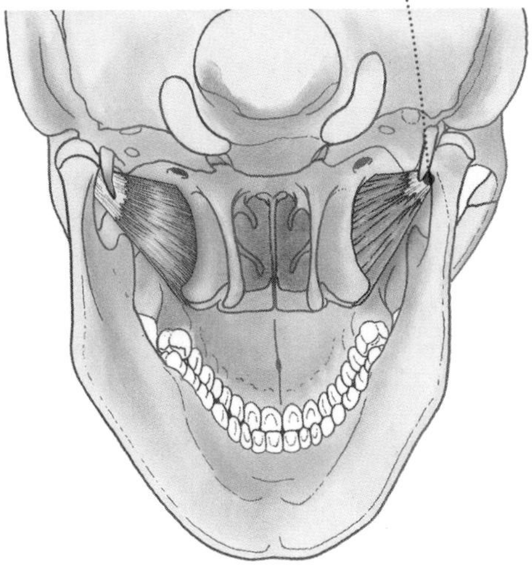

Its Action

It moves the mandible forward (propulsion).

Common Action of the Pterygoids

These muscles move the jaw laterally when they contract on one side. The jaw moves toward the side opposite the contraction (for example, if the pterygoids contract on the right side, the jaw moves to the left).

The pterygoids are most often involved in mastication, which is not only an up/down movement (raising or lowering the mandible) but also a lateral movement.

Relaxing the Jaw

In vocal work, we often give the cue to relax the muscles that elevate the mandible (masseter, temporalis, pterygoid). In fact, these muscles contribute to keeping your mouth closed all the time, and they are, therefore, difficult to release (for some people, they remain contracted even during sleep). Their excessive contraction often spreads to their neighbors, such as the muscles of the soft palate, pharynx, or tongue, hindering the ease of their mini-movements. However, when you are standing, if these muscles relax completely, your mouth will open wide (see the next page). Therefore, you must moderate this relaxation, finding just the right amount of contraction. You must also understand that to really relax those muscles (which is sometimes essential in prevocal exercises), standing is not appropriate: you have to be lying down, preferably on your side and even possibly supporting your jaw with one hand.

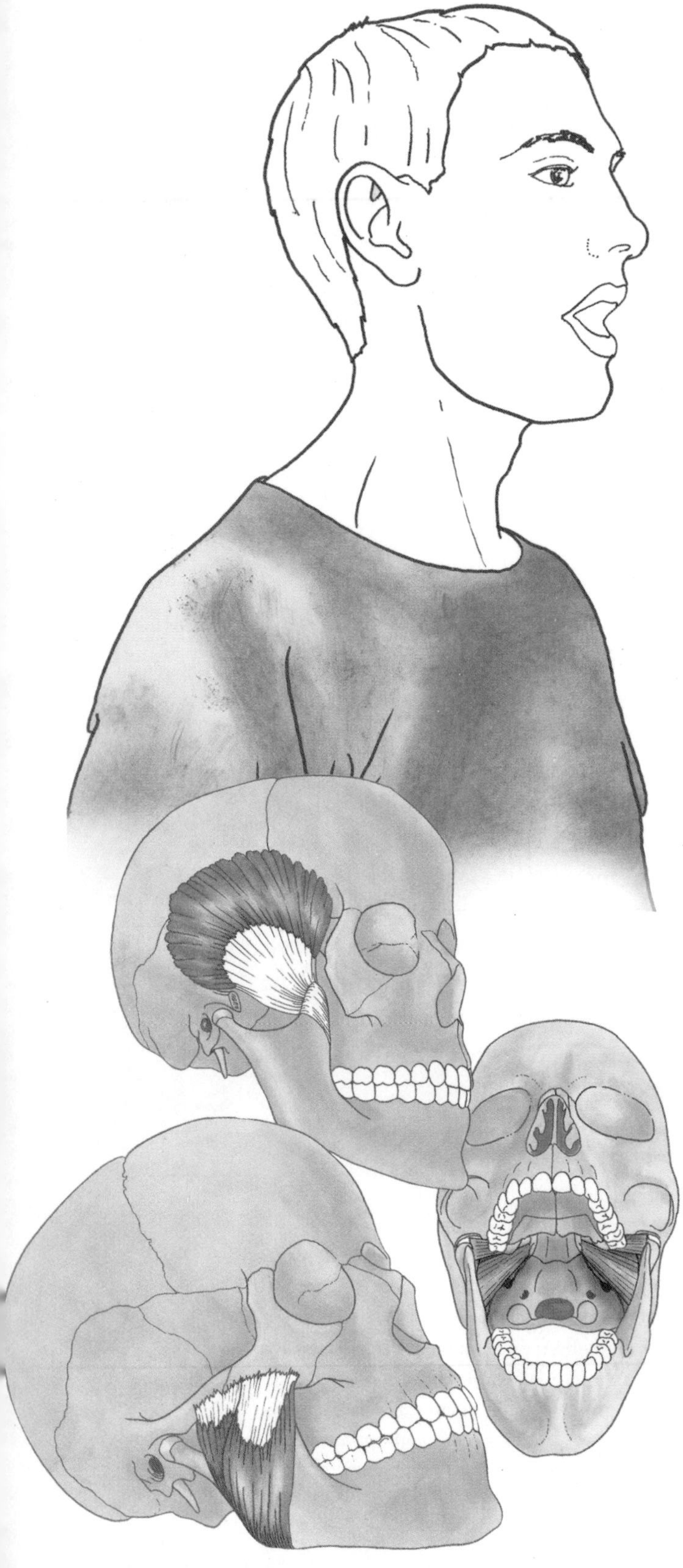

Gravity Is Partially Responsible for Movements of the Mouth

In a standing position, the jaw drops under the effect of gravity.

We can add to that the weight of the tongue, the weight of the bone of the mandible, and also the weight of everything that is suspended from the jaw: the larynx, hyoid bone, and trachea.

Therefore, when we are in a vertical position, we must work to keep the mouth closed at all times. This closure is assured by the whole of the elevator muscles.

In vocal work we are often asked to relax our jaw to free up the movement of the mouth. However, the complete relaxation of the muscles of the mandible leaves the mouth wide open. What we actually want is an adjustment of the tonus of the elevator muscles, which are often too contracted.

It is interesting to alternate the recruitment of these muscles:

- Use the pterygoids to keep your mouth closed (the feeling is internal, inside the mouth).
- Use the temporal muscles to keep your mouth closed (the feeling is in the temples).
- Use the masseter to keep your mouth closed (the feeling is around the jaw).

The position of the head and neck causes opening/closing of the mandible.

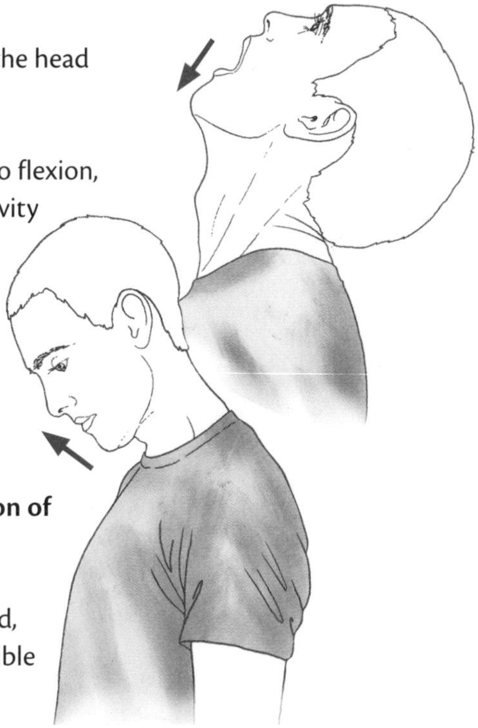

Placing the neck/head in extension, or even rocking the head backward, can cause the mouth to open.

At a certain point when we are bringing the head into flexion, the jaw will start to elevate because of the pull of gravity on the head.

Therefore, in positioning the head for vocal work, the position of the head on the atlas, more or less in extension, and relatively symmetrical, influences the placement of the jaw.

The position of the head and neck causes translation of the mandible.

If we keep our mouth closed with our teeth separated, and we mobilize the neck, we can see that the mandible moves like a "drawer" under the superior maxilla:

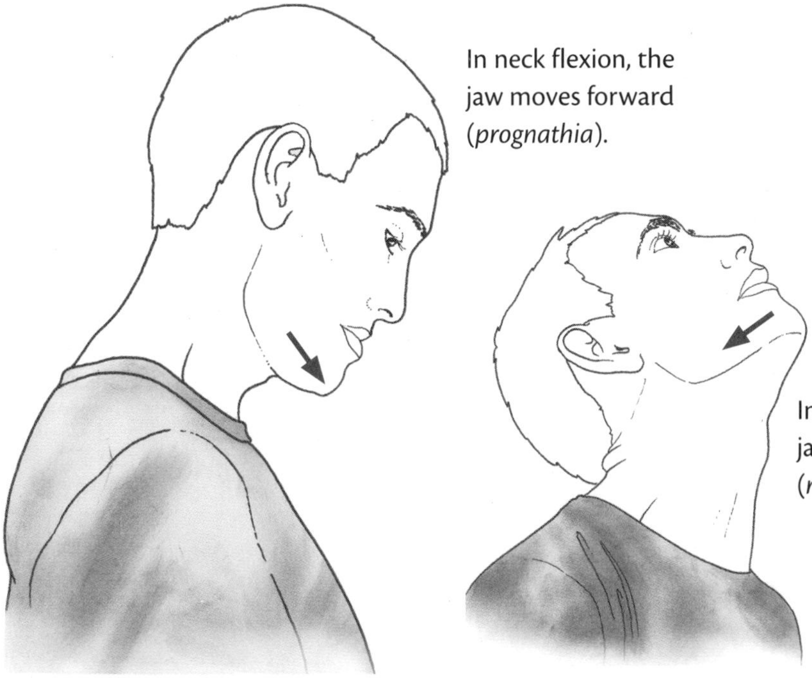

In neck flexion, the jaw moves forward (*prognathia*).

In neck extension, the jaw moves backward (*retrognathia*).

These tendencies, which are very obvious in large movements of the head and neck, are also present in small movements.

The Mandible: Articulation and Resonance

The Mandible and the Articulation of Vowels

The opening of the mouth (lowering of the mandible), which we also call "aperture," varies depending on the vowel produced. Here we will look at the vowel sounds *ee*, *ou*, and *ah*.

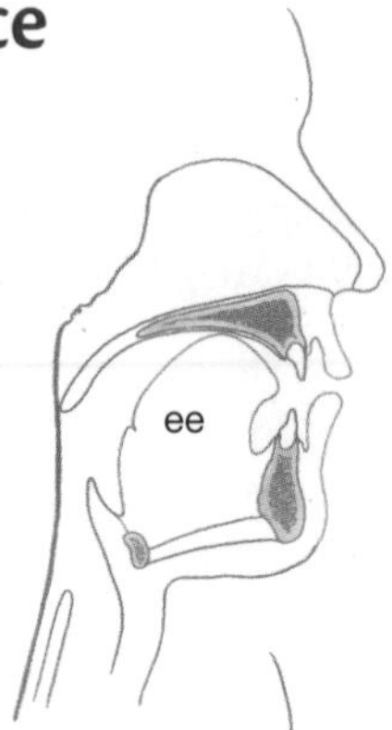

For the vowel sound *ee*, the mouth is slightly open. The jaws are close together. The mandible is kept high by the action of the masseter, temporalis, and pterygoids.

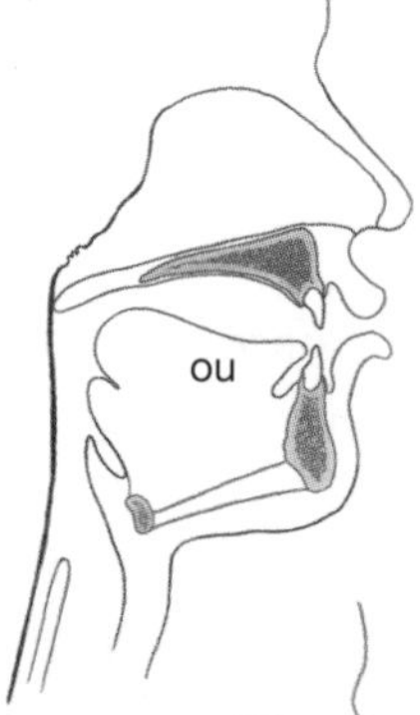

For the vowel sound *ou*, the mouth is almost completely closed. The jaws are very close. The mandible is kept high by the action of the masseter, temporalis, and pterygoids.

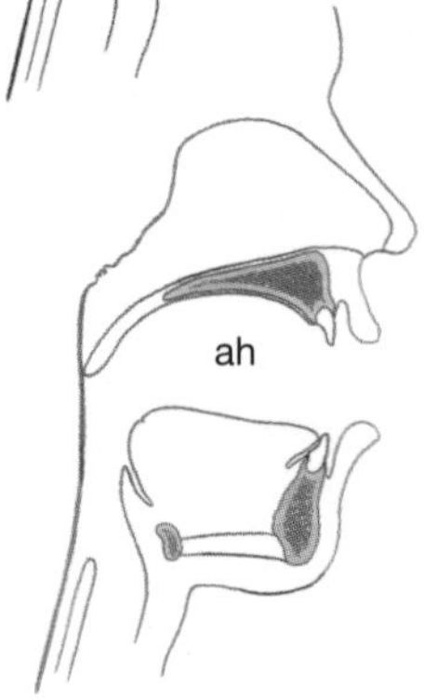

For the vowel sound *ah*, the mouth is open. The jaws are separated. The elevator muscles are less contracted, letting gravity lower the mandible. To this we can add action of the hyoid muscles to actively lower the mandible.

To feel the mandible lower gradually, we can pronounce the vowel sounds *ee*, *ey*, *eh*, and *ah* in succession.

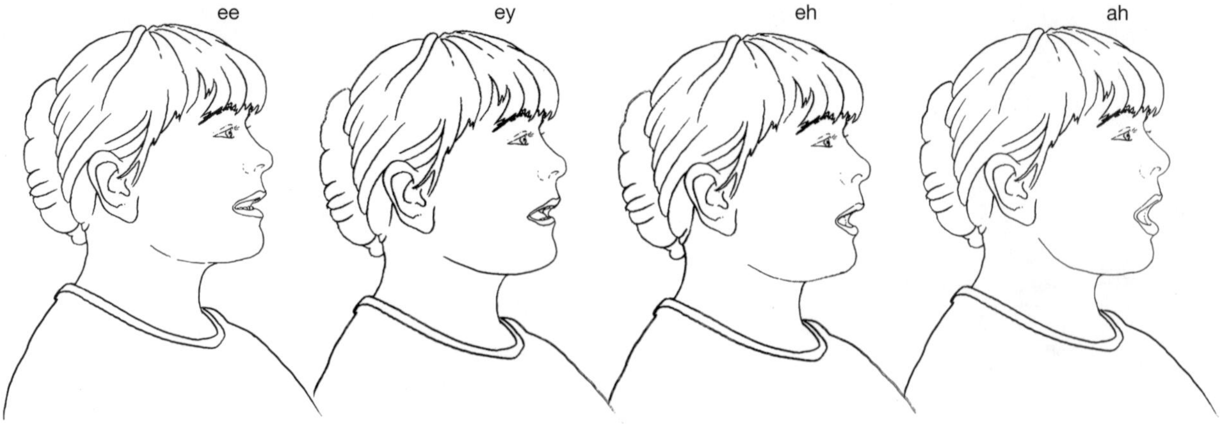

The Mandible and the Articulation of Consonants

Jaw movements are involved in forming certain consonants.

The **explosive consonants** (*p, b, t, d, k, g*) are made by two successive movements of the mandible:

- These start out with a lift of the mandible to create an occlusion (lips, tongue/teeth, tongue/palate), an action that is performed by the elevator muscles: masseter, temporalis, pterygoids.
- Then the mandible is lowered abruptly to release the pressure and allow an "explosion" of sound. Gravity is responsible for this second action, as is relaxation of the elevator muscles. It can also be done by a contraction of the depressors: the hyoid muscles.

Fricative consonants (for example, *f, s*) are made by keeping the mandible lifted. This is accomplished by activation of the elevator muscles: masseter, temporalis, pterygoids.

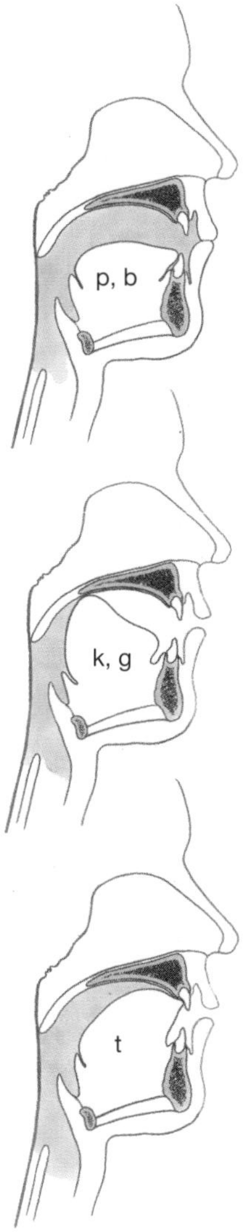

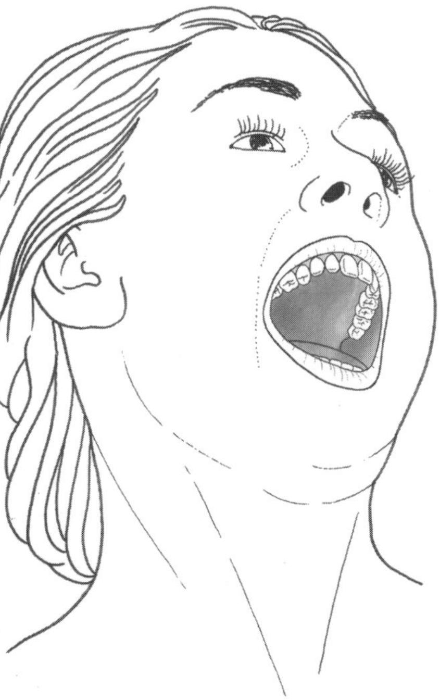

The Jaw Is Involved in the Resonance of the Voice

In the domain of the voice, a sound is called "open" when it is emitted with the mouth more open in the front (at the level of the lips or mandible) than it is in the back (at the level of the oropharynx). Taken to the extreme, a very open voice can become what we call "colorless," or a voice without sonority.

The Soft Palate

We don't see it, and we don't always feel it, because it's located at the back of the mouth, but it is a centerpiece of our vocal instrument. Extending the hard palate, the soft palate is a muscular and fibrous partition at the rear of the nasal cavity (above it) and mouth (below it).

In everyday life, the soft palate is primarily used when we swallow, to prevent food, especially liquids, from going up into the nose from the rear. When it is weak, it is also the source of snoring. It is a place from which we can influence the opening of the Eustachian tube, putting the pharynx and the middle ear in communication.

In the voice, the soft palate is essential to the articulation of certain consonants and vowels (see pages 245–46). But its movements are also among the most important "shapers" of timbre above the larynx and can considerably enrich the resonance of the sound.

The Uvula, the Uvula . . .

Several words refer to the soft palate and will be found in the following pages:

- The word *velum* and the prefix *velo-* refer to the palate.
- The word *palatine* and the prefix *palato-* refer to the palate.
- The word *staphylin* and prefix *staphylo-* refer to the uvula (the word *uvula* derives from the Greek for "bunch of grapes," which matches the description of the uvula itself).

Description of the Soft Palate

The soft palate is a sheet approximately quadrilateral in shape of about 4 to 5 square centimeters, and about 0.5 centimeter thick. It is an extension of the hard palate. Its superior surface extends the floor of the nasal cavity, and its inferior surface is oriented more or less forward and downward depending on the position.

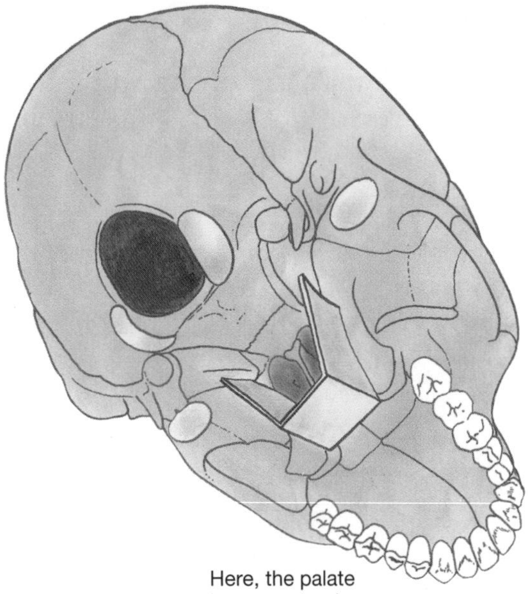

Here, the palate is represented schematically.

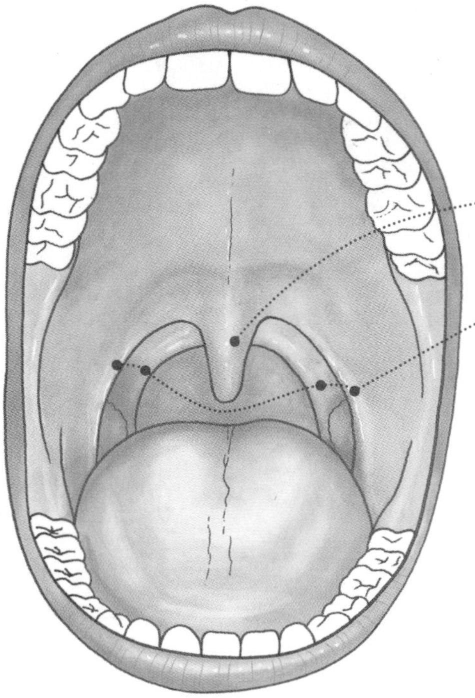

Its posterior extremity, or posterior border, is extended by a hanging medial zone: the **uvula**. Seen from the mouth, on either side of the uvula are borders that extend downward as two vertical recesses: the **arches** of the soft palate. There is a front arch and a rear arch on either side.

The front and rear arches form, with the base of the tongue and soft palate, a limit called the **isthmus of the fauces**. The isthmus is the boundary between the mouth (in front) and the pharynx (in back).

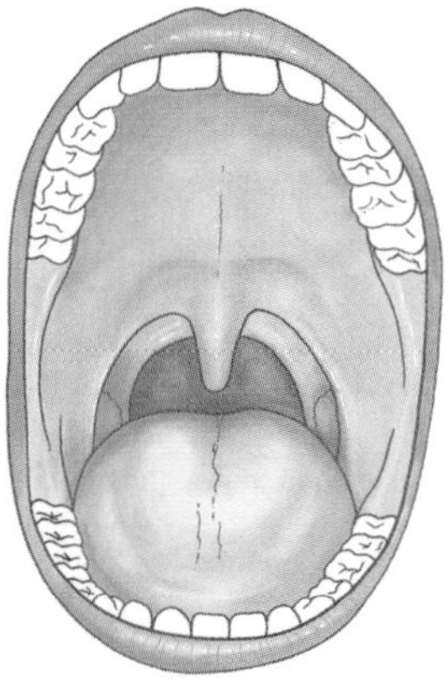

An Arch That Tugs

The word *arch* conveys a false image that suggests the soft palate would be held up by this element. In fact, the opposite is true: this element tugs the palate downward.

The Fascia of the Soft Palate

The soft palate is built around a fibrous structure: the **palatine aponeurosis.**

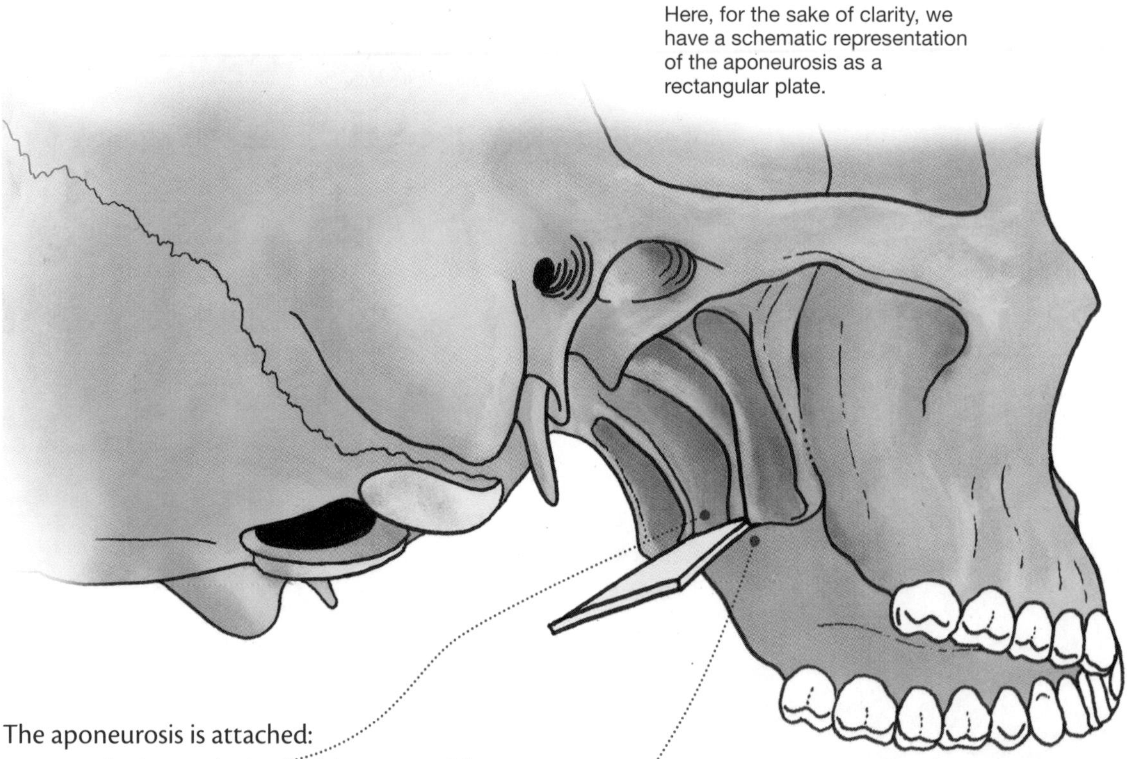

Here, for the sake of clarity, we have a schematic representation of the aponeurosis as a rectangular plate.

The aponeurosis is attached:

- to the internal wings of the pterygoid process (on the "hooks")
- to the posterior border of the hard palate, on the palatine bone

On top of this fascia attach the muscles that raise the palate as well as a muscle that lowers it: the palatopharyngeus. On the bottom is another depressor muscle: the palatoglossus.

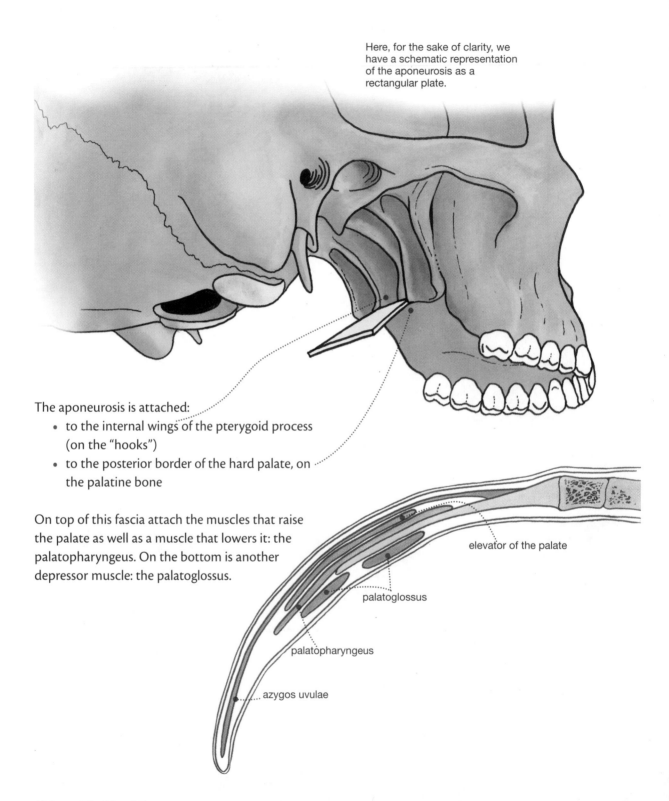

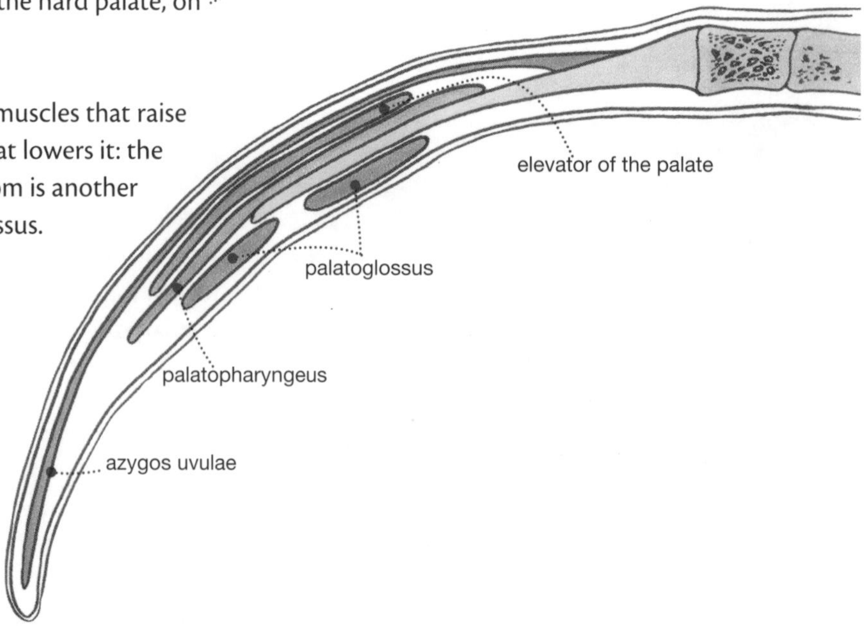

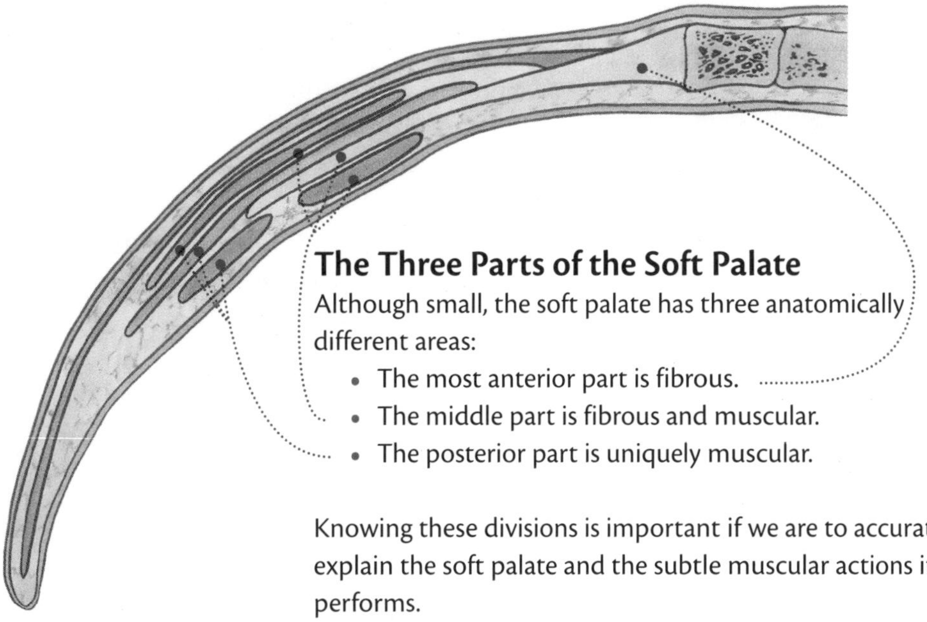

The Three Parts of the Soft Palate

Although small, the soft palate has three anatomically different areas:

- The most anterior part is fibrous.
- The middle part is fibrous and muscular.
- The posterior part is uniquely muscular.

Knowing these divisions is important if we are to accurately explain the soft palate and the subtle muscular actions it performs.

A Better Differentiated Soft Palate

In very deep and precise vocal work, these three parts need to be increasingly differentiated:

- sensorily and spatially (we know better and better where to locate each part)
- functionally (we know better and better what each part does; see page 247)

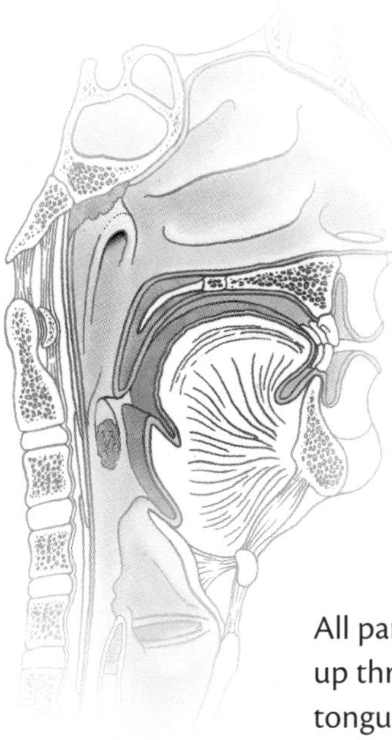

All parts are lined with mucosa, which continues up through the nasal cavity to the hard palate and tongue and down through the pharynx.

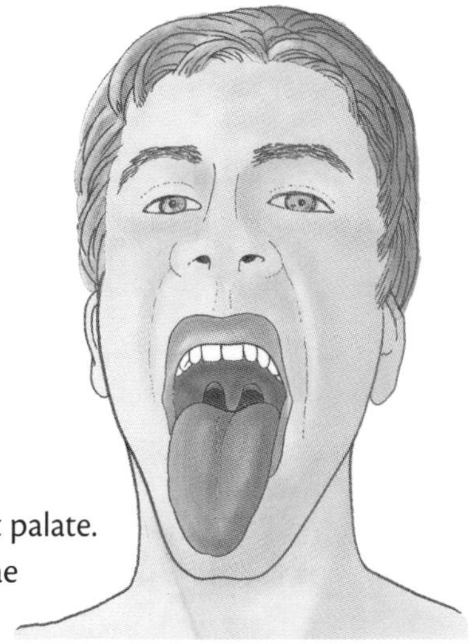

The Uvula

This is the median and posterior portion of the soft palate. It is the location of a small muscle: the azygos uvulae (see page 238).

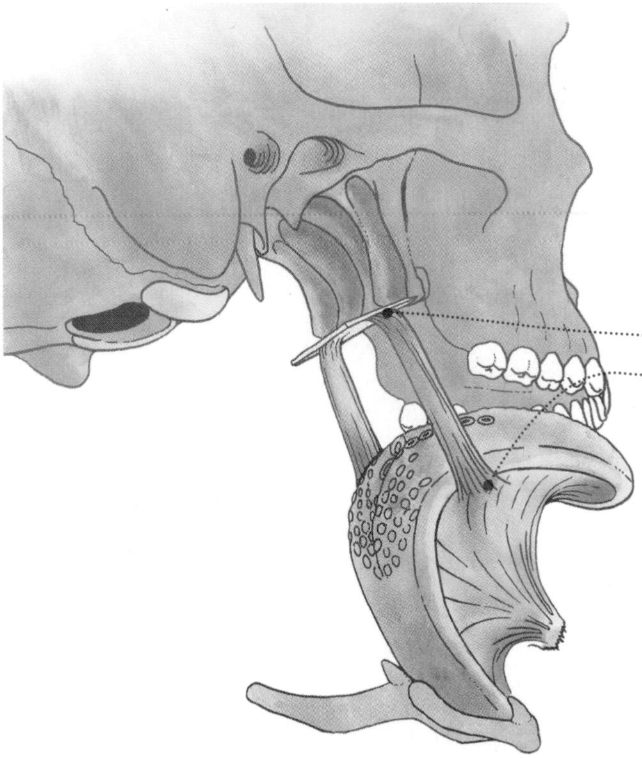

The Muscles of the Soft Palate

Two muscles depress the soft palate:

- the palatoglossus
- the palatopharyngeus

The Palatoglossus (Glossopalatinus or Palatoglossal) Muscle

This muscle attaches at the top against the underside of the palatine aponeurosis. It descends vertically and terminates along the lateral edge of the tongue.

It forms the anterior arch of the palate, which runs in front of the palatine tonsil.

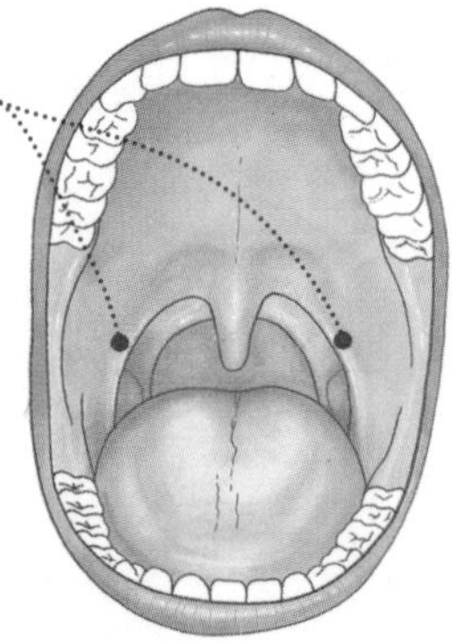

Its Action

It lowers the palate while raising the back of the tongue (see page 254). It contributes to closure of the isthmus of the throat (fauces). It has an important action in the formation of vowels and nasal consonants.

When singing, it's interesting to stretch this muscle that lifts the palate while it's working, making it more accurate and tonic. You can practice this while doing a "nasal yawn."

Feel the Palatoglossus

1. In the joint, we are working this muscle when we transition from the *ey* sound to *in*.
2. We can feel the muscle on the edges of the tongue when we start making an *mm* sound softly and move toward a crescendo.

It's good to be familiar with the action of this muscle, but don't put it into play constantly, because if you work it too hard (especially producing nasal consonants) you can create a narrow isthmus, which is not conducive to keeping an "open throat."

The Palatopharyngeus (Pharyngopaltinus or Palatopharyngeal) Muscle

This muscle attaches at the top against the superior surface of the **palatine aponeurosis**. It descends toward the back. The fibers of one part terminate on the **thyroid cartilage** (superior and posterior edges). The fibers of the other part join those of the **inferior pharyngeal constrictor** and intersect like a scarf on the posterior part of the pharynx.

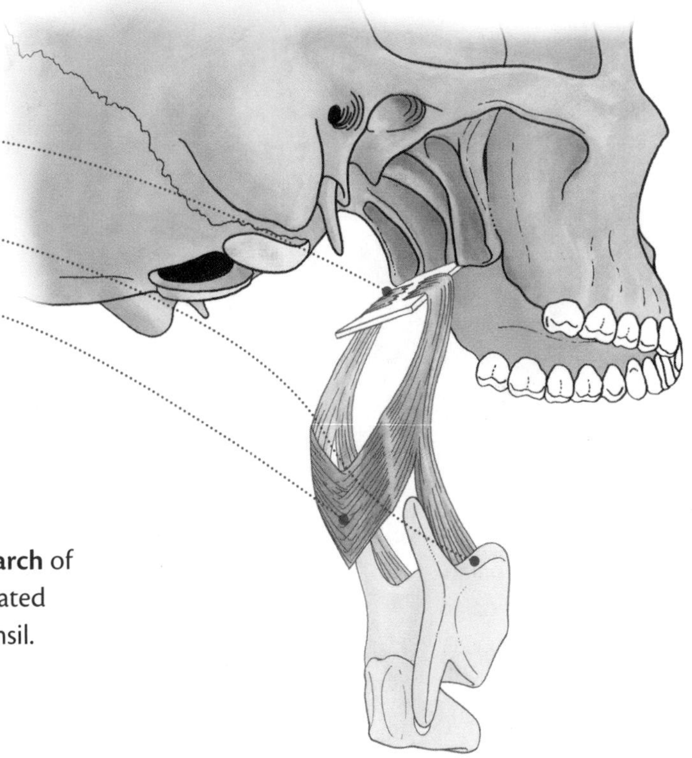

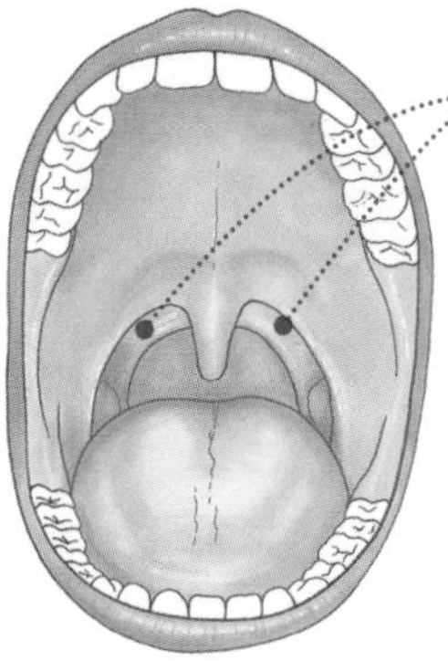

It forms the **posterior arch** of the palate, which is located behind the palatine tonsil.

Its Action

It lowers the palate while contributing to the closure of the posterior part of the isthmus of the fauces. It elevates the pharynx and larynx. It has an important action in producing vowels and nasal consonants.

The actions of the palatopharyngeus and internal levator palati can be commingled easily. If this happens when sound is produced, it can give the sound a certain pharyngeal hardness.

Feel the Palatopharyngeus

1. We use this muscle when we go from making an *ou* sound to an *on* sound.
2. We can feel the action of the muscles that lower the palate when we pronounce *rrr* a little to the posterior.

It is affected by the Eustachian tubes (see page 283), attaching secondarily on their inferior borders.

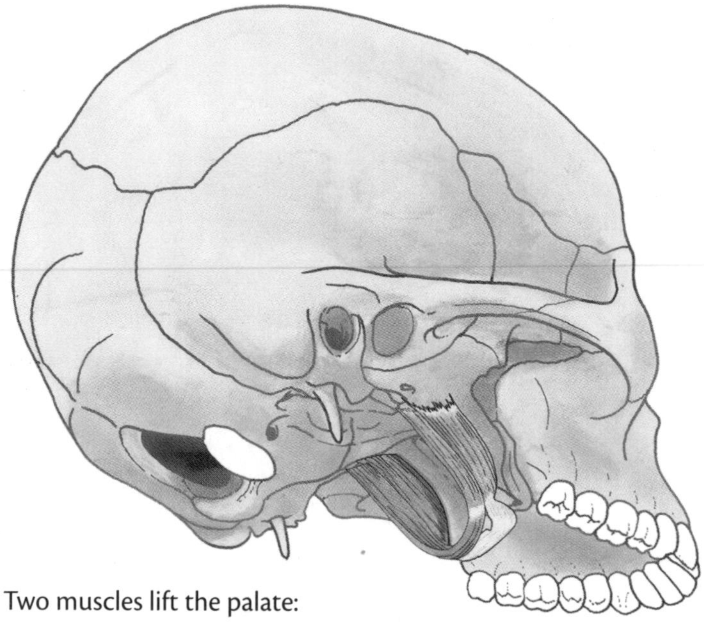

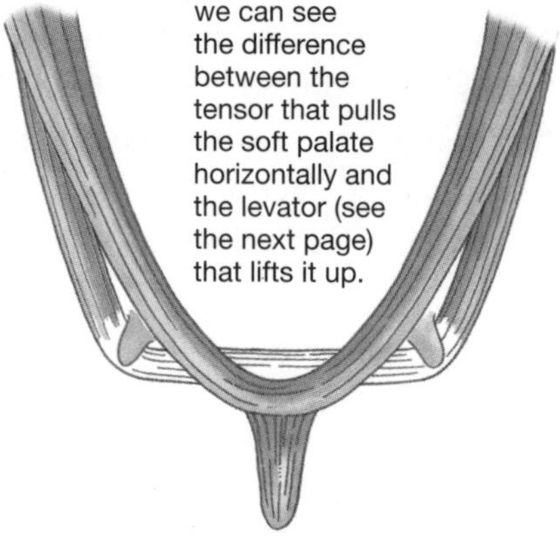
In this drawing, we can see the difference between the tensor that pulls the soft palate horizontally and the levator (see the next page) that lifts it up.

Two muscles lift the palate:

- the tensor palati
- the levator palati

The Tensor Palati

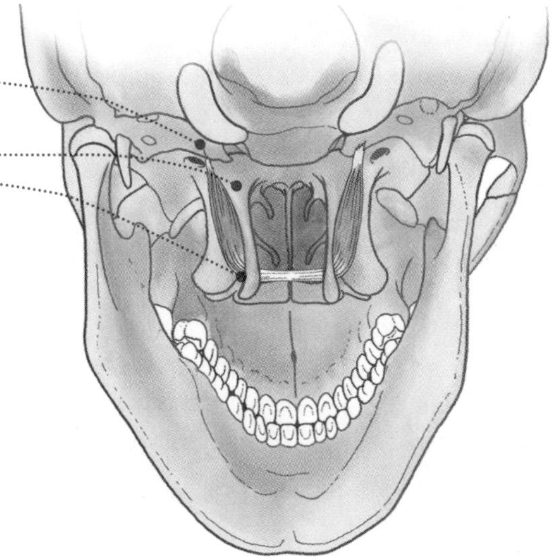

This muscle attaches at the top to the sphenoid body and on the external surface of the Eustachian tube. It descends along the medial lamina of the pterygoid process and then bends around the hamulus (the "hook" of the process; see page 71). Its fibers then join those of the symmetrical muscle.

Its Action

It puts tension on the palatine aponeurosis, pulling it laterally so that it stretches and becomes more horizontal. Along with the levator, it opens the Eustachian tube (see page 283).

A Reactive Muscle

This muscle is very rich in proprioceptive muscle spindles that react to the muscle stretching and can adapt very quickly by contracting and reducing its length.

Feel the Tensor Palati

You can feel this muscle working when you move from making an *in* sound to an *e* sound.

The Levator Palati

At the top this muscle attaches to the underside of the sphenoid body (to the inside of the tensor palati) and on the internal surface of the Eustachian tube. It descends toward the middle, under the extremity of the Eustachian tube. It terminates on the palatine aponeurosis, behind the tensor palati, merging its fibers with those of the symmetrical muscle to form the **median raphe of the palate**.

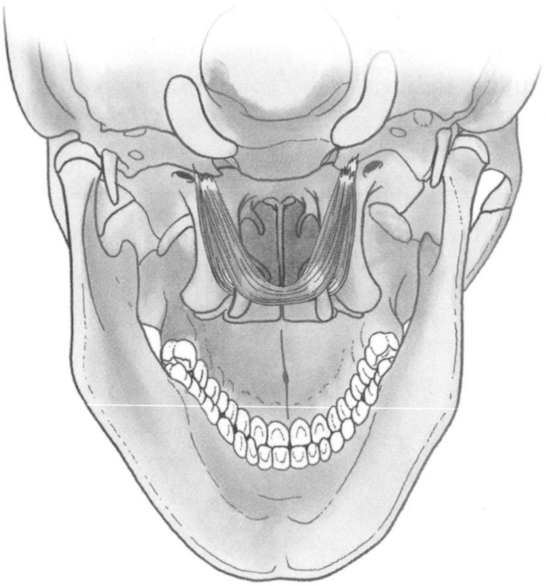

Its Action

It lifts the posterior part of the fascia of the palate like a hammock.

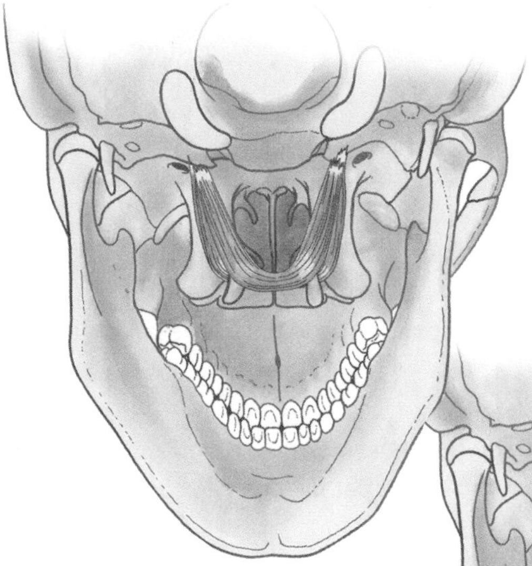

In daily life, we contract this muscle when we yawn, especially the part of the yawn that opens the back of the mouth.

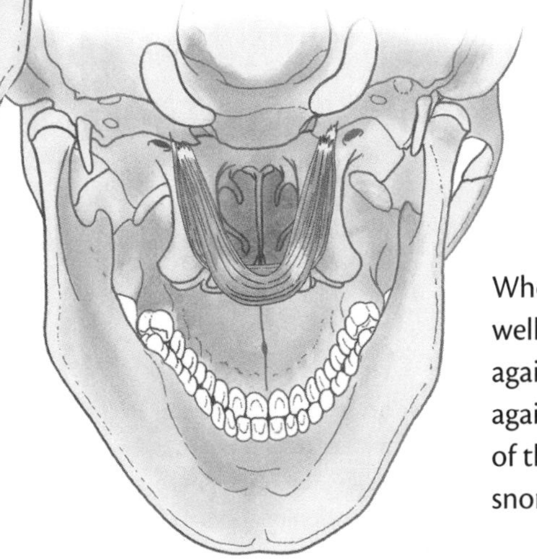

When this muscle is not well toned, it can vibrate against the tongue or against the posterior wall of the larynx and cause snoring.

Feel the Levator Palati

When using the voice, we can feel this muscle at the back of the mouth when we move from making the *on* sound to the *ou* sound.

The Soft Palate and Respiration

When you breathe through your nose, the palate is lowered. The air passes through the nasopharynx, oropharynx, hypopharynx, and finally the larynx. It doesn't pass at all through the mouth.

This does not require you to close your mouth: you can breathe through your nose, inhaling as well as exhaling, and keep your mouth wide open at the same time. This means that you can watch the palate lower at the back of your throat by looking in a mirror.

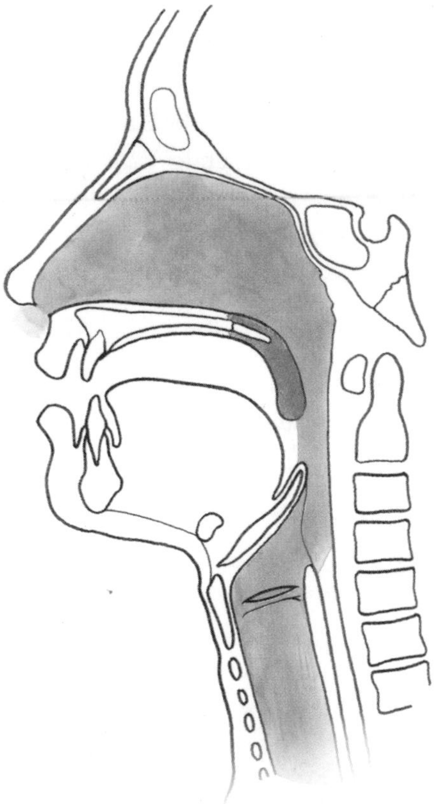

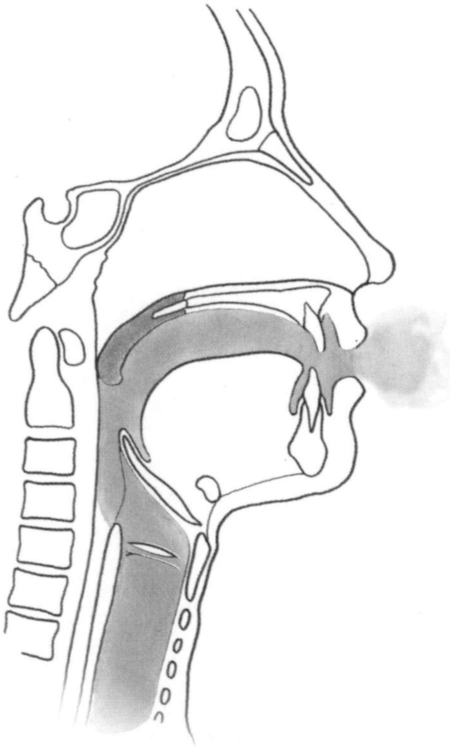

When you breathe through your mouth, the mouth is open and the palate is lifted. The air passes through the mouth, followed by the oropharynx, the hypopharynx, and finally the larynx. It does not pass through the nose.

When you're breathing through your mouth, you can see the palate rise at the back of the throat by looking in a mirror.

Breathing Practice

It's interesting to practice breathing while paying attention to the soft palate. Explore inhalation/exhalation combinations and oral/nasal breathing. For example, inhale through the mouth and exhale through the nose several times, then do the opposite. You can then develop a recognition of and better sensory awareness in the region of the palate, which can be an aid in vocal work.

It is possible to mix these two air pathways by lowering the palate only partially and keeping the mouth open: you can then breathe through both at once, partially through the mouth and partially through the nose.

The Soft Palate: Articulation and Resonance

The Palate and the Articulation of Vowels

Because the palate can be lowered and raised, it is involved in numerous vocal sounds.

When we are standing, it's the force of gravity that lowers it: the palate falls by itself at the back of the tongue with different degrees of force. We can lower it more actively and with more force by using the depressors of the palate: the palatoglossus and the palatopharyngeus. If we want to lift the palate, we use the tensor palati and/or the levator palati.

Independent of this, the muscles that lift and depress the palate often act to moderate the actions of each other, giving the palate the mobility to adapt in an instant to any vocal situation.

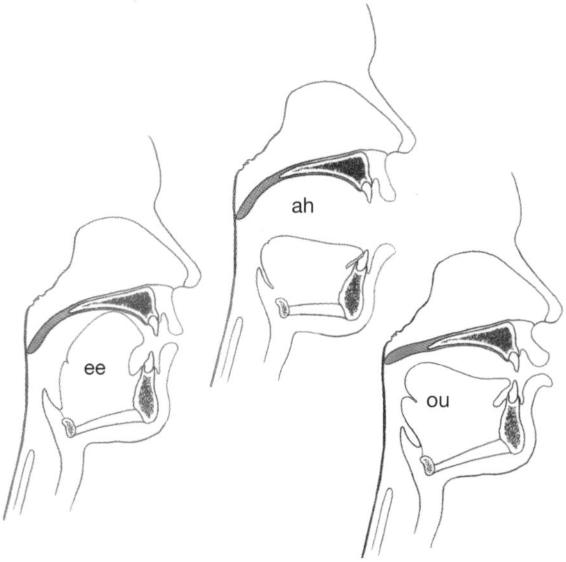

The palate is lifted high or halfway up to produce the "oral" vowels—for example, *ah, ee, ou*. This lifting of the palate is accomplished by the action of the elevators.

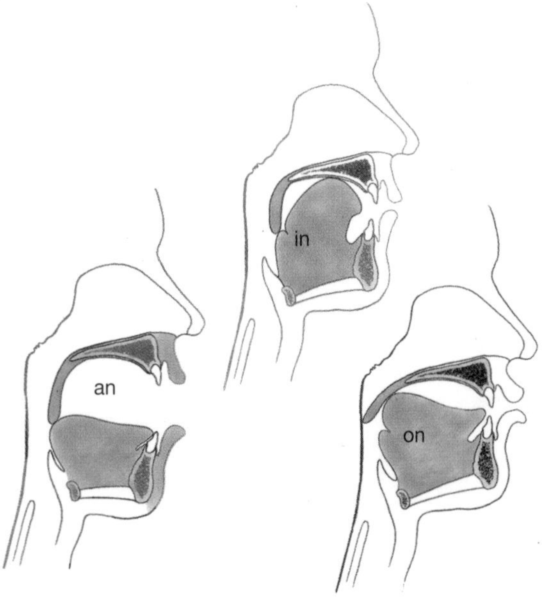

The palate lowers itself, like a lid, to promote nasal resonance and produce the so-called nasal vowels—for example, *an, in, on*. The action is due to the action of the depressors, in addition to gravity.

The Palate and the Articulation of Consonants

1. Like lowering a lid, the lowering of the palate encourages both the exit of air through the nose and nasal resonance, which allows for the production of the nasal consonants.

We call a consonant *nasal* if the soft palate is lowered to produce it—for example, *m, n, ny*. The air exits through the mouth and perhaps a bit through the nose.

This action of lowering the palate is performed by the palatoglossus and the palatopharyngeus acting symmetrically.

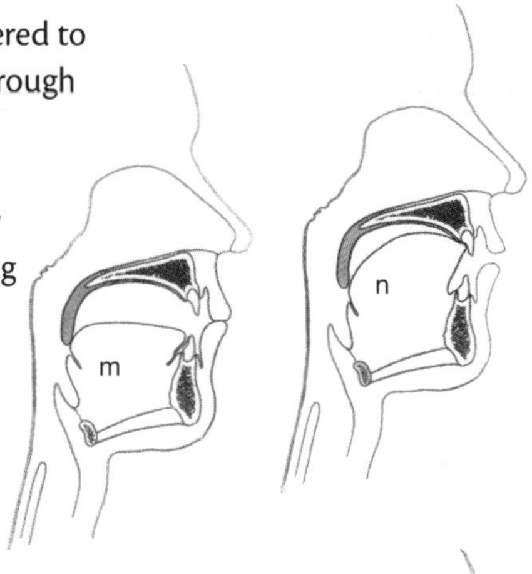

We say that a vowel or consonant is *oral* if the palate is lifted to produce it. The air exits only by the mouth.

2. The back of the tongue meets the soft palate to produce the "rolling" *r*—that is, an *r* sound produced at the back of the mouth.

The top of the tongue is placed against the hard palate or the front of the soft palate when producing this "French" *r*.

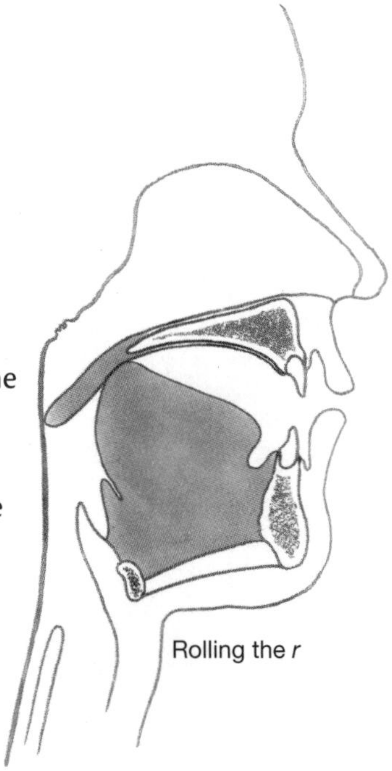
Rolling the *r*

Practicing with the *R*

It is interesting to practice articulating the *r* sound carefully to activate the front of the soft palate. It is also interesting when doing this to add a yawn to the active lowering of the palate, which will work the muscles that both lift and lower the palate at the same time, and both actions will be reinforced.

The Palate Participates in the Resonance of the Voice

The movements of the soft palate are among the most important "shapers" of timbre above the larynx. The flexible and mobile tissue at the junction of the oropharynx and nasopharynx allows us to pass air through the nose, through the mouth, or through both at the same time, and in adjustable proportions.

Perhaps most important, the palate can balance the resonance between two of the principal resonating chambers of the vocal tract: the junction of the nasopharynx and oropharynx and the angled junction between the pharynx (vertical part) and the mouth/nose (horizontal part). It can simultaneously modify the size of these two areas, resulting in a change in resonance.

We can distinguish two different actions for the palate that affect resonance:

Lifting the Palate

This is the action of elevating the palate. The feeling is in the "posterior border" of the palate. Lifting the palate immediately makes the voice "brighter." This is why, when we are singing, we may be asked to lift the palate as if we were yawning at the back of the mouth, as if we "had a hot potato in the back of the mouth," or "like we're trying to hide a yawn" (yawning without opening mouth). In this position the resonating effect is fairly easy to find. However, if this concept is applied exclusively, it may ultimately hinder the adaptability of the palate and the neighboring areas. Therefore, it should not be practiced long-term.

The Inner Smile

This action is quite different, and it uses the action of the tensor to pull the palate wider and more horizontal. The feeling is less posterior on the palate. Additionally, we can distinguish the two zones if we pass successively from this action ("inner smile") to the previous one ("inner yawn"). This action will allow us to find a middle position between the ascent and total descent. We talk also of "inhaling as if we're sniffing a fragrance," knowing that this inspiratory position can often be held for the vocal exhalation.

The Singer Lifts the Palate

The singer brings the palate into a lifted position, closing access to the nasal cavity (which consumes a portion of the airflow without producing adequate resonance). Moreover, with the soft palate lifted like this, there will be more area in the mouth allotted for volume and resonance, and the orifice of the oropharynx will be cleared.

The Tongue

The tongue is a key part of the vocal instrument. Often we forget that fact because it is first and foremost an essential organ of mastication, swallowing, and taste. But the tongue, with its ability to make numerous, fine-tuned movements, is one of the most important "sound shapers" above the larynx.

When it comes to the tongue, it's easy to just think of the tip of the tongue and how it reacts with the rest of the mouth in sound production. But the bulk of the tongue is behind and below the tip, and here, too, it is capable of transforming the voice in various ways.

The fact that the tongue can "steer" sound is one of the keys to creating resonance.

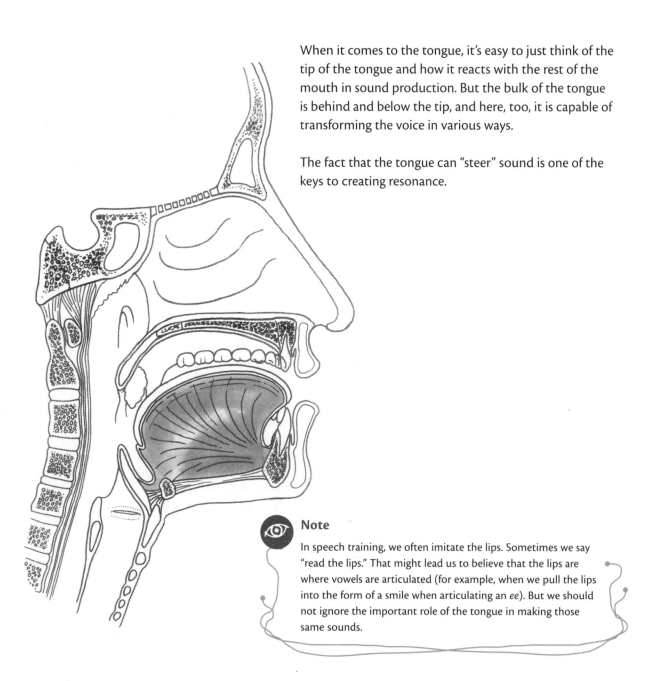

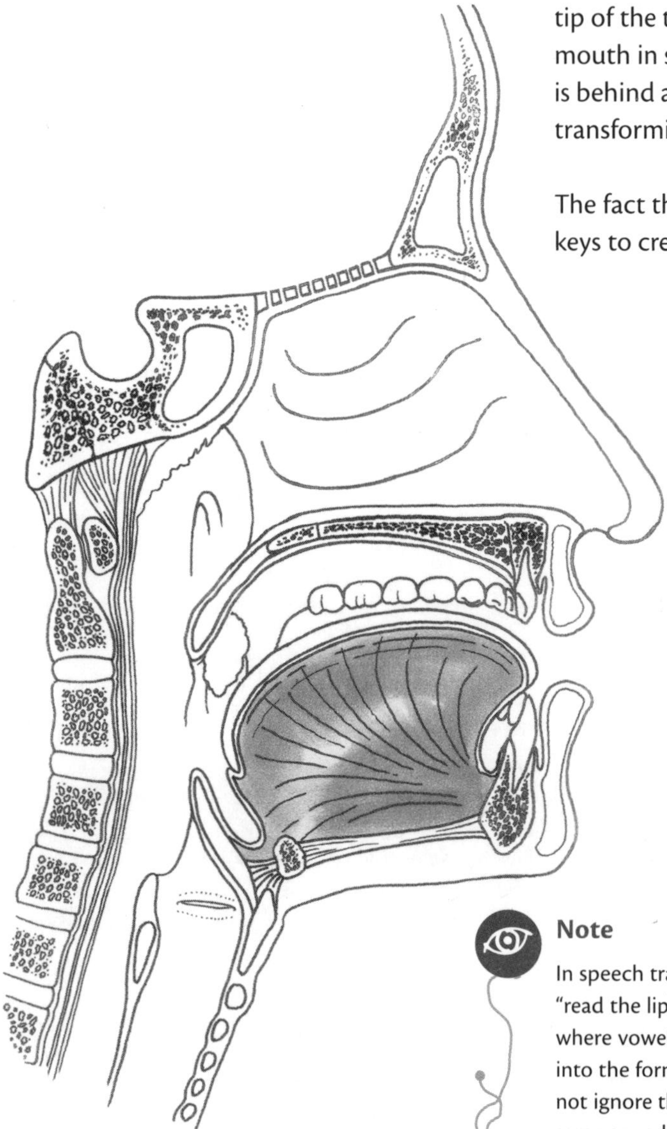

Note

In speech training, we often imitate the lips. Sometimes we say "read the lips." That might lead us to believe that the lips are where vowels are articulated (for example, when we pull the lips into the form of a smile when articulating an *ee*). But we should not ignore the important role of the tongue in making those same sounds.

Description of the Tongue

Bordered by the dental arch of the inferior maxilla (see page 86), the tongue occupies the lower part of the mouth. It constitutes both the floor and the principal mass. Its mass is much larger than we generally imagine; it's actually about the size of a fist.

The tongue moves freely superiorly and anteriorly. It is made up of the following:

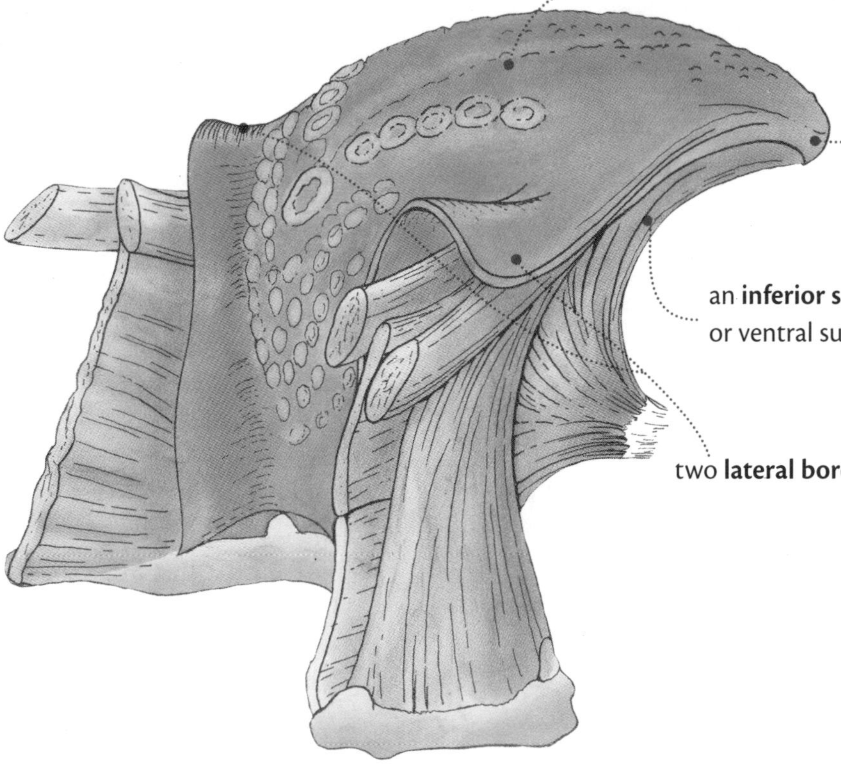
a dorsal surface called the **dorsum** of the tongue
a **tip**
an **inferior surface,** or ventral surface
two **lateral borders**

The areas at the back and at the bottom, called the **base** of the tongue, are much less free. Here the tongue is connected by many muscles to:

- the inferior and superior maxillae
- the hyoid bone
- the soft palate
- the pharynx
- the base of the skull

The Skeleton of the Tongue

Although the tongue is a soft mass, it is structured at its base by rigid elements, which we call the **skeleton of tongue,** although they are not all bone.

The Mandible

This bone was seen in detail on page 80. It serves as a platform for the insertion of multiple tongue muscles.

The Hyoid Bone

This small bone was seen in detail on page 88. It also serves as a platform for the insertion of multiple tongue muscles.

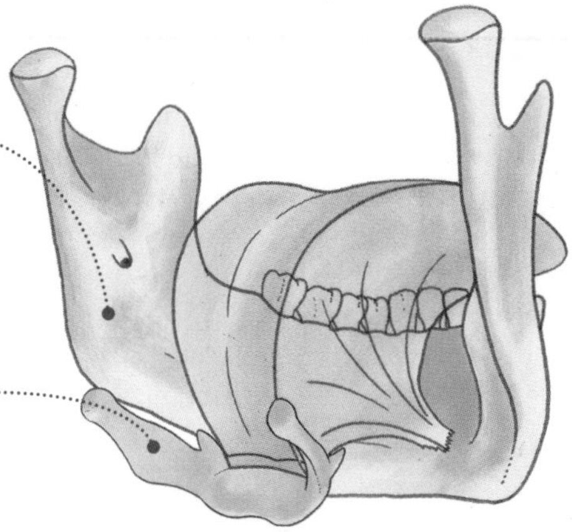

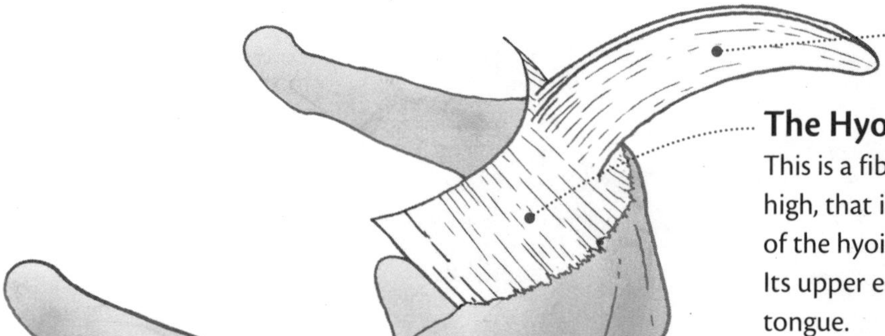

The Hyoglossal Membrane

This is a fibrous band, about 1 centimeter high, that is attached to the upper edge of the hyoid bone and the lesser horns. Its upper edge is lost in the mass of the tongue.

The Lingual Septum

This is a somewhat rigid, fibrous layer that is implanted perpendicularly on the front of the hyoglossal membrane. It is sickle-shaped, convex upward, and concave downward. It runs upward and to the front. It forms a fibrous skeleton on each side on which the muscles of the tongue are arranged.

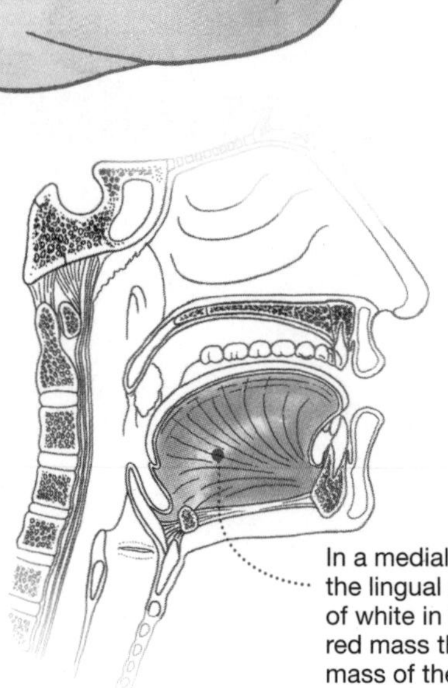

In a medial profile, we see the lingual septum as an arc of white in the middle of a red mass that is the muscle mass of the tongue.

The Muscles of the Tongue

Around the fibrous skeleton, the mass of the tongue is formed mainly by muscles:

- eight muscle pairs; seven are visible in the illustration below
- one unpaired muscle

Every muscle pair is symmetrical, with one muscle on the right and one on the left.

Some of these muscles are involved in other parts of the mouth as well, but they are described here in relation to their action that is specific to the tongue.

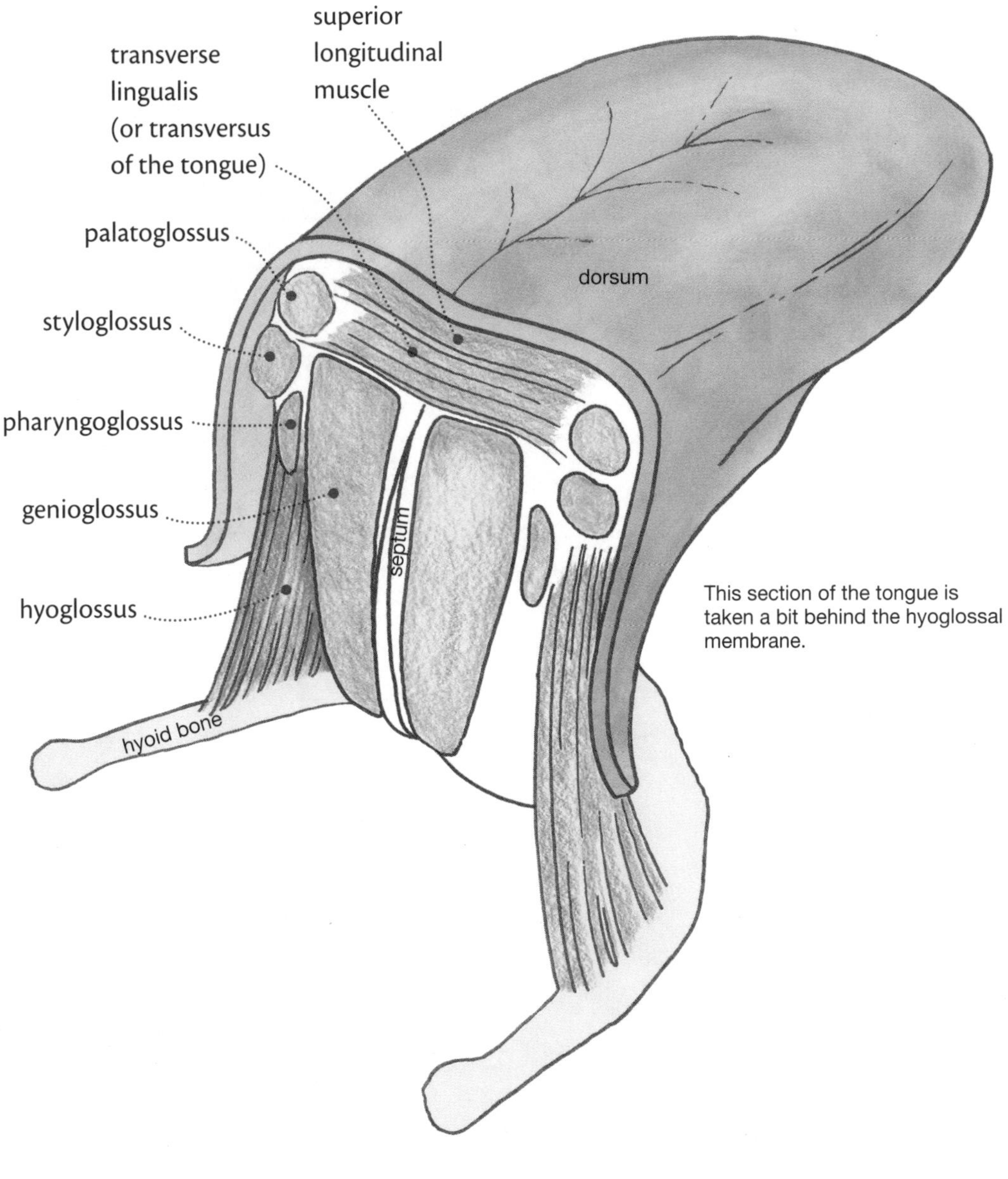

The Biggest Muscle: The Genioglossus

This muscle alone makes up 80 percent of the volume of the tongue. It is a paired muscle (there are two genioglosses, right and left, though they are intertwined).

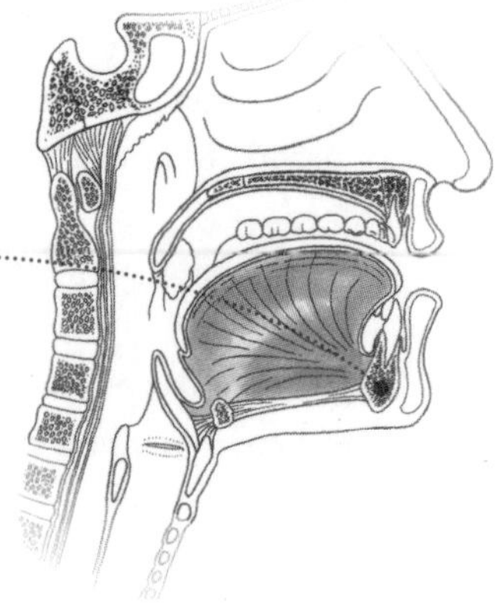

It attaches to the deep surface of the inferior maxilla, at the lower part, and the superior mental spines (see the illustration at the bottom of this page). From there, its fibers fan out in three directions. At this point, they have very different arrangements and different functions:

The **anterior fibers** run toward the tip, that is to say, upward and to the front.

The **middle fibers** run toward the dorsum of the tongue, that is to say, up and backward.

The **posterior fibers** run upward toward the back or downward toward the hyoid bone.

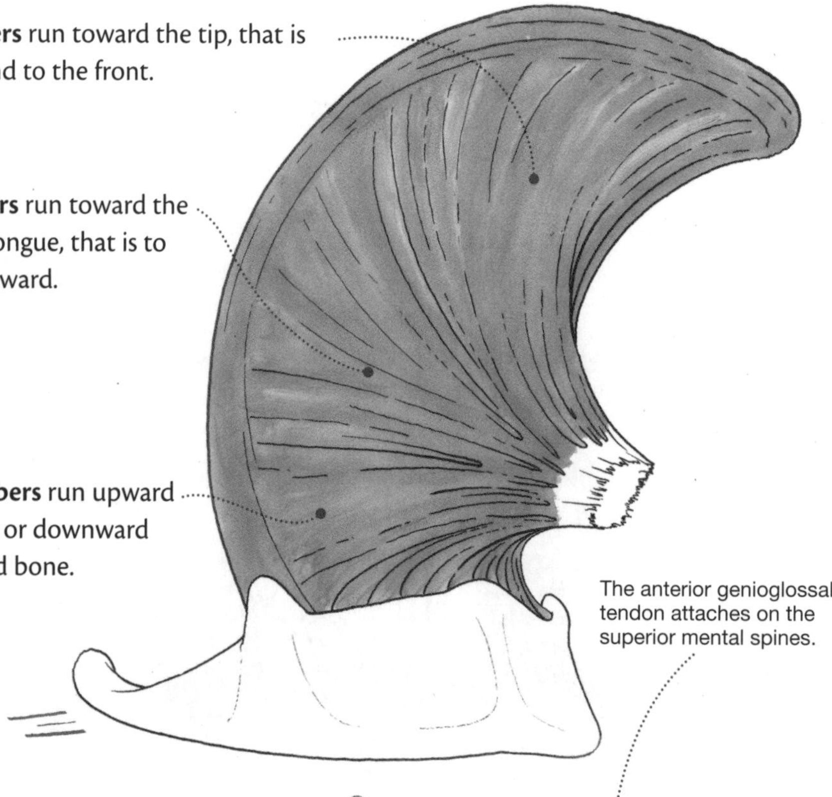

The anterior genioglossal tendon attaches on the superior mental spines.

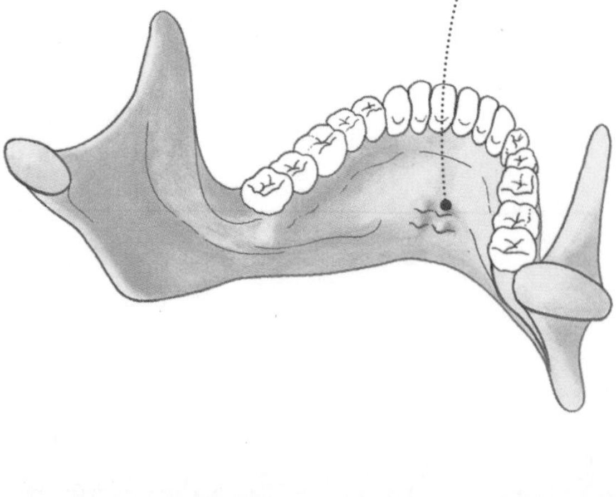

Here, the inferior maxilla is seen from the back, showing the mental spines.

Its Action

In the voice, the genioglossus constantly intervenes. It contracts by zone, coordinating its action with that of other muscles to bring one part or another of the back of the tongue against a particular part of the palate.

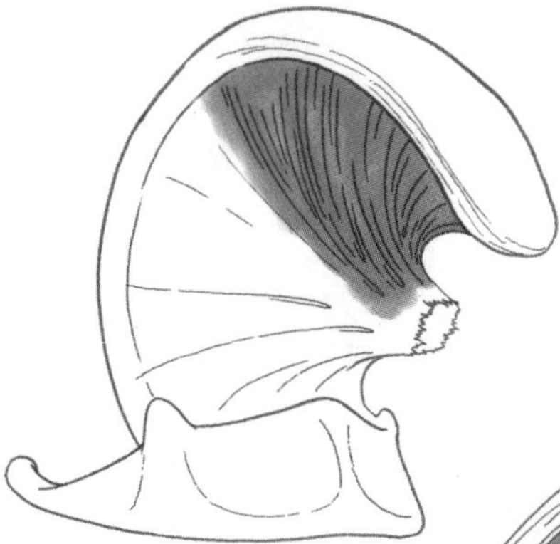

The **anterior fibers** move the tip of the tongue backward and downward.

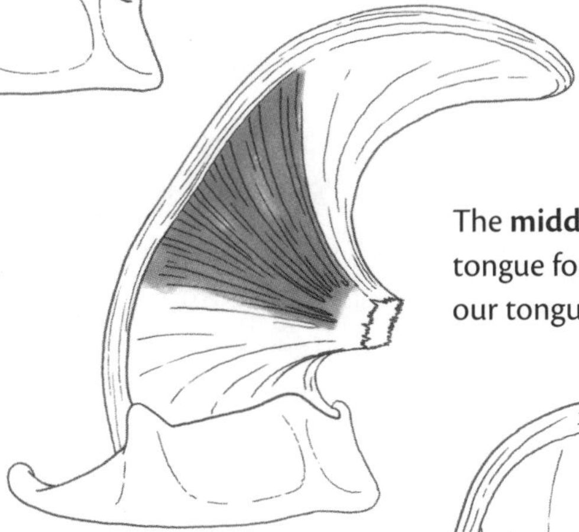

The **middle fibers** move the mass of the tongue forward, as when we “stick out” our tongue.

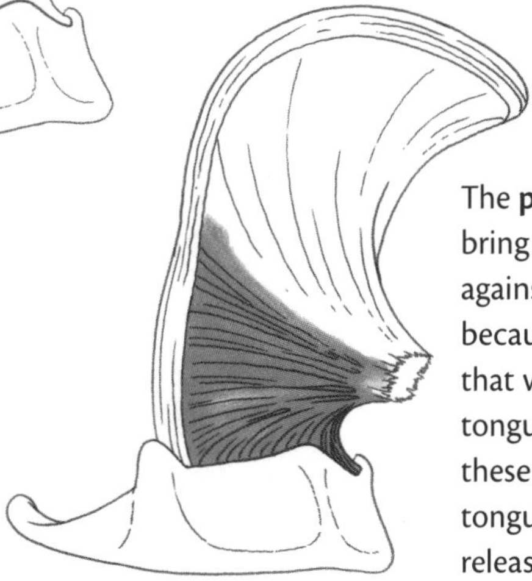

The **posterior fibers** bring the tongue against the palate. It is because of these fibers that we can click our tongue; the action of these fibers lifts the tongue before the release.

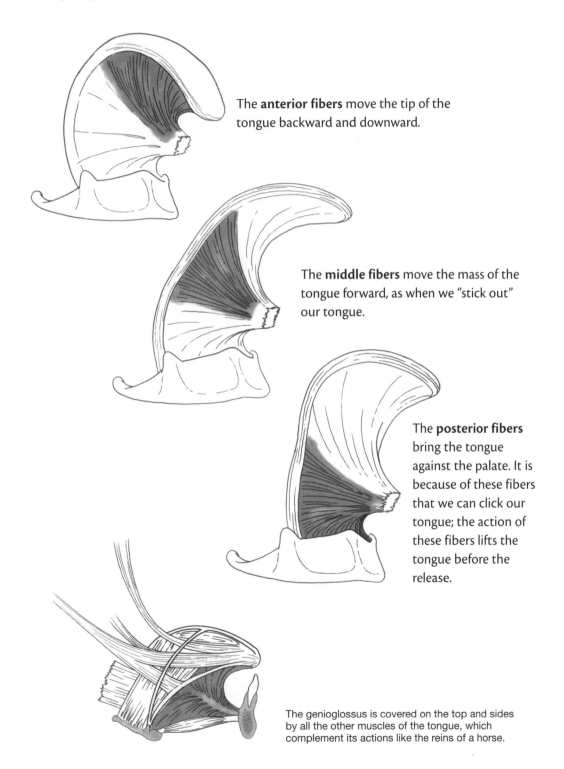

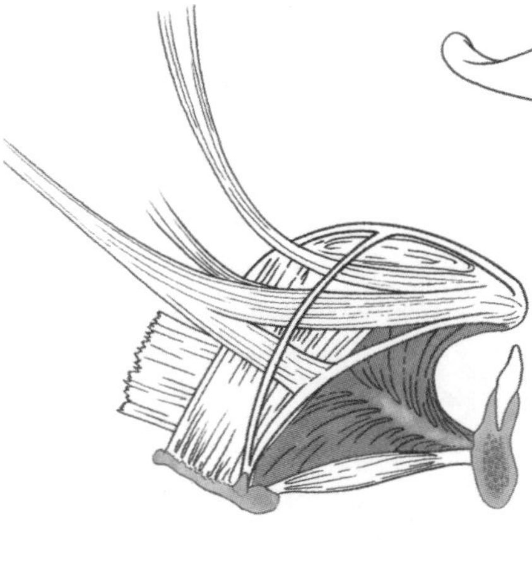

The genioglossus is covered on the top and sides by all the other muscles of the tongue, which complement its actions like the reins of a horse.

The following muscles are located on the sides of genioglossus, but they come from above; they are like the "reins of the tongue."

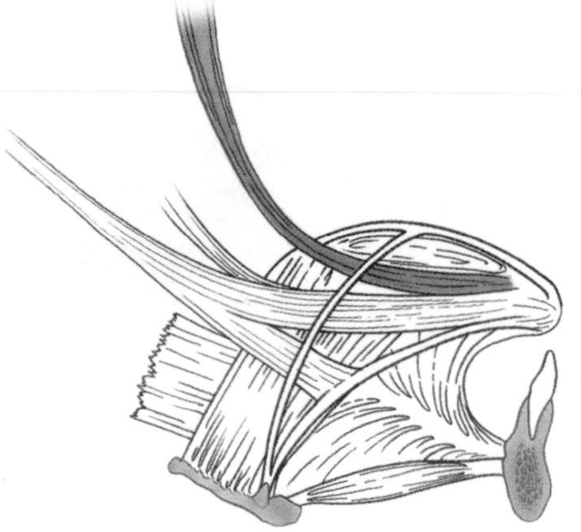

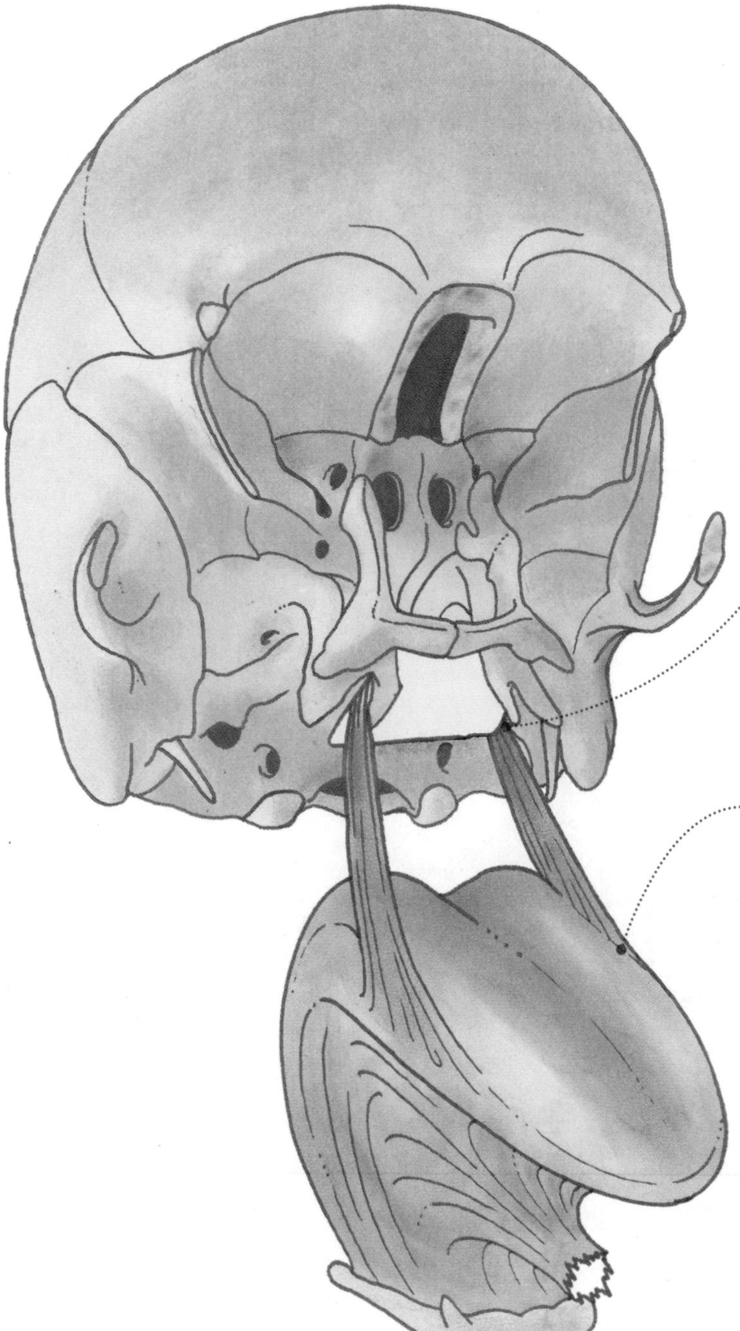

The Palatoglossus

This muscle attaches, at the bottom, on the thick lateral part of the tongue. It runs upward and to the back, terminating on the inferior surface of the aponeurosis of the soft palate.

It forms the anterior arch of the soft palate, in front of the palatine tonsil. (See more details about this muscle on page 240.)

Its Action

On the side where it is located, it pulls the tongue up and back. As it forms the anterior arch of the soft palate, it narrows the isthmus of the fauces (see page 237) in two ways:

- It brings this arch toward the midline, making the isthmus narrower.
- It lowers the soft palate, making the isthmus shorter.

It participates in the formation of the vowel sound *ew*.

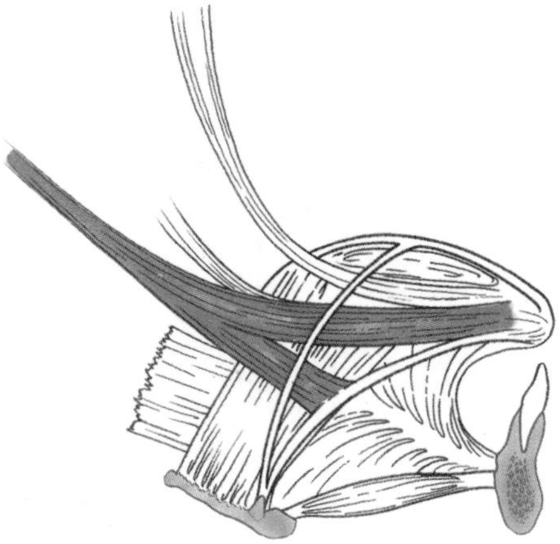

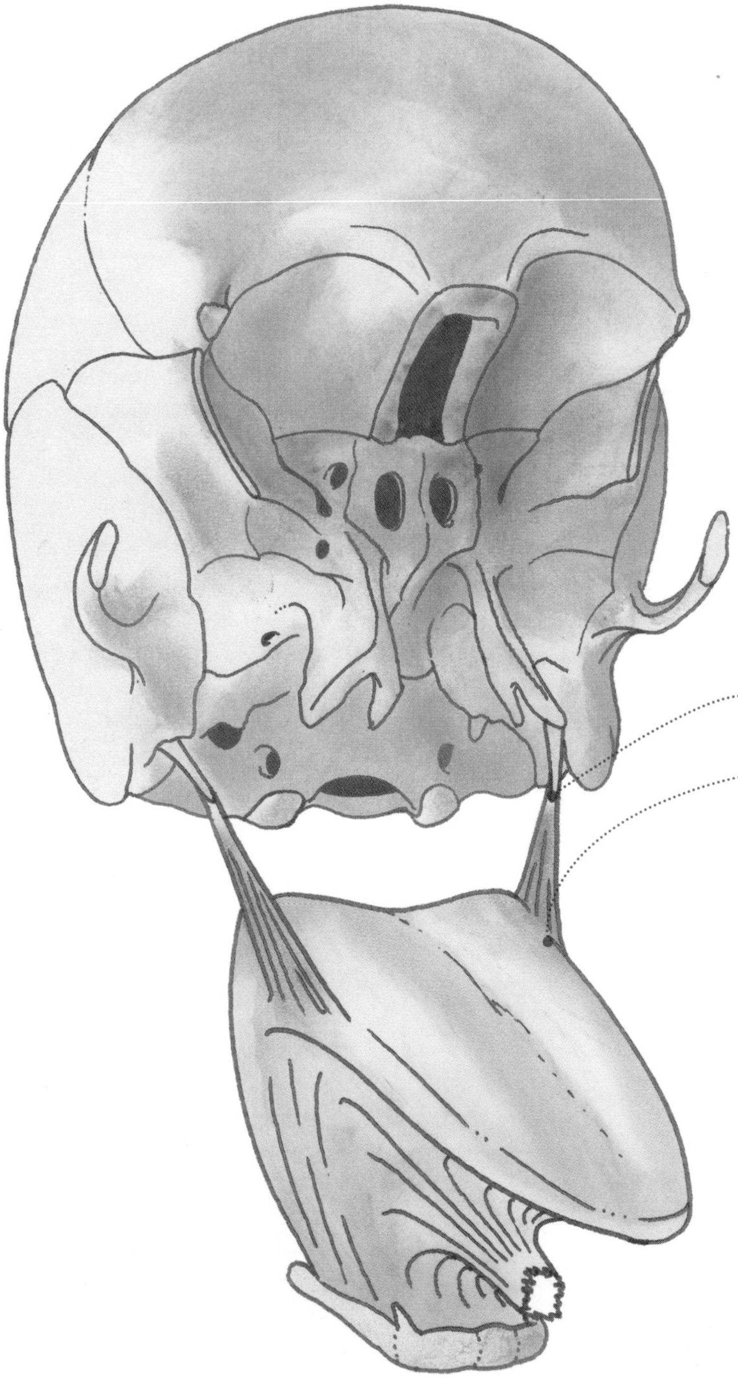

The Styloglossus

This muscle attaches, in the front and at the bottom, by fanlike insertions at the side of the dorsum.

Toward the middle, its posterior fibers join the inferior part of the lingual septum. The most anterior fibers extend to the front of the tongue.

Its muscle fibers ascend toward the back to terminate at the tip of the styloid process of the temporal bone (see page 73).

Its Action

On the side where it is located, it pulls the base of the tongue upward and toward the back as the base widens slightly.

It may participate in the formation of the vowel *ou*.

Feel the *Ew/Ou*

We can feel the base of the tongue become narrowed or wider when we pronounce *ew* or *ou*.

The following muscles are found at the sides of the genioglossus.

The Pharyngoglossus

The muscle fibers arise from the lateral edge of the tongue. They run downward and toward the front, where they terminate on the superior pharyngeal constrictor. (For more details on these muscles, see page 222.)

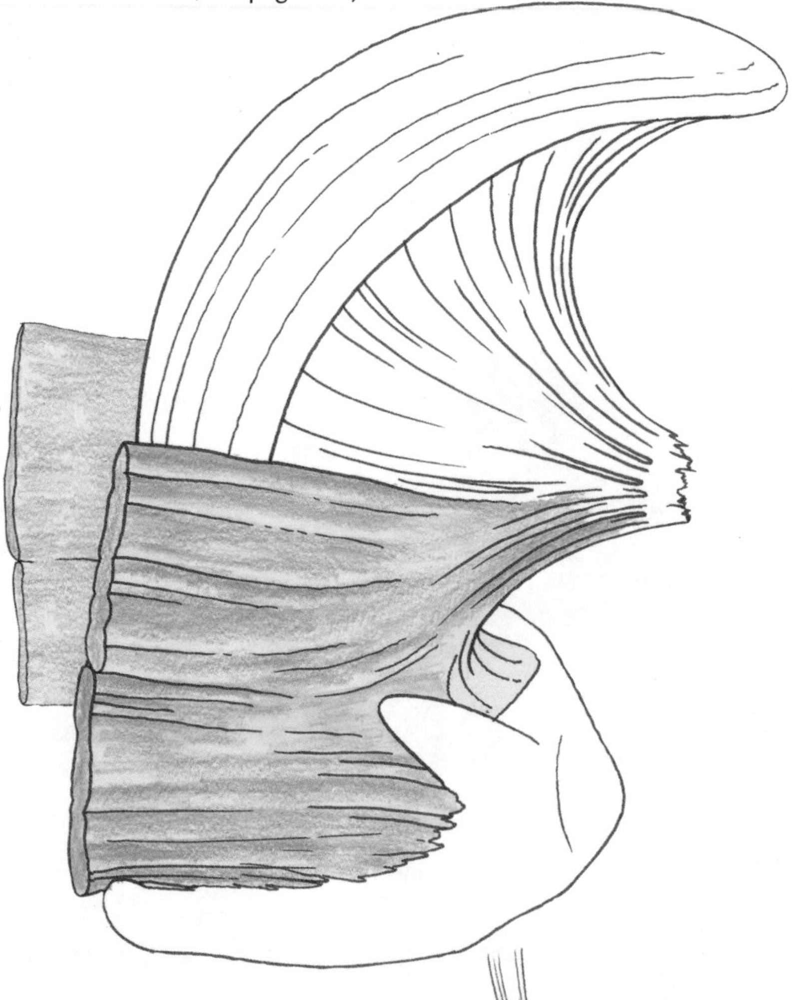

Its Action

On the side where it is located, it pulls the tongue back and up a bit.

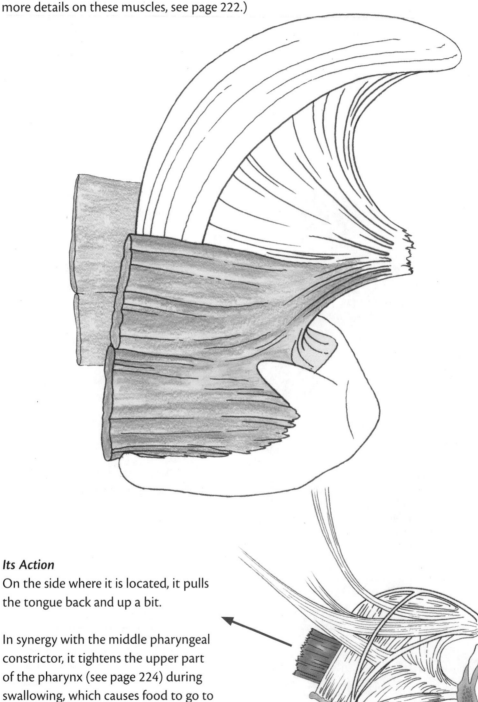

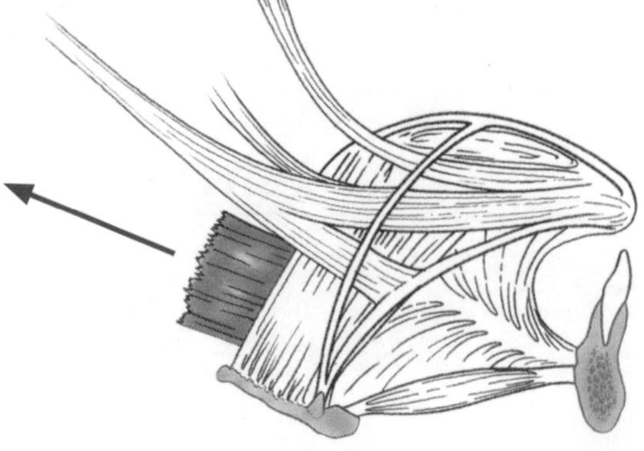

In synergy with the middle pharyngeal constrictor, it tightens the upper part of the pharynx (see page 224) during swallowing, which causes food to go to the back of the mouth to the pharynx.

The Hyoglossus

This muscle attaches high on the lingual septum and in the muscle mass of the genioglossus. Its fibers flare outward as they descend, terminating on the great horn of the hyoid bone.

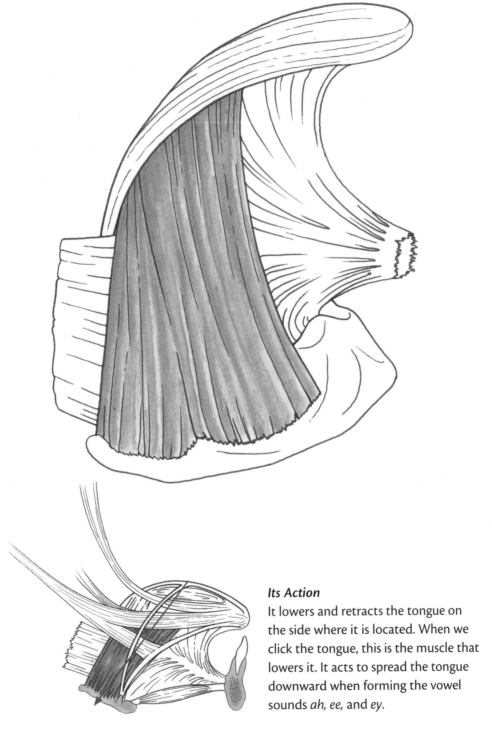

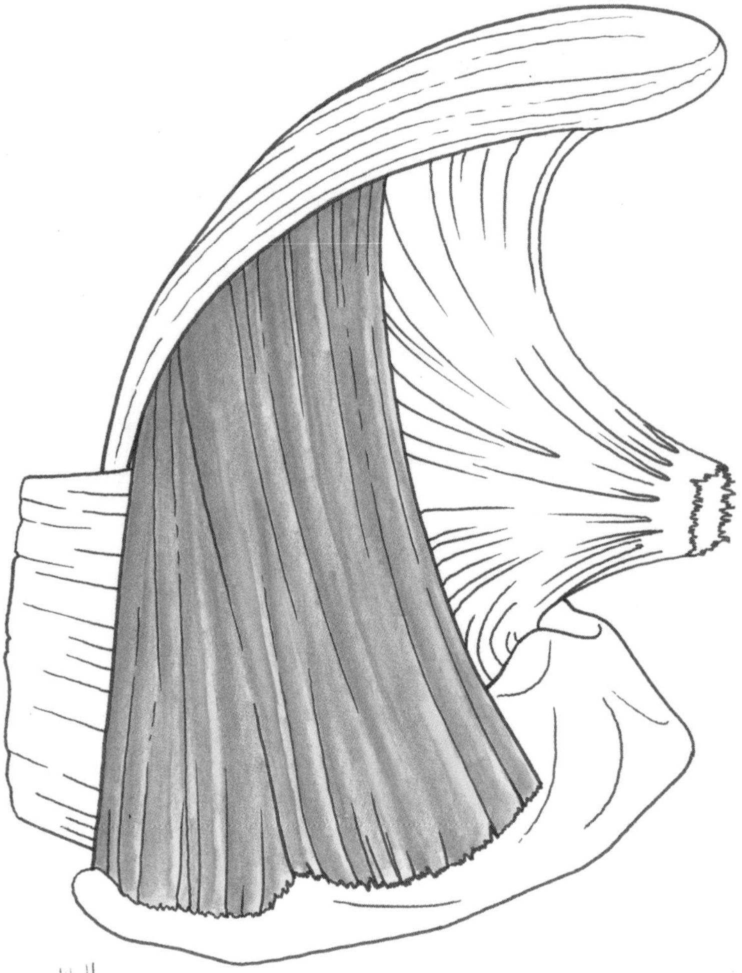

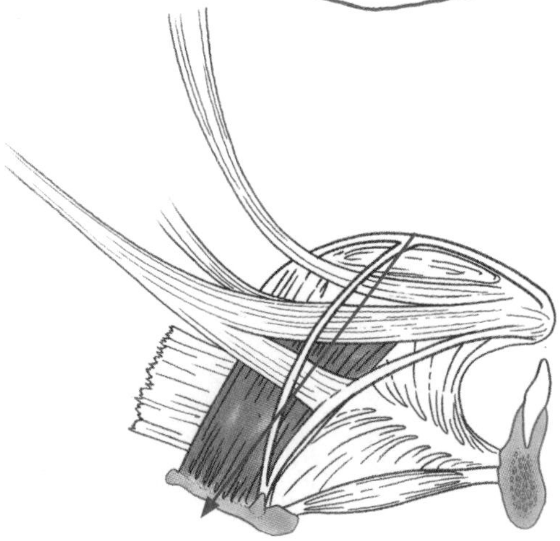

Its Action

It lowers and retracts the tongue on the side where it is located. When we click the tongue, this is the muscle that lowers it. It acts to spread the tongue downward when forming the vowel sounds *ah, ee,* and *ey*.

The Transverse Lingualis (Transverse Muscle of the Tongue)

This muscle is formed by transverse fibers that run from the lateral surfaces of the lingual septum to the mucosa at the lateral borders of the tongue.

Its Action

On the side where it is located, it can hollow and narrow the top of the tongue, or at least prevent it from collapsing.

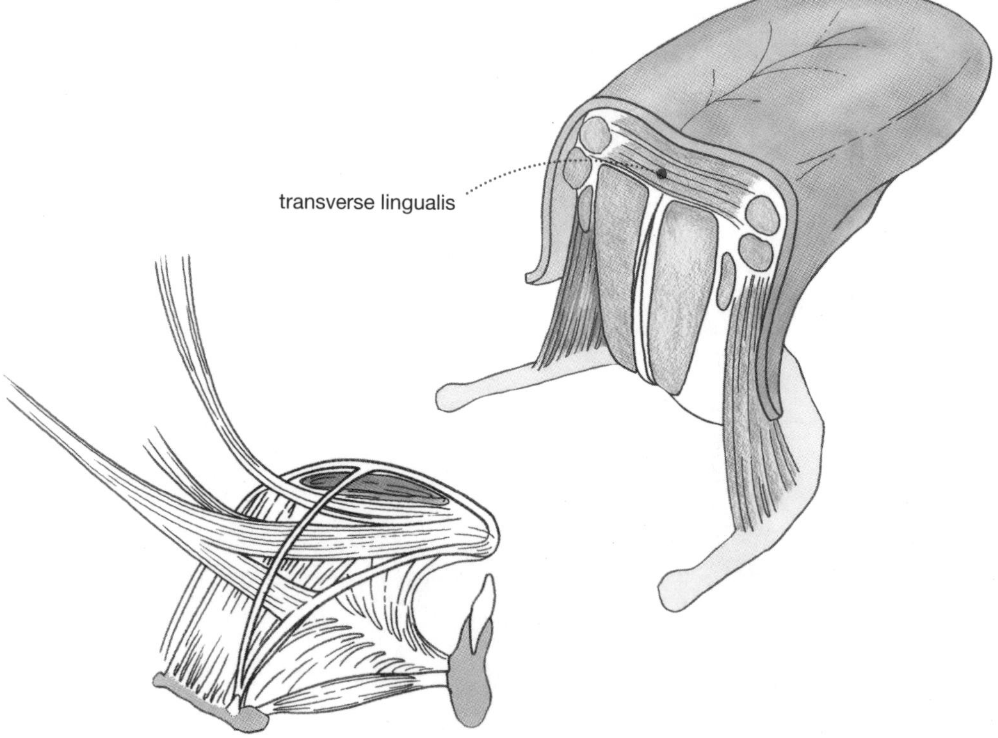

The Amygdaloglossus

This muscle arises from within the pharyngoglossus on the fibrous layer of the pharynx. It terminates on the outer surface of the capsule of the palatine tonsil.

Its Action

On the side where it is located, it raises the base of the tongue. It contributes to forming the vowel sounds *ou* and *ew*.

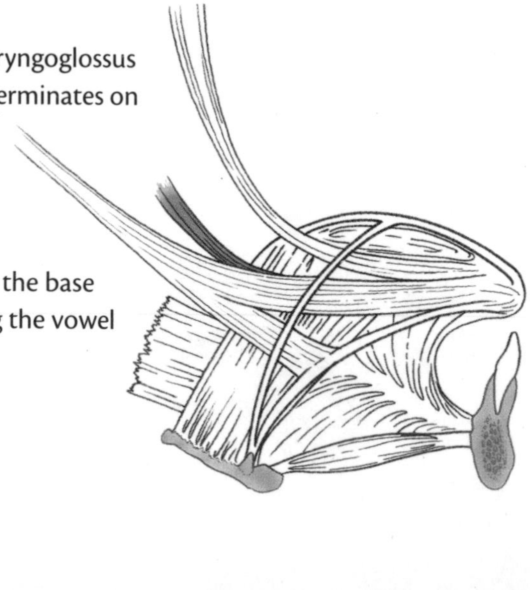

Inferior Lingual Muscle

This muscle is attached at the bottom to the lesser horn of the hyoid bone. It extends from the root to the apex of the tongue.

Its Action

On the side where it is located, it retracts and lowers the tongue. It acts to spread the tongue downward when forming "open" vowel sounds such as *ah, ee,* and *ey*. It lifts the hyoid bone.

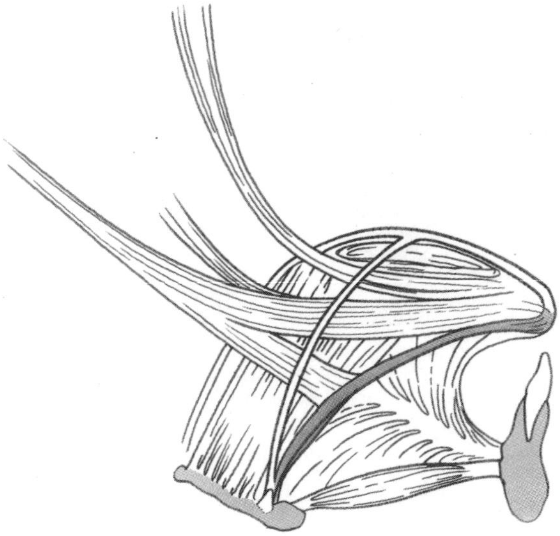

Superior Lingual Muscle

This is the only unpaired tongue muscle. It attaches at the back on the base of the epiglottis and by two lateral bundles of fibers on the lesser horns of the hyoid bone. It forms a long medial sheet that underlies the mucosa of the dorsum of the tongue.

Its Action

It lowers and shortens the upper part of the tongue. This action contributes to the formation of the *ah* vowel sound.

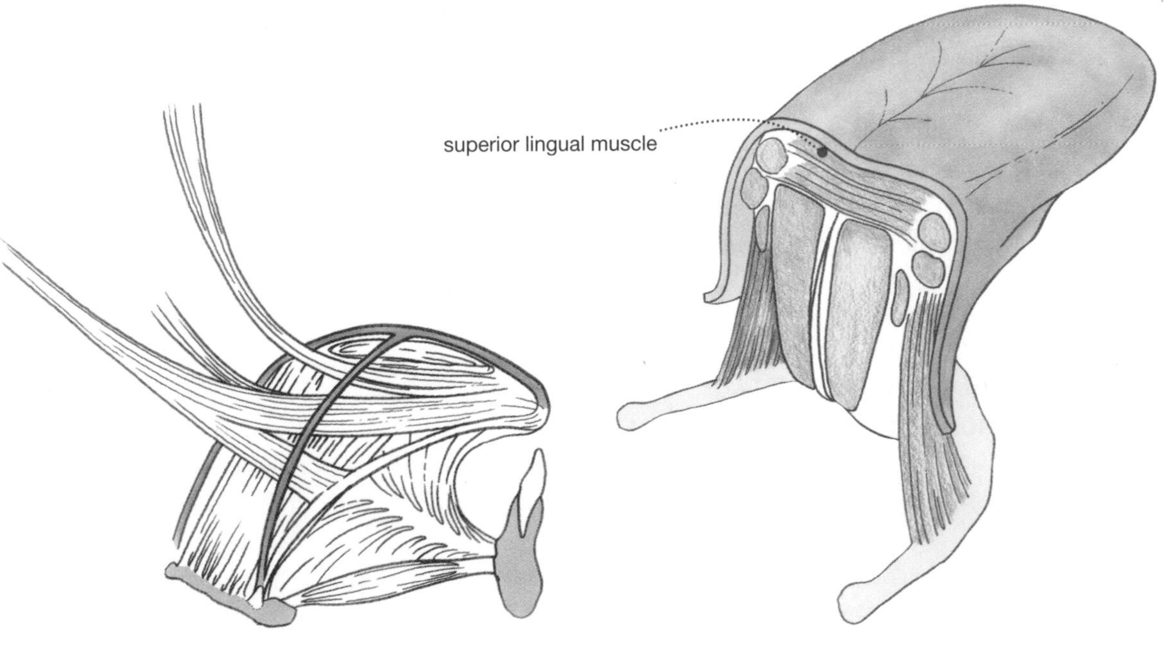

The Dynamics of the Tongue

The Movements of the Tongue Are Influenced by Gravity

If we are in an upright position with our mouth closed, and we bring our neck into flexion, our tongue will fall forward in our mouth.

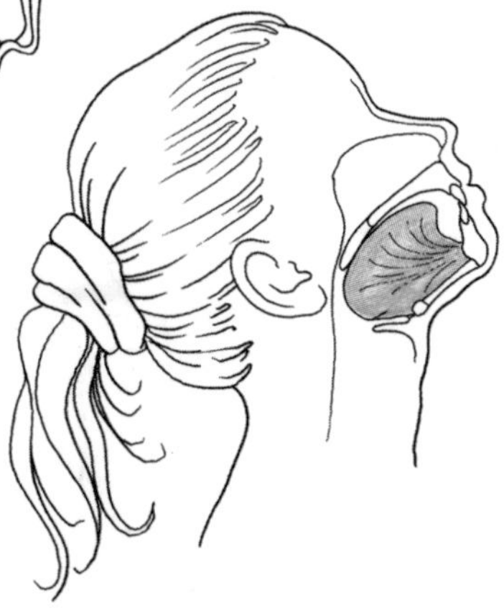

If we bring our head backward into extension, the tongue will fall toward the back of our mouth.

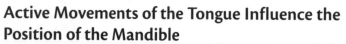

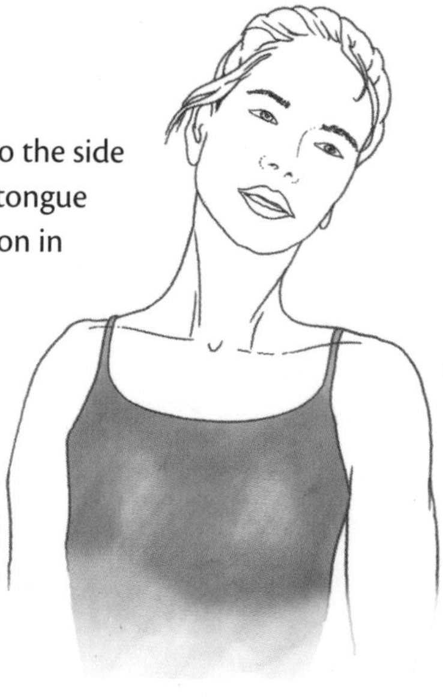

If we lean our head to the side (lateral flexion), the tongue will fall in the direction in which we're leaning.

These tendencies are very clear when we make large movements of the head and neck, but they also take place when we make smaller movements and even mini-adjustments. With this in mind, we can see how the placement of the head before beginning vocal work is important, as it determines the starting position of the tongue.

Active Movements of the Tongue Influence the Position of the Mandible

If we pull our tongue backward, we can feel how the mandible tends to pull backward as well (for example, when we sing the vowel sound *ooh*, the jaw tends to go backward).

To a lesser extent, if the tongue advances, the mandible tends to move forward (this can be seen in the anterior vowels/consonants—for example, when we say *ba ba*).

The Rest Position of the Tongue

With the mouth closed, the tongue contacts the entire surface of the hard palate:

- The tip is on the palatal bumps behind the incisors.
- The side edges are raised and come into contact with the alveolar process.
- The back of the tongue is grooved and slightly hollowed from front to back.

In this situation, the mass of the tongue is positioned relatively forward in the mouth, and the back of the tongue (the base) is not too far back in the mouth. This "rest" position allows the tongue to mobilize vertically, supported by the palate, specifically for swallowing.

The Articulatory Position of the Tongue

When preparing to speak, we start by lowering the mandible a bit; the teeth separate a little, and the lips start to open. The tip of the tongue moves from the upper incisors toward the lower incisors. The sides of the tongue remain in contact with the upper molars. This is the starting position from which the movements of the tongue are minimal.

In this situation, the mass of the tongue is positioned relatively forward in the mouth and the back of the tongue (the base) is not moved toward to the pharynx, so that the larynx is more free.

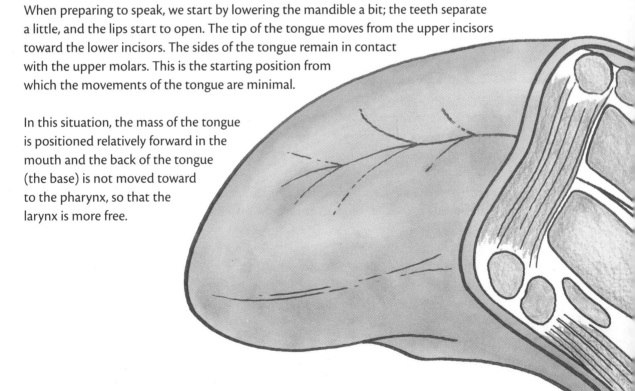

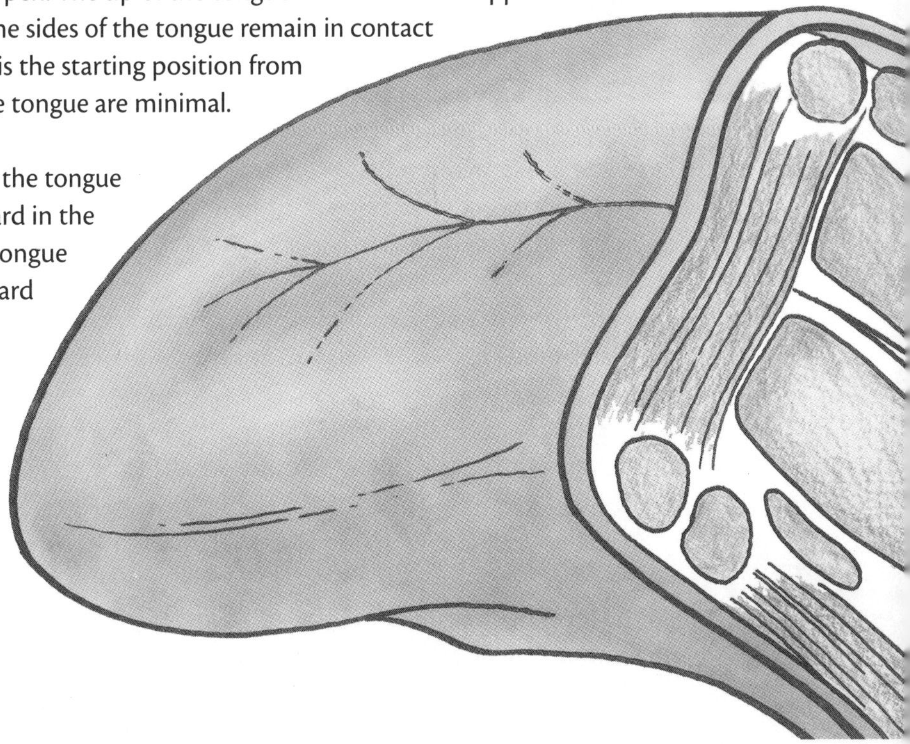

The Tongue: Articulation and Resonance

The Tongue and Articulation of Vowel Sounds

Thanks to the action of its muscles, the tongue can change shape and participate in the articulation of vowels. Vowels are said to be *anterior* or *posterior*, depending on where the mass of the tongue is located on the palate. Here we look at only three vowels sounds: *ee*, *ah*, and *ou*.

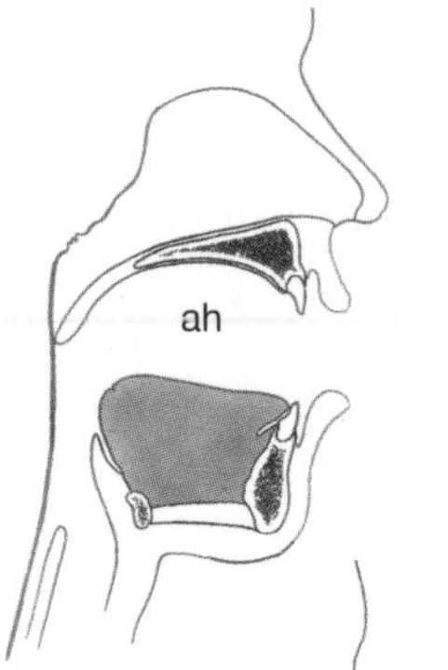

With the *ah* sound, the dorsum of the tongue is lowered.

When pronouncing the *ee* sound, the dorsum of the tongue is raised, and at the same time the mass of the tongue is brought forward.

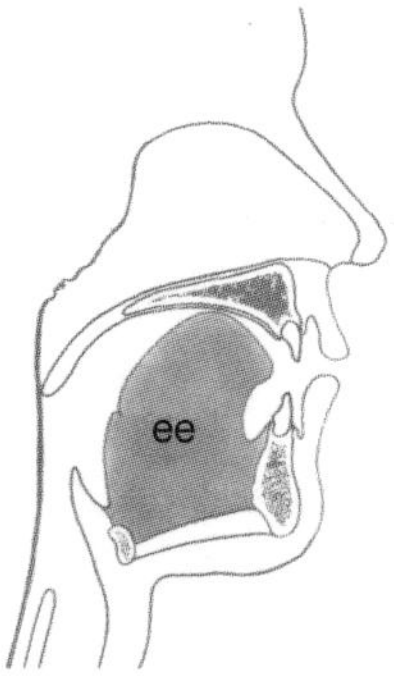

When making the *ou* sound, the dorsum of the tongue is lifted and the mass of the tongue is toward the back.

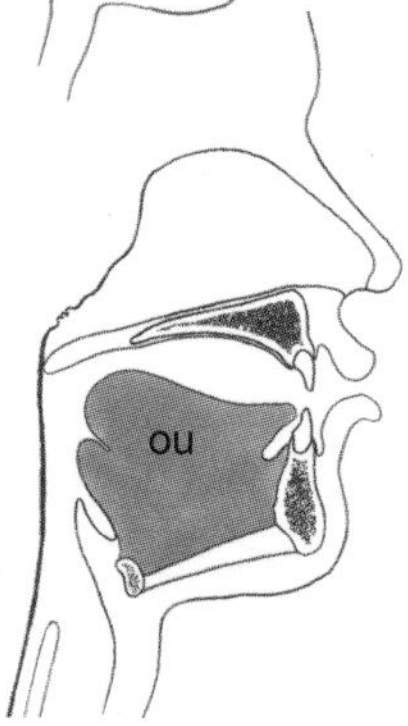

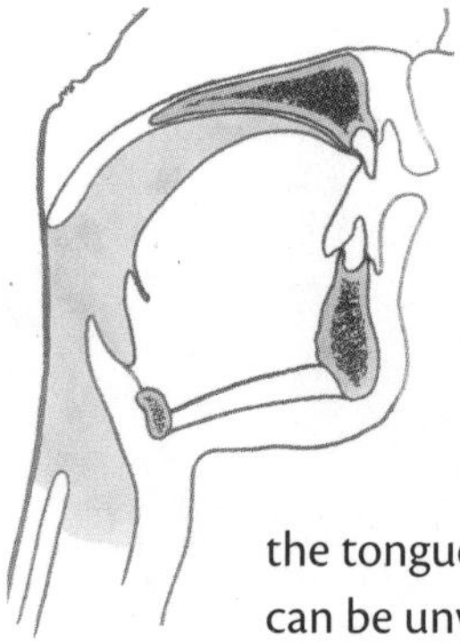

The Tongue and Articulation of Consonants

Dental Consonants

Many consonants are called *dental* because they involve contact between the tongue and teeth. The tip of the tongue can make contact with the rear of the incisors and completely prevent the expiratory air from passing. If at the same time the soft palate is lifted, the air can't escape through the nose either. Pressure then builds in the oral cavity. From there, if we separate the tongue from the teeth suddenly, we create an *explosive dental consonant*, which can be unvoiced (for example, *t*, without the vibration of the larynx) or voiced (*d*, with laryngeal vibration).

We can have the same type of contact with the tongue and the teeth and lower the soft palate at the same time. The air then exits through the nose. This creates the *occlusive nasal consonant n*.

The tip of the tongue can be placed against the teeth, with the middle portion allowing for a "whistle" of air to pass. This creates the dental fricative consonant, which can be unvoiced (for example, *s*; see page 181) or voiced (*z*; see page 165).

The tip of the tongue can be placed against the teeth with the lateral areas of the tip allowing enough air to pass to make a slight whistle, while at the same time there is a vibration at the laryngeal level. This forms a *lateral liquid consonant* (such as *l*).

Palatal Consonants

Palatal consonants involve contact between the tongue and the hard palate.

The lateral edges of the dorsum of the tongue can come in contact with the lateral edges of the hard palate. This creates a *fricative palatal consonant,* which can be unvoiced (for example, *sh,* without a laryngeal vibration; see page 181) or voiced (*zh,* with laryngeal vibration; see page 165).

The dorsum of the tongue can come in contact with the hard palate, letting air pass on both sides. From there, if we separate the tongue from the palate, we create a *palatal sonorous consonant* (for example, *ny*).

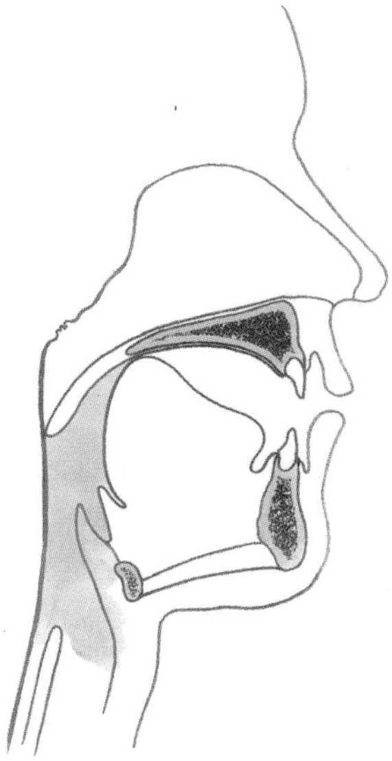

Velar Consonants

Velar consonants involve contact between the tongue and the soft palate.

The dorsum of the tongue can come into contact with the soft palate and completely prevent the escape of expiratory air. If at the same time the soft palate is lifted, the air can't escape through the nose either. Pressure then builds in the oral cavity. From there, if we separate the tongue from the soft palate abruptly, we can create an *explosive velar consonant,* which can be unvoiced (for example, *k,* without a laryngeal vibration) or voiced (*g,* with a laryngeal vibration).

The dorsum of the tongue can make contact with the palate and let the air escape intermittently, creating a vibration. This creates a *vibrating velar consonant,* which can be unvoiced (for example, *r,* without a laryngeal vibration) or voiced (again, *r,* but with a laryngeal vibration).

The Tongue Participates in Resonance of the Voice

With the Tongue Flat behind the Lower Incisors

We look for this position so that the oral cavity isn't filled and encumbered by the mass of the tongue.

With the Mass of the Tongue Lifted Posteriorly

When we want to enhance the bass spectrum, we give preference to this position.

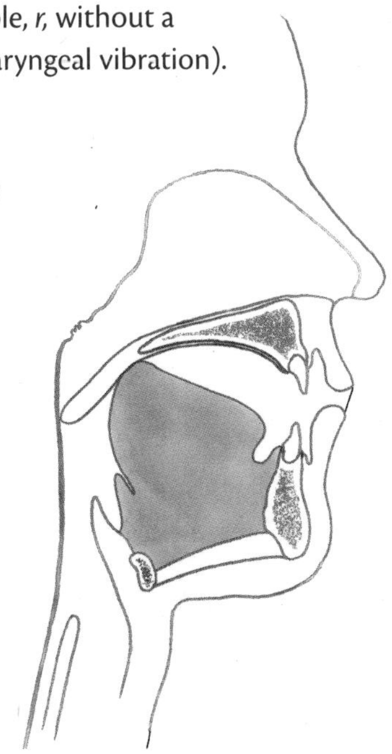

The Lips

A very important component of the vocal instrument, the lips are two muscular and membranous folds that define the limits of the orifice of the mouth. They are the last area of the vocal tract where sound can be influenced before "leaving the body."

In everyday life, we use our lips both for chewing and for sucking. They are a place of elective facial expression, and therefore communication. Regarding the voice, the lips are among the most important players in terms of articulation and modifying the resonance.

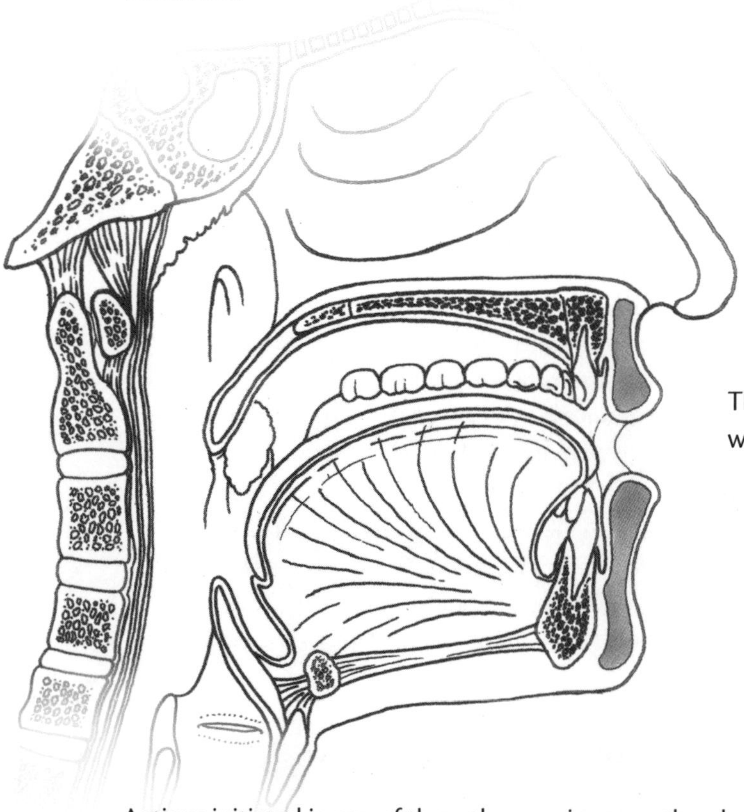

The lips are bordered by three regions with which they interact:

- the region of the cheeks
- the region of the nose
- the region of the chin

Actions initiated in any of these three regions can then have an effect on the lips, and action in the lips can affect any or all of these three regions. In vocal work, this linkage is important. For example, a change in the cheeks due to facial expression can produce a change in the shape of the lips, which in turn transforms the resonance of the voice. All this can be voluntary or involuntary.

Movements of the lips are often complemented by those of the mandible or jaw, and they are easily mistaken for each other. When we are using the voice, it is often important to distinguish movement that is happening in one place or the other, or how movements are influencing each other. For example, when we close our mouth, the movement is made by the jaw, and it is completed more or less by a movement made by the lips.

Description of the Lips

There are two lips, the superior (upper) and inferior (lower). Each has two parts:

- a **cutaneous area,** sometimes covered in men by a mustache on the top and a beard on bottom
- a **mucosal area,** thicker or thinner depending on the person, with many anteroposterior folds

In the cutaneous area of the upper lip there is a vertical groove called the **philtrum**. Where it joins with the lip mucosa, it is shaped like an arc or bow; this is called **Cupid's bow,** and it is particularly noticeable in children.

On each side, the lips unite at the corners, forming the labial **commissures:**

- With the mouth open, they frame the orifice of the mouth on the sides.
- With the mouth closed, they form the **slit** of the mouth, which is about 45 to 55 millimeters long.

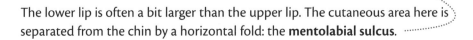

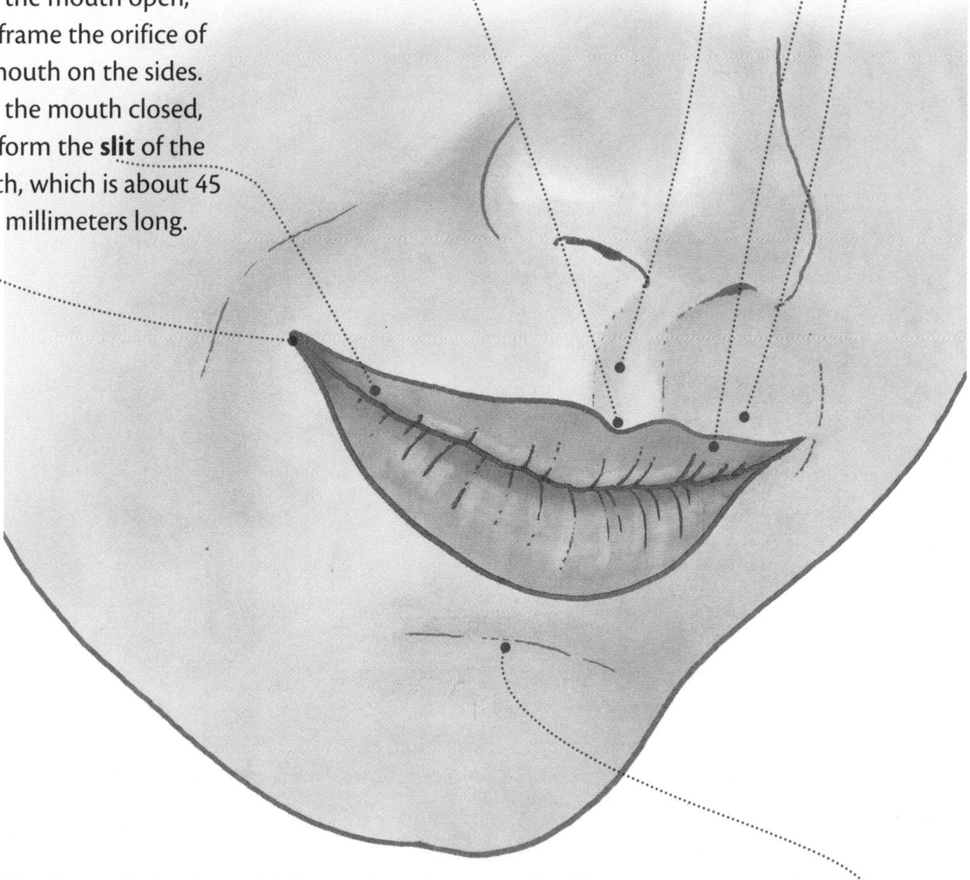

The lower lip is often a bit larger than the upper lip. The cutaneous area here is separated from the chin by a horizontal fold: the **mentolabial sulcus**.

The Muscles of the Lips

The structure of the lips is not supported by a bony framework. They are soft regions. However, they can change their form and tone thanks to the many muscles that lie under the skin and mucous membranes.

The Orbicularis Oris Muscle Closes the Buccal Orifice

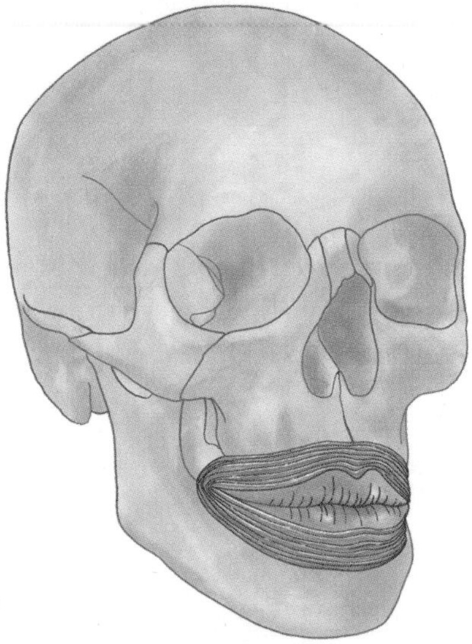

This muscle has two parts.

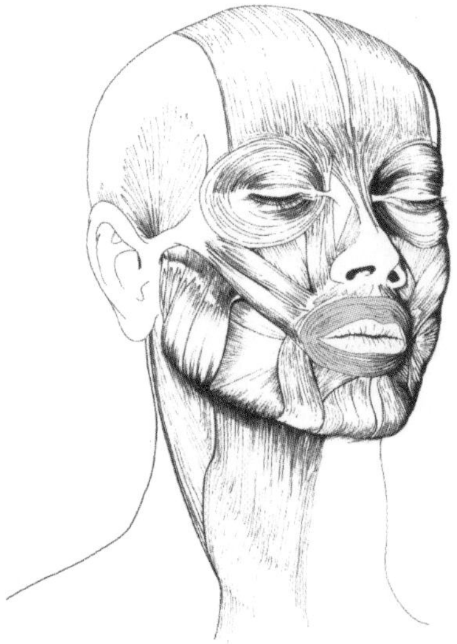

The main portion, called the **internal orbicularis,** is located along the free border of the lips. The fiber bundles of the superior and inferior lips intersect at the commissures, then terminate on the deep surface of the skin and mucosa.

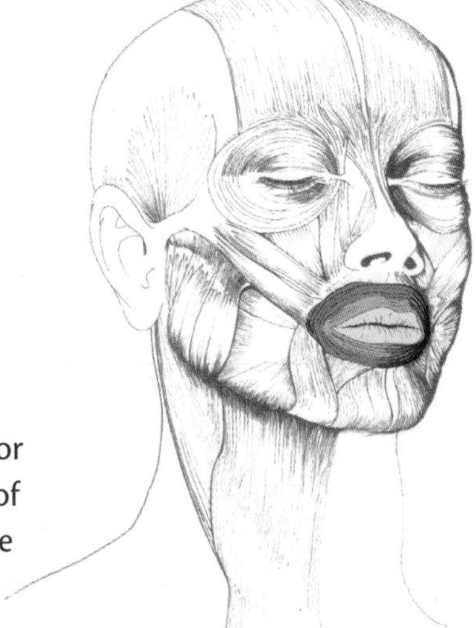

An outer peripheral portion called the **external orbicularis** is composed of two types of fibers:

- Some are terminal muscle fibers that connect to the commisssures.
- Others are called **incisive muscles.** There are two superior and two inferior, and they run from the alveolar border of the maxilla to the lips, mingling with the terminal muscle fibers.

Its Action

The orbicularis oris is the main constrictor of the buccal orifice: it closes the lips. Each part of the orbicularis brings a specific action to this closure:

The internal orbicularis allows us to press our lips together.

The external orbicularis lets us pucker our lips, projecting them forward.

In the production of sound, the action of the orbicularis will often be combined with the action of its neighboring antagonist muscles (see the following pages).

A Small Muscle Completes the Action of the Orbicularis

Aeby's muscle (not shown) is composed of small fibers that run from the skin of each lip to the deep surface of the mucosa.

Its Action

It compresses the lips from back to front.

Around the Orbicularis, Three Groups of Muscles Open the Buccal Orifice

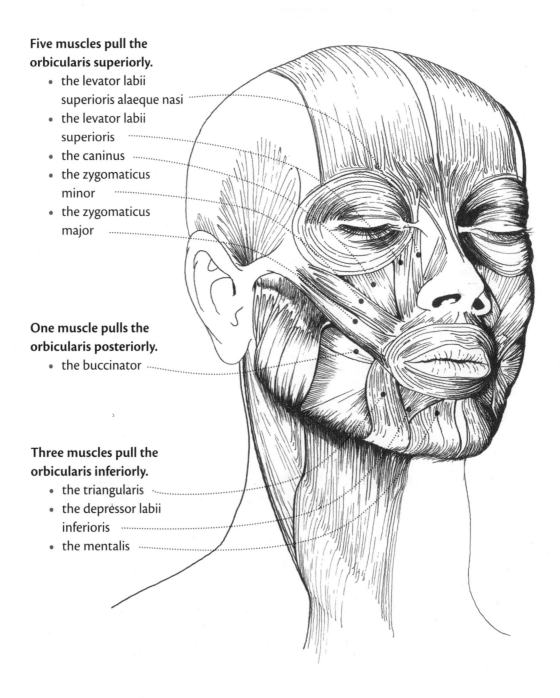

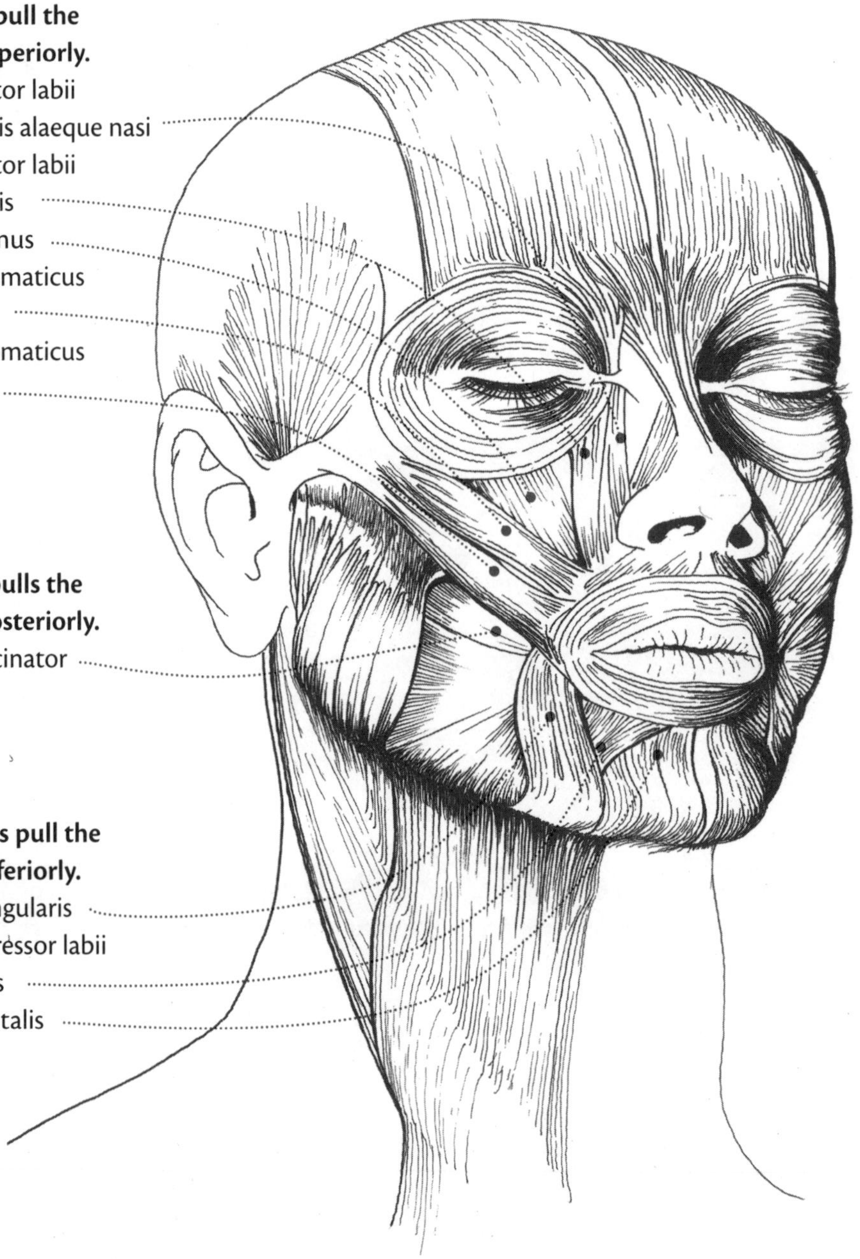
Five muscles pull the orbicularis superiorly.
- the levator labii superioris alaeque nasi
- the levator labii superioris
- the caninus
- the zygomaticus minor
- the zygomaticus major

One muscle pulls the orbicularis posteriorly.
- the buccinator

Three muscles pull the orbicularis inferiorly.
- the triangularis
- the depressor labii inferioris
- the mentalis

The Buccinator Pulls the Lips to the Sides

This deep muscle fills the space between the upper and lower maxillae. It attaches to the commissures of the mouth, and further back on the outer face of each maxillary alveolus.

Some superior and inferior fibers intertwine toward the back, terminating on the pterygomaxillary ligament and buccinator crest of the mandible.

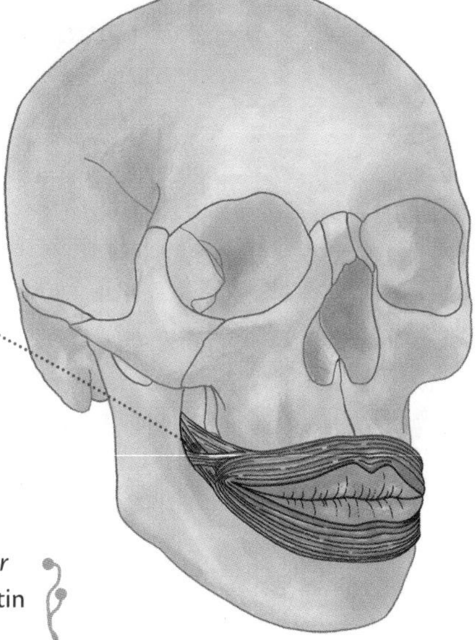

Details

The word *buccinator* derives from the Latin for "trumpeter."

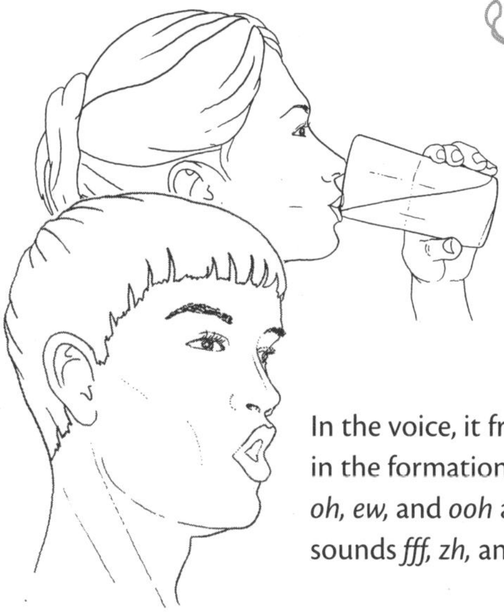

Its Action

It pulls the commissure posteriorly, expanding the lip laterally. But it is also a deep muscle of the cheek, which it flattens. We can see its action when we whistle or drink from a glass or bottle.

In the voice, it frequently participates in the formation of the vowel sounds *oh*, *ew*, and *ooh* and the consonant sounds *fff*, *zh*, and *sh*.

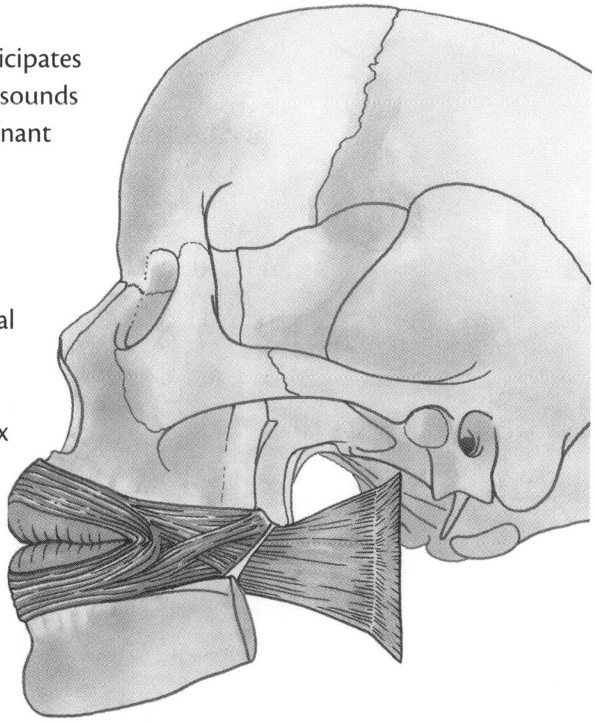

In the back, the buccinator meets the superior pharyngeal constrictor muscle; the two muscles attach to the pterygomaxillary ligament, forming a muscular chain. It contributes to the closure of the cheeks and nasopharynx when we swallow, which pushes the food bolus toward the oropharynx. It can also help close the pharynx when the cheeks contract. In preparatory vocal work, it is interesting to separate these two actions and learn, for example, how to relax the pharynx at the back of the throat and, at the same time, tone the cheeks to articulate vowels sounds such as *oh* and *ooh*.

The Levator Labii Superioris Alaeque Nasi

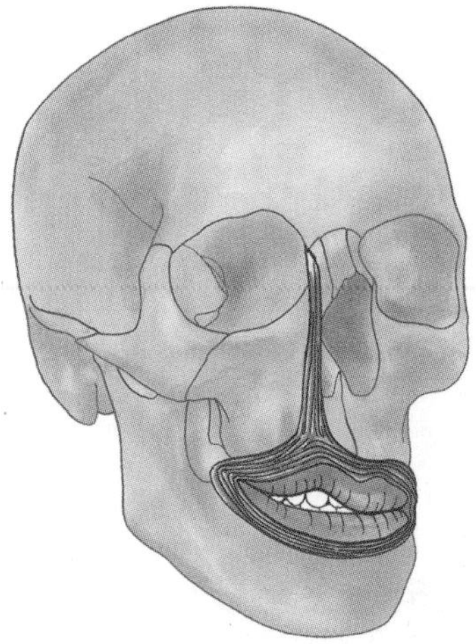

This muscle attaches at the bottom on the deep surface of the skin of the upper lip and the skin of the superior border of the wing of the nose. It rises to terminate on the orbital process of the frontal bone and the ascending branch of the superior maxilla.

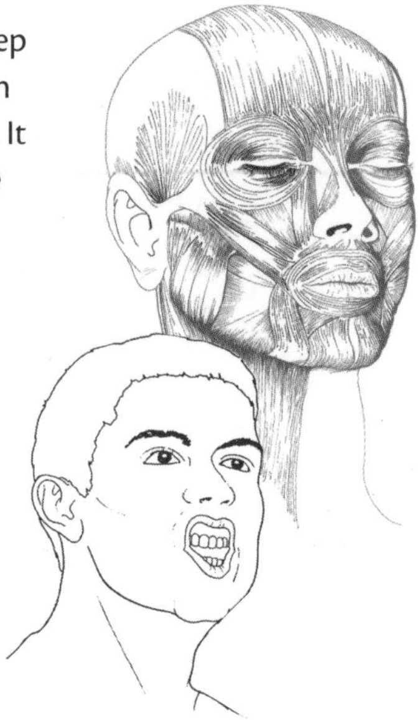

Its Action

It lifts the upper lip. In the medial part, it also lifts the wing of the nose.

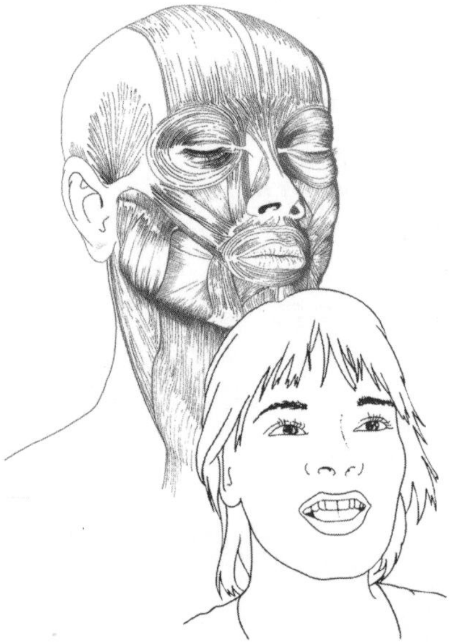

The Levator Labii Superioris

The levator arises on the deep surface of the skin on the border of the upper lip, a bit outside the levator labii superioris alaeque nasi. Its fibers ascend to the inside and terminate at the inferior border of the orbit.

Its Action

It lifts the part of the upper lip that we find just outside of the area that the levator labii superioris alaeque nasi lifts.

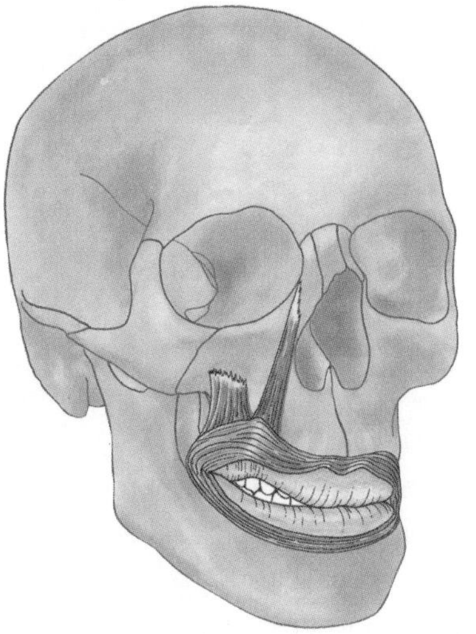

The Caninus

The caninus muscle arises a little to the outside of the levator labii superioris alaeque nasi. Its fibers rise inward and terminate on the outer surface of the superior maxilla.

Its Action

It raises the labial commissure.

The Zygomaticus Minor

Located at the top of the cheek, this muscle, which is actually quite small, originates on the deep surface of the skin of the upper lip, a little to the inside of the corner. It ascends to the outside and terminates on the external surface of the malar bone.

Its Action

It lifts the upper lip a little to the outside of the levator of the lip. This pulls the upper lip outward. It's the "smile muscle."

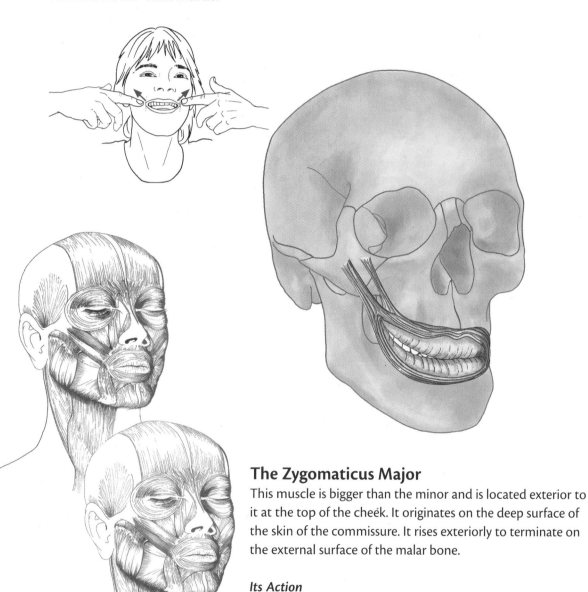

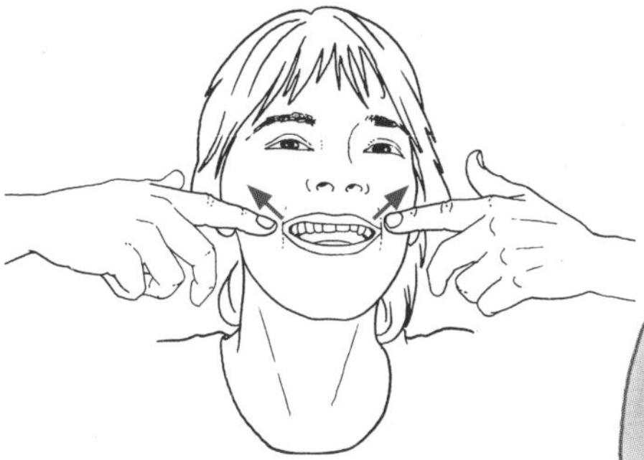

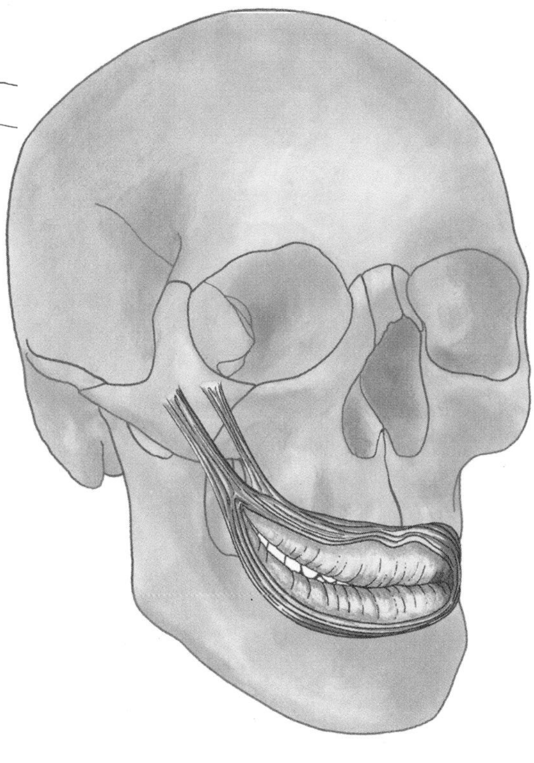

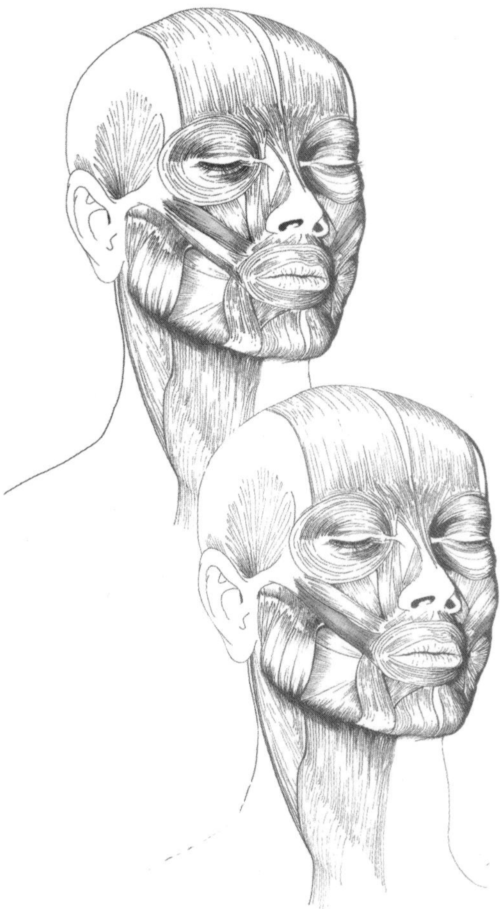

The Zygomaticus Major

This muscle is bigger than the minor and is located exterior to it at the top of the cheek. It originates on the deep surface of the skin of the commissure. It rises exteriorly to terminate on the external surface of the malar bone.

Its Action

It elevates the corner of the mouth toward the back and participates in opening the orifice of the mouth. This leads to a curvature of the groove between the lip and cheek. It is, like the zygomaticus minor, the "smile muscle."

The Triangularis

This muscle inserts in the skin of the commissure and the inferior external border of the orbicular muscle of the lips. It flares as it descends to terminate on the outer surface of the mandible.

Its Action

It lowers the commissure (the corner of the mouth).

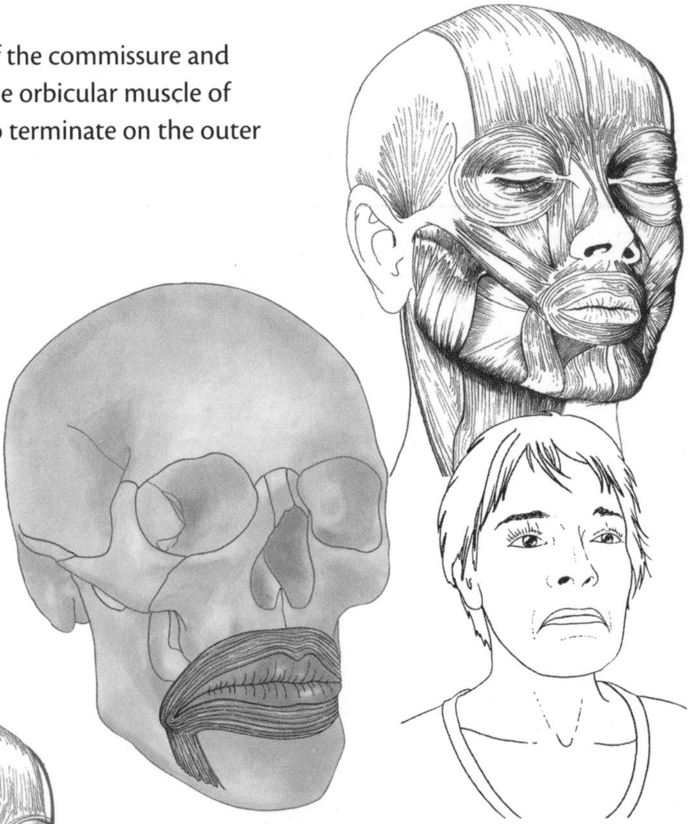

The Mentalis Muscle

This muscle attaches at the top to the inferior border of the orbicular muscle of the lips. It descends slightly to the inside to terminate at the bottom on the mandible, at about the median line of the chin.

Its Action

It lowers the bottom lip.

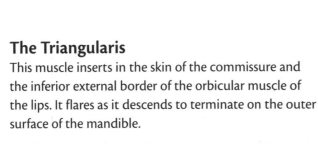

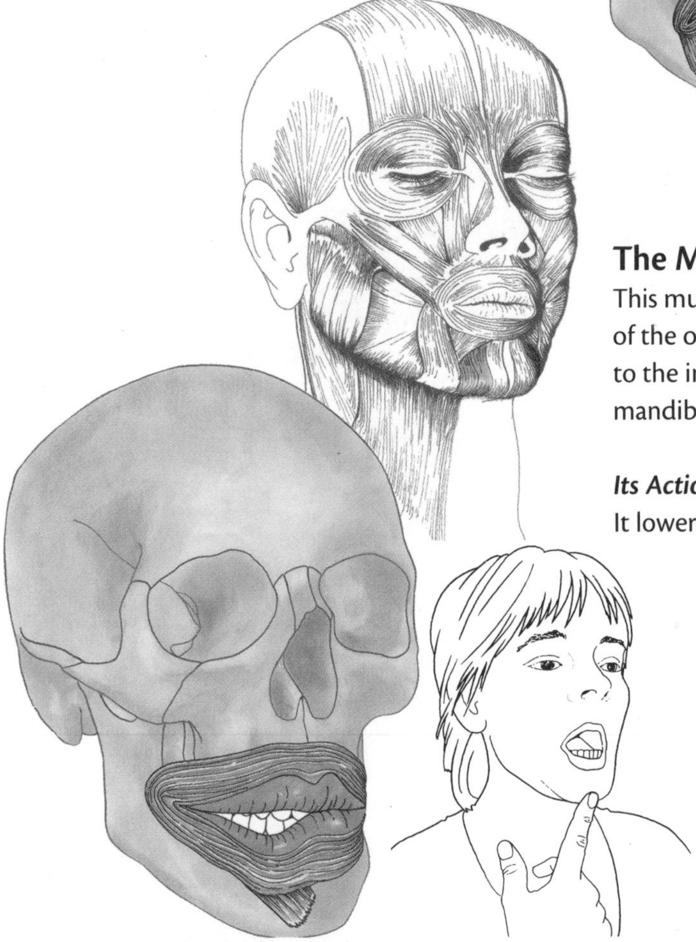

The Depressor Labii Inferioris

This muscle attaches to the deep surface of the skin of the lower lip and to the inferior fibers of the orbicular muscle of the lips. It descends and runs a bit outward to terminate on the mandible at the lower end.

Its Action

It lowers the lower lip, while pulling the borders outward. This muscle can interfere when we try to make a very open *aaaah* sound.

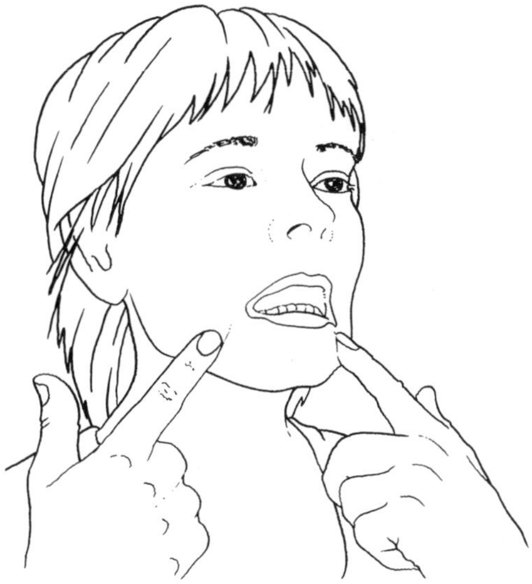

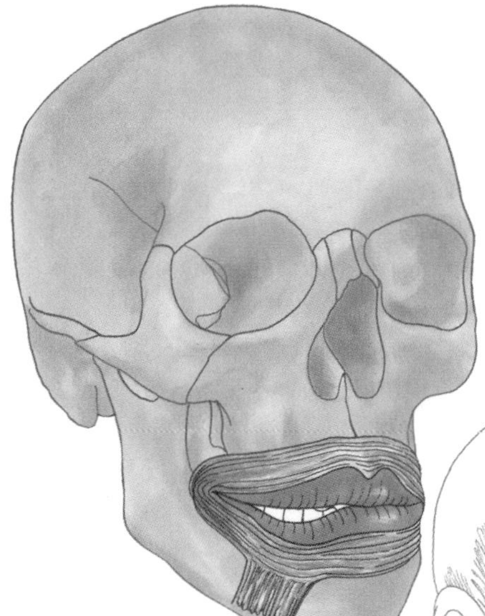

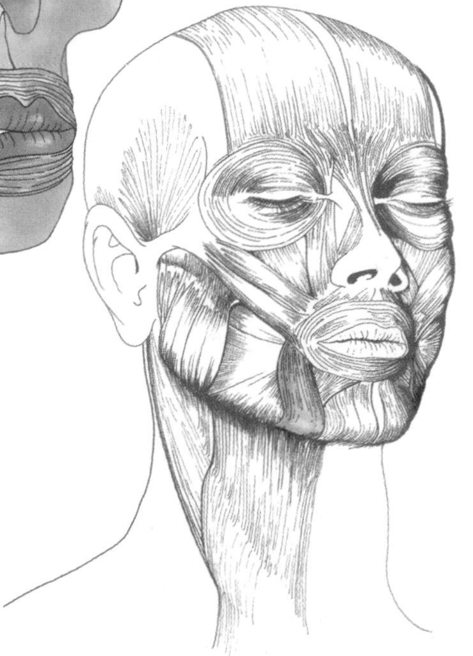

The Lips: Articulation and Resonance

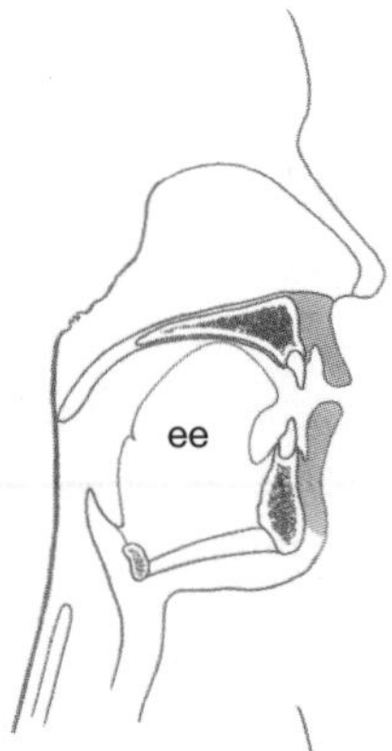

The Lips and the Articulation of Vowels

The lips can spread their corners (an action made mainly by the buccinator muscles) for the formation of certain vowels such as *ee, ey,* and *eh*.

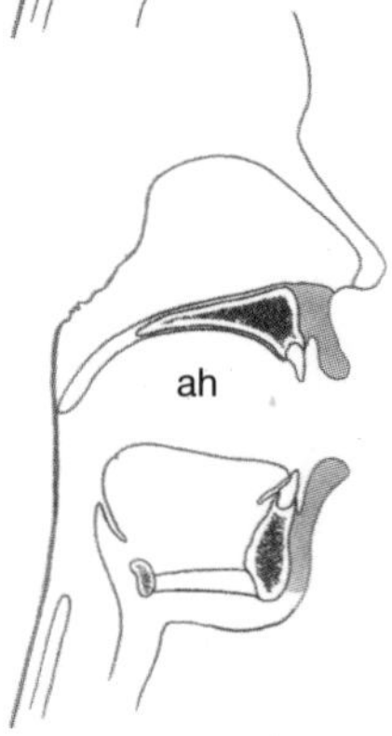

With the *ah* sound, the spreading of the lips is very slight.

If the soft palate is lowered, the spreading of the lips contributes to the production of the nasal vowel sound *in*.

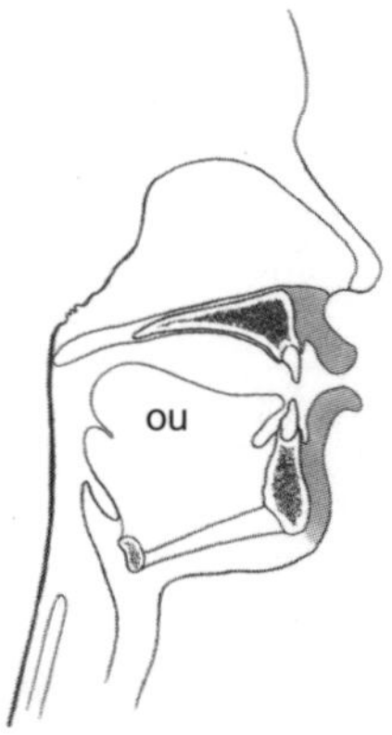

The lips can, while being slightly separated, take a circular form (by the simultaneous action of the orbicularis and its antagonists) for the formation of "rounded" vowel sounds, also known as labialized vowels, such as *oh, ew,* and *ou*.

If the soft palate is lowered, the rounding contributes to the production of nasal vowel sounds such as *on* and *un*.

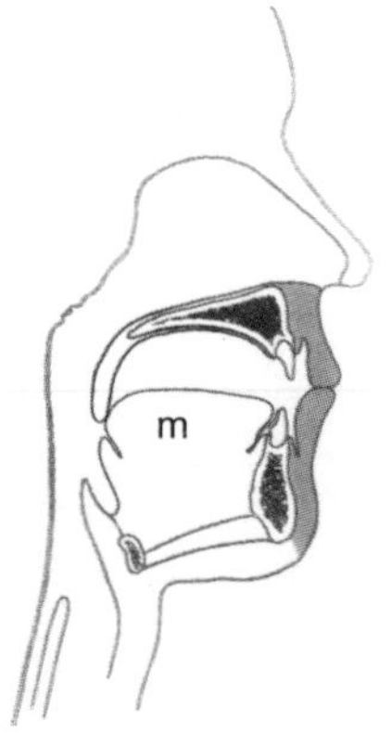

The Lips and the Articulation of Consonants

Many consonants involve the action of tightening the lips, made by the orbicular muscle of the lips.

The lips can join completely while, at the same time, the soft palate is lowered. The air then exits through the nose. This creates the nasal occlusive consonant *m*.

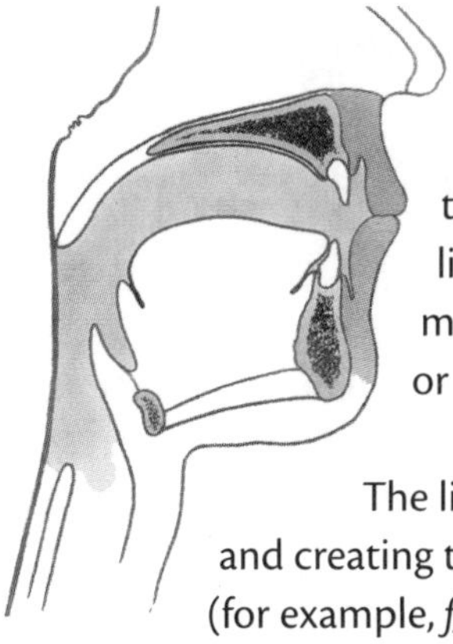

The lips can join completely and prevent any expiratory air from passing between them. If at the same time the soft palate is lifted, air cannot get out through the nose. Pressure then builds in the oral cavity. From there, if we separate the lips suddenly, we create an *explosive bilabial consonant,* which may be unvoiced (for example, *p,* without laryngeal vibration) or voiced (*b,* with laryngeal vibration).

The lips may be incompletely joined, allowing some air to pass and creating the *bilabial fricative consonant,* which can be unvoiced (for example, *f,* without laryngeal vibration) or voiced (*v,* with laryngeal vibration).

The Lips Participate in Vocal Resonance

We can project the lips forward while rounding them (make the vowel sound *ou* in an exaggerated way) to create a resonance cavity, with the lips anterior to the oral cavity. (Warning: The lips can't be joined or the effect is lost.) This favors the creation of deep harmonics.

We can do the contrary and pull the lips backward, opening them at the top and bottom (which will expose the teeth). In this case, the oral cavity is not extended by the labial resonator. This promotes the opposite effect, creating high harmonics.

The Nose and the Nasal Cavities

The most visible part of the vocal tract, the nose follows the pharynx and is the second possible passageway for phonatory air.

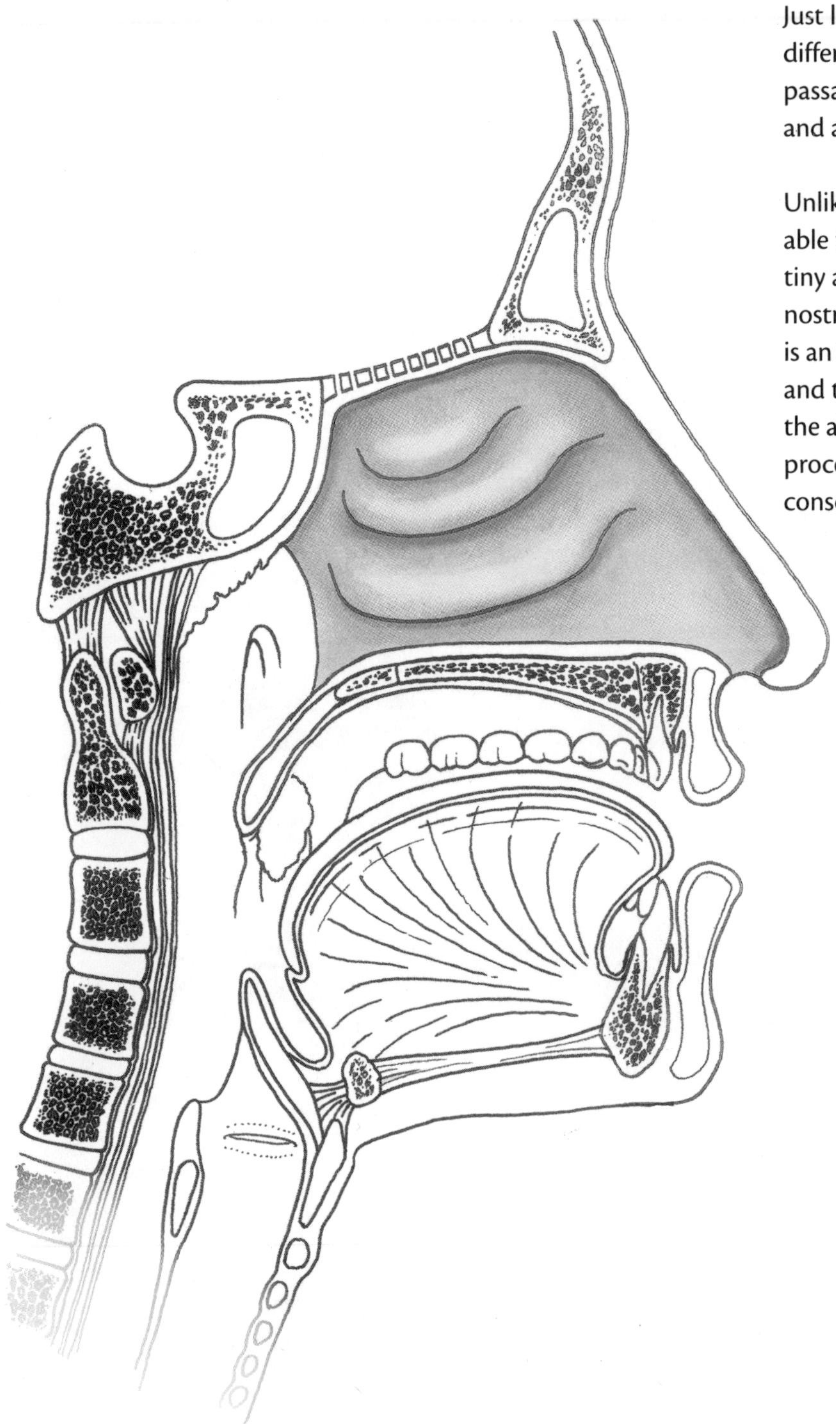

Just like the mouth, it performs many different functions: it is one of the air passageways of the respiratory system and also the olfactory organ.

Unlike the mouth, the nose is not able to change shape, except for tiny adjustments in the size of the nostrils. Nevertheless, for the voice, it is an important place for resonance and the articulation of sound. When the air exits through the nose in the process of phonation, the vowels and consonants are called *nasal*.

Description of the External Part of the Nose

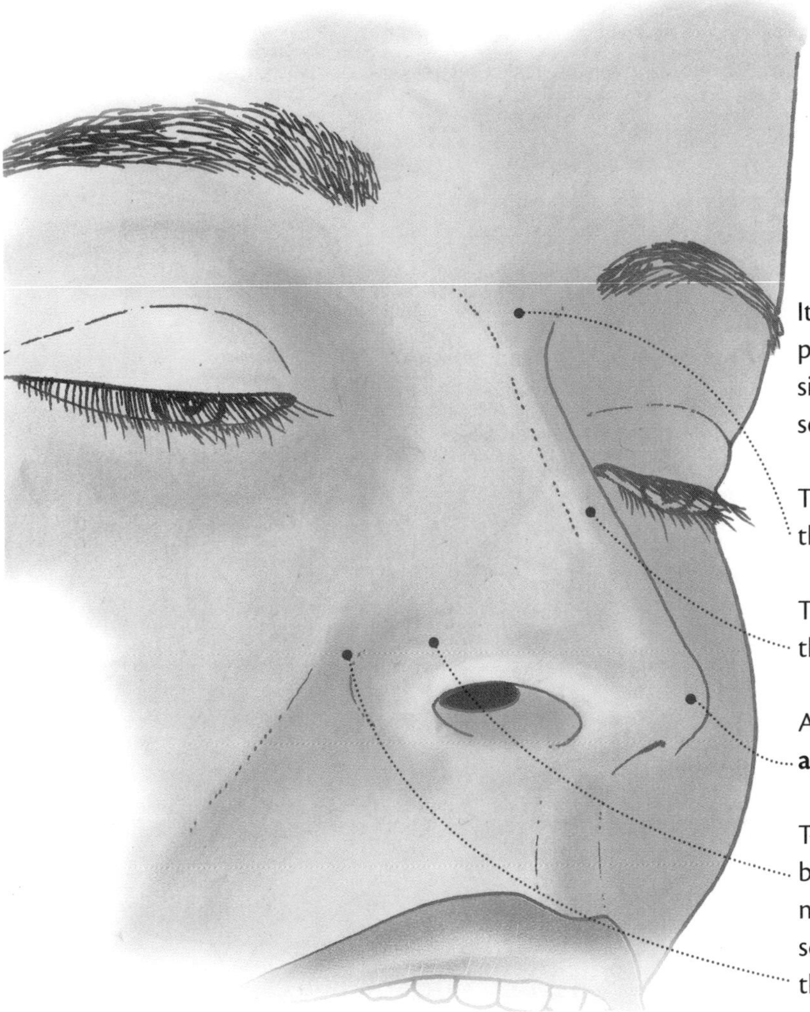

It is shaped like a three-sided pyramid. The two nostrils are situated at the base and are separated by the nasal septum.

The top of the pyramid is called the **nasal root.**

The front of the nose is called the **bridge.**

At the tip of the bridge is the **apex**.

The nostrils are surrounded by the **alae,** or wings, of the nose, which are themselves separated from the cheeks by the **nasogenian grooves**.

This visible anterior part of the nose is called the **vestibule**. It gets its structure from several bones: the maxilla bones, the nasal bones, and the frontal bone, which are extended downward by cartilage under the skin.

But the nose is much deeper than it appears to be from just the outside. Inside, it is composed of two cavities or nasal fossa.

The Nasal Fossa

The fossa are the two hollow areas located inside the nose. Each is high, narrow, and elongated from front to back. The flexible anterior area, part of the external and visible part of the nose, is smaller than the posterior part, which is enclosed in the mass of the facial skull. Each nasal cavity opens in front at the nostril and in back at an orifice called the **nasal choana**. In each we find four walls and two extremities.

The **superior wall** (the "ceiling") is formed by an assemblage of several bones:
- the posterior surface of the nasal bone
- the inferior surface of the nasal spine of the frontal bone (see page 74)
- the cribriform lamina of the ethmoid bone (see page 75)
- the anterior and inferior surfaces of the sphenoid body (see page 70)

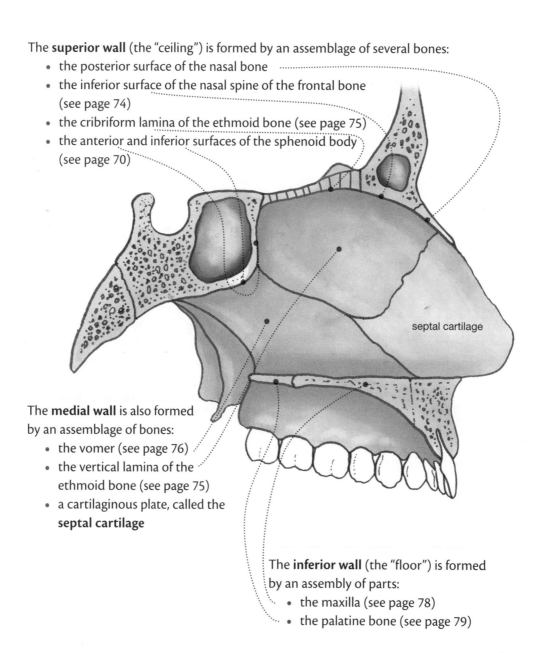

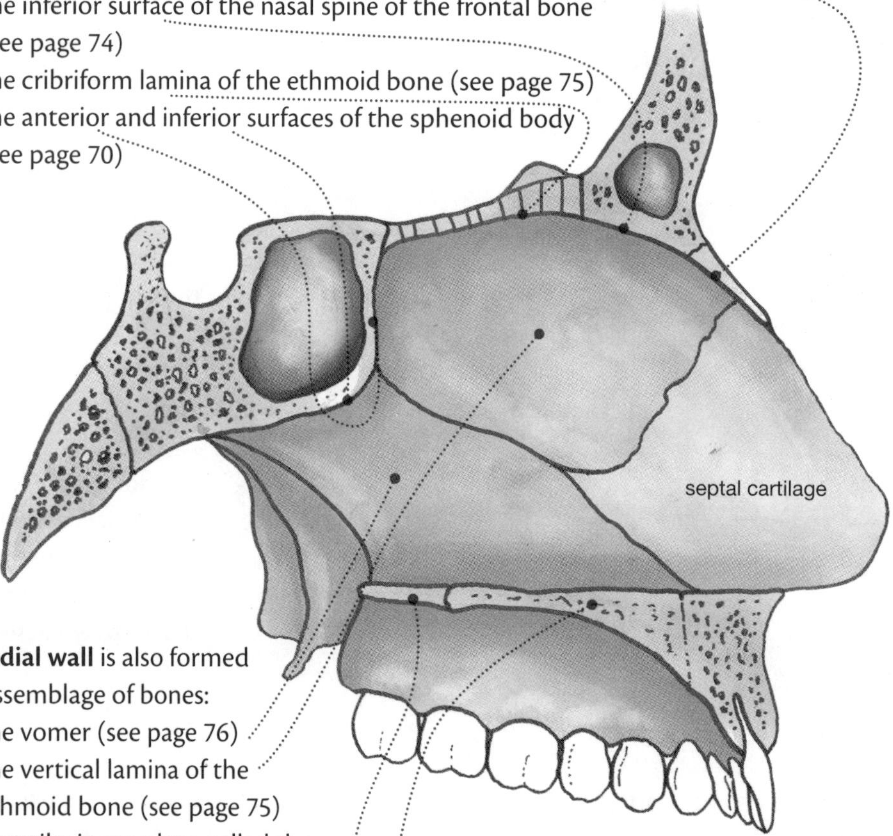

The **medial wall** is also formed by an assemblage of bones:
- the vomer (see page 76)
- the vertical lamina of the ethmoid bone (see page 75)
- a cartilaginous plate, called the **septal cartilage**

The **inferior wall** (the "floor") is formed by an assembly of parts:
- the maxilla (see page 78)
- the palatine bone (see page 79)

The **lateral wall** is made up of an assemblage of many bones in three planes:

The **deep plane** is formed by the medial aspect of the maxilla, pierced by the orifice of the maxillary sinus (see page 280), and the medial lamina of the pterygoid process (see page 71).

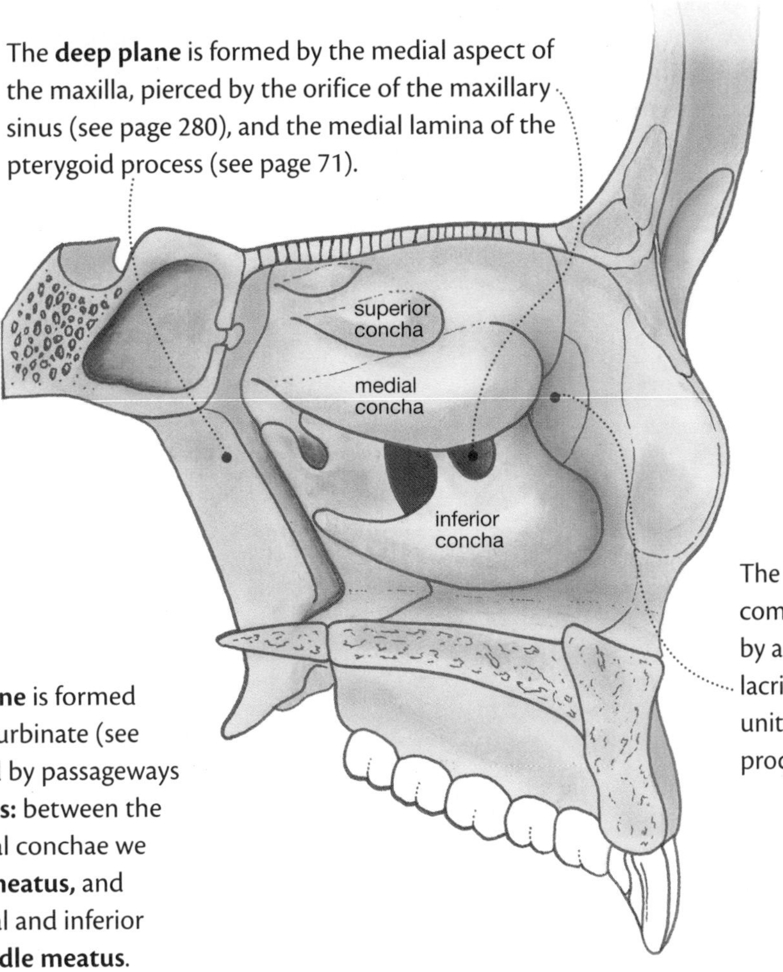

The **medial plane** is composed in the front by a small flat bone: the lacrimal bone, which unites with the frontal process of the maxilla.

The **superficial plane** is formed by the concha, or turbinate (see page 77), separated by passageways called the **meatuses:** between the superior and medial conchae we find the **superior meatus,** and between the medial and inferior conchae is the **middle meatus**.

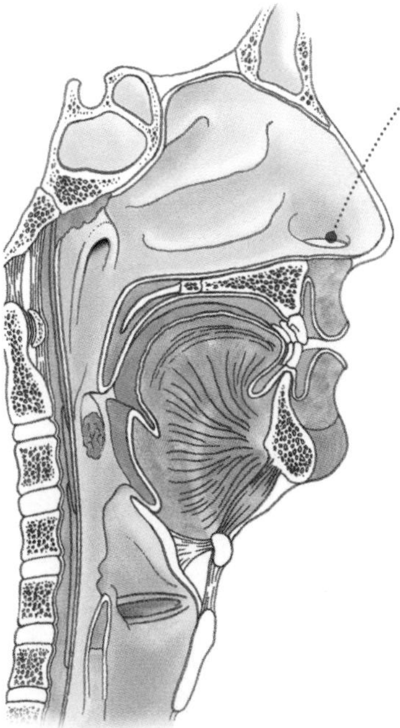

The anterior extremity is the **nostril,** the orifice that links the nasal cavity to the outside.

The posterior extremity is the **choana,** an orifice that links the nasal cavity with the nasopharynx.

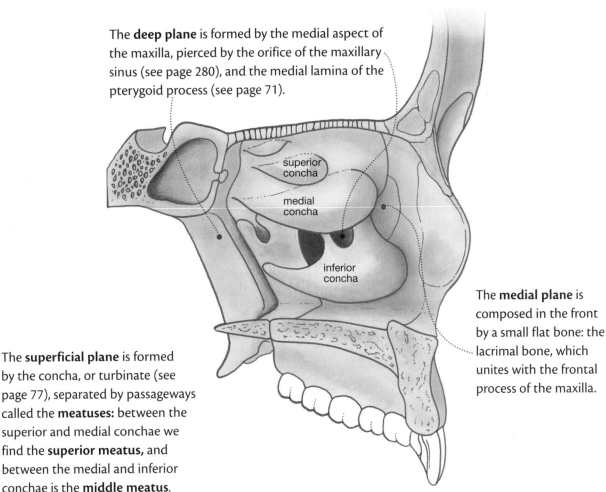

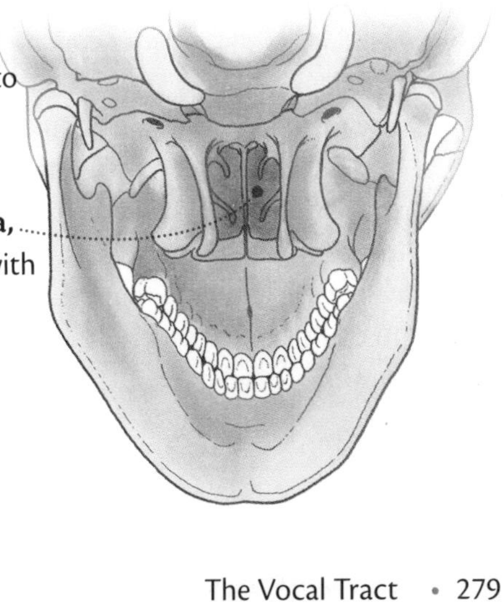

The Paranasal Sinuses

The sinuses are air cavities connected to the nasal cavity in almost all directions. Some are hollow areas in the bones of the skull that are linked to the nasal cavity by a small conduit called the **ostium**.

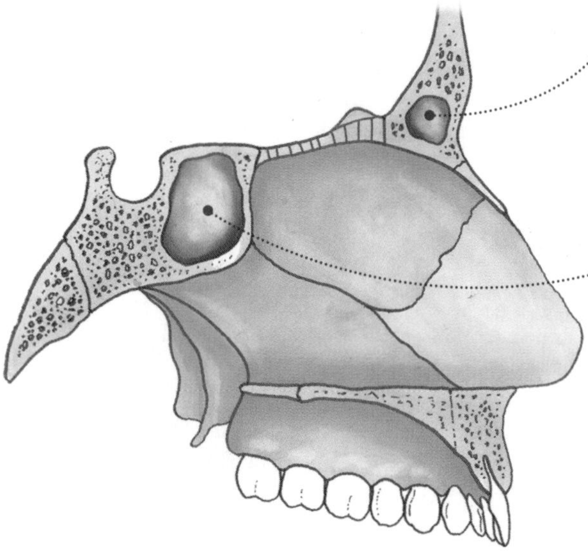

The **frontal sinus** is a hollow area in the mass of the frontal bone, above the nasal cavity and behind the brow ridges. Its ostium opens onto the middle meatus.

The **sphenoid sinus** is a hollow area in the body of the sphenoid. Its ostium opens onto the posterior part of the nasal fossa. There are often two adjacent sinuses.

The **ethmoid sinus** is a hollow area in the lateral mass of the ethmoid bone. It is made up of tiny cavities that are joined together, the ethmoid "cells," of which there are six to nine on each side. The most anterior cells open, via the small ostia, beneath the middle concha (or turbinate), and the most posterior cells, under the superior concha.

The **maxillary sinus** is a hollow area in the body of the maxilla bone. It is above both molars and the premolars. Its ostium opens into the nasal cavity below the middle turbinate. At about 10 cubic centimeters, it's the largest sinus.

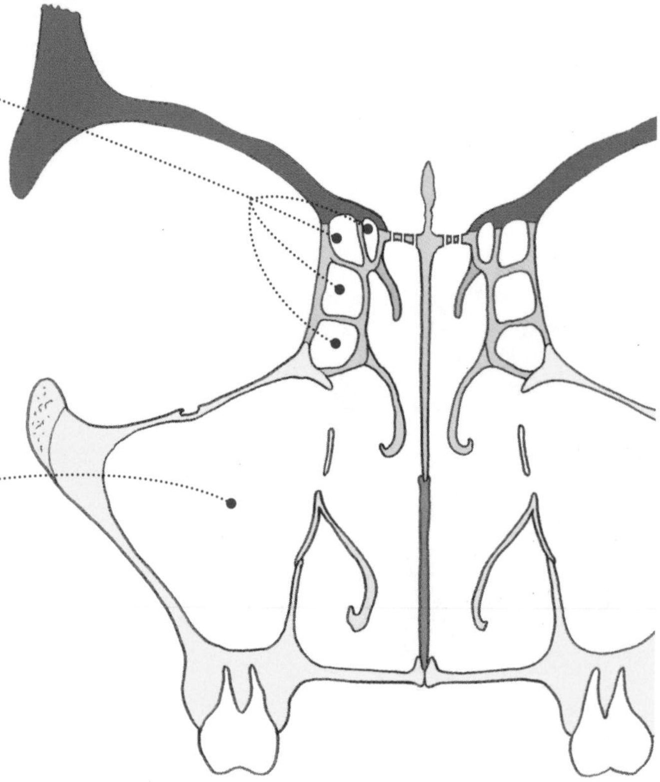

The Nasal Mucosa

The entire nasal cavity is lined with a mucous layer that covers all of the nasal folds; it even continues via the ostia to the walls of each of the sinuses. This lining is warm and humid, and it transmits these qualities to the air that passes through the nasal cavity.

The mucosa contains:

- hairlike projections: cilia, which block dust and clean the air coming through the passage
- a sticky mucus that contains an antibacterial enzyme

When we inhale through the nose, the air that passes into the airway below is of high quality, and this is important for the larynx and the vocal cords. However, when using the voice, circumstances don't always allow us to inhale through the nose. Sometimes it's necessary to inhale quickly, before continuing to speak, recite, or sing, and it is faster to breathe in through the mouth. It is important to take a subsequent breath through the nose as soon as possible. This helps prevent dryness of the mucous membranes.

The mucosa folds in and around each of the nasal concha, transforming the lateral surface of the nasal fossa into the mucosal lining that is easily recognizable in a sagittal cut of the nose.

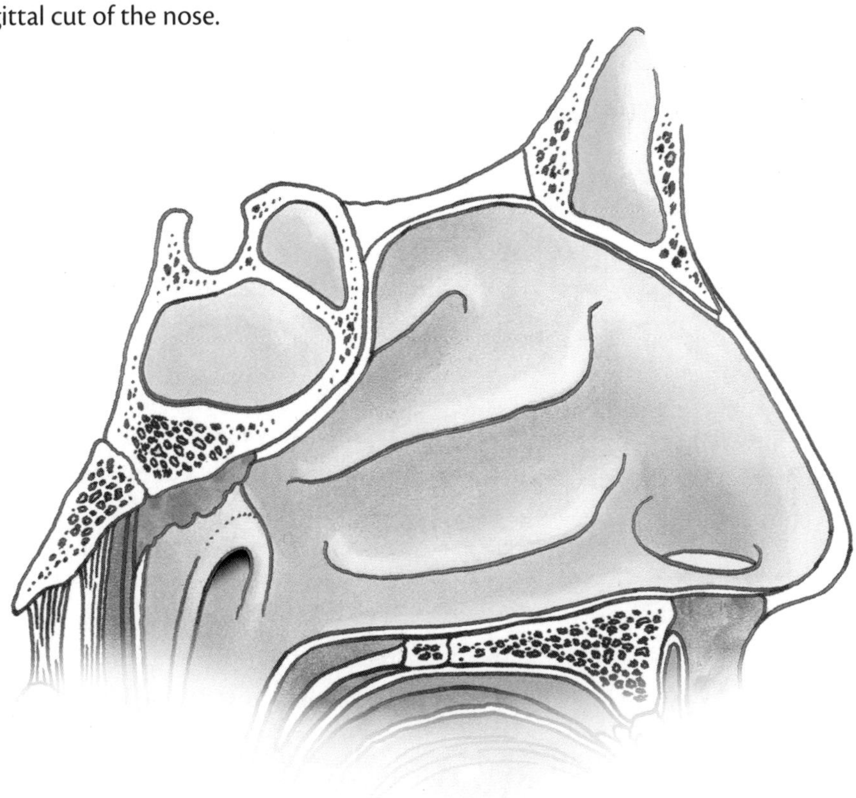

The Ears

The ears have a special place in vocal work, since audition of a sound makes it possible for us to adjust it. This is called the *audio-phonation loop*. We'll just touch upon the anatomy of the ears here.

The ear is made up of three parts. We refer to them as the outer ear, the middle ear, and the inner ear.

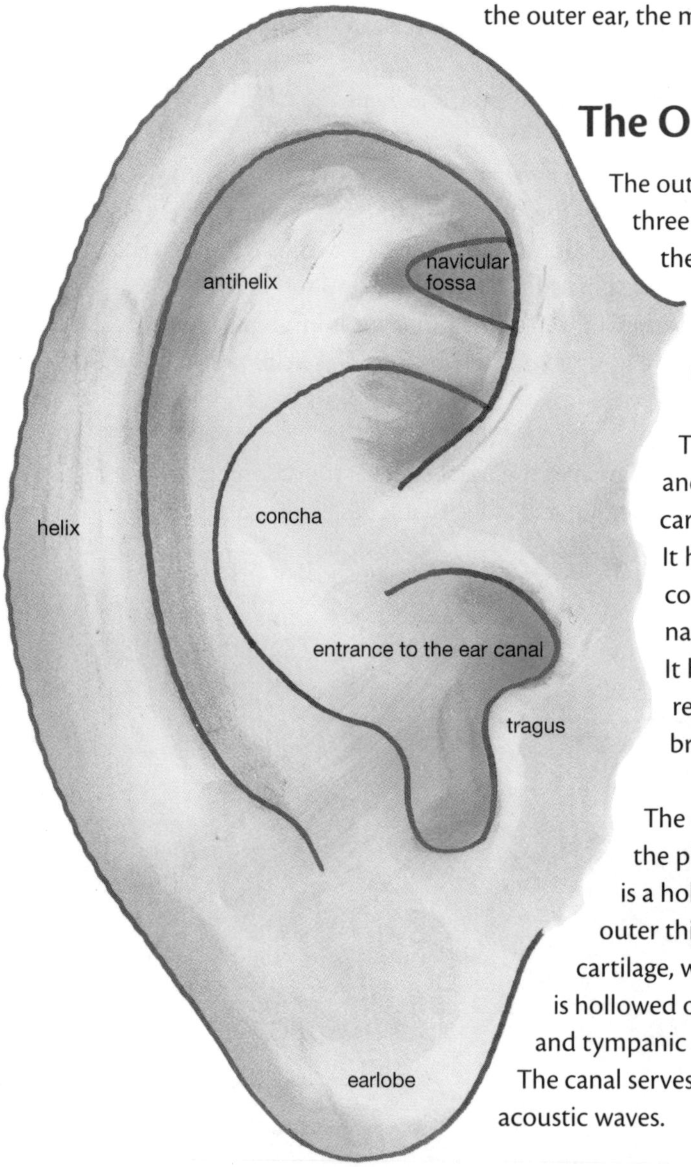

The Outer Ear

The outer ear is itself made up of three parts: the pinna (or auricle), the ear canal (or external acoustic meatus), and the eardrum (or tympanic membrane).

The pinna is oval in shape and is made up of elastic cartilage covered with skin. It has many folds: the tragus, concha, helix, antihelix, navicular fossa, and earlobe. It has a role as an acoustic receptor that captures and brings together sound waves.

The external ear canal connects the pinna with the eardrum. This is a hollow cylinder whose two outer thirds are made up of elastic cartilage, while the inner third is hollowed out of the temporal bone and tympanic cavity (see pages 72–73). The canal serves as a passage for the acoustic waves.

The tympanic membrane (or eardrum) is a rounded membrane that is vibrated by acoustic waves. It separates the outer ear from the middle ear chamber.

The Middle Ear

The middle ear is a small chamber: the tympanic cavity. It contains air. Here we find the auditory ossicles, three tiny interconnected bones: the stirrup, anvil, and hammer.

It has six surfaces. We'll touch on three of them here: The external surface is occupied mainly by the eardrum. The inferior surface, which has a hole and opens onto the auditory or Eustachian tube, connects the middle ear with the nasopharynx. This regulates the air pressure of the tympanic cavity. The medial surface is where the stirrup is in contact with the labyrinth of the inner ear. The vibrations received by the eardrum are transmitted by the ossicles to the cochlea.

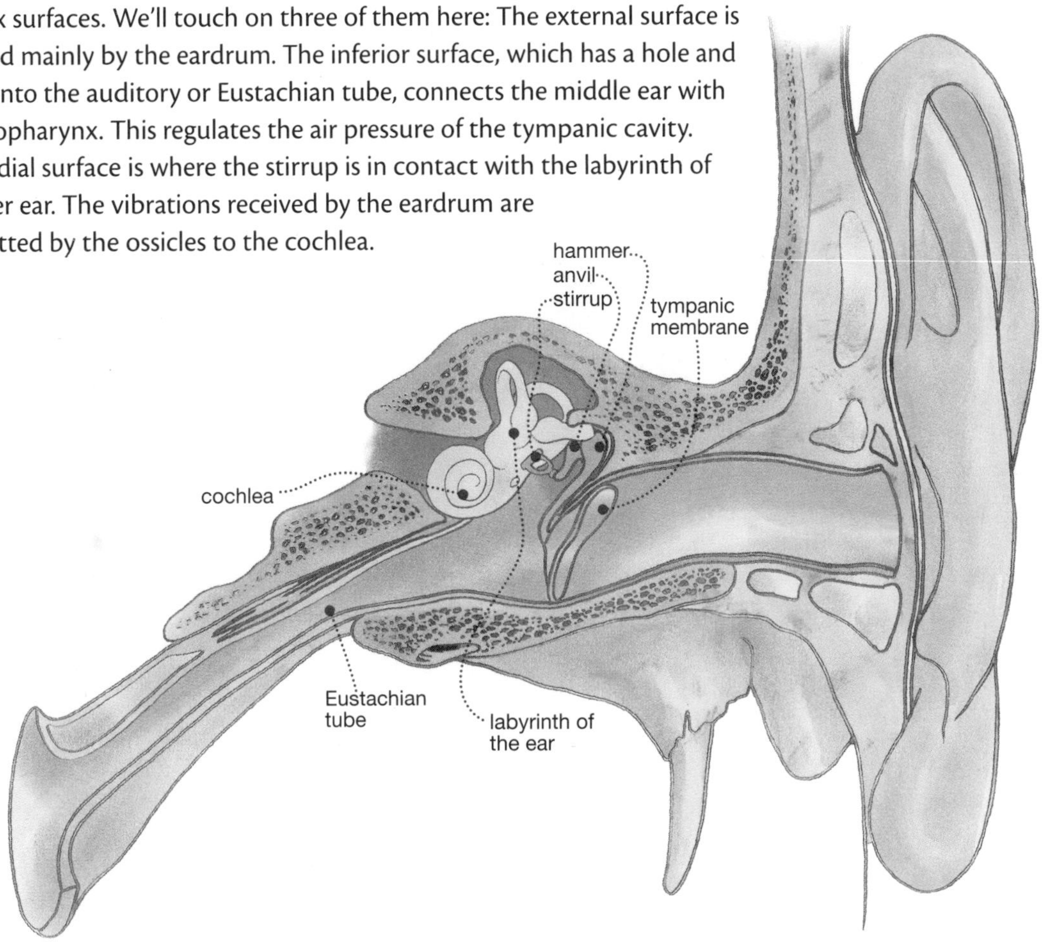

The Inner Ear

The inner ear is located in the mass of the temporal bone (see page 72). It consists of two parts:

- the bony labyrinth, which contains a fluid, the perilymph, that bathes the membranous labyrinth (which itself also contains a fluid, called endolymph)
- the membranous labyrinth, which includes three regions: the vestibule, the semicircular canals, and the cochlea.

The vestibule and semicircular canals contribute to our sense of balance, and we won't discuss them here. The cochlea is the organ of hearing. It is a spiral-shaped cavity in the bony labyrinth. It contains the organ of Corti, which is made up of a membrane and cilia-lined sensory cells that are stimulated by vibration of the membrane.

6
Some Terms Used in the Vocal Professions

Matter

This part is devoted to giving simple and accessible explanations of some terms that are frequently used in the study of the voice (and the rest of the book): a *vibration*, a *harmonic*, a *form*, et cetera.

We focus primarily on the links between these terms, which belong to the world of physics, and the anatomical structures and their function.

What Is Matter?

Everything that makes up the concrete and tangible objects in our world (the planet, the air, the human body), we call matter.

We might say that matter is the opposite of a vacuum (when there is a vacuum, there is no material, and when there is matter, there is no vacuum).

What Is Matter Composed Of?

At the base of all matter are atoms. Atoms are the basic units that can be assembled to form molecules, which themselves can be assembled to form the matter that we can perceive.

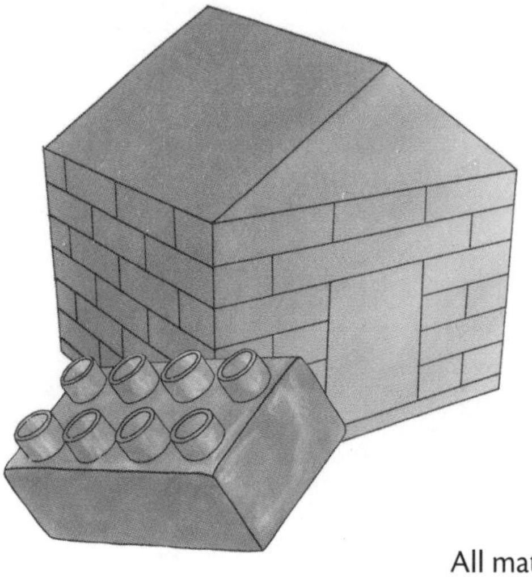

In the same vein, if we build a Lego building, it is an assemblage of Lego bricks.

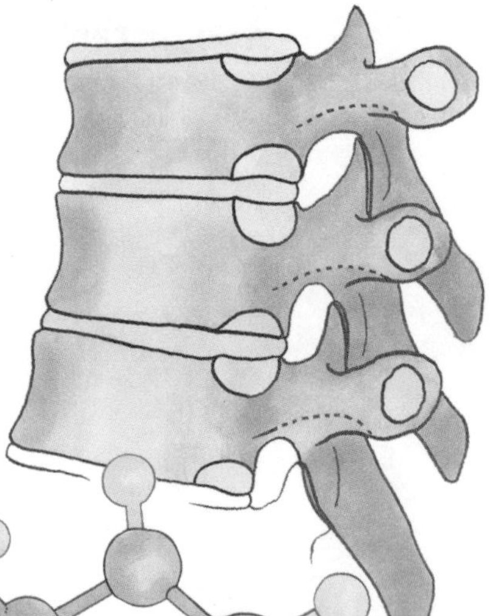

All matter is an assemblage, more or less complex, of molecules . . .

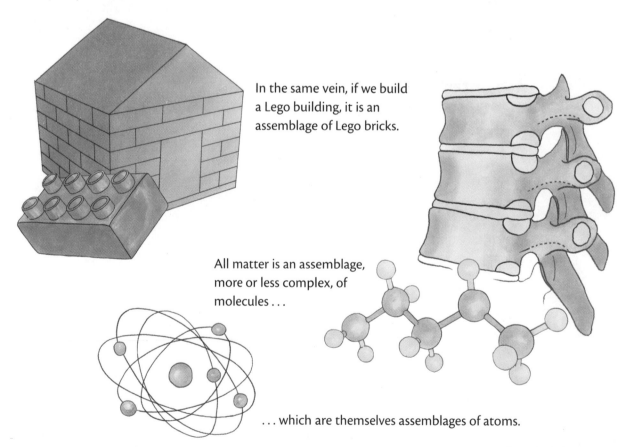

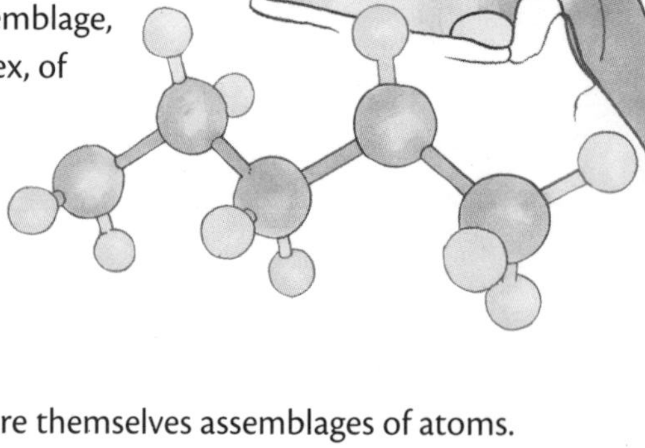

. . . which are themselves assemblages of atoms.

The Three States of Matter

Molecules can be assembled in different ways. There are three different types of assemblages, called *states*.

Solid State

The bonds between the molecules are strong and the molecules move very little in relation to one another.

A solid has a specific form and a specific volume.

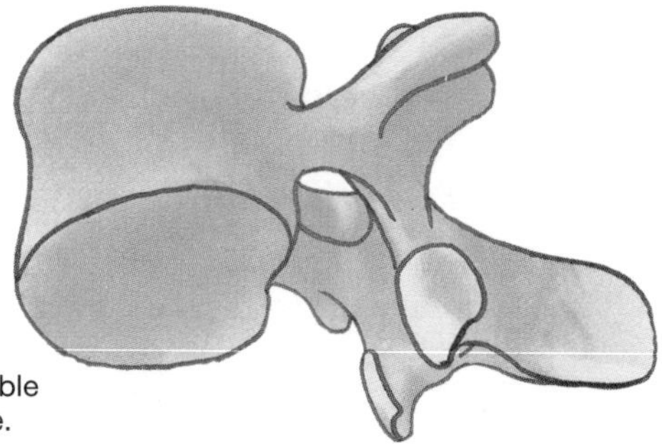

A solid has a stable form and volume.

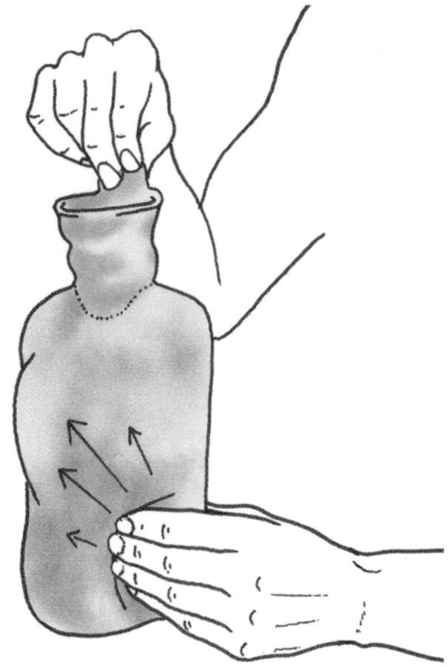

Liquid State

The bonds between the molecules are very weak. In this case, the molecules can slide over one another but they can't separate. A liquid has a specific volume but not a specific form.

The volume of a liquid stays the same, but we are able to manipulate the form.

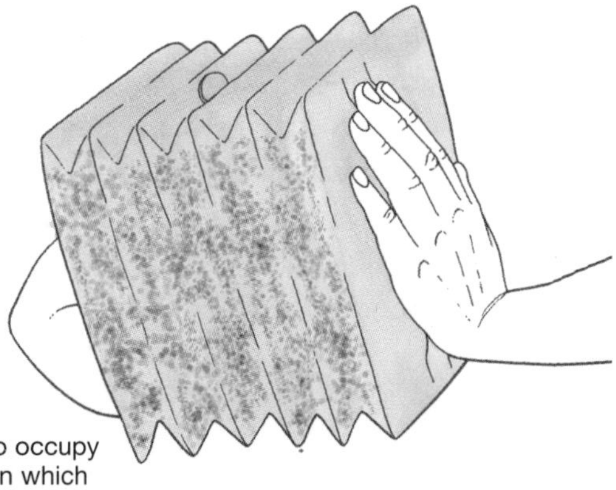

Gaseous State

The molecules are independent (not connected). Because they are free, they can move, collide with their neighbors, and be repelled, so they tend to move away from one another. For this reason, we say that a gas has no specific volume or form, and it tends to occupy all the space in which it is contained.

A gas tends to occupy all the space in which it is contained.

Gas and Pressure

What Is Pressure?

Pressure is a physical concept that quantifies the fact that in a gas, the molecules are closer together or farther apart and that they repel each other. The closer together they are, the more they repel each other, and the higher the pressure. This pressure can be expressed in Pascal or bar units (1 bar unit is 100,000 Pascal units).

Air Is a Gas under Pressure

The air we breathe is a gas composed of nitrogen and oxygen molecules that are held close to the earth by the force of gravity. Its average pressure is 1 bar.

Pressure and Volume

For the same amount of air, the greater the volume of the container, the lower the pressure, and vice versa.

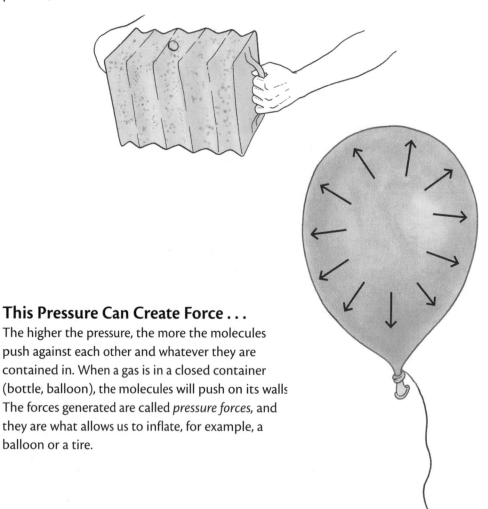

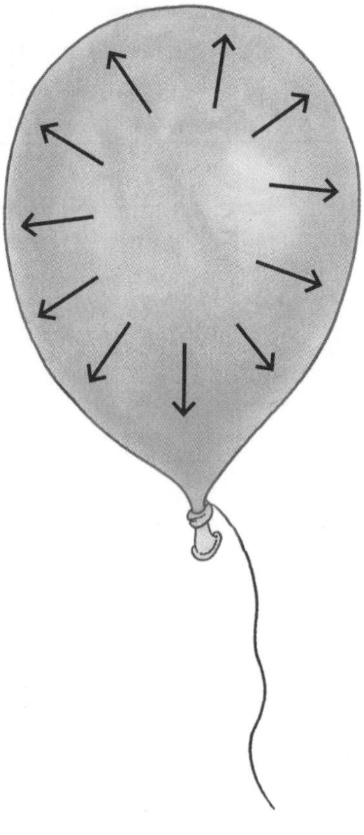

This Pressure Can Create Force . . .

The higher the pressure, the more the molecules push against each other and whatever they are contained in. When a gas is in a closed container (bottle, balloon), the molecules will push on its walls The forces generated are called *pressure forces*, and they are what allows us to inflate, for example, a balloon or a tire.

. . . the Forces Are Often Colossal

For example, the pressure generated by the force of ambient air on 1 square meter of glass is approximately 10 tons. This is the combined weight of two African elephants.

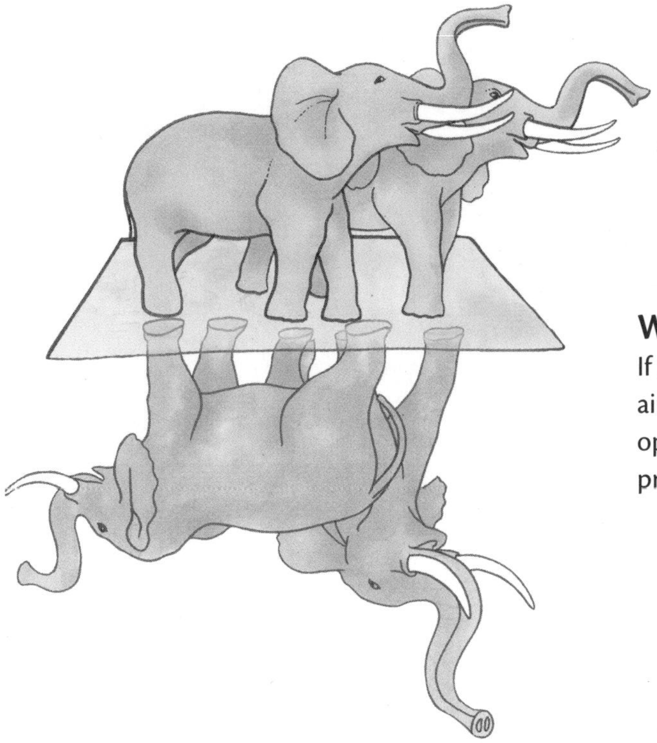

We Live in a World under Pressure

If the glass does not shatter, this is because the air on the other side exerts a similar force in the opposite direction. This balance of pressure is very present and necessary in our world.

Nature Abhors a Vacuum

We say that a vacuum pulls air in. This is because a volume filled with "nothing" is subject to the pressure of the outside air, which pushes against its walls and can distort its form (a vacuum cannot exert a counter force). This is what happens when you remove the air from a bag of coffee.

From Pressure to Sound

We have seen how the "generator" can put the thoracic rib cage under pressure (see page 97), and then how some of that air can escape through the vocal cords (see page 154), creating a zone of localized pressure (peak pressure).

It then causes the following phenomenon: the molecules of "peak pressure" will push against the surrounding molecules, creating a new pressure zone that will in turn push against other areas.

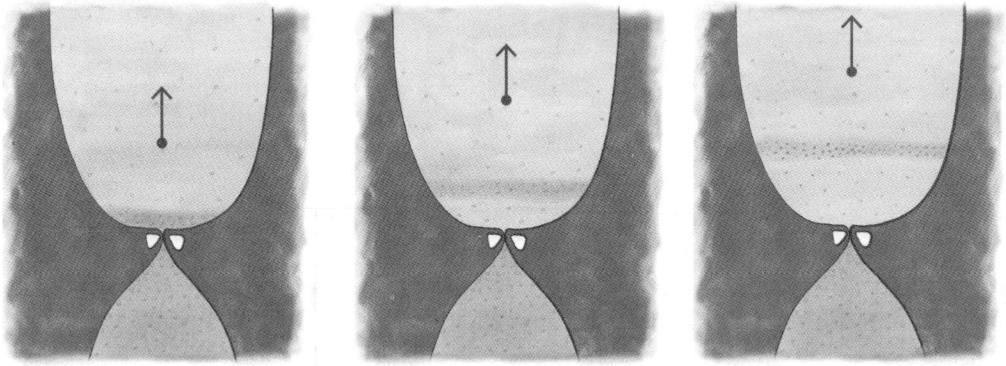

The zone of peak pressure moves without displacing the matter around it; this creates a **pressure wave**.

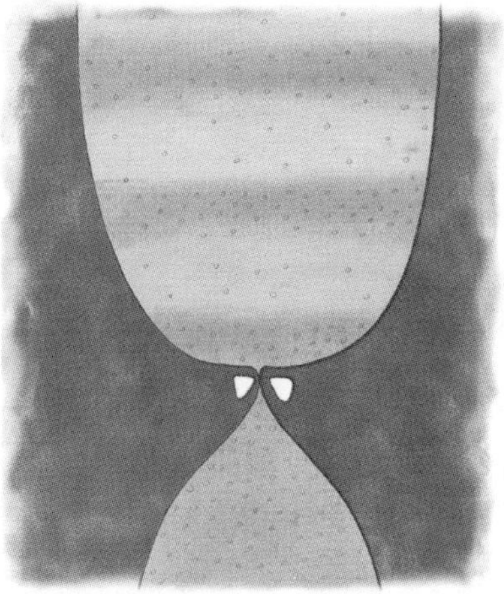

Note that in general peak pressure is never created alone—that is to say, a succession of peaks is produced, following one another very closely (in vocal work, this corresponds to several open/close cycles of the vocal cords). We talk of **sound waves** when describing a succession of peaks.

A sound is therefore a sound wave. It is created by an **emitter** that can create an imbalance of pressure. It then spreads through the air before being picked up by a **receptor** (ear, microphone) capable of interpreting these pressure fluctuations.

Graphical Representation

Imagine that we could place at the opening of the mouth a sensor that could register in graph form the variations in pressure over time. The result would look something like the following chart.

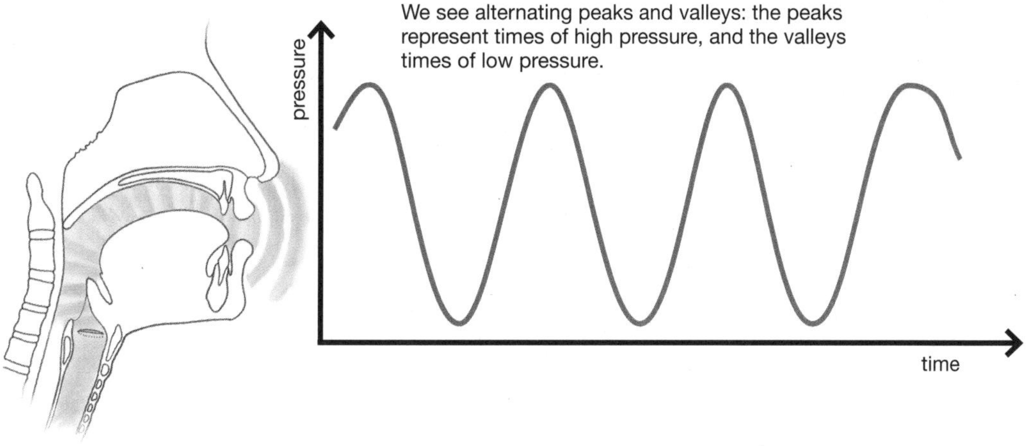

We see alternating peaks and valleys: the peaks represent times of high pressure, and the valleys times of low pressure.
pressure
time

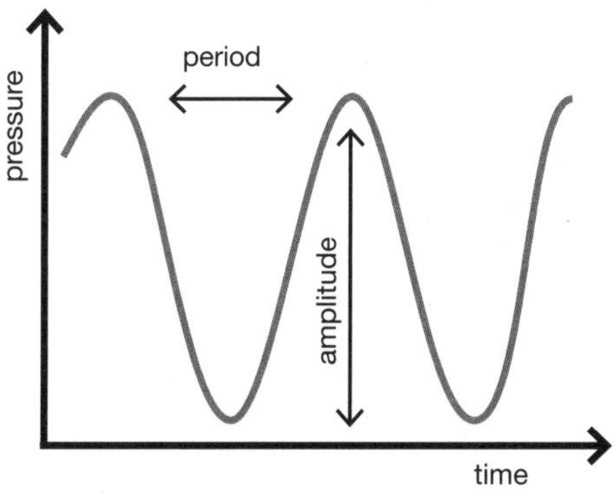

pressure
period
amplitude
time

Some Important Qualities

Amplitude

This is the difference between the lowest and highest pressure.

Period

The length of time between two pressure peaks.

Frequency

The number of peaks picked up by the sensor in a second.

Pitch, Intensity, and Duration of Sound

What Is a Pure Sound?

Pure sound is sound devoid of harmonics (see page 294). This is a special sound close to that produced by a tuning fork. It's quite easy to study because it depends on only two parameters: period (frequency) and amplitude.

It has the following form (called *sinusoidal*):

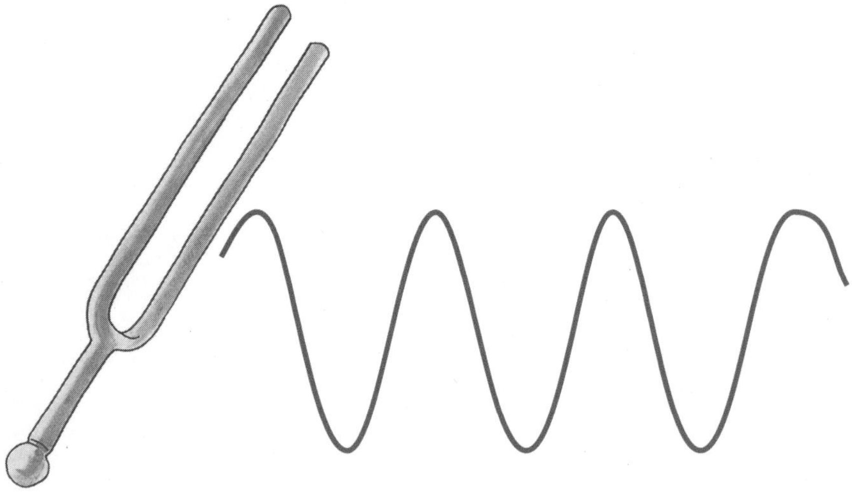

What Is a Musical Note?

A musical note is a sound that can be characterized by four components: pitch, duration, intensity, and timbre.

The Pitch

For listeners, the pitch is responsible for the sensation that allows them to identify the note as low or high.

Different scales allow us to classify notes according to their pitch. The most common scale used in the West is the chromatic scale (divided into octaves, which are themselves divided into twelve semitones). The pitch of a note is determined by the frequency of its sound wave: the higher the frequency, the higher the note (which is why, for example, the sounds on a cassette are higher when we speed it up). Frequency is expressed in hertz (Hz).

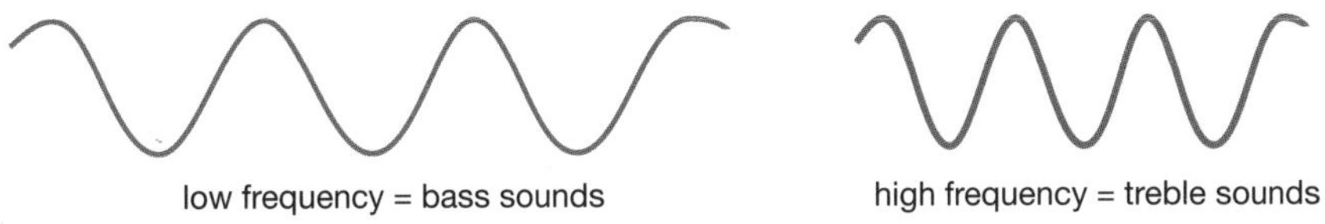

The Duration

The duration is the length of time for which we can perceive one note. This is quantitatively defined by the number of its vibrations (number of peaks). The more numerous they are, the longer the note.

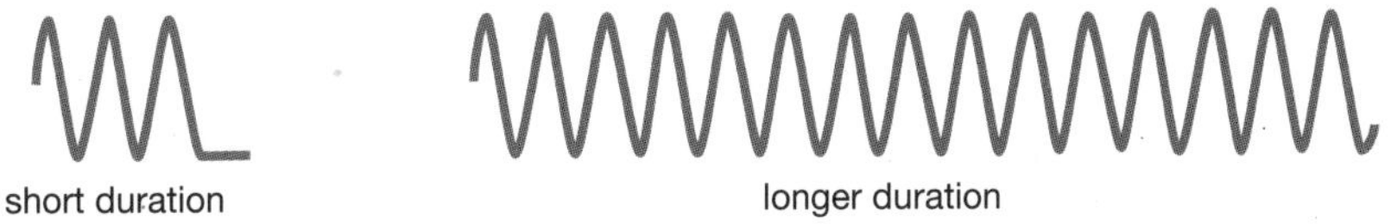

The Intensity

The intensity is a value that allows us to describe the strength or weakness of a note. We can also talk about sound volume, or nuance. The intensity of a note is a psychoacoustic phenomenon associated with the amplitude of the pressure wave associated with it. The greater the amplitude of the wave, the stronger and louder the note. The intensity is measured in decibels.

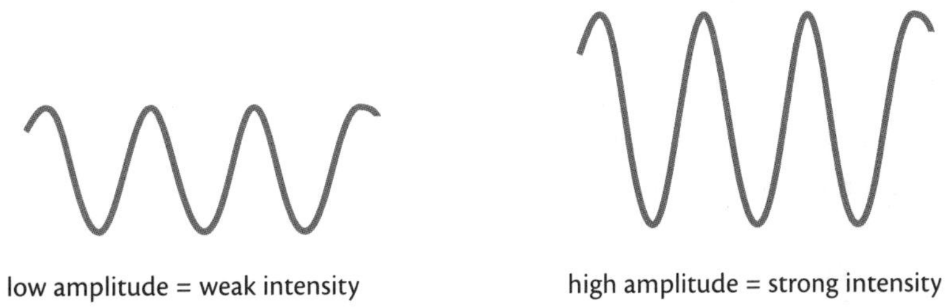

Timbre

The timbre is what allows us to identify a unique sound. The notes produced by two different instruments (piano and harpsichord, for example) can have the same pitch and the same intensity, but they will never have the same timbre.

Real-World Sounds Are Not Pure

In nature, pure tones are very rare. Pitch (frequency) and intensity are not sufficient to describe the waveform of an actual sound, which can vary greatly from one instrument to another.

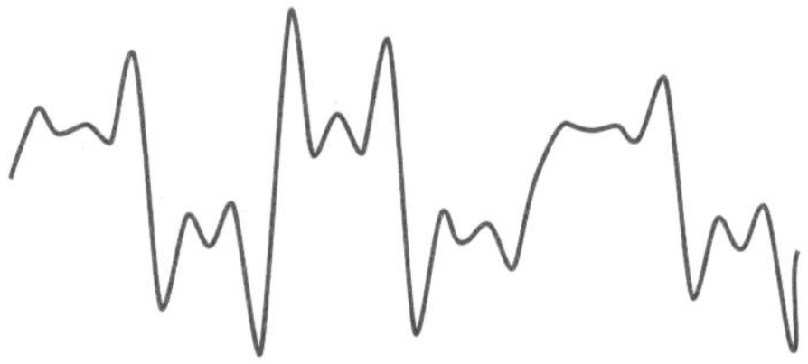

actual sound 1

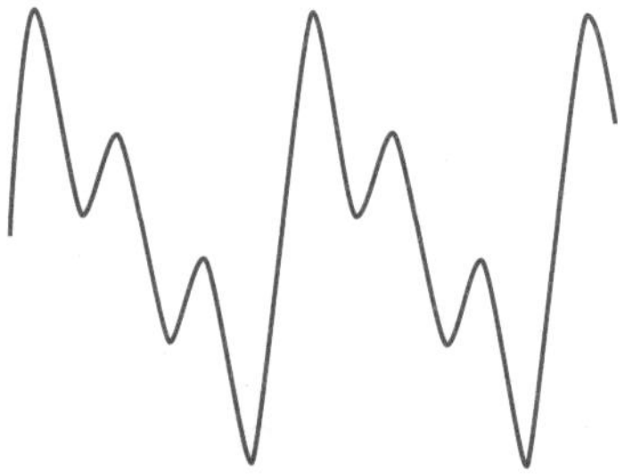

actual sound 2

A Real Sound Is the Sum of Its Pure Tones

The work of Joseph Fourier (nineteenth-century physicist) showed that any sound is the sum of its pure tones, called *harmonics:*

- Each harmonic is defined by its frequency and intensity.
- The harmonic with the lowest frequency is called the *fundamental.*
- The frequency of each harmonic is an integer multiple (1, 2, 3 . . .) of the fundamental frequency.

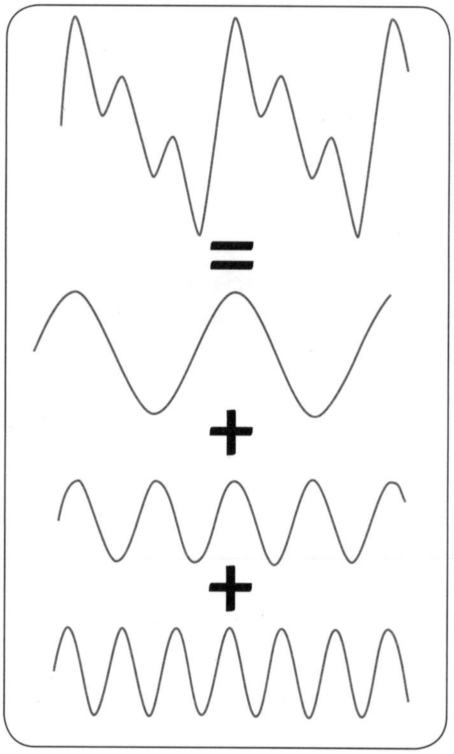

The fundamental is important because it defines the pitch of the reconstructed sound. For example, a note of 40 Hz has a fundamental of 40 Hz and harmonics of 80 Hz, 120 Hz, and 160 Hz.

Graphical Representation (Spectral)

We call the totality of all of the component harmonics of a sound the *spectrum*. Because they depend only on the frequency and amplitude, it is convenient to represent the spectrum using these two parameters (see below).

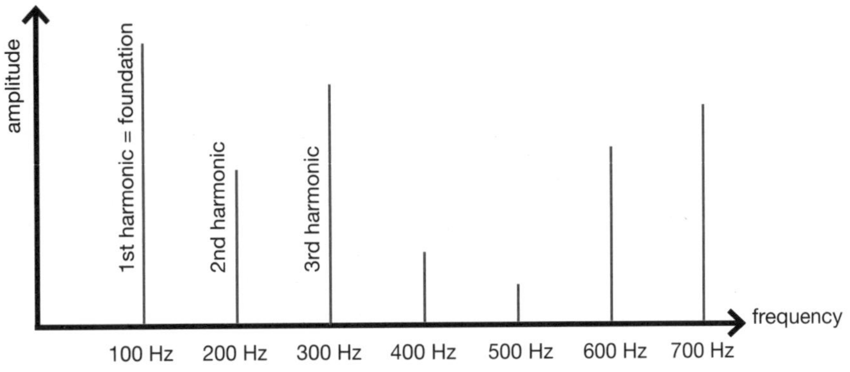

Consonance, Dissonance, and Harmony

A chord is a set of several notes played simultaneously. Depending on the notes, this chord can be harmonious, pleasant to the ear (octave, fifth), or, on the contrary, dissonant (second, seventh).

This is because each note has its harmonics, which may or may not be juxtaposed with the others. The more they are juxtaposed, the more the sound will be unified and consonant. On the other hand, they might fight each other, which makes the sound dissonant.

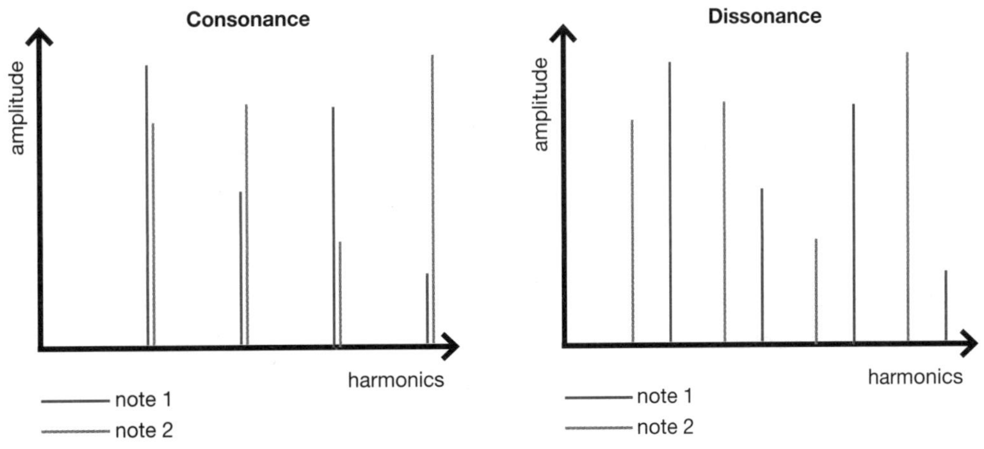

Bibliography

Amy de La Bretèque, Benoît. *À l'origine du son: le soufflé.* Marseille, France: Solal, 2000.

———. *L'équilibre et le rayonnement de la voix.* Marseille, France: Solal, 1999.

———. *Le chant: contraintes et libertés.* Courlay, France: Fuzeau, 1991.

Barthélémy, Yva. *La voix libérée.* Paris: Robert Laffont, 1985.

Bouchet, A., and J. Cuilleret. *Anatomie topographique, descriptive et fonctionnelle.* Paris: Simep, 1990.

Brizon, J., and J. Castaing. *Les feuillets d'anatomie.* Paris: Maloine, 1996.

Calais-Germain, Blandine. *Anatomy of Breathing.* Seattle, Wash.: Eastland Press, 2006.

———. *Anatomy of Movement.* Revised edition. Seattle, Wash.: Eastland Press, 2007.

———. *The Female Pelvis: Anatomy & Exercises.* Seattle, Wash.: Eastland Press, 2003.

———. *No-Risk Abs: A Safe Workout Program for Core Strength.* Rochester, Vt.: Healing Arts Press, 2011.

Clemente, Carmine D. *Anatomy: A Regional Atlas of the Human Body.* 6th ed. Baltimore: Lippincott Williams & Wilkins, 2010.

Cornut, Guy. *La mécanique respiratoire dans la parole et dans le chant.* Paris: PUF, 1959.

———. *La Voix.* Paris: PUF, 1983.

———. *Moyens d'investigation et pédagogie de la voix chantée.* Lyon, France: Symétrie, 2001.

Dinville, Claire. *La voix chantée.* Paris: Masson, 1982.

Drake, Richard L., A. Wayne Vogl, and Adam W. M. Mitchell. *Gray's Anatomy for Students.* Third edition. Philadelphia: Churchill Livingstone, 2015.

Feldenkrais, Moshe. *Awareness Through Movement.* Reprint. New York: HarperCollins, 2009.

Fournier, Cécile. *La voix, un art et un métier.* Éditions Comp'Act, 1999.

Habermann, Günther. *Stimme und spräche.* Stuttgart, Germany: Thieme, 1978.

Kahle, W., H. Leonhard, and W. Platze. *Anatomie.* Paris: Flammarion, 1986.

Legent, François, Léon Perlemuter, and Claude Vandenbrouck. *Cahiers d'anatomie O.R.L.* Paris: Masson, 1976.

Le Huche, François. *Anatomie et physiologie des organes de la voix et de la parole.* Paris: Masson, 1984.

Miller, Richard. *La structure du chant.* Paris: IPMC, 1990.

Netter, Frank H. *Atlas of Human Anatomy.* 6th edition. Philadelphia: Saunders, 2014.

Ormezzano, Yves. *Le guide de la voix.* Paris: Odile Jacob, 2000.

Paire, Yvonne. *Ouf! Je respire . . .* Paris: Fleurus, 2012.

Pfauwadel, Marie-Claude. *Respirer, parler, chanter.* Paris: Le Hameau, 1981.

Piron, A. *Techniques ostéopathiques appliquées à la phoniatrie.* Lyon, France: Symétrie, 2007.

Stanislavski, Constantin. *An Actor Prepares.* Reprint. New York: Routledge, 1989.

Rondeleux, Louis-Jacques. *Trouver sa voix.* Paris: Seuil, 1977.

Scotto di Carlo, Nicole. "L'arme secrète des chanteurs d'opéra." *Revue la Recherche* 218 (Feb. 1990).

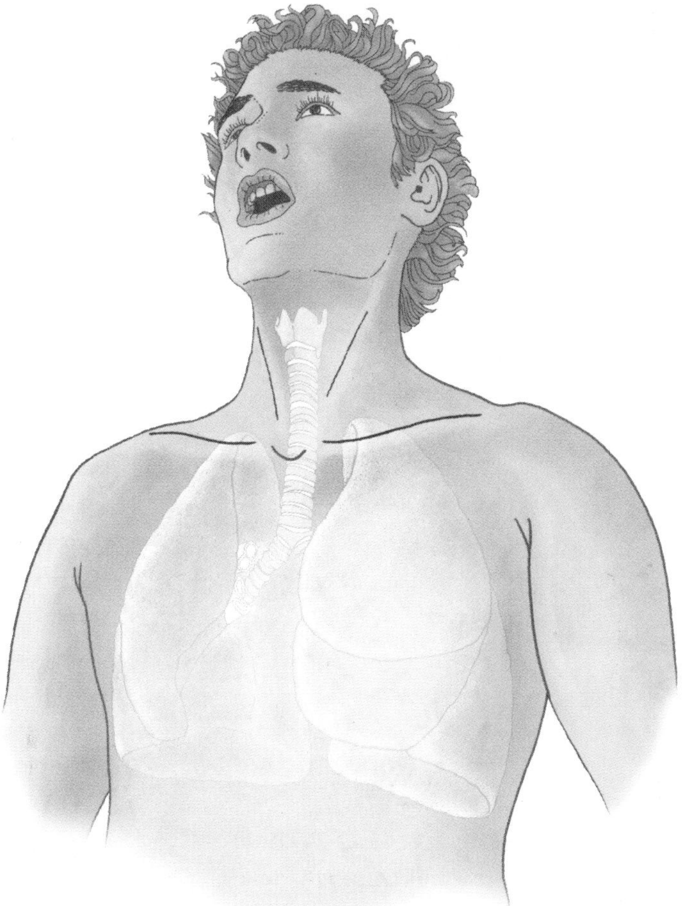

Index

Books of Related Interest

No-Risk Abs
A Safe Workout Program for Core Strength
by Blandine Calais-Germain

No-Risk Pilates
8 Techniques for a Safe Full-Body Workout
by Blandine Calais-Germain and Bertrand Raison

Preparing for a Gentle Birth
The Pelvis in Pregnancy
by Blandine Calais-Germain and Núria Vives Parés

Seed Sounds for Tuning the Chakras
Vowels, Consonants, and Syllables for Spiritual Transformation
by James D'Angelo

The Healing Power of the Human Voice
Mantras, Chants, and Seed Sounds for Health and Harmony
by James D'Angelo

Chakra Frequencies
Tantra of Sound
by Jonathan Goldman and Andi Goldman

Healing Sounds
The Power of Harmonics
by Jonathan Goldman

The Soundscape
Our Sonic Environment and the Tuning of the World
by R. Murray Schafer